Start Here,
Go Anywhere.

WILEY

I dedicate this book to my new granddaughters, Gianna Leigh Madden and Taralyn Kayhomas.
—CONNIE ALLEN

I dedicate this book to my husband Chuck, who has supported and encouraged me, my chlren Scott and Kate, who were patient when I worked long hours, and my nieces Jessica and KristinFatigati, who are interested in studying anatomy.
—VALERIE HARPER

5th
EDITION

Laboratory Manual for Anatomy and Physiology

CONNIE ALLEN
Edison State College

VALERIE HARPER
Colorado Mesa University

WILEY

VP & EXECUTIVE PUBLISHER	Kaye Pace
EXECUTIVE EDITOR	Bonnie Roesch
ASSOCIATE EDITOR	Lauren Elfers
EDITORIAL ASSISTANT	Grace Bagley
MARKETING MANAGER	Maria Guarascio
PRODUCTION MANAGER	Juanita Thompson
PRODUCTION EDITOR	Barbara Russiello
SENIOR DESIGNER/COVER DESIGNER	Madelyn Lesure
SENIOR PHOTO EDITOR	MaryAnn Price
ILLUSTRATION EDITOR	Claudia Volano
SENIOR PRODUCT DESIGER	Linda Muriello
MEDIA SPECIALIST	Svetlana Barskaya
COVER	IIlustration by Bryan Christie Design

This book was typeset in 10/12 Times Roman by Aptara and printed and bound by Courier Kendallville Press. The cover was printed by Courier Kendallville.

This book is printed on acid free paper.

Founded in 1807, John Wiley & Sons, Inc. has been a valued source of knowledge and understanding for more than 200 years, helping people around the world meet their needs and fulfill their aspirations. Our company is built on a foundation of principles that include responsibility to the communities we serve and where we live and work. In 2008, we launched a Corporate Citizenship Initiative, a global effort to address the environmental, social, economic, and ethical challenges we face in our business. Among the issues we are addressing are carbon impact, paper specifications and procurement, ethical conduct within our business and among our vendors, and community and charitable support. For more information, please visit our website: www.wiley.com/go/citizenship.

The procedures in this text are intended for use only by students with appropriate faculty supervision. In preparing the text, care has been taken to identify potentially hazardous steps and to insert safety precautions where appropriate. The authors and publisher believe the procedures to be useful tools if performed with the materials and equipment specified, in careful accordance with the instructions and methods in the text. However, these procedures must be conducted at one's own risk. The author and publisher do not warrant or guarantee the safety of individuals using these procedures and specifically disclaim any and all liability resulting directly or indirectly from the use or application of any information contained in this book.

ISBN 978-1-118-34440-8

Printed in the United States of America.

10 9 8 7 6 5 4 3 2 1

Preface

Anatomy and physiology is a challenging course, and this laboratory manual is written to help students meet that challenge. It is written for students interested in allied health fields, such as nursing; physical, respiratory, cardiovascular, or occupational therapy; radiology; and dental hygiene. This manual may be used with any two-semester anatomy and physiology textbook.

The design of this laboratory manual is based on the authors' experience as anatomy and physiology instructors and uses three learning styles: visual, auditory, and kinesthetic. When students label diagrams, they focus on the structure rather than just the dot at the end of a line. Writing out the structure's name and pronouncing it reinforces learning. Also, having students become subjects of laboratory exercises personalizes the learning process. Animal dissections give students an opportunity to physically manipulate structures, comparing location and texture, and to observe how structures are supported, protected, and attached by connective tissue.

Special features incorporated in this laboratory manual include:

- This lab manual can be used for online anatomy and physiology classes. New lab activities have been added that can be performed by students at home or used in the laboratory. Online students can also use the Real Anatomy Virtual Dissection program and PowerPhys physiology activities to enhance their learning.
- Just enough text is provided to introduce concepts in each section and to set up and support the laboratory section. The exercises are written so students do not need their textbooks to complete the laboratory activities.
- New material is divided into small segments, starting with simple diagrams, illustrating the basic concepts and building up to more complex diagrams. Subsequent activities add to the students' knowledge in a stepwise fashion. This is especially noticed in the skeletal and muscular exercises.
- Each exercise contains a list of objectives, materials needed for the exercise, and easily identifiable laboratory activity sections.
- Unlabeled four-color drawings, photographs, and photomicrographs are included for students to label either at home or in the laboratory. Students first write out the name of the structure to help learn it. Then the completed diagrams will be used to identify structures on models.
- Word derivatives for bolded terms are given in the text.
- Phonetic pronunciation is included as new words are introduced.

- Physiology experiments use students as subjects and can be completed with either simple, inexpensive equipment and materials or more complex lab setups.
- Experimental report sections after physiology experiments where students are asked to make predictions, collect and analyze data, and write simple lab reports.
- Discussion Questions are within the activities to make the students think about the material presented.
- An Answer Key is provided at the end of the laboratory manual for the activities in each exercise. Students receive immediate feedback, and they are not dependent on the instructor for the correct answers.
- "Reviewing Your Knowledge" and "Using Your Knowledge" sections follow the activities at the end of each exercise. "Reviewing Your Knowledge" provides a thorough review of the material in the exercise, whereas "Using Your Knowledge" requires students to apply information learned. Either or both of these sections may be handed in to the instructors for a grade, because neither section has answers in the back of the laboratory manual. Answers to these two sections are provided on the Instructor Companion site.

New Features to the Fifth Edition

- Exercises are now organized into Before Going to Lab and Lab Activities. Before Going to Lab sections are to be completed at home or outside of class.
- Completely revised Exercise 15: Surface Anatomy and Exercise 26: Blood Components and Blood Tests.
- Suggested Real Anatomy activities are included throughout the lab manual.
- PowerPhys 3.0 is now tablet enabled and includes three new modules.
- New Biopac Laboratory Guide Experiments are available online for several Exercises. Activities include:
 ○ Recruitment and Fatigue
 ○ Isometric and Isotonic Contractions
 ○ Effect of Physical and Mental Distractions on Patellar Reflex (Knee Jerk) Response
 ○ The Effect of Mental and Sensory Stimulations on Brain Wave Patterns
 ○ Observing ANS Responses to Lying
 ○ Observing ANS Responses Following Meditation
 ○ Effect of Body Position and Exercise on ECG
 ○ Heart Sounds and Events of the Cardiac Cycle
 ○ Effect of Body Position on Resting Blood Pressure
 ○ Effect of Exercise on Blood Pressure
 ○ Effect of Exercise on Lung Volumes and Capacities
 ○ Comparison of Predicted and Measured FVC and FEV_1

PowerPhys 3.0 is Tablet Enabled and Fully Integrated with The New Edition

Each Lab Manual contains online access to *Power-Phys 3.0* for students. *PowerPhys 3.0* is physiological simulation software for the A & P laboratory that allows students to explore physiology principles through 14 self-contained activities. Each activity follows the scientific method containing objectives with illustrated and animated review material, pre-lab quizzes, pre-lab reports (including predictions and variables), data collection and analysis, and a full lab report with discussion and application questions. Experiments contain real data that is randomly generated, allowing users to experiment multiple times, but still arrive at the same conclusions. These activities focus on core physiological concepts and reinforce techniques experienced in the laboratory. www.wiley.com/go/powerphys

- Three New Activities:
 - Hematocrit and Hemoglobin Concentration and Blood Typing
 - Acid-Base Balance
 - Effect of Dietary Fiber on Transit Time and Bile
- Now available on mobile devices

Dissection Manuals

- Depending on your needs, a Cat Dissection Manual or Fetal Pig Dissection Manual is available to accompany the main Lab Manual. Both dissection manuals contain dissection activities for use in the lab accompanied by full color photos and figures.
- All cat and pig photographs have been updated in the new editions.
- New cat and pig dissection videos accompany this edition, available in *WileyPLUS*. Dissections performed by Shawn Miller and photographs taken by Mark Nielsen of the University of Utah.

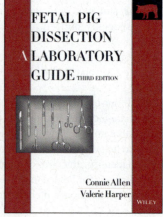

Resources That Power Success

Anatomy Drill and Practice in
WileyPLUS

Anatomy Drill and Practice lets you test your knowledge of structures with simple-to-use drag and drop labeling exercises, or fill-in-the-blank labeling. You can drill and practice on these activities using illustrations from the text, cadaver photographs, histology micrographs, or anatomical models, so you can practice labeling these models outside of the lab.

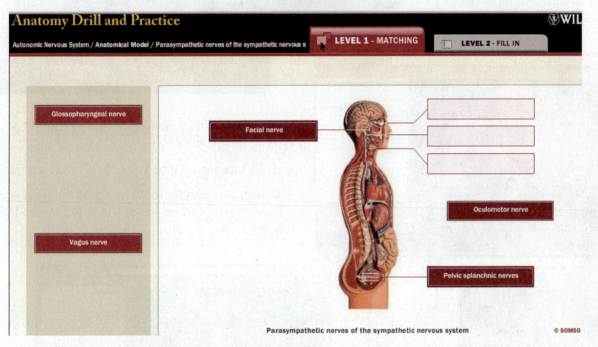

Parasympathetic nerves of the sympathetic nervous system © SOMSO

Student and Instructor Companion Websites

www.wiley.com/college/allen

Resources for Students:

- Audio Glossary
- Rat Dissection video
- Pig Dissection video
- Answers to figure questions
- Interactions Review Sheets
- Real Anatomy Review Sheets
- Lab Safety Guidelines
- Experimental Reports
- Blood Activities

Resources for Instructors:

- Anatomy and Physiology Visual Library
- Editable PowerPoints that include all figures from the Lab Manual
- All tables from the Lab Manual in PowerPoint
- All images from the Cat Dissection Manual
- All images from the Pig Dissection Manual
- Answer Keys for "Reviewing Your Knowledge"
- Answer Keys for "Using Your Knowledge"
- Interactions Review Sheets
- Real Anatomy Review Sheets

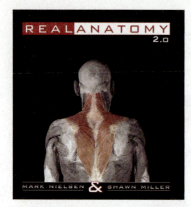

Real Anatomy 2.0

Mark Nielsen and Shawn Miller, *University of Utah*

Real Anatomy is 3-D imaging software that allows you to dissect through multiple layers of a three-dimensional real human body to study and learn the anatomical structures of all body systems.

NEW to Real Anatomy 2.0

• Now available on the Web, accessible by iPad and Android tablets.

• All possible highlight structures on an image are now accessible via a drop-down list as well as being searchable.

• New crumb trail navigation shows context of system, image, structure.

• Fully integrated into *WileyPLUS* for Anatomy.

• Dissect through up to 40 layers of the body and discover the relationships of the structures to the whole.

• Rotate the body as well as major organs to view the image from multiple perspectives.

• Use a built-in zoom feature to get a closer look at detail.

• A unique approach to highlighting and labeling structures does not obscure the real anatomy on view.

• Snapshots can be saved of any image for use in Power-Points, quizzes, or handouts

• Audio pronunciation of all labeled structures is readily available

• Related Images provide multiple views of structures being studied.

• Snapshots can be saved of any image for use in PowerPoints, quizzes, or handouts.

• View histology micrographs at varied levels of magnification with the virtual microscope.

• Audio pronunciation of all labeled structures is readily available.

REAL ANATOMY

Photographic Atlas of Human Anatomy, First Edition

Mark Nielsen and Shawn Miller, *University of Utah*

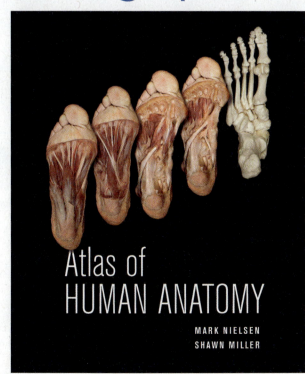

This atlas, filled with outstanding photographs of meticulously executed dissections of the human body, has been developed to be a strong teaching and learning solution, not just a catalog of photographs. Organized around body systems, each chapter includes a narrative overview of the body system followed by detailed photographs that accurately and realistically represent the anatomical structures. Histology is included. *Photographic Atlas of Human Anatomy* will work well in your laboratories, as a study companion to your lab manual, and as a print companion to the Real Anatomy DVD.

WileyPLUS

WileyPLUS is a research-based online environment for effective teaching and learning.

WileyPLUS builds students' confidence because it takes the guesswork out of studying by providing students with a clear roadmap:

- **what to do**
- **how to do it**
- **if they did it right**

It offers interactive resources along with a complete digital textbook that help students learn more. With *WileyPLUS*, students take more initiative so you'll have greater impact on their achievement in the classroom and beyond.

Now available for

Bb

Blackboard

For more information, visit www.wileyplus.com

WileyPLUS

ALL THE HELP, RESOURCES, AND PERSONAL SUPPORT YOU AND YOUR STUDENTS NEED!

www.wileyplus.com/resources

1st DAY OF CLASS ... AND BEYOND!

2-Minute Tutorials and all of the resources you and your students need to get started

WileyPLUS

Student Partner Program

Student support from an experienced student user

Wiley Faculty Network

Collaborate with your colleagues, find a mentor, attend virtual and live events, and view resources
www.WhereFacultyConnect.com

WileyPLUS

Quick Start

Pre-loaded, ready-to-use assignments and presentations created by subject matter experts

Technical Support 24/7 FAQs, online chat, and phone support
www.wileyplus.com/support

© Courtney Keating/iStockphoto

Your *WileyPLUS* Account Manager, providing personal training and support

WileyPLUS and You

WileyPLUS for Anatomy and Physiology is an innovative, research-based online environment—designed for both effective teaching and learning. Utilizing *WileyPLUS* in your lab course provides your students with an accessible, affordable, and active learning platform and the tools and resources to efficiently build presentations for a dynamic laboratory experience and to create and manage effective assessment strategies. The underlying principles of **design, engagement,** and **measurable outcomes** provide the foundation for this powerful, new release of *WileyPLUS*.

Design

- New research-based design helps students manage their time better and develop better study skills
- Course Calendars help track assignments for both students and teachers
- New Course Plan makes it easier to assign pre- and post-lab activities and assessment. Simple drag-and-drop tools make it easy to assign the course plan as-is or in any way that best reflects your course syllabus

The new design makes it easy for students to know **what** it is they need to do, boosting their confidence and preparing them for greater engagement in lab.

Engagement

- Complete online version of the Lab Manual for seamless integration of all content
- Online version of the Lab Manual includes interactive figure labeling exercises for students
- Relevant student study tools and learning resources ensure positive learning outcomes

- Immediate feedback boosts confidence and helps students see a return on investment for each study session
- Resources like Cadaver Videos, Anatomy Drill, and Practice help students study for laboratory practicals
- Course materials, including editable PowerPoint stacks and Wiley's Visual Library for Anatomy and Physiology, help you personalize lessons and optimize your time

Concept mastery in this discipline is directly related to students keeping up with the work and not falling behind. The new Concept Modules, Activities, Self Study, and Progress Checks in *WileyPLUS* will ensure that students know **how** to study effectively so they will remain engaged and stay on task.

Measurable Outcomes

- Progress check enables students to hone in on areas of weakness for increased success
- Self-assessment and remediation for all learning objectives lets students know exactly how their efforts have paid off
- Instant reports monitor trends in class performance, use of course materials, and student progress towards learning objectives
- New gradable versions of the PowerPhys 2.0 laboratory reports allow you to assign and grade these within *WileyPLUS*

With new detailed reporting capabilities students will **know** that they are doing it right. With increased confidence, motivation is sustained so students stay on task, and success will follow. Please contact your Wiley representative for details about these and other resources or visit our website at *www.wiley.com/college/sc/aandp* and click on the laboratory text cover to explore the assets more fully.

Acknowledgments

We deeply appreciate the support, instruction, and encouragement from the members of our editorial, production, and marketing team at Wiley: Bonnie Roesch, Lauren Elfers, Barbara Russiello, MaryAnn Price, and Maria Guarascio. We also wish to thank Gerard Tortora and Bryan Derrickson for producing a wonderful textbook that provided many illustrations and ideas for our laboratory manual. A special thank you to Susan Baxley for reviewing all the exercises, making suggestions and to Bob Clemence for allowing us to use his figure of the Respiratory Volumes and Capacities. A special thanks to Charles Harper for answering many clinical questions. We also wish to thank Kierstan Hong at Imagineering Art for the artwork she provided for our laboratory manual.

Thank you to our colleagues at Edison State College: Bob Clemence, Colleen Swanson, Jody Gootkin, Richard McCoy,

Jeff Davis, Dick Felden, Lyman O'Neil, Kitty Gronlund, Tony Contino, Cheryl Black, Jed Wolfson, Jay Koepke, and Roy Hepner who encouraged us, answered our questions, and provided critiques of exercises. We also wish to thank Nicole Yarbrough George for her critique of the skeletal muscle chapter. Thank you to Chaim Jay Margolin of Regional Radiology Associates and David Michie of Clinical Physiology Associates for providing images for this manual. Special thanks to SOMSO for providing images for our online Anatomy Drill and Practice: Anatomical Models section. Thanks to contributors Jerri Lindsey, *Tarrant County College*, and Terry Thompson, *Wor-Wic Community College*.

Contents

Anatomical Language

OBJECTIVES

1 Describe the anatomical position
2 Use anatomical and directional terms correctly
3 Identify the various body planes and sections

MATERIALS

- human models or anatomical charts
- apples (1 per group) and plastic knives or scalpels
- plastic tubing (eight-inch piece per group) or plastic straw
- 5 sheep brains (for class demonstration)

Anatomical terms describe body positions, body regions, specific body areas, and landmarks. Most of these words are derived from Latin or Greek and are often part of the names of muscles, bones, nerves, and blood vessels. Learning these terms at this time will help you throughout the course.

A. Anatomical Terms

The **anatomical position** is the reference position anatomists and people in medical fields use to describe the location of body parts or regions. In the anatomical position, the body is erect (vertical) and facing forward; the arms are straight and at the sides of the body with the palms facing forward; the legs are straight with the feet facing forward and flat (Figure 1.1).

In the **supine position**, the body is horizontal and lying on the back. In the **prone position**, the body is horizontal and lying on the stomach.

Body regions refer to specific areas of the body. There are common misperceptions about the boundaries of a few body regions.

Two common misconceptions are that the arm is the area between the shoulder and wrist and that the leg includes the thigh. Actually, the **arm** is located between the shoulder and elbow, and the **forearm** is located from the elbow to wrist. The **thigh** is located between the groin and knee, and the **leg** is located between the knee and ankle.

The **chest** is the area of the trunk between the neck and diaphragm and contains the heart and lungs. The **abdomen** is the area of the trunk between the diaphragm and hip bones and contains the digestive organs. The **pelvis** is the area of the trunk below the hip bones and contains internal reproductive organs and the urinary bladder.

The **groin** is the area on the anterior (front) surface marked by a crease where the lower limb attaches to the pelvis.

Many anatomical terms have one or more word roots with a prefix and/or a suffix added. For example, in the word *antecubital*, *ante-* is a prefix meaning before or in front of, the word root *cubit-* means elbow, *-al* is a suffix meaning pertaining to. Table 1.1 contains anatomical terms with four different suffixes, all of which mean pertaining to. These suffixes are *-al*, *-ic*, *-ar*, and *-ary*. When suffixes like these are added to word roots they form adjectives, whereas nouns have different endings such as *-um*, *-us*, *-is*, and *-a*. For example, *stern-* is a word root meaning chest; *sternum* is the noun and *sternal* is the adjective. Anatomical terms and their definitions are found in Table 1.1. Word roots and their definitions are found in Appendix A, as well as nouns and adjectives formed from the word roots.

Before Going to Lab

1 Label Figure 1.1 with the appropriate anatomical terms for each body region or area. Refer to Table 1.1.

2 Refer to Appendix A to review how word roots, suffixes, and prefixes are combined to form nouns and adjectives.

LAB ACTIVITY 1 **Anatomical Terms**

1 Use anatomical terms to identify the specific body regions or areas on models, anatomical charts, or yourself. ■

TABLE 1.1 Anatomical Terms

TERM	DEFINITION	TERM	DEFINITION
AXIAL	Pertaining to the central part of the body, the head and trunk	**APPENDICULAR**	Pertaining to the extremities or limbs
Cephalic (se-FAL-ik)	Pertaining to the head	**Upper Limb (Appendage)**	
• Cranial	Pertaining to the portion of the skull surrounding the brain	• Acromial (a-KROM-ee-al)	Pertaining to the highest point of the shoulder
• Facial	Pertaining to the face	• Axillary (AX-il-ary)	Pertaining to the armpit
• Frontal	Pertaining to the forehead	• Brachial (BRAY-key-ul)	Pertaining to the arm
• Orbital	Pertaining to the eye	• Antecubital (an-teh-KYOO-bi-tul)	Pertaining to the anterior (front) surface of the elbow
• Otic (OH-tik)	Pertaining to the ear	• Olecranal (oh-LEK-ra-nul)	Pertaining to the posterior (back) surface of the elbow
• Nasal	Pertaining to the nose	• Antebrachial	Pertaining to the forearm
• Buccal (BUCK-al)	Pertaining to the cheek	• Carpal	Pertaining to the wrist
• Oral	Pertaining to the mouth	• Manual	Pertaining to the hand
• Mental	Pertaining to the chin	• Palmar	Pertaining to the palm of the hand
• Occipital (ox-SIP-i-tal)	Pertaining to the back of head	• Digital	Pertaining to the digits (fingers)
Cervical	Pertaining to the neck	**Lower Limb (Appendage)**	
Thoracic	Pertaining to the chest	• Inguinal (ING-won-ul)	Pertaining to the groin where the thigh attaches to the pelvis
• Sternal	Pertaining to the breast bone	• Gluteal (GLUE-tee-ul)	Pertaining to the buttocks
• Pectoral	Pertaining to the chest	• Femoral (FEM-or-ul)	Pertaining to the thigh
• Mammary	Pertaining to the breast	• Patellar (pa-TEL-ur)	Pertaining to the anterior (front) surface of the knee
Abdominal	Pertaining to the abdomen	• Popliteal (pop-lih-TEE-ul)	Pertaining to the posterior (back) surface of the knee
• Umbilical (um-BIL-ih-cal)	Pertaining to the navel	• Crural (CROO-rul)	Pertaining to the anterior (front) surface of the leg
• Coxal (COX-al)	Pertaining to the hip	• Fibular (FIB-you-lur) or peroneal (peh-RONE-ee-ul)	Pertaining to the lateral side of the leg
Pelvic	Pertaining to the pelvis	• Sural (SIR-ul)	Pertaining to the posterior (back) surface of the leg
• Pubic (PYOO-bik)	Pertaining to the genital area	• Tarsal (TAR-sul)	Pertaining to the ankle
Dorsal	Pertaining to the back	• Pedal	Pertaining to the foot
• Scapular	Pertaining to the shoulder blade region	• Plantar	Pertaining to the sole of foot
• Vertebral (ver-TEE-brul)	Pertaining to the spinal column	• Calcaneal (kal-KANE-ee-ul)	Pertaining to the heel
• Lumbar	Pertaining to the area of the back between the lowest rib and buttocks.	• Digital	Pertaining to the digits (toes)

(a) Anterior view

(b) Posterior view

FIGURE 1.1 **Anatomical terms.**

(a) Anterior View

1 _____
2 _____
3 _____
4 _____
5 _____
6 _____
7 _____
8 _____
9 _____
10 _____
11 _____

12 _____
13 _____
14 _____
15 _____
16 _____
17 _____
18 _____
19 _____
20 _____
21 _____
22 _____

23 _____
24 _____
25 _____
26 _____
27 _____
28 _____
29 _____
30 _____
31 _____

(b) Posterior View

32 _____
33 _____
34 _____
35 _____
36 _____
37 _____
38 _____
39 _____
40 _____
41 _____
42 _____

B. Directional Terms

Directional terms are used to describe the location of body structures relative to other structures. An example of a directional term is *inferior*, which means below. It would be correct to say that the neck is inferior to the head but incorrect to say that the neck is inferior. The directional terms are listed in Table 1.2, along with an example of how they are used. Note that opposite terms are paired.

The directional terms *proximal* and *distal* apply to the point of attachment of a limb to the torso or the point of origin of a structure such as a blood vessel or nerve. These terms refer to the location of structures relative to the point of attachment or point of origin, whether they are closer (proximal) or farther away (distal).

More than one directional term can apply to describe the location of a body structure. For example, the ears are posterior and lateral to the nose.

Before Going to Lab

1 Label Figure 1.2 with the directional terms from the bulleted list by writing the term in the appropriate numbered blank.

LAB ACTIVITY 2 Directional Terms

1 With your partner, complete the sentences using the appropriate directional term from Table 1.2. Refer to the anatomical terms in Table 1.1 and Appendix A as needed.

a. The sternum is _____ to the vertebrae.

b. The nose is _____ and _____ to the eyes.

c. The heart is _____ to the lungs.

d. The wrist is _____ to the arm.

e. The right lung and right kidney are _____ .

f. The skin is _____ to the bones.

- anterior or ventral 1 _____
- distal 2 _____
- inferior 3 _____
- posterior or dorsal 4 _____
- proximal 5 _____
- superior 6 _____

FIGURE 1.2 Directional terms.

TABLE 1.2 Directional Terms

DIRECTIONAL TERM	DEFINITION	EXAMPLE OF USE
Superior	Above	The head is superior to the neck.
Inferior	Below	The neck is inferior to the head.
Anterior (Ventral)	Closer to front of body	The lips are anterior to the teeth.
Posterior (Dorsal)	Closer to back of body	The teeth are posterior to the lips.
Medial	Closer to midline of body	The nose is medial to the eyes.
Lateral	Farther from midline of body	The eyes are lateral to the nose.
Intermediate	Between two structures	The elbow is intermediate between the shoulder and wrist.
Ipsilateral	On same side of body	The right arm and right leg are ipsilateral.
Contralateral	On opposite sides of body	The right arm and left arm are contralateral.
Proximal	Nearer to point of attachment of limb to trunk	The elbow is proximal to the wrist.
Distal	Farther from point of attachment of limb to trunk	The wrist is distal to the elbow.
Superficial	Closer to surface of body	The skin is superficial to the muscles.
Deep	Farther from surface of body	The muscles are deep to the skin.

C. Body Planes and Sections

Planes are flat surfaces that divide the body or organs in order to expose internal structures (Figure 1.3). The exposed surfaces produced by planes are called **sections. Sagittal** (*sagitta* = arrow) **planes** pass vertically through the body or organs and divide them into right and left sections (**sagittal sections**).

If a plane passes vertically through the midline and divides the body into equal right and left halves, the plane is a **midsagittal plane,** but if a plane divides the body into unequal right and left portions, it is a **parasagittal plane.**

A **frontal** or **coronal plane** passes vertically through the body or organs and produces anterior and posterior sections (**frontal sections**). A **transverse plane** passes horizontally through the body and produces superior and inferior sections (**transverse sections** or **cross-sections**). **Oblique planes** pass through the body at an angle forming oblique sections.

We often look at sections of individual organs, such as blood vessels, intestines, or long bones. Sections that are produced by a plane running along the long axis of a long narrow structure are called **longitudinal sections.** Sections that are produced by a plane running perpendicular to the long axis are called **cross-sections.** Because blood vessels and intestines twist and bend, one body plane may produce longitudinal sections, cross-sections, and oblique sections of these structures.

> **CLINICAL NOTE:** *Transverse sections observed with computed tomography (CT) scans or magnetic resonance imaging (MRIs) are called **axial sections**.*

Before Going to Lab

1 Label the planes in Figures 1.3(a) and the sections in Figure 1.3(b) with the terms in the accompanying bulleted list by writing the term in the appropriate numbered blank.

2 Identify the type of sections of the human brain in Figure 1.4.

(a) Right anterolateral view

(b) Longitudinal and cross-sections

- cross-section 1 _____

- frontal plane 2 _____

- longitudinal section 3 _____

- midsagittal plane 4 _____

- oblique plane 5 _____

- parasagittal plane 6 _____

- transverse plane 7 _____

FIGURE 1.3 **Body planes and sections.**

(a)

(b)

(c)

- frontal

- midsagittal

- transverse

a _____

b _____

c _____

FIGURE 1.4 Human brain sections.

LAB ACTIVITY 3 Body Planes and Sections

1 Observe sagittal, frontal, and transverse sections using an apple.
- Working in a group, draw a face on the apple.
- Cut sagittal, frontal, and transverse planes through the apple to make sagittal, frontal, and transverse sections.
- Compare the appearance of the apple core in each section. Describe any difference in shape, size, and number of seed chambers.
- Keep sections together to form a whole apple to show to your instructor.

2 Observe longitudinal sections and cross-sections using plastic tubing or plastic straw.
- Observe a demonstration provided by your instructor of a tube cut along its longitudinal axis to produce a longitudinal section and a tube cut perpendicular to its longitudinal axis to produce a cross-section.
- Obtain an eight-inch piece of plastic tubing or plastic straw and twist it so you can visualize <u>one plane</u> that would simultaneously divide one area of the tube

into a longitudinal section and another area into a cross-section.
- Do not cut the tube unless instructed to do so.
- Show your instructor where a cut would produce both a longitudinal section and a cross-section.

3 Identify sagittal, frontal, transverse, and oblique sections on sheep brains.
- Your instructor will display five sheep brains—one whole brain and four brains that have been cut into different sections.
- Determine the anterior, posterior, superior, and inferior surfaces of the brains.
- Decide which brain has been cut into sagittal, frontal, transverse, or oblique sections.
- Compare the appearance of the different sections.

Brain 1—Whole brain

Brain 2 _____ section

Brain 3 _____ section

Brain 4 _____ section

Brain 5 _____ section

Reviewing Your Knowledge

A. Body Regions

Complete the following sentences.

_____ 1. The leg is to the lower limb as the _____ is to the upper limb.

_____ 2. The arm is to the upper limb as the _____ is to the lower limb.

_____ 3. The armpit is to the upper limb as the _____ is to the lower limb.

_____ 4. The tarsal bones are to the lower limb as the _____ bones are to the upper limb.

_____ 5. The elbow is to the upper limb as the _____ is to the lower limb.

_____ 6. The shoulder is to the upper limb as the _____ is to the lower limb.

_____ 7. True or False. The hand includes the wrist and fingers and the foot includes the ankles and toes.

_____ 8. True or False. The bones of the face are also part of the skull.

B. Anatomical Terms

Write the anatomical terms that the phrase or word describes. Phrases or words referring to nouns are indicated. All other phrases refer to adjectives.

_____ 1. Navel (noun)

_____ 2. Pertaining to the area between the neck and abdomen

_____ 3. Pertaining to the ear

_____ 4. Pertaining to the palm of hand

_____ 5. Pertaining to the high point of the shoulder

_____ 6. Pertaining to the anterior surface of the elbow region

_____ 7. Pertaining to the face; anterior portion of the head

_____ 8. Pertaining to the nose

_____ **9.** Pertaining to the neck

_____ **10.** Pertaining to the posterior surface of the knee

_____ **11.** Wrist (noun)

_____ **12.** Pertaining to the area between the elbow and wrist

_____ **13.** Back (noun)

_____ **14.** Armpit area (noun)

_____ **15.** Pertaining to the mouth

_____ **16.** Pertaining to the anterior surface of the knee

_____ **17.** Breast bone (noun)

_____ **18.** Pertaining to the hip

_____ **19.** Pertaining to the side of the leg

_____ **20.** Pertaining to the calf

_____ **21.** Pertaining to the area between the shoulder and elbow

_____ **22.** Pertaining to the fingers or toes

_____ **23.** Pertaining to the hand

_____ **24.** Pertaining to the breast

_____ **25.** Pertaining to the cheek

_____ **26.** Pertaining to the heel

_____ **27.** Pertaining to the sole of the foot

_____ **28.** Pertaining to the groin where the thigh attaches to the pelvic region

_____ **29.** Pertaining to the head

_____ **30.** Pertaining to the chin

_____ **31.** Pertaining to the foot

_____ **32.** Pertaining to the eye

_____ **33.** Pertaining to the genital area

_____ **34.** Pertaining to the area between the hip and knee

_____ **35.** Pertaining to the area that includes the bones enclosing the brain

_____ **36.** Pertaining to the forehead

_____ 37. Pertaining to the spinal column

_____ 38. Pertaining to the inferior back of the head

_____ 39. Pertaining to the anterior surface of the leg

_____ 40. Pertaining to the area of the lower back or loin

_____ 41. Pertaining to the trunk below the abdomen

_____ 42. Pertaining to the area of the back that contains the shoulder blades

_____ 43. Pertaining to the posterior surface of the elbow

_____ 44. Arm (noun)

_____ 45.⎫
 ⎬ Pertaining to the chest
_____ 46.⎭

C. Body Planes and Sections

Write the name of the plane that the phrase describes.

_____ 1. Divides body or organ into unequal right and left sections

_____ 2. Divides body or organ into anterior and posterior sections

_____ 3. Divides body or organ into superior and inferior sections

_____ 4. Divides body into right and left halves

_____ 5.⎫
 ⎬ Which planes when passed through the body would result in two sections, with each
 ⎪ section containing a piece of the heart and a piece of each lung?
_____ 6.⎭

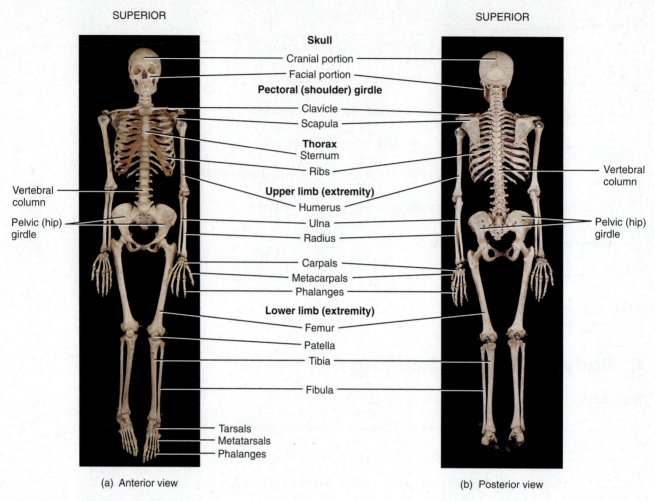

FIGURE 1.5 Human skeleton.

D. Directional Terms

Complete the sentences using directional terms. Use Figure 1.5 for reference.

1. The clavicle is _____ to the ribs.

2. The ribs are _____ to the sternum.

3. The humerus is _____ to the radius.

4. The ulna is _____ to the radius.

5. The tibia is _____ to the femur.

6. The right humerus and the right radius are _____.

7. The pelvic girdle is _____ to the ribs.

8. The sternum is _____ to the vertebral column.

9. The scapula is _____ to the clavicle.

10. The right fibula and left fibula are _____.

Using Your Knowledge

A. Body Regions, Anatomical Terminology, and Directional Terms

1. A 55-year-old male presented with an irregularly shaped and abnormally pigmented mole in the left scapular region, just lateral to the vertebrae. Indicate on Figure 1.6 where this mole is likely to be found.

2. A 37-year-old female presented to the emergency room with a severe burn (3rd degree) on the right brachial region just proximal to the antecubital region. Indicate on Figure 1.6 where the laceration is likely to be found.

3. A 19-year-old female was identified by a tattoo on the fibular surface of the right leg just proximal to the tarsal region. Indicate on Figure 1.6 where the tattoo is likely to be found.

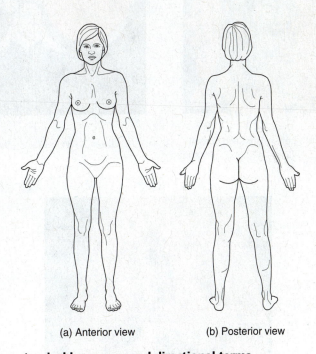

(a) Anterior view (b) Posterior view

FIGURE 1.6 Body regions, anatomical language, and directional terms.

Questions 4–7 have italicized words that are derived from word roots used to form the adjectives in Table 1.1. Using the locations suggested by the italicized words, answer questions 4–7.

4. Is the *popliteal* artery proximal or distal to the *femoral* artery?

5. Is the *pectoralis* major muscle anterior or posterior to the *subscapularis* muscle?

6. Is the *sternocleidomastoid* muscle superior or inferior to the *rectus abdominis* muscle?

7. Are the *thoracic* vertebrae medial or lateral to the *scapulae*?

B. Body Planes and Sections

Figure 1.7 contains three different sections through the thorax. Indicate which section (view a, b, or c) is a

8. Frontal section _____

9. Sagittal section _____

10. Transverse (axial) section _____

(a) (b)

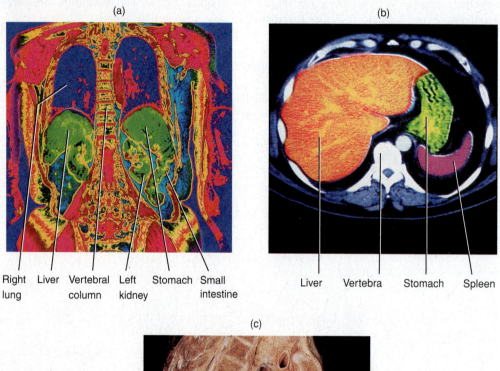

Right Liver Vertebral Left Stomach Small Liver Vertebra Stomach Spleen
lung column kidney intestine

(c)

Spinal Vertebral Heart
cord column Trachea Sternum

FIGURE 1.7 Sections through the thorax.

Organic Systems and Body Cavities

Organ Systems and Body Cavities

OBJECTIVES

1 Name the organ systems and describe the functions of each

2 Name and identify the major organs of each organ system

3 Describe the location of the body cavities and name the organs they contain

4 Describe the structure, location, and function of the serous membranes

5 Identify the abdominopelvic quadrants and regions and the major organs found in each

MATERIALS

- human torso models or charts
- male and female human reproductive models or charts
- paper or plastic large enough to outline student torsos, markers
- articulated skeleton
- one-gallon zippered plastic bags (1 per group)
- masking tape
- rat dissection video in the Wiley Student Companion Site

Organ systems are like different departments within a company. Within a company, departments work together to keep the company functioning. Within the body, organ systems work together to keep the body alive. In this exercise, you will learn the basic function and location of each organ system.

Before Going to Lab

1 Observe the organs in Figures 2.1(a) and (b). Refer to your textbook for a list of organ systems, their function, and the major organs in each organ system.

2 Write each labeled organ under the appropriate organ system. There may be organ systems that don't have any organs in these figures, and some organs may function with more than one system.

A. Overview of Organ Systems and Major Organs

An **organ system** is a group of organs performing a common function. All organ systems cooperate to maintain an optimal environment for body cells through a process called **homeostasis** (*homeo-* = same; *stasis* = standing). Failure to maintain homeostasis results in disorders, disease, and possibly death.

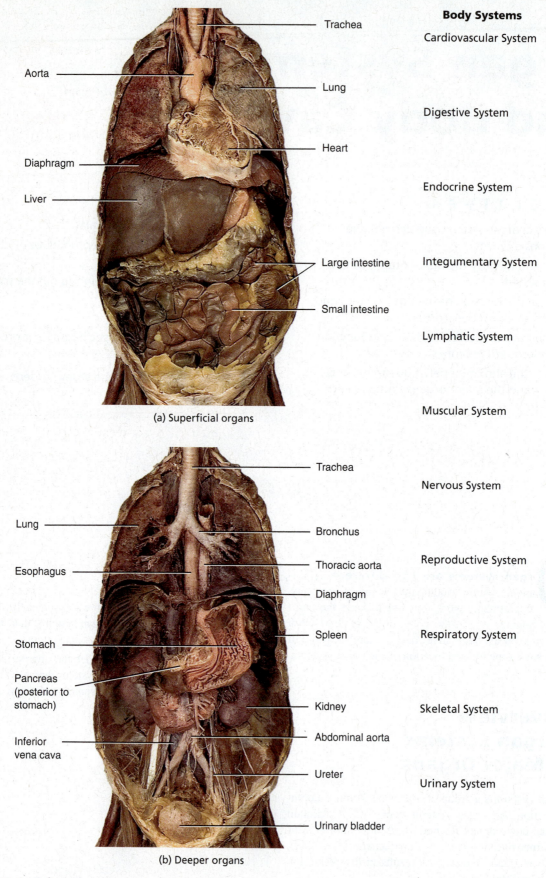

(a) Superficial organs

(b) Deeper organs

FIGURE 2.1 **Selected organs and organ systems.**

LAB ACTIVITY 1 Identification of Organs on Torso

You will be identifying organs from anterior to posterior on a torso model and answering questions concerning their position relative to the organs around them.

1 Identify the following organs on the anterior surface of a torso model. Identify all the organs without removing any organs from the model.
- brain
- trachea
- heart
- lungs
- liver
- stomach (torso's left side)
- small intestine
- large intestine (colon)

2 Remove the lungs, heart, liver, and stomach. Locate the gallbladder on the inferior surface of the liver.

3 Identify the following organs on the human torso model or chart:
- esophagus
- bronchi (right and left)
- inferior vena cava
- pancreas (posterior to stomach)
- spleen

4 Remove the small intestine and large intestine. Locate the appendix at the inferior right end of the large intestine.

5 Identify the following organs on the human torso model:
- abdominal aorta
- adrenal glands (superior surface of kidneys)
- kidneys
- ureters
- urinary bladder

6 Identify the female reproductive organs on a female reproductive model or chart. Observe the position of the urinary bladder relative to the uterus.
- ovaries
- uterus
- urinary bladder

7 Identify the male reproductive organs on a male reproductive model or chart.
- penis
- scrotum (skin covering testes)
- testes

8 Answer the following questions about the position of each organ on the torso model or Figure 2.1.

1. The stomach is _____ to the small intestine.
 a. superior b. inferior c. medial d. lateral

2. The liver is _____ to the lungs.
 a. superior b. inferior c. medial d. lateral

3. The lungs are _____ to the heart.
 a. superior b. inferior c. medial d. lateral

4. The trachea is _____ to the esophagus.
 a. medial b. inferior c. anterior d. posterior

5. The pancreas is _____ to the stomach.
 a. superior b. anterior c. lateral d. posterior

6. The large intestine is _____ to the stomach.
 a. superior b. inferior c. posterior d. lateral

7. The stomach is _____ to the spleen.
 a. lateral b. medial c. superior d. inferior

8. The abdominal aorta and inferior vena cava are _____ to the kidneys.
 a. medial b. lateral c. superior d. inferior

9. The kidneys are _____ to the small intestine.
 a. anterior b. posterior c. superior d. inferior

10. The urinary bladder is _____ to the kidneys.
 a. posterior and superior b. medial and inferior
 c. medial and superior d. lateral and posterior ■

LAB ACTIVITY 2 Organ Location

1 Draw the outline of a full-size torso on paper or plastic.

2 Using a marker, draw life-size outlines of all superficial organs in the appropriate place on the paper or plastic torso. ■

B. Body Cavities

Many of the body's organs are found within body cavities. The *cranial cavity* contains the brain, and it is continuous with the *vertebral* (*vertebra* = back) canal that contains the spinal cord.

The **thoracic cavity** is a space enclosed by the ribs, sternum, and vertebral column. This cavity contains three small cavities: the **pericardial cavity** (*peri-* = around; *-cardia* = heart) and two **pleural cavities** (*pleuro-* = side or rib). The pericardial cavity surrounds the heart, and each pleural cavity contains a lung. The **mediastinum** (*media-* = middle; *-stinum* = partition), a central area within the thoracic cavity, extends from the neck to the diaphragm and from the sternum to the vertebral column. The organs located in the mediastinum are the heart, thymus gland, esophagus, trachea, blood vessels, and bronchi. The pleural cavities are located on either side of the mediastinum. The **diaphragm** separates the thoracic cavity from the abdominopelvic cavity.

The **abdominopelvic cavity** consists of two continuous cavities: the abdominal cavity and the pelvic cavity. The **abdominal cavity** is the superior portion located between the diaphragm and the brim of the pelvis (hip bones). This cavity contains the stomach, liver, gallbladder, pancreas, spleen, small intestine, kidneys, appendix, and part of the large intestine. Within the abdominal cavity is the

peritoneal cavity. Although most abdominal organs are positioned within the peritoneal cavity, a few organs are **retroperitoneal** (*retro-* = backward), or located posterior to the peritoneum. These organs are the pancreas, kidneys, adrenal glands, and portions of the large intestine, small intestine, aorta, and inferior vena cava. The **pelvic cavity** is the inferior portion of the abdominopelvic cavity. The pelvic cavity contains part of the large intestine, rectum, urinary bladder, female reproductive organs (ovaries, uterine tubes, uterus, vagina), and male reproductive organs (prostate, and part of ductus deferens). It is important to note that the testes and penis are not located in the pelvic cavity but are located inferior to it.

Before Going to Lab

1 Label the major body cavities and the diaphragm on Figure 2.2(a) and (b).

LAB ACTIVITY 3 Body Cavities

1 Locate the major body cavities on a skeleton and torso model. Identify the organs located in each body cavity.

2 Locate the mediastinum (meed-ee-uh-STINE-um) on a torso model or on Figure 2.1. Identify the organs located within the mediastinum. ■

- abdominal cavity
- cranial cavity
- diaphragm
- pelvic cavity
- thoracic cavity
- vertebral canal

1 _____

2 _____

3 _____

4 _____

5 _____

6 _____

(a) Right lateral view (b) Anterior view

FIGURE 2.2 Body cavities.

C. Serous Membranes

Most of the organs in the ventral body cavity are covered with thin **serous** (*serum* = any clear, watery fluid) **membranes,** which are composed of two layers: a visceral layer and a parietal layer. The **visceral** (*viscera* = internal organs) **layer** covers the organ, whereas the **parietal** (*paries* = wall) **layer** attaches to and covers the ventral body wall. These two layers make up one continuous sheet that folds to form a sac. Between the two layers is a potential cavity containing a small amount of serous fluid secreted by the membranes. The clear, watery **serous fluid** prevents friction as the organs move within the ventral body cavity. For example, the heart has movement within the thoracic cavity as it fills with and ejects blood.

Serous membranes are named for the cavities they surround. Thoracic serous membranes include the **pleura,** which covers the lungs, and the **pericardium,** which covers the heart. The serous membrane that covers the abdominal organs is the **peritoneum** (*peri-* = around; *teinein* = to stretch).

Before Going to Lab

1 In Figure 2.3, observe how the serous pericardium folds to form a double layer.

2 Label the two layers of the serous pericardium in Figure 2.3.

- parietal
- visceral

1 _____

2 _____

FIGURE 2.3 Serous pericardium folds to surround the heart.

LAB ACTIVITY 4 Serous Membranes

1 Make a replica or model of a serous membrane with your lab group.
- Obtain a 1-gallon zippered plastic bag.
- Push all the air out of the bag and zip the bag.
- Have a lab partner place a fist (simulating an organ) on the bottom edge of the bag and push up into the bag so the bag surrounds the fist.
- Remove the fist, unzip the bag, and add about 40 to 50 mL of water to the bag. Push out the extra air before rezipping the bag.
- Now have the same lab partner place a fist (simulating an organ) on the bottom edge of the bag and push up into the bag so the bag surrounds the fist.

2 Clean up as directed by your instructor.

3 Answer the Discussion Questions with your lab group.

DISCUSSION QUESTIONS
Serous Membranes

1 In the bag with water, what is the name of the simulated serous membrane layer that is touching the fist (organ)?

2 In the same bag, what is the name of the simulated outer serous membrane layer?

3 What does the water represent?

4 Was it easier to push a fist into the bag with no water or into the bag with water?

5 Based on your observations, does the presence of serous fluid make it easier for organs to move? Explain.

D. Organ Systems, Body Cavities, and Serous Membranes in the Rat

The organ systems, body cavities, and serous membranes of the rat are similar to those of humans. The rat dissection will allow you to see the relationship of organs to each other, organ location within body cavities, and serous membranes.

LAB ACTIVITY 5 Rat Dissection Video

Go to the Wiley Student Companion Site to view the rat dissection video.

E. Abdominopelvic Regions and Quadrants

Anatomists divide the abdominopelvic cavity into nine **regions** using two vertical and two horizontal lines in a tic-tac-toe grid so that the location of any organ is simple to describe. The two vertical lines are drawn mid-clavicular (mid-collar bone) and just medial to the nipples, beginning at the diaphragm and extending inferiorly through the pelvic area. The upper horizontal line is drawn across the abdomen, inferior to the ribs and across the inferior portions of the liver and stomach. The lower horizontal line is drawn slightly inferior to the superior portion of the pelvic bones. These nine regions from the top right to the lower left are **right hypochondriac** (*hypo-* = under; *chondro-* = cartilage), **epigastric** (*epi-* = upon; *gastro-* = stomach), **left hypochondriac, right lumbar** (*lumbar* = loin), **umbilical, left lumbar, right inguinal** or **iliac** (*inguinal* = groin), **hypogastric** or **pubic**, and **left inguinal** or **iliac**. Clinicians are more apt to divide this cavity into four **quadrants** that are formed by transverse and sagittal planes running through the umbilicus (navel). These quadrants are useful clinically when one is trying to describe abnormalities or to determine which organ may be the cause of pain. The four quadrants are **right upper quadrant (RUQ), left upper quadrant (LUQ), right lower quadrant (RLQ),** and **left lower quadrant (LLQ).**

> **NOTE:** *Right and left always refer to the model's or specimen's own right and left.*

Before Going to Lab

1 Draw lines on Figure 2.4(a) separating the abdominopelvic cavity into quadrants and label the quadrants.

2 Draw lines on Figure 2.4(b) separating the abdominopelvic cavity into regions and label the regions.

LAB ACTIVITY 6 Abdominopelvic Quadrants and Regions

1 Using a piece of masking tape, mark the location of the diaphragm on a human torso or on yourself.

2 Using two pieces of masking tape, divide the abdominopelvic cavity into quadrants on a human torso or on yourself.

3 Using the torso model or your textbook, identify in which abdominopelvic quadrant(s) each organ is *primarily* located. Use the abbreviations RUQ, LUQ, RLQ, and LLQ.

a. appendix _____

b. large intestine or colon _____

c. liver _____

d. ovaries _____

e. pancreas _____

f. small intestine _____

g. spleen _____

h. stomach _____

4 Using four pieces of masking tape, divide the abdominopelvic cavity into regions on a human torso or on yourself.

5 Using the torso model or your textbook, identify in which abdominopelvic region each organ is *primarily* located.

a. appendix _____

b. gallbladder _____

c. left ovary _____

d. bifurcation of the abdominal aorta _____

e. spleen _____

f. stomach (majority of) _____

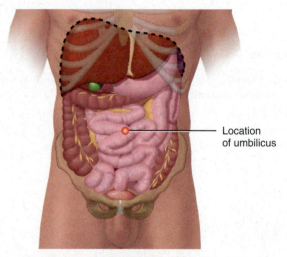

Location of umbilicus

(a) Quadrants

Location of umbilicus

(b) Regions

FIGURE 2.4 Abdominopelvic cavity.

Reviewing Your Knowledge

A. Functions and Identification of Organ Systems

Identify the organ system by its function as described below.

_____ 1. Maintains blood oxygen and carbon dioxide levels

_____ 2. Controls muscles and glands by electrical impulses; helps control homeostasis

_____ 3. Causes movement of bones

_____ 4. Waterproof barrier that blocks the entrance of pathogens into the body and prevents the loss of water from the body

_____ 5. Transports nutrients, oxygen, and carbon dioxide throughout the body

_____ 6. Changes food into absorbable nutrients; expels wastes

_____ 7. Regulates composition of blood by eliminating nitrogenous wastes, excess water, and minerals

_____ 8. Uses hormones to control cell function; helps control homeostasis

_____ 9. Provides framework for the body and protects body organs

_____ 10. Produces gametes (sperm and egg)

_____ 11. Returns fluid to the bloodstream and provides protection against pathogens that have entered the body

B. Organ System Identification

Identify the correct organ system for the following organs.

_____ 1. spleen

_____ 2. liver

_____ 3. trachea

_____ 4. blood vessels

_____ 5. hair

_____ 6. kidney

_____ 7. uterus

_____ 8. pituitary gland

_____ 9. spinal cord

_____ 10. testes (2 systems)

_____ **11.** prostate gland _____ **14.** adrenal gland

_____ **12.** large intestine _____ **15.** thyroid

_____ **13.** pancreas (2 systems)

C. Body Cavities

Identify all the cavities for each organ as follows: cranial (C), vertebral (V), thoracic (T), pleural (PL), pericardial (PC), peritoneal (PT), abdominal (A), or pelvic (P).

_____ **1.** brain _____ **7.** spinal cord

_____ **2.** small intestine _____ **8.** liver

_____ **3.** heart _____ **9.** kidneys

_____ **4.** lungs _____ **10.** uterus

_____ **5.** bronchi _____ **11.** urinary bladder

_____ **6.** stomach _____ **12.** ovaries

D. Abdominopelvic Quadrants and Regions

Name the quadrant(s) (RUQ, LUQ, RLQ, and LLQ) and region(s) (right hypochondriac, epigastric, left hypochondriac, right lumbar, umbilical, left lumbar, right inguinal or iliac, hypogastric or pubic, and left inguinal or iliac) that the following organs predominantly occupy.

_____ **1.** liver _____ **5.** appendix

_____ **2.** stomach _____ **6.** left kidney

_____ **3.** spleen _____ **7.** right ovary

_____ **4.** gallbladder _____ **8.** uterus

E. Serous Membranes

Write the term the phrase describes.

_____ **1.** Attaches the heart to the body cavity

_____ **2.** Covers the surface of the lungs

_____ **3.** Covers the surface of abdominal organs

_____ **4.** The lubricating liquid in serous cavities

_____ **5.** Circle the organs that are found within the peritoneal cavity: pancreas, liver, kidney, spleen, adrenal glands, abdominal aorta, inferior portions of vena cava, stomach

Using Your Knowledge

A. Homeostatic Imbalances of Organ Systems

Using your textbook, identify the organ system that is homeostatically imbalanced in the following diseases or disorders.

_____ **1.** muscular dystrophy

_____ **2.** hypothyroidism

_____ **3.** myocardial ischemia

_____ **4.** infectious mononucleosis

B. Body Cavities and Serous Membranes

Identify *all* the cavities entered for each procedure, beginning with the largest cavity and ending with the most specific body cavity. Use these abbreviations for the body cavities: abdominal (A), cranial (C), pelvic (P), pericardial (PC), pleural (PL), peritoneal (PT), thoracic (T), and vertebral (V).

_____ **5.** coronary bypass surgery

_____ **6.** cholecystectomy (gallbladder removal)

_____ **7.** spinal tap

C. Abdominopelvic Quadrants

_____ **8.** A 44-year-old male went to the emergency room complaining of severe pain in his RLQ. The doctor palpated the area and determined that the pain was originating from an organ in that quadrant. Which organ might be involved?

 (a) liver (b) appendix (c) gallbladder (d) spleen (e) stomach

_____ **9.** A 23-year-old female went to the doctor with the chief complaint of RLQ pain. Which organ is most likely the cause?

 (a) adrenal gland (b) ovary (c) gallbladder (d) pancreas (e) kidney

D. Organ Identification

Identify the organs in the color-enhanced medical images in Figure 2.5.

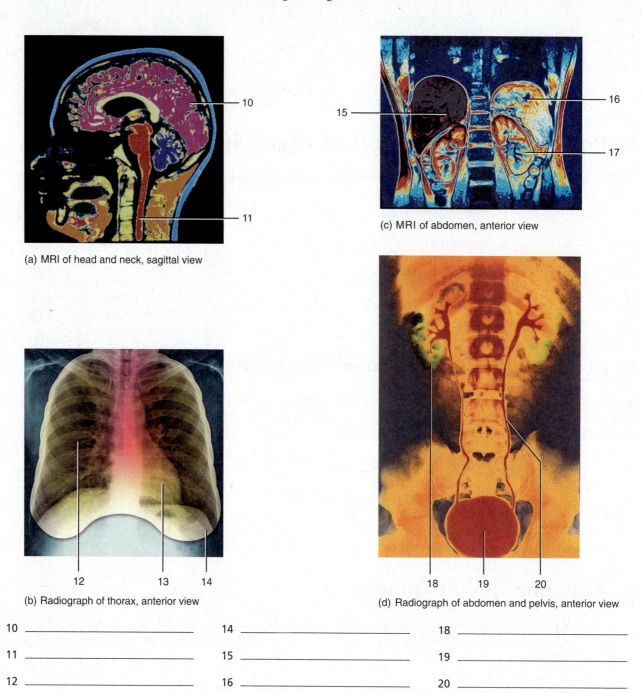

(a) MRI of head and neck, sagittal view

(b) Radiograph of thorax, anterior view

(c) MRI of abdomen, anterior view

(d) Radiograph of abdomen and pelvis, anterior view

10 _____ 14 _____ 18 _____

11 _____ 15 _____ 19 _____

12 _____ 16 _____ 20 _____

13 _____ 17 _____

FIGURE 2.5 Identification of organs on medical images.

Compound Light Microscope

OBJECTIVES

1 Describe and demonstrate how to carry, clean, use, and store a compound light microscope

2 Identify the parts of a compound light microscope and describe their function

3 Calculate total magnification for each objective lens

4 Demonstrate how to view an object with the microscope using all magnifications

5 Demonstrate how to measure the field of view

6 Measure the diameter of a cell

7 Prepare a wet-mount slide

MATERIALS

- compound light microscopes, lens paper, immersion oil
- thin, clear plastic rulers
- prepared microscope slides of the letter "e"
- prepared microscope slides of the trachea (or other organ)
- wet mount of cheek cells; clean microscope slides, coverslips, lens paper, flat toothpicks, and dropper bottle of dilute methylene blue, 0.9% saline solution, 10% bleach solution

A compound light microscope is used to observe small structures such as cells and tissues. The term *compound* refers to the two types of lenses (ocular and objective) that are used simultaneously to magnify the image. The term *light* refers to the necessity of using a light source to view the object. Most human cells must be magnified to be seen by the unaided human eye. The compound light microscope can magnify images up to approximately 1,000 times, depending on the magnifying power of the lenses.

Microscopic examination of cells and tissues allows students to observe how cell and tissue structure determines function. Changes in normal cell and tissue structure cause changes in organ function that lead to a disorder or disease. Tissue biopsies are performed to observe whether normal cellular structure has changed, which would indicate the absence or presence of a disorder or disease.

A. Transporting the Microscope

The compound light microscope is an expensive, precision instrument that must be handled appropriately. Demonstrate care in transporting, cleaning, using, and storing the microscope.

- Pick up the microscope with two hands, one holding the arm and the other supporting the base with the cord in a secure position.
- Carry the microscope upright so that a lens or eyepiece does not fall out, and carefully place the microscope on the lab table in front of you.

B. Parts of the Microscope

- **Base**—The wide bottom part that supports the microscope.
- **Arm**—The straight or curved vertical part that connects the base to the head.
- **Head** (or body tube)—The upper part of the microscope that extends from the arm and contains the ocular lens(es) and the rotating nosepiece with the objective lenses.

NOTE: *All other microscope parts attach to the base, arm, and head—the three basic parts of the framework.*

- **Ocular lens(es)**—Removable eyepieces used to observe the microscope slide. Microscopes with one ocular lens are called **monocular** (*mono-* = one; *ocu-* = eye), and those with two ocular lenses are called **binocular** (*bi-* = two). Typically, these lenses magnify an object tenfold (10×). Look at an ocular lens and record the magnification power. _____ One of the ocular lenses may have a **pointer** used to identify a specific area on the slide. A **micrometer,** used to measure the field of view and object size, may also be present in one ocular lens. State whether your microscope has a pointer and/or a micrometer. If it has a pointer or micrometer, give the ocular lens (right or left) in which each is found.

 Pointer _____

 Micrometer _____

- **Objective lenses**—A microscope will usually have three or four objective lenses mounted on a **revolving nosepiece.** Most microscopes have objective lenses that magnify an object 4× (scanning), 10× (low-power), 40× (high-dry), and 100× (oil immersion). List the magnification powers of the objective lenses on your microscope. _____ As the barrel of the objective lens increases in length, the magnifying power also increases.

- **Stage**—The flat platform located beneath the objective lenses on which the microscope slide is placed. The stage has a hole in the middle, the light aperture, through which light is focused on the slide. The slide may be held onto the stage with either two spring clips or a mechanical stage clamp. Does your microscope have 2 spring clips or a mechanical stage? _____

- **Mechanical stage**—Holds the slide securely in place with a spring clamp for viewing and can be moved with precision by using the **adjuster knobs.** One knob moves the slide side to side, and the other forward and backward.

- **Coarse focus knobs**—On each side of the microscope toward the base is a large knob or dial that may or may not have a smaller knob in the middle. Only one of the large knobs needs to be used, depending on whether one is right- or left-handed. The large knob is used for coarse focusing and either moves the stage up and down quickly or moves the objective lenses up and down quickly. This knob is to be used with scanning or low-power lenses. Does your stage or objective lens move? _____.

- **Fine focus knob**—The smaller knob on each side of the microscope that is used for precision focusing.

- **Condenser**—Located just below the stage is a lens that condenses light through the specimen on the slide above. If the condenser has an **adjustment knob** that raises and lowers the condenser, it usually needs to be in the highest position to focus the most light on the specimen.

- **Iris diaphragm**—Located beneath the condenser, the iris diaphragm works similarly to the iris of the eye. By adjusting its **lever,** the aperture changes diameter and regulates the amount of light that passes through the condenser. Decreasing the aperture size decreases the amount of light on the specimen and increases contrast.

- **Substage light**—The light source is usually built into the base of the microscope and typically has a dial or sliding bar on one side to control the light intensity.

LAB ACTIVITY 1 Parts of the Microscope

1 Identify the parts of your microscope as shown in Figure 3.1.

2 Compare your microscope with the one in Figure 3.1 and identify any differences with your lab group. ■

C. Calculating Magnification

Total magnification is determined by multiplying the ocular lens power times the objective lens power. *Example:* Ocular lens power = 10×; Objective lens power = 4×; Total magnification = 40×.

LAB ACTIVITY 2 Calculating Magnification

1 Calculate the total magnification by multiplying the magnifying powers of your microscope lenses.

 scanning lens ____ × ocular lens ____ = ____ total magnification

 low-power lens ____ × ocular lens ____ = ____ total magnification

 high-dry lens ____ × ocular lens ____ = ____ total magnification

 oil immersion lens ____ × ocular lens ____ = ____ total magnification ■

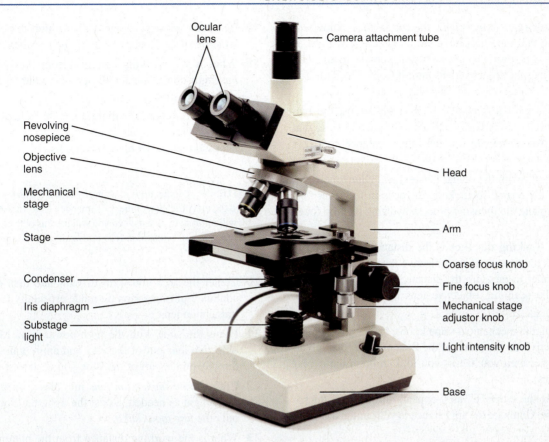

FIGURE 3.1 **Parts of the microscope.**

D. Using the Microscope

- Clean up your lab area and put nonessentials away so you will have plenty of room to use the microscope.

- Unwind the cord and plug it in.

- Clean the ocular, objective lenses, and condenser lenses only with the *special lens paper* (optical safe) provided by your instructor. Whenever the image on the slide cannot be focused clearly, it may be that the ocular, objective lens, or slide is dirty and needs additional cleaning. If all else has failed, consult your instructor.

- Turn on the light and adjust it to the lowest light setting feasible for good visibility and color to reduce eye strain. The scanning (4×) and low-power (10×) lenses will not need as much light as the high-dry (40×) and oil immersion (100×) lenses.

- *Trouble-shooting:* If no light comes on initially, check two things before consulting your instructor. (a) Turn the light dial to a higher setting. (b) Check the safety switch on the electrical outlet by pushing in the reset button. If the light still does not work, plug your microscope into an outlet that you know works.

LAB ACTIVITY 3 **Using the Microscope**

1 Move the **scanning objective lens** into place so it is over the light aperture on the stage. This objective lens has the shortest barrel. Make sure you feel the lens click into position or your field of view will be black.

2 Obtain a prepared slide with the letter "e" from your instructor and place it on the stage, securing it with either the mechanical stage clamps or slide clips. Draw the letter "e" as it appears on the stage without looking in the ocular lens. _____

3 Without looking into the ocular lens, practice moving the slide from side to side in addition to backward and forward using the mechanical stage knobs (or your hands if your stage has slide clips).

4 Using the mechanical stage knobs (or your hands if your stage has slide clips), position the letter "e" over the light hole in the stage.

5 Check to see that the condenser lens is raised completely up to the stage.

6 If your *stage is moveable,* the coarse focus knob will move the stage. Raise the stage as far as it will go. If your *objective is moveable,* use the coarse focus knob to lower the objective lens until it stops. The slide and the scanning objective lens will not actually touch.

7 If you have a binocular microscope, adjust the two ocular lenses as you would a pair of binoculars so that the two lenses are a comfortable distance apart for your eyes.

8 Look through the ocular lens(es) and adjust the light. Use the coarse focus knob to focus in the letter "e." Complete the focusing process by using the fine focus knob.

9 The **working distance** is the distance a specimen is from the bottom of the objective lens. Use a millimeter ruler to measure the distance between the bottom of the scanning objective lens and your specimen. _____ mm

10 Using the mechanical stage knobs, bring the letter "e" directly into the center of the **field of view** (the lighted circular area you see as you look through the ocular lenses).

11 Draw the letter "e" as it appears through the microscope. Compare the appearance to your initial drawing. _____

12 While observing the letter "e" through the ocular lens(es), describe the movement that you observe as you move the slide:

- to the left _____
- to the right _____
- forward _____
- backward _____

13 Reposition the letter "e" directly in the middle of the field of view and switch to the **low-power objective lens.**

14 Most microscopes are **parfocal** so that when you move to a different magnification the specimen is almost, but not quite, in focus. You will need only the fine focus knob to focus the image. Center the specimen because the previously centered object is usually not in the center.

15 What is the working distance from the bottom of the low-power objective lens to your specimen? _____ mm

16 Use the iris diaphragm lever to adjust the amount of light and improve the contrast of your image.

CAUTION: *Do not use the coarse focus knob with high-dry or oil immersion lenses.*

17 Repeat the above procedure with the **high-dry objective lens.** Be sure to focus *only* with the *fine focus knob.*

18 What is the working distance from the bottom of the high-power objective lens to your specimen? _____ mm

19 Describe the change in diameter of the field of view as one switches from the scanning lens to the low-power lens and then to the high-power lens.

NOTE: Most slides used in anatomy and physiology do not need the magnification of the oil immersion lens. Your instructor will inform you if and when you will use this objective lens. *Only use the oil immersion lens if instructed to do so.*

20 Center the area you want to view and then obtain a container of immersion oil made especially for the oil immersion lens.

21 Focus the slide with the high-power lens. Move the objective lens out of the way and apply a drop of oil directly on the part of the slide you wish to study.

22 Click the *oil immersion lens* into place, open the iris diaphragm as needed, adjust the light, and focus with only the *fine focus knob.*

23 What is the working distance from the bottom of the oil immersion lens to your specimen? _____ mm

24 When you finish, move the scanning power objective lens back into place.

25 Move the stage as far from objectives as possible by either lowering the stage or raising the objectives. Remove the slide and clean the oil from the oil immersion lens with lens paper. (Also clean the high-dry objective lens if you passed it through the oil.) Your instructor may ask you to use an additional cleaner.

26 Clean the slide with a new lens paper. If necessary, clean the stage as well.

E. Measuring the Field of View

The **field of view** is the area on the slide that is being observed and is inversely proportional to the magnification (the field of view decreases in size with increasing magnification). Once you know the diameter of the field of view in millimeters (mm) at various magnifications, you will be able to estimate the size of cells or other structures in the field of view. The object being viewed should be in the center of the field of view when you are switching to a higher objective lens (higher magnifying power).

Measuring the field of view at different magnifications demonstrates the advantage of scanning at a lower magnification to find a structure of interest before working up to a higher magnification.

LAB ACTIVITY 4 **Measuring the Field of View and Estimating Object Size**

1 Move the scanning objective lens in place.

2 Place a clear plastic ruler over the light opening in the stage, or use the micrometer in the ocular lens or grid slide.

3 Look through the ocular lens, and move the ruler or grid slide so that a line touches the left edge of the field and count the number of millimeter intervals that can be seen. Record. _____ mm

4 Switch to the low-power objective lens and repeat this procedure to count the number of millimeter intervals that can be seen. Record. _____ mm

5 Switch to the high-dry objective lens and repeat this procedure to count the number of millimeter intervals that can be seen. Record. _____ mm. The closeness of this lens to the slide may not allow a ruler to be added.

6 Move the scanning power objective lens in place and move the stage as far as possible from the objectives before removing the ruler.

7 Obtain a prepared slide with the letter "e" and place it on the stage. Using the scanning power objective lens, estimate the diameter of the letter e. If it occupies ½ of the field of view, then it is ½ times the measured diameter of the field of view of the scanning objective lens. Record the diameter of the letter "e." _____ mm

8 Move the stage as far as possible from the objectives before removing the ruler. ■

F. Microscopic Structure of an Organ

An organ is composed of a variety of cells and tissues. This activity starts with an observation of a section of a whole organ with the scanning lens to get the "big picture" and then moves to higher magnifying powers to see additional detail.

LAB ACTIVITY 5 **Observation of an Organ**

1 Obtain a prepared slide of the trachea or other organ supplied by your instructor.

2 Begin your observation using the scanning lens. Center and focus the organ, and note that you can see several different tissues (stained different colors) present at this magnifying power.

3 Switch to the low-power lens. Center and focus, and note the additional tissue detail that can be discerned at this magnifying power. Make a drawing of the tissues in Figure 3.2(a).

4 Switch to the high-power lens. Center and focus, and note that you can now observe the cells that constitute the various tissues in this organ. Make a drawing of the cells in Figure 3.2(b).

5 Estimate the diameter of 3 different types of cells.

_____ mm

_____ mm

_____ mm

6 Move the scanning power objective lens into place and move the stage as far as possible from the objectives before removing the slide. ■

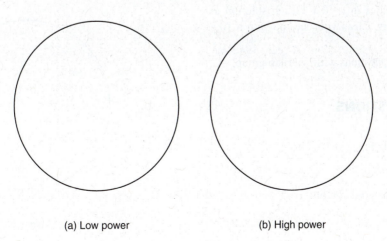

(a) Low power (b) High power

FIGURE 3.2 **Student drawings of tissues and cells.**

G. Wet Mount of Cheek Cells

Cells that line the interior of the mouth fit closely together like floor tiles and form a thick layer of thin cells that protect the underlying tissue from abrasion and microbes (bacteria and viruses). The superficial cells continually slough off and are replaced by underlying cells. Gently scraping the lining of the cheek removes the superficial cells that are about to slough off.

LAB ACTIVITY 6 **Wet Mount of Cheek Cells**

1 Prepare a cheek smear slide and observe it under the compound microscope.
- Obtain a toothpick, a clean microscope slide, and a coverslip.
- Place a drop of 0.9% saline on the microscope slide.
- Gently scrape the flat end of the toothpick (no blood, please!) on the inner lining of your cheek only one time. Do not scrape hard enough to hurt.
- To apply the cells to the slide, rotate the toothpick between your thumb and forefinger to dislodge the cells into the saline.
- Dispose of the toothpick as your instructor directs.
- Add *one drop* of dilute methylene blue to the cells on your slide.
- Cover the sample with a coverslip as directed by your instructor.
- Using low-power, locate the blue-stained cells. Switch to high-power to observe more cellular detail.

2 Compare the cells on your slide with Figure 3.3(a). Draw a picture of your cells in Figure 3.3(b).

3 Estimate the diameter of the cheek cells. _____ mm

4 Move the scanning power objective lens in place and remove the slide.

5 Place microscope slides in a 10% bleach solution as your instructor directs. Clean your lab top with a 10% bleach solution.

6 Answer Discussion Questions with your lab group.

DISCUSSION QUESTIONS
Cheek Smear

1 Why was stain added to the cheek cells?

2 What cellular structures did you observe?

H. Storing the Microscope

It is important to put the microscope away properly.

- Check that the scanning objective lens is in place and that the slide is removed from the stage.
- Depending on the type of microscope, either lower the stage or raise the objective to put maximum distance between the objective and the stage.
- Center the mechanical stage.
- Turn off the substage light. The bulb life is extended if it cools before moving the microscope.
- Clean the ocular and objective lenses with lens paper.
- Coil the cord neatly according to your instructor's directions.
- Place the dust cover over the microscope.
- Using both hands and the proper carrying technique, return the microscope to the appropriate cabinet.

Cells folded over • Nucleus • Cytoplasm • Overlapping cells

400×

(a) Photomicrograph of cheek cells

(b) Student drawing

FIGURE 3.3 **Cheek cells.**

Reviewing Your Knowledge

A. Care and Use of the Microscope

Correct each statement by crossing out the incorrect word(s) and inserting the correct word(s).

1. One hand is to be used to transport the microscope.

2. Tissue paper can be used to clean the microscope lenses and prepared slides.

3. The microscope should be put away with the high-dry lens in position.

4. The coarse focusing knob should be used when using the high-dry lens.

5. The iris diaphragm should be completely open to obtain maximum contrast.

6. The condenser should be in the lowest position (far from the stage) to focus the most light on the specimen.

B. Parts of the Microscope

Write the term that the phrase describes.

_____ 1. Large knob that moves the stage or objective lens a great distance. Used with scanning or low-power objective lenses only.

_____ 2. Flat platform beneath the objective lens on which the microscope slide is placed.

_____ **3.** Removable lenses that you look through to observe the microscope slide.

_____ **4.** Small knob that moves the stage or objective lens a very small distance and is used for precision focusing.

_____ **5.** Extends from the arm and contains the ocular lenses and rotating nosepiece.

_____ **6.** Lens that condenses light through the specimen and is located below the stage.

_____ **7.** Light from specimen passes through these lenses first. These lenses are located in the rotating nosepiece.

_____ **8.** Wide bottom part that supports the microscope.

_____ **9.** Regulates the amount of light passing through the condenser.

_____ **10.** Vertical portion that connects the base to the head.

C. Total Magnification and Field of View

1. Calculate the total magnification of an object viewed with a 10× ocular and a 60× objective lens.

2. Does the size of the field of view increase or decrease when going from a lower- to higher-power objective lens?

Cell Structure and Cell Cycle

OBJECTIVES

1 Identify cellular components on a model or diagram
2 Describe the function of the plasma membrane and cellular organelles
3 Identify cells and observable cellular structures on prepared microscope slides or on photomicrographs
4 Identify the stages of mitosis
5 Describe the events of each stage of mitosis

MATERIALS

- model or diagram of a cell
- compound microscopes and lens paper, prepared slides of human skeletal muscle cells, pseudostratified ciliated columnar epithelium (trachea), nonciliated simple columnar epithelium with microvilli (small intestine), motor neurons, sperm, and blood or **Real Anatomy** (Histology)
- 3-dimensional models of mitosis
- whitefish blastula slides

The human body contains over a trillion cells. These cells form the organs of the human body and are responsible for organ function. Cells take in nutrients delivered to them by the blood and use these nutrients to make carbohydrates, proteins, lipids, and nucleic acids. Cells use these macromolecules to make cellular and extracellular structures, repair themselves, and perform the tasks required for organ function.

A. Cell Structure

Cells are the smallest structural and functional units of living organisms. They are enclosed by a plasma membrane that controls the movement of substances into and out of the cell. The interior of the cell is filled with cytoplasm that contains cytosol (a viscous fluid) and organelles (little organs). Like an automobile, a cell has different parts or organelles that perform different functions. A "generalized" animal cell is shown in Figure 4.1, and functions of cellular organelles are described in Table 4.1.

LAB ACTIVITY 1 Cell Structure

1 Point to each cell structure shown in Figure 4.1 on a cell model or chart.
2 Describe the function of each organelle in Figure 4.1(a).
3 Answer the Discussion Question with your lab group.

DISCUSSION QUESTIONS
Cell Structures

1 Which cell structures from Table 4.1 are not found in most human cells?

(a) Sectional drawing

1 mitochondrion	7 microvillus	13 nucleolus
2 peroxisome	8 flagellum	14 nucleus
3 smooth endoplasmic reticulum	9 cilium	15 cytoplasm
4 lysosome	10 secretory vesicle	16 rough endoplasmic reticulum
5 plasma membrane	11 chromatin	17 ribosome
6 centrioles	12 nuclear membrane	18 Golgi complex

(b) Transmission electron micrograph 4000×

FIGURE 4.1 Generalized animal cell.

TABLE 4.1 **Function of Cell Structures**

STRUCTURE	FUNCTION
Plasma Membrane	Controls movement of substances into and out of the cell
Microvilli	Folds of the plasma membrane that increase the surface area of the cell to increase absorption or secretion
Nucleus	Contains DNA molecules and nucleolus
Nucleolus	Assembly site for ribosomes
Chromatin	Long thin strands within nucleus. Each strand composed of one DNA molecule and associated proteins.
Cytoplasm	Area of the cell that includes the cytosol and organelles
Cytosol	Fluid portion of cytoplasm that surrounds organelles
Organelles	
• Mitochondria	Makes ATP via aerobic cellular respiration
• Ribosomes	Site of protein synthesis
• Rough endoplasmic reticulum (RER)	Processes and transports proteins made at attached ribosomes; synthesizes phospholipids
• Smooth endoplasmic reticulum (SER)	Fatty acid and steroid synthesis; detoxifies toxic substances
• Golgi complex	Receives and modifies proteins from RER; sorts and transports them
• Secretory vesicles	Secrete substances outside the cell by exocytosis
• Lysosomes	Enzymes digest and recycle worn-out organelles and substances entering the cell; can digest the cell
• Peroxisomes	Produce hydrogen peroxide; detoxify harmful substances
• Cytoskeleton	Three kinds of protein filaments; maintain cell shape and involved in cell movement and movement of organelles
• Centrosomes (centrioles)	Form mitotic spindle; needed to form cilia and flagella
• Cilia	Abundant, hair-like cell projections that move fluids and particles along the cell surface
• Flagella	Long cell projection; whip-like motion moves sperm

B. Cell Specialization

The human body contains over 200 different types of cells with different functions. These differences in function are reflected in cell structure. Cells of the human body differ from the generalized animal cell in shape, size, or number and type of organelles present. In the next activity you will observe cells of skeletal muscle, pseudostratified ciliated columnar epithelium, nonciliated simple columnar epithelium with microvilli, motor neurons, sperm, and blood.

- **Skeletal muscle** cells are long, cylindrical cells that contain specialized proteins (contractile proteins) that enable them to contract (shorten in length) to move bones. The contractile proteins are organized into repeating units that can be observed in the light microscope as striations.

- **Pseudostratified ciliated columnar epithelial cells** have cilia that move substances like mucus along the surface of the cells. Mucus is produced by specialized cells called goblet cells.

- **Nonciliated simple columnar epithelium with microvilli.** Microvilli increase the surface area of the plasma membrane which provides a larger area for absorption of nutrients along the gastrointestinal tract or secretion of product from glands.

- **Motor neurons** are nervous tissue cells with many processes (cell extensions) that receive information from other neurons and send electrical signals to muscle cells causing them to contract.

- **Sperm cells** (sperm) are small, oval cells with a flagellum that propels them through the female reproductive tract.

- **Red blood cells** do not have a nucleus (anucleate) but contain large amounts of hemoglobin, a red pigment that binds oxygen.

- **White blood cells** have nuclei with different shapes and defend the body from pathogens and cancerous cells.

LAB ACTIVITY 2 Cell Specialization

1 Observe each prepared slide (skeletal muscle, pseudostratified ciliated columnar epithelium, motor neuron, sperm, blood cells, and nonciliated simple columnar epithelium with microvilli) and identify the cells and cell components shown in Figure 4.2 or **Real Anatomy** (Histology).

2 Using Figure 4.2, describe each cell's shape and list the cell structures that can be seen with the light microscope in each cell type.

a. skeletal muscle cell: _____

b. pseudostratified ciliated columnar epithelial cell:

c. motor neuron: _____

d. sperm cell: _____

e. red blood cell _____

f. white blood cell _____

g. nonciliated simple columnar epithelium:

Width of skeletal muscle cell Nucleus

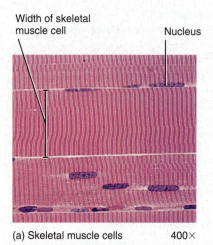

(a) Skeletal muscle cells 400×

Goblet cell Nucleus Cilia

(b) Ciliated cells 630×
(Pseudostratified ciliated
columnar epithelium)

Processes Nucleus Cell body

(c) Motor neuron

Nucleus Flagellum

(d) Sperm cells 400×

Red blood cells Nucleus of white blood cell

(e) Blood cells

Nucleus Microvilli

(f) Nonciliated simple columnar 500×
epithelium with microvilli

FIGURE 4.2 Cell specialization.

C. Somatic Cell Division: Mitosis and Cytokinesis

Somatic (*soma-* = body) **cell division** occurs when one cell divides to produce two genetically identical cells. Cell division is needed for growth of the individual and cell replacement.

The **cell cycle,** a period during which a cell grows and divides into two genetically identical cells (daughter cells), begins when a cell is produced by cell division and ends when the cell divides (Figure 4.3). The length of the cell cycle differs according to the type of cell, with some cells dividing more frequently than others. The cell cycle can be divided into two basic periods: **interphase,** a long period during which the cell conducts its normal activity, grows, and prepares for cell division; and the **mitotic phase,** when the cell is dividing. The mitotic phase consists of **mitosis,** or nuclear division, and **cytokinesis,** or cytoplasmic division. The four stages of mitosis are prophase, metaphase, anaphase, and telophase (Table 4.2).

To observe interphase and the stages of mitosis, you will examine a prepared microscope slide containing several sections of a whitefish blastula. The **blastula** is an early embryonic stage in which cells are dividing rapidly, providing many cells in different stages of mitosis.

Before Going to Lab

1 Using Table 4.2, identify interphase, each phase of mitosis, and cytokinesis in Figure 4.4(a)–(e).

LAB ACTIVITY 3 Mitotic Phases

1 Observe the 3-dimensional models of the mitotic phases, noting the changes in each phase.

2 Obtain a prepared whitefish blastula slide and hold it up to the light. Notice that there are many blastula sections on each slide. It will be necessary to view several of these sections to find all of the phases.

3 Using a compound microscope, begin looking at your slide with the low-power objective lens. Use the high-power objective lens to identify interphase, the four stages of mitosis, and cytokinesis. ■

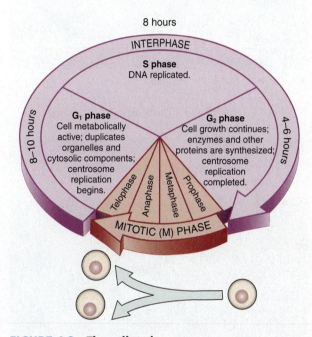

FIGURE 4.3 The cell cycle.

TABLE 4.2 Phases of Somatic Cell Cycle

PHASE	DESCRIPTION OF ACTIVITY
Interphase (*inter-* = between)	Normal cell work; cell metabolically active and growing; DNA replicates
Mitotic Phase	Cell division
Mitosis (*mitos-* = thread)	Nuclear division
• Prophase (*pro-* = first)	Nucleolus and nuclear membrane disappear; chromatin condenses into chromosomes; centrioles move to opposite poles; spindle fibers form
• Metaphase (*meta-* = next)	Chromosomes line up at metaphasal plate; spindle fibers attach to centromeres of chromatids
• Anaphase (*ana-* = apart)	Chromatids of chromosomes separate; move to opposite poles
• Telophase (*telo-* = end)	Cell reverses prophase activities
Cytokinesis (*cyto-* = cell; kinesio-* = movement)	Cytoplasmic division into two genetically identical daughter cells

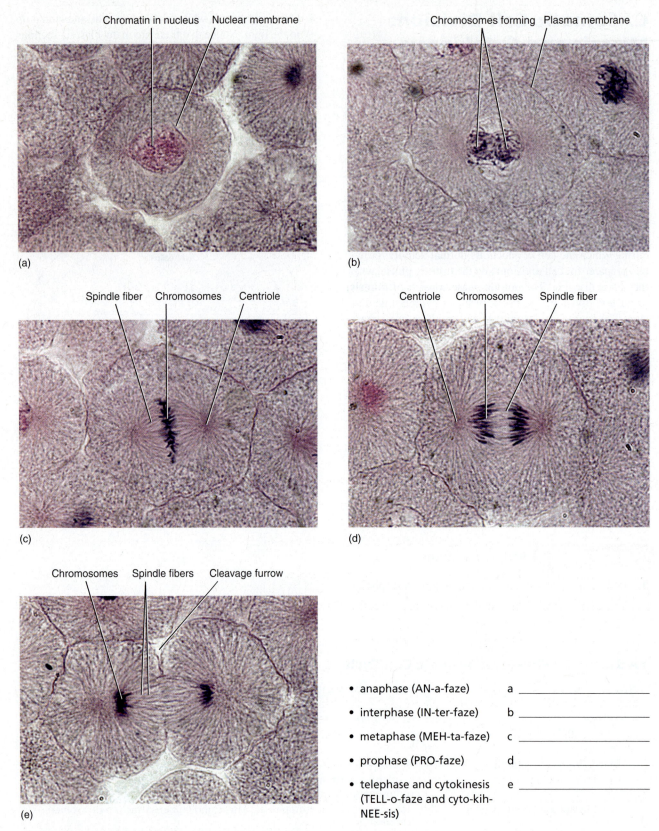

FIGURE 4.4 **Mitotic phases.**

- anaphase (AN-a-faze) a _____
- interphase (IN-ter-faze) b _____
- metaphase (MEH-ta-faze) c _____
- prophase (PRO-faze) d _____
- telephase and cytokinesis (TELL-o-faze and cyto-kih-NEE-sis) e _____

Reviewing Your Knowledge

A. Cellular Structure

Fill in the blank with the name of the cell structure that fits the description.

_____ 1. short, hair-like projections for movement of substances along cell surface

_____ 2. intracellular fluid

_____ 3. site of energy production by cellular respiration

_____ 4. site of protein synthesis

_____ 5. site of steroid and fatty acid synthesis

_____ 6. small vesicle with digestive enzymes

_____ 7. organelles needed to form cilia and flagella

_____ 8. thread-like strand of DNA with associated proteins

_____ 9. site of secretory and membrane protein synthesis

_____ 10. site where protein products are stored, packaged, and exported

_____ 11. contains DNA that control cellular activities

_____ 12. site of ribosome synthesis

_____ 13. gives the cell shape, support, movement, and holds organelles in position

_____ 14. controls movement of substances into or out of the cell

_____ 15. folds of the plasma membrane that increase the cell's surface area

_____ 16. detoxifies harmful substances, produces hydrogen peroxide, and oxidizes amino acids

_____ 17. double membrane that separates the nucleus from the cytoplasm

_____ 18. a small membranous sac that delivers proteins to the plasma membrane to exit the cell

B. Phases of the Cell Cycle

Write the phase of the cell cycle that the phrase describes.

_____ **1.** cytoplasmic division

_____ **2.** cell performing normal functions; longest phase

_____ **3.** nuclear division

_____ **4.** chromatid pairs line up at equatorial plate

_____ **5.** chromatin condenses into chromosomes

_____ **6.** spindle fibers break up; nucleus reappears

_____ **7.** centromeres divide; chromosomes move to opposite poles

_____ **8.** nuclear membrane disassembles and disappears

_____ **9.** chromosomes unravel to form chromatin

_____ **10.** mitotic spindle forms

_____ **11.** DNA replicates

Using Your Knowledge

A. Cellular Organelles and Their Function

Write the letter for the correct answer in the blank.

_____ 1. A cell makes and secretes a protein-based hormone. This particular cell would have a great amount of RER and:

 (a) SER (b) Golgi complex (c) mitochondria (d) lysosomes

_____ 2. Testes and ovaries that make steroids (lipids) would have a larger amount of:

 (a) SER (b) RER (c) mitochondria (d) lysosomes

_____ 3. Muscle cells that need large amounts of ATP would have many:

 (a) Golgi complexes (b) ribosomes (c) SER (d) mitochondria

_____ 4. Cells that line the small intestine are specialized for absorption and secretion. The plasma membrane structure they have to accomplish this is:

 (a) centrioles (b) cilia (c) flagella (d) microvilli

_____ 5. Immune cells that destroy bacteria with chemicals need an abundance of:

 (a) SER (b) ribosomes (c) lysosomes (d) centrioles

B. The Cell Cycle and Mitosis

Answer each question with a short answer.

6. Explain the role of somatic cell division as a person ages from infancy to adulthood.

7. Explain the role of cell division in wound healing.

8. List the cell structures involved in mitosis.

9. Can red blood cells undergo mitosis? Explain.

10. Sperm and eggs have one-half the number of chromosomes of the somatic cells that divided to form them. Are sperm and eggs formed by mitosis? Explain.

11–14. Name four cellular organelles that are membrane bound.

 11. _____ **13.** _____

 12. _____ **14.** _____

15–17. Which cellular organelles are not membrane bound?

 15. _____

 16. _____

 17. _____

18–20. Identify the three structures indicated in Figure 4.5.

 18. _____

 19. _____

 20. _____

TEM 45,000×

FIGURE 4.5 **Transmission electron micrograph of a cell section.**

Transport Across the Plasma Membrane

OBJECTIVES

1 Describe diffusion and osmosis
2 Compare hypotonic, hypertonic, and isotonic solutions
3 Observe simple diffusion, diffusion and osmosis across a dialysis membrane, osmosis across egg viteline membrane, and osmosis in living red blood cells

MATERIALS

- **Simple Diffusion:** 2 agar Petri dishes per group, small millimeter rulers, forceps, methylene blue crystals, potassium permanganate crystals

- **Diffusion and Osmosis Across a Dialysis Membrane (per group):** dialysis tubing 12 cm long, 2 dialysis clips, scissors, Congo red or red food coloring, 40% sucrose solution, 500-mL beakers, distilled water, gram scale

- **Osmosis Across the Egg Vitelline Membrane:** uncooked egg, vinegar, Karo syrup or 25% sucrose solution, water, 500-ml beaker or glass container with lid, gram scale

- **Osmosis in Living Red Blood Cells:** blood (uncoagulated), disposable gloves, safety glasses, clean microscope slides and coverslips, 4 medicine droppers per group, compound microscope, filter paper

The plasma membrane with its unique design is responsible for discriminately allowing substances into and out of a living cell. **Active processes** require cellular energy (ATP) to transport substances against their concentration gradients across the plasma membrane. **Passive processes** do not require the cell to expend energy because the kinetic energy of the particles causes them to move from an area of their higher concentration to an area of their lower concentration. In this exercise, we will look at the passive processes of diffusion and osmosis.

A. Simple Diffusion

Diffusion can occur in solids, liquids, or gas, and across the plasma membrane. The net movement of substances from a region of their greater concentration to a region of their lesser concentration is called moving substances "down the concentration gradient." A **concentration gradient** indicates there is a difference in the concentration of molecules or ions inside the cell (intracellular) compared to outside the cell (extracellular).

In the following activity, diffusion of two substances through a solid (agar) will be studied. Methylene blue has a molecular weight of 320, and potassium permanganate has a molecular weight of 158. The two substances move at different rates through the agar, which is made up mostly of water.

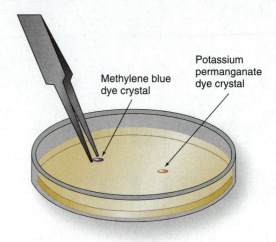

Methylene blue dye crystal

Potassium permanganate dye crystal

FIGURE 5.1 **Diffusion in agar plate setup.**

LAB ACTIVITY 1 Experiment: Simple Diffusion

1 Prediction: With your lab group, predict which substance will move faster. (*Hint: Use the molecular weights.*) Circle your answer: *methylene blue* or *potassium permanganate*.

2 Materials: Obtain the materials for simple diffusion.

3 Data Collection: Measure the diffusion rates of methylene blue and potassium permanganate.
- Decide who will set up the experiment, who will time, who will measure, and who will record.
- Carefully using forceps, place a large crystal of methylene blue on the surface of an agar Petri dish (Figure 5.1). Be careful not to drop any extra crystals on the agar surface.
- Using the same technique, place a similar size crystal of potassium permanganate on the other side of the Petri dish.
- Using a millimeter (mm) ruler, measure each substance's diameter of diffusion at 15-minute intervals for at least 1 hour (or longer if desired).
- After each observation, record the diffusion diameter of each substance in millimeters (mm) in Table 5.1.

4 Clean up as directed by your instructor.

5 Data Analysis: Calculate the diffusion rates for each time period for the two substances using Table 5.1.

6 Complete the Experimental Report with your lab group.

EXPERIMENTAL REPORT
Simple Diffusion

Results: Which substance moved faster?

Discussion: Discuss why the diffusion rates of the two substances differed.

Conclusion: State how molecular weight affects diffusion rate.

TABLE 5.1 Simple Diffusion Results

TIME (min)	DIFFUSION OF METHYLENE BLUE		DIFFUSION OF POTASSIUM PERMANGANATE	
	DIFFUSION DIAMETER (mm)	DIFFUSION RATE (mm/min)	DIFFUSION DIAMETER (mm)	DIFFUSION RATE (mm/min)
15				
30				
45				
60				
75				

B. Diffusion and Osmosis Across a Dialysis Membrane

Osmosis is the diffusion of a solvent (dissolving medium, which is water in living organisms) across a selectively permeable membrane that occurs in response to differences in solute (substance dissolved in solvent) concentrations. Water diffuses from an area of higher water concentration but lower solute concentration (**hypotonic solution**) to an area of lower water concentration but higher solute concentration (**hypertonic solution**). The terms hypotonic (*hypo-* = deficient; *-tonos* = stretching) solution or hypertonic solution are used only when two solutions are compared. A solution's *tonicity* is a measure of the solution's ability to change the volume of a cell by changing water content. Water will move into the hypertonic solution until the solute concentrations of the two solutions equalize to become **isotonic solutions** (*iso-* = same) or if enough pressure is applied to stop the flow of water.

In the following activity, a sucrose solution is added to a dialysis membrane bag (semipermeable membrane), which is placed in distilled water. Dialysis membranes contain small pores that allow water and small solutes to cross. If water enters the bag, the weight of the bag will increase, and if water leaves the bag, its weight will decrease.

The dialysis membrane used in this experiment contains pores that are large enough to allow diffusion of sucrose and red dye molecules into and out of the dialysis membrane bag.

Before Going to Lab

1 Label Figure 5.2.

LAB ACTIVITY 2 **Experiment: Diffusion and Osmosis Across a Dialysis Membrane**

NOTE: *This experiment should be set up at the beginning of class so that changes can be observed and recorded throughout the lab time. The results become more dramatic the longer this experiment runs.*

1 Prediction: With your lab group, predict the direction of water and sucrose movement in Figure 5.2 by circling the correct italicized choice.

- The net movement of water (measurable by weight) will be *into* or *out of* the dialysis bag (see Figure 5.2).
- The net movement of sucrose will be *into* or *out of* the dialysis bag (see Figure 5.2).

2 Materials: Obtain materials for osmosis and diffusion across a dialysis membrane.

3 Data Collection: Observe osmosis and diffusion through a dialysis membrane and record your results in Table 5.2.

- Decide who will set up the experiment, who will time, who will weigh and observe the color, and who will record.

- hypertonic solution 1 _____
- hypotonic solution 2 _____

FIGURE 5.2 Dialysis bag setup.

- Cut a 12-cm piece of dialysis tubing and *soak the bag in water* for 3 minutes to make it easier to open.
- Fold over one end of the dialysis tubing and secure it with a dialysis clip.
- Rub the open end of the dialysis tubing between your thumb and finger to separate the sides, and fill the dialysis tubing more than three-quarters full with the red 40% sucrose solution.
- Pushing the air out of the bag, fold over the end of the bag and clip it with a dialysis clip. The dialysis bag now looks like a large "cell."
- Check to see that no liquid is leaking out of either end of the dialysis bag.
- Rinse the bag to remove excess sucrose solution, dry the bag, and weigh it. Record the beginning weight in Table 5.2.
- Submerge the dialysis bag in a beaker of distilled water.
- Dry and weigh the dialysis bag after 15-minutes, 30-minutes, and 45-minutes. If your lab lasts longer, you may also do a 60-minute measurement.
- Record each measurement in Table 5.2.

4 Clean up as directed by your instructor.

5 Complete the Experimental Report with your lab group.

TABLE 5.2 Dialysis Bag Results

TIME (min)	COLOR OF BEAKER WATER	WEIGHT OF DIALYSIS BAG (grams)
0 min		Start weight:
15 min		
30 min		
45 min		
60 min		

EXPERIMENTAL REPORT
Dialysis Bag

Results:

- Describe how the weight of the bag changed over time.

- Describe how the color of the beaker water changed over time.

Discussion:

- Identify the solutes and solvent in this experiment.

- Why was the red dye added to the sucrose solution?

- Which direction did osmosis occur?

- Which direction did diffusion of solutes occur?

- Did net osmosis and diffusion of solutes stop? Did the two solutions become isotonic? Explain.

- In this experiment, what represented the plasma membrane, intracellular fluid, and extracellular fluid?

Conclusion: State what drives the osmosis and diffusion of solutes; when does the net movement of both stop?

C. Osmosis Across the Egg Vitelline Membrane

The vitelline membrane of the egg is a thin, semipermeable membrane found on the underside of the shell in a raw egg. The chicken egg is not a single cell, and the vitelline membrane is not equivalent to the cell plasma membrane. In this activity, a raw egg is soaked in vinegar for 24–48 hours to dissolve the eggshell and leave the vitelline membrane. The egg surrounded by the vitelline membrane is then placed in different solutions to observe osmosis.

LAB ACTIVITY 3 **Experiment: Osmosis Across the Egg Vitelline Membrane**

1 **Prediction:** With your lab group, predict the direction of water movement by circling the correct italicized choice.
 - When the egg is placed in distilled water, the net movement of water will be *into* or *out of* the egg.
 - When the egg is placed in 25% sucrose solution (or Karo syrup), the net movement of water will be *into* or *out of* the egg.

2 **Materials:** Obtain materials for osmosis across the egg vitelline membrane. The eggs used in this experiment are raw eggs that have been soaked in vinegar for 24–48 hours.

3 **Data Collection:** Observe osmosis across the vitelline membrane of the egg and record your results in Table 5.3.
 - Decide who will set up the experiment, who will time, who will weigh, and who will record.
 - Remove two eggs from the vinegar solution and gently rinse the eggs in water. Blot water off the eggs and then weigh the eggs. Record the weight in Table 5.3.
 - Place one egg in a 25% sucrose solution (Karo syrup or maple syrup can also be used) and the other egg in a distilled water solution.
 - Dry and weigh the eggs after 15-minutes, 30-minutes, 45-minutes, and 60-minutes. Observe any change in appearance of the eggs over time.
 - Record each measurement in Table 5.3.
 - Break the vitelline membrane and observe the appearance of the egg. Is it similar to a raw egg or a cooked egg?

4 Clean up as directed by your instructor.

5 Complete the Experimental Report with your lab group.

TABLE 5.3 Osmosis Across the Vitelline Membrane

TIME (min)	WEIGHT OF EGG IN SUCROSE SOLUTION (grams)	WEIGHT OF EGG IN WATER (grams)
0 min	Start weight:	Start weight:
15 min		
30 min		
45 min		
60 min		
Change in Appearance of Egg over Time		
Egg in sucrose solution		
Egg in water		

EXPERIMENTAL REPORT
Egg Vitelline Membrane

Results:
- Describe how the weight of the egg in 25% sucrose solution (or syrup) changed over time.

- Describe how the appearance of the egg in 25% sucrose solution (or syrup) changed over time.

- Describe how the weight of the egg in distilled water solution changed over time.

- Describe how the appearance of the egg in distilled water solution changed over time.

Discussion:
- When the egg was in the 25% sucrose solution (or syrup), in which direction did osmosis occur?

- When the egg was in distilled water solution, in which direction did osmosis occur?

- Compare the appearance of the egg in 25% sucrose solution (or syrup) at the last weighing with the appearance of the egg in distilled water solution at the last weighing. Explain the difference in appearance.

- Is the 25% sucrose solution (or syrup) a hypertonic, hypotonic, or isotonic solution?

- Is water a hypertonic, hypotonic, or isotonic solution?

Conclusion:
- Explain why osmosis occurs across the vitelline membrane of the egg.

D. Osmosis Across a Living Plasma Membrane

Red blood cells (RBCs) are good examples to use for this osmosis experiment because their shape changes dramatically when they are exposed to hypotonic or hypertonic solutions. Their normal shape is a round biconcave disc with a smooth plasma membrane. The plasma membrane does not allow salts in the RBC cytoplasm to leave the cell, but water can freely enter or leave the cell through the plasma membrane.

Under the microscope, RBCs look two-toned, with the center being lighter because of its concavity and thinness. If the cell loses most of its water by osmosis when put in a hypertonic solution, it becomes **crenated** or shriveled with spiked edges. If the cell gains a significant amount of water by

being placed in a hypotonic solution, it swells and may eventually burst—a process called **hemolysis** (*hemo-* = blood; *-lysis* = break down). As a basis for comparing various solutions to blood, the salt content (NaCl) of blood is 0.9% and is the same salt content as a physiologic saline solution.

> **NOTE:** *Remember . . . salt shrivels and water swells (a cell).*

> **SAFETY NOTE:** *Wear safety glasses and disposable gloves when handling body fluids or animal blood.*

LAB ACTIVITY 4 **Experiment: Osmosis in Living RBCs**

1. With your lab group, discuss and identify which solutions for this experiment are hypotonic, hypertonic, or isotonic to the RBCs (Figure 5.3). The isotonic solution is also called physiologic saline. The arrows in Figure 5.3 indicate the direction of water movement.

 - 0.9% saline solution _____
 - 5% saline solution _____
 - distilled water _____

2. In Figure 5.3, identify the shape of the RBC (normal, swollen, or crenated) and the type of solution into which the RBC is immersed (isotonic, hypertonic, or hypotonic).

3. **Prediction:** With your lab group, predict the net movement of water in all three types of solutions. Circle your answer, from the italicized choices, for all three situations:
 - In the isotonic solution, the RBCs will *swell, crenate,* or *not change shape.*
 - In the hypotonic solution, the RBCs will *swell, crenate,* or *not change shape.*
 - In the hypertonic solution, the RBCs will *swell, crenate,* or *not change shape.*

4. **Materials:** Obtain materials to observe osmosis in RBCs.

5. **Data Collection:**
 - Decide who will set up and initially observe each slide. Every member of your group should observe each slide.

 Slide #1: 0.9% saline solution
 - Place a drop of animal blood on a slide with a medicine dropper. With a different medicine dropper, add a drop of 0.9% saline solution to the blood. Tilt the slide to intermix the two solutions, and cover with a coverslip.
 - Using high power, observe the slide for changes in cell shape. Blood cells are very tiny and difficult to see under low power.
 - Record shape of RBCs and type of solution into which they were placed.

 shape _____ type of solution _____

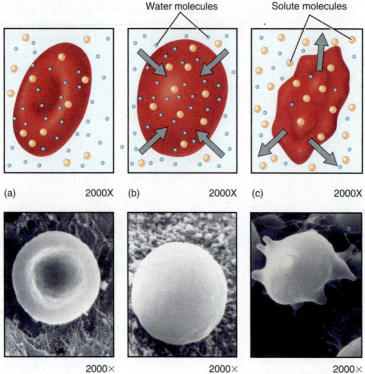

Water molecules Solute molecules

(a) 2000X (b) 2000X (c) 2000X

(a) shape _____

 solution _____

(b) shape _____

 solution _____

(c) shape _____

 solution _____

2000× 2000× 2000×

FIGURE 5.3 Osmosis in RBCs.

Slide #2: 5% saline solution
- Place a drop of animal blood on a second slide and cover with a coverslip. Observe the RBCs using high power. On one side of the coverslip, add one drop of 5% saline solution with a clean medicine dropper. Place filter paper or a small piece of paper towel on the opposite side of the coverslip to absorb the liquid and pull the 5% saline solution into the RBCs.
- Immediately observe the second slide under high power and watch for changes in cell shape. You may have to wait a few minutes for cell changes to occur.
- Record shape of RBCs and type of the solution into which they were placed.
 shape _____ type of solution _____

Slide #2 again: distilled water
- Using slide #2 again, add one drop of distilled water to the same edge of the coverslip that was used for the saline solution.
- Hold a piece of filter paper at the opposite edge of the coverslip, observe the paper absorb the saline solution, and watch as the distilled water is drawn under the coverslip and into the RBCs.
- Immediately observe the slide under high power and watch if the RBCs change shape.
- Record shape of RBCs and type of solution into which they were placed.
 shape _____ type of solution _____

6 Clean up as directed by your instructor.
- Place blood-stained items (slides, coverslips, and droppers) in a 10% bleach solution or as directed by your instructor.

- Place gloves in an autoclavable bag or location indicated by your instructor.
- Wash down the lab counters with 10% bleach solution in a squirt bottle and wipe with paper towels wet with 10% bleach solution. Allow to air dry.
- Wash your hands with soap and water before leaving the lab area.

7 Complete the Experimental Report with your lab group.

EXPERIMENTAL REPORT
Osmosis in Living RBCs

Results: Describe how the RBC shape changes when placed in each solution.

- 0.9% saline solution _____
- 5% saline solution _____
- distilled water _____

Discussion:
1. Describe how extracellular solute concentration affects osmosis across the plasma membrane.

2. Describe how water movement affects the cell shape.

Conclusion: State which solution(s) caused osmosis in RBCs.

Reviewing Your Knowledge

A. Transport Across the Plasma Membrane

Match the definition with the term.

 a. to shrink or shrivel
 b. water moving through selectively permeable membrane
 c. substance dissolved in a solution
 d. to burst a red blood cell
 e. difference between solute concentrations across a membrane
 f. same solute concentration on both sides of membrane
 g. lower concentration of solutes than in cytosol of cell
 h. a fluid that contains dissolved substances
 i. higher concentration of solutes than in cytosol of cell
 j. random movement of particles from their greater concentration to their lesser concentration

____ **1.** concentration gradient

____ **2.** crenate

____ **3.** diffusion

____ **4.** hemolysis

____ **5.** hypertonic solution

____ **6.** hypotonic solution

____ **7.** isotonic solution

____ **8.** osmosis

____ **9.** solute

____ **10.** solvent

B. Diffusion

Select the correct lettered answer.

_____ **1.** In the dialysis bag experiment, sucrose and red dye molecules are:

(a) solutes (b) solvents

_____ **2.** If there is no concentration gradient, a substance will not have net movement.

(a) true (b) false

_____ **3.** Passive processes use _____ to move substances across a plasma membrane.

(a) ATP (b) kinetic energy

C. Osmosis

Select the correct lettered answer.

_____ **1.** An isotonic solution will _____ an RBC.

(a) crenate (b) hemolyze (c) cause no change to

_____ **2.** A hypertonic solution will _____ an RBC.

(a) crenate (b) hemolyze (c) cause no change to

_____ **3.** A hypotonic solution will _____ an RBC.

(a) crenate (b) hemolyze (c) cause no change to

Using Your Knowledge

A. Diffusion

Select the correct lettered answer in questions 1–3.

____ **1.** White blood cells engulf bacteria. This is an example of diffusion.

 (a) true (b) false

____ **2.** Food cooking on the stove in the kitchen can be smelled in the living room. This is an example of diffusion.

 (a) true (b) false

____ **3.** Oxygen in the lungs moves into the bloodstream and carbon dioxide moves in the opposite direction. This is an example of diffusion.

 (a) true (b) false

B. Osmosis

Select the correct lettered answer in questions 4–8.

____ **4.** A test tube with blood in it has a particular solution added to it. After several minutes, the solution is not clear anymore but becomes red. Which solution was added to the blood to obtain this result?

 (a) 0.9% saline (b) 5% saline (c) distilled water

____ **5.** A 0.8% saline solution would be ____ to the cytosol of a cell.

 (a) hypotonic (b) hypertonic (c) isotonic

____ **6.** If a 50% sugar solution had been used in the dialysis bag in Activity 2, there would be a faster rate of osmosis.

 (a) true (b) false

____ **7.** If you placed a peeled apple or potato in a 5% salt solution, it would ____

 (a) gain weight (b) lose weight (c) stay the same weight

_____ **8.** A person's hands become wrinkled after spending a long, relaxing time in the tub. Tub water does not have as many solutes in it compared with the human body. The hands look wrinkled because _____.

(a) the tub water is *hypotonic* to body cells and water *enters* the cells
(b) the tub water is *hypotonic* to body cells and water *leaves* the cells
(c) the tub water is *hypertonic* to body cells and water *enters* the cells
(d) the tub water is *hypertonic* to body cells and water *leaves* the cells

C. Application

9. When a person becomes dehydrated, the amount of water in extracellular fluids such as blood decreases, causing the solute concentration of these fluids to increase. State whether osmosis results in water entering or leaving cells.

10. Severe vomiting and diarrhea cause a loss of water and solutes from extracellular fluids. If a person was given only water, what effect would this have on the solute concentration of extracellular fluids? Would osmosis result in water entering or leaving cells?

Tissues

OBJECTIVES

1 Name the four primary tissue types
2 Compare and contrast primary tissue structure and function
3 Identify examples of epithelial, connective, muscular, and nervous tissue types on prepared microscope slides or photomicrographs and describe their location and function

MATERIALS

- compound microscope and lens paper
- prepared epithelial tissue slides or **Real Anatomy** (Histology): whole mount of mesothelium, simple squamous (lung), simple cuboidal (kidney), simple columnar nonciliated (small intestine), stratified squamous nonkeratinized (esophagus), transitional (relaxed urinary bladder), and ciliated pseudostratified columnar (trachea)
- prepared connective tissue slides or **Real Anatomy** (Histology): areolar (spread), reticular, adipose, dense regular (tendon), dense irregular (skin), elastic, hyaline cartilage (trachea), elastic cartilage, fibrocartilage, dried compact bone, cancellous (spongy) bone, and blood
- prepared muscle tissue slides or **Real Anatomy** (Histology): skeletal muscle, cardiac muscle, and smooth muscle
- prepared motor neuron slide or **Real Anatomy** (Histology)
- Dissection: 1 fresh chicken leg per group, dissection equipment, disposable gloves, dissecting microscope, compound microscope, lens paper

There are four primary tissue types in the human body: epithelial, connective, muscle, and nervous. **Epithelial tissue** covers surfaces, lines cavities, and forms glands. **Connective tissue,** the most abundant primary tissue in the body, connects different tissues, provides a framework, resists pulling forces, and protects other tissues. **Muscle tissue** causes movement, and **nervous tissue** receives and generates nerve impulses. Organs are formed from two or more different tissues working together to perform a specific function. **Histology** is the study of the microscopic anatomy of cells and tissues.

A. Epithelial Tissue

Epithelial tissues or epithelia (epithelium, sing.) exhibit **cellularity;** that is, most of the tissue consists of cells packed together in an orderly fashion with little extracellular material (matrix) between cells. Epithelial tissues are avascular and receive nutrients from the vascular underlying connective tissue. A **basement membrane** separates epithelial and connective tissues.

There are two categories of epithelial tissues: covering and lining epithelia, and glandular epithelia. **Covering and lining epithelia** cover the surface of the body and some organs, and line all hollow body structures. Tight junctions between neighboring cells enable this type of epithelium to form barriers that protect and control what substances can cross into adjacent tissues. **Glandular epithelia** form glands that produce and secrete products needed by the body.

Covering and lining epithelia face exterior or interior spaces. The epithelial cell surface adjacent to the space is called the **apical** (*apex* = tip) **surface,** whereas the epithelial cell surface adjacent to the basement membrane is the **basal surface.** When viewed from the apical surface, one can observe how epithelial cells fit together like floor tiles to form a lining or barrier (Figure 6.1). When viewed in cross-section (side view of cells), one can observe differences in structure of the apical and basal surfaces. The differences in structure are associated with differences in function.

Epithelial tissue types are classified according to the shape of the cell in the apical cell layer and the number of epithelial cell layers. Epithelial tissue function is determined by the cell type and number of cell layers.

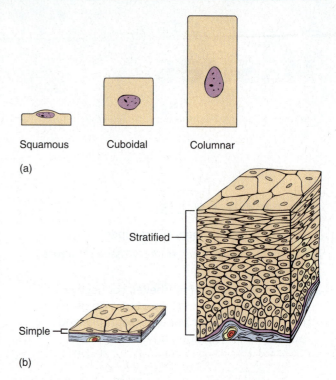

(a)

(b)

FIGURE 6.2 **Epithelial tissue classification.**

Epithelial cells have four shapes: squamous, cuboidal, columnar, and transitional. Cell shapes are best seen in side views of the cells [Figure 6.2(a)]. **Squamous cells** are the thinnest cells, and they have a flattened nucleus. **Cuboidal cells** are cube-like with a round nucleus in the center of the cell. **Columnar cells** are tall with an oval nucleus close to the base of the cell. **Transitional cells** change shape; the apical cells are cuboidal when the tissue is relaxed and squamous when the tissue is stretched.

Epithelial tissue with only one cell layer is **simple epithelium,** while epithelial tissue with two or more cell layers is **stratified epithelium** [Figure 6.2(b)]. **Simple epithelium** provides a selective barrier allowing diffusion, filtration, secretion, or absorption of selected substances. There are three types of simple epithelium: simple squamous, simple cuboidal, and simple columnar. **Stratified epithelium** is thicker, subject to wear and tear, and forms a protective barrier. Multiple cell layers make the tissue more resistant to damage, thereby preventing pathogens and foreign materials from crossing into underlying tissues. There are four types of stratified epithelium: stratified squamous, stratified cuboidal, stratified columnar, and transitional. **Pseudostratified** (*pseudo-* = false) **columnar epithelium** gives the illusion of several different layers of cells but is only one cell layer thick. All the cells touch the basement membrane, although not all cells reach the apical surface. Therefore, there are cells of different shapes and heights, and their nuclei are at different levels. The tallest cells are narrow where they touch the basement membrane but have a columnar shape toward the outer surface, while the shorter cells do not reach the apical surface.

FIGURE 6.1 **Epithelial tissue lining the mouth.**

NOTE: *Observe each tissue first with a scanning or low-power objective lens, focusing with the coarse focus knob. Before switching to a more powerful objective lens, move the slide so that the area you want to observe is in the center of the field of view. After changing to a more powerful lens, use the fine focus knob only to bring the tissue into focus. Use this procedure with all tissue slides.*

Before Going to Lab

1 Examine the photomicrograph of a surface view of simple squamous epithelium from mesothelium in Figure 6.3.
 • This photomicrograph shows cells viewed from the apical surface.
 • Observe how the epithelial cells fit close together to form a good barrier.

2 Examine the photomicrographs of cross-sections of epithelial tissue types in Figures 6.4–6.9. Stratified cuboidal and stratified columnar are not included because they are less common.
 • On the survey photomicrograph (40×) for each tissue:
 * Observe that the epithelial and connective tissue layers stain differently.
 * Locate where the basement membrane forms the border between the epithelial and connective tissue layers.

• Note whether there is one epithelial cell layer (simple epithelium) or two or more epithelial cell layers (stratified epithelium).
• On the higher magnification photomicrograph (400×) for each tissue:
 * Examine the shape of the epithelial cell at the apical surface. This determines the tissue type as squamous, cuboidal, or columnar epithelium.
• Note any structural differences between apical surface and basal surface of epithelial tissue.

3 Label the photomicrographs in Figures 6.3–6.4.

4 Review the location and function for each epithelial tissue type in Tables 6.1–6.6.

LAB ACTIVITY 1 Microscopic Examination of Epithelia

1 Examine a prepared microscope slide of a whole mount of mesothelium. Use Figure 6.3 to help you locate the major structures. Draw the tissue in the space provided and label the major structures.

2 Examine prepared microscope slides or use **Real Anatomy** (Histology) to observe tissue cross-sections. Use the survey photomicrographs in Figures 6.3–6.9 to help you locate the epithelia at low power. Use the photomicrograph at 400× to help you locate the major structures in each tissue.

3 Draw each tissue in the space provided and label the major structures.

LM 243x

Simple squamous epithelium

• cytoplasm
• nucleus
• plasma membrane

FIGURE 6.3 **Surface view of mesothelium.**

Student drawing

1 _____

2 _____

3 _____

Alveoli

Pleura

165×, H&E
(a) Survey photomicrograph

Student drawing

1

2
3

4

400×, H&E, human
(b) Simple squamous epithelium

Student drawing

- connective tissue
- nucleus of simple squamous epithelial cell in alveolar wall
- nucleus of simple squamous epithelial cell in visceral layer of pleura
- simple squamous epithelium

1 _____

2 _____

3 _____

4 _____

FIGURE 6.4 **Sectional view of lung.**

TABLE 6.1 Location and Function of Selected Simple Squamous Epithelia

LOCATION	FUNCTION
Mesothelium (epithelial layer of serous membranes)	**Secretion** of serous fluid into serous cavity.
Alveoli (air sacs of lungs)	Single layer of squamous cells creates a short distance for **diffusion** of oxygen and carbon dioxide.
Glomerular capsule (part of filtration membrane in kidney)	**Filtration** of blood to form urine filtrate (substance that is converted into urine).
Endothelium of capillaries	Single layer of squamous cells creates a short distance for **diffusion** of substances between blood and interstitial fluid (tissue fluid).

40×
(a) Survey photomicrograph

Student drawing

400×, H&E, human
(b) Simple cuboidal epithelium

- apical surface of simple cuboidal epithelial cell
- lumen of kidney tubule
- nucleus of simple cuboidal epithelial cell
- simple cuboidal epithelium

Student drawing

1 _____

2 _____

3 _____

4 _____

FIGURE 6.5 Sectional view of kidney tubules.

TABLE 6.2 Location and Function of Selected Simple Cuboidal Epithelia

LOCATION	FUNCTION
Walls of kidney tubules	Modify urine filtrate by **absorption** of substances from the filtrate and **secretion** of other substances into the filtrate.
Glands	**Secretion** of products made by the simple cuboidal epithelial cells.

40×
(a) Survey photomicrograph

Villus
Lumen

Student drawing

400×, H&E, human
(b) Simple columnar epithelium

Goblet cell
1
2
3
4

Student drawing

- connective tissue
- microvilli on apical surface of simple columnar epithelial cell
- nucleus of simple columnar epithelial cell
- simple columnar epithelium

1 _____

2 _____

3 _____

4 _____

FIGURE 6.6 Sectional view of small intestine.

TABLE 6.3 Location and Function of Selected Simple Columnar Epithelia

LOCATION	FUNCTION
Lining of stomach and intestines	**Secretion** of digestive juices by simple columnar cells and secretion of mucus by goblet cells. In the small intestine the simple columnar epithelial cells have **microvilli** (*micro-* = small; *villi* = shaggy hair) to increase surface area for **absorption.**
Uterine tubes (fallopian tubes)	Simple columnar epithelial cells have **cilia** that help **move** the egg to the uterus.
Central canal of spinal cord	Ciliated cells **move** cerebrospinal fluid.

Mucous glands

Lumen

Epithelial tissue

Connective tissue

Smooth muscle

40×
(a) Survey photomicrograph

Student drawing

1

2

3

4

200×, H&E, human
(b) Nonkeratinized stratified squamous epithelium

- connective tissue
- nucleus of epithelial cell in basal layer of epithelium
- nucleus of squamous epithelial cell
- stratified squamous epithelium

Student drawing

1 _____

2 _____

3 _____

4 _____

FIGURE 6.7 **Sectional view of esophagus.**

TABLE 6.4 **Location and Function of Selected Stratified Squamous Epithelia**

LOCATION	FUNCTION
Surface of skin (keratinized stratified squamous epithelium)	Epithelial layer of skin is a tough, dry, waterproof outer surface that forms a **protective barrier.**
Lining of mouth, esophagus, anus, and vagina (nonkeratinized stratified squamous epithelium)	Moist epithelial layer that forms a **protective barrier** in areas subject to abrasion and friction.

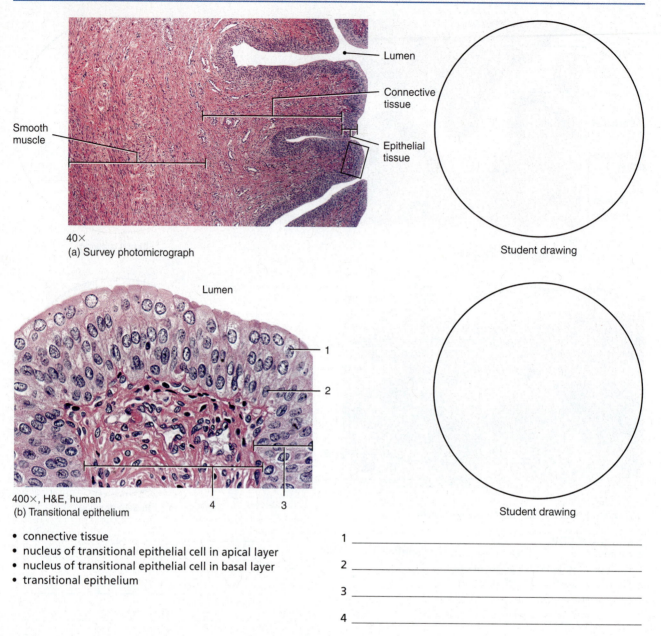

40×
(a) Survey photomicrograph

Lumen

Connective tissue

Epithelial tissue

Smooth muscle

Student drawing

Lumen

1

2

4 3

400×, H&E, human
(b) Transitional epithelium

Student drawing

- connective tissue
- nucleus of transitional epithelial cell in apical layer
- nucleus of transitional epithelial cell in basal layer
- transitional epithelium

1 _____

2 _____

3 _____

4 _____

FIGURE 6.8 Sectional view of a relaxed urinary bladder.

TABLE 6.5 Location and Function of Transitional Epithelia

LOCATION	FUNCTION
Lining of urinary bladder and parts of the ureters and the urethra	Provides a **protective barrier** that permits **distension.**

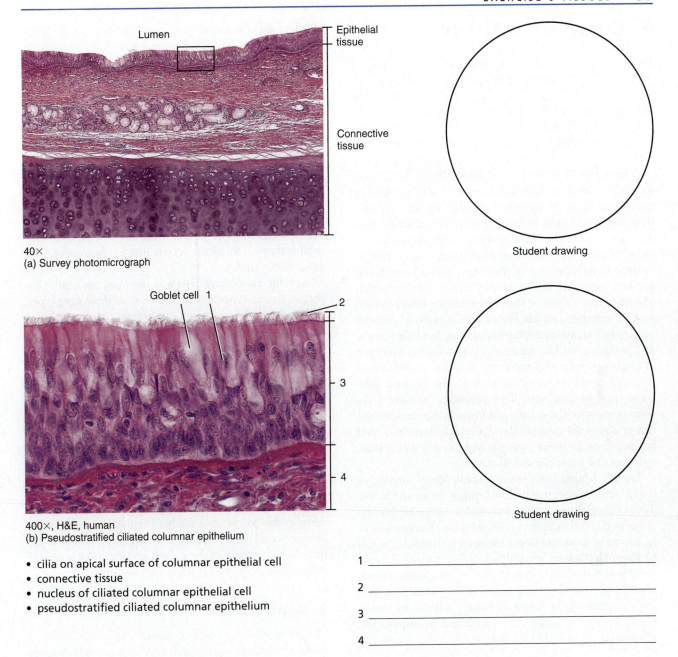

40×
(a) Survey photomicrograph

Lumen

Epithelial tissue

Connective tissue

Student drawing

Goblet cell 1

2

3

4

400×, H&E, human
(b) Pseudostratified ciliated columnar epithelium

- cilia on apical surface of columnar epithelial cell
- connective tissue
- nucleus of ciliated columnar epithelial cell
- pseudostratified ciliated columnar epithelium

Student drawing

1 _____

2 _____

3 _____

4 _____

FIGURE 6.9 **Sectional view of the trachea.**

TABLE 6.6 **Location and Function of Selected Pseudostratified Columnar Epithelia**

LOCATION	FUNCTION
Lining of nasal cavity, trachea, and bronchi	**Secretion** of mucus by goblet cells. The columnar cells have **cilia,** which **move** mucus toward the pharynx.

B. Connective Tissue

Connective tissue is the most abundant primary tissue in the body and has a variety of functions. It connects epithelial tissue to other tissues, forms the internal framework for soft organs, and forms tendons (connect muscle to bone) and ligaments (connect bone to bone). Bone, the hardest connective tissue, **protects** body organs and provides a framework for movement of muscles. Adipose tissue (fat) **insulates** body tissues and **stores lipids,** and blood provides a liquid medium for transportation of substances throughout the body. Connective tissues are not very cellular, containing more extracellular matrix than cells (Figure 6.10). Connective tissue function is determined by the properties of the extracellular matrix components. **Extracellular matrix** consists of **fibers** and **ground substance** that are synthesized and secreted by connective tissue cells. The three major types of fibers are **collagen fibers** (which provide strength), **elastin fibers** (which provide strength and elasticity), and **reticular fibers** (fine, branching fibers that provide a net-like framework). The number and type of fibers present contribute to the strength, elasticity, and structure of the extracellular matrix. The **ground substance** may be fluid, semi-fluid, gelatinous, or hard. It secures connective tissue cells and fibers in the extracellular matrix and is the medium through which interstitial fluid diffuses between blood and cells and through which macrophages and white blood cells move.

Mature connective tissues include loose connective tissue, dense connective tissue, cartilage, bone, and blood. The connective tissue types have different connective tissue cells that form the extracellular matrix. The extracellular matrix of **loose and dense connective tissues** is formed by **fibroblasts** (-*blast* = early, developing stage); the extracellular matrix of **cartilage** is formed by **chondroblasts** (*chondro-* = cartilage) and maintained by **chondrocytes;** and the extracellular matrix of **bone** is formed by **osteoblasts** (*osteo-* = bone) and maintained by **osteocytes.** **Plasma** is the extracellular matrix in which blood cells and platelets are suspended; however, plasma is not made by these cells.

1. Loose Connective Tissue

Loose connective tissue includes areolar, reticular, and adipose tissues. In **loose connective tissue,** the fibers in the extracellular matrix are loosely arranged. Collagen, elastic, and reticular fibers in this tissue provide strength, elasticity, and support. Ground substance is semi-fluid (viscous), but interstitial fluid can easily diffuse through it. Fibroblasts and adipocytes (lipid-storing cells) permanently reside in the tissue. Cells that are involved in body defense, such as macrophages, mast cells, and white blood cells, enter connective tissue from blood vessels that traverse through the extracellular matrix.

Areolar connective tissue is the most abundant connective tissue. It contains fibroblasts, all three fiber types, a semi-fluid (viscous) ground substance, and a variety of cells involved in body defenses. **Reticular** (*reticulo-* = net-like) **fibers** are the dominant fiber type in **reticular connective tissue. Reticular cells** synthesize the reticular fibers and the ground substance. Reticular cells are actually fibroblasts, but they synthesize more reticular fibers than collagen fibers. **Adipose** (*adipo-* = pertaining to fat) **tissue,** like areolar connective tissue, contains fibroblasts, fibers, ground substance, and adipose cells. However, adipose tissue has a greater number of **adipocytes** (lipid-storing cells) and very little extracellular matrix. Lipid within the adipose cells occupies most of the cell volume and pushes cytoplasm and organelles to the periphery. Some slides do not have the lipid stained, or the lipid has been removed during tissue preparation. In those slides, the nucleus and cytoplasm are stained, and the space inside the cell normally occupied by lipid appears "empty."

2. Dense Connective Tissue

The extracellular matrix of dense connective tissue is packed with fibers and contains very little ground substance and few fibroblasts. In **dense regular connective tissue,** the extracellular matrix is packed with parallel bundles of collagen fibers that run in the direction of the pulling forces applied to the tissue. Fibroblasts are squeezed between the bundles, and the cytoplasm extends between and around collagen bundles. Collagen fibers give unstained tissue a silvery white appearance, which is why this tissue is often called **white fibrous tissue.**

Dense irregular connective tissue, like dense regular connective tissue, has little ground substance and few fibroblasts. The extracellular space is also packed with bundles of collagen fibers with fibroblasts squeezed between these bundles. In dense irregular connective tissue, however, the bundles of collagen fibers are irregularly arranged. This reflects the direction of the pulling forces to which this tissue is exposed, usually from many different directions.

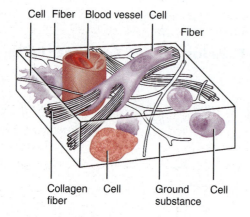

Cell Fiber Blood vessel Cell

Fiber

Collagen fiber Cell Ground substance Cell

FIGURE 6.10 Connective tissue (areolar).

In **elastic connective tissue,** the extracellular matrix is packed with elastic fibers, and fibroblasts are found in the spaces between these fibers. Elastic fibers allow the tissue to be stretched and then regain its original size and shape (recoil).

3. Cartilage

Cartilage is a specialized form of connective tissue. Its extracellular matrix consists of collagen and elastic fibers embedded in a firm gelatinous ground substance. Collagen fibers give the tissue its strength, and the ground substance, which binds water to form a firm gel, gives cartilage its resiliency. **Chondroblasts** (*chondro-* = cartilage) secrete the fibers and ground substance, become isolated in spaces called **lacunae** (little lakes), and transform into **chondrocytes.** Since the tissue is avascular, the chondrocytes receive their nutrients through diffusion from adjacent vascular tissues.

There are three types of cartilage: hyaline (*hyalos* = glass), elastic, and fibrocartilage. These cartilage types differ in the amount and kinds of fibers and ground substance molecules present in their extracellular matrix. **Hyaline cartilage** is the most predominant cartilage in the body and contains fine collagen fibers in the extracellular matrix. This cartilage type appears glassy to the naked eye, and under the compound microscope its matrix looks smooth and homogeneous. It contains collagen fibers that are thin and not visible with a compound microscope. The matrix of hyaline cartilage is resilient and firm. **Elastic cartilage** is similar to hyaline cartilage, except the matrix is packed with elastic fibers. The many elastic fibers allow this cartilage to be much more flexible. **Fibrocartilage** has fewer lacunae and chondrocytes, and its extracellular matrix is packed with thick collagen fibers that give it tensile strength similar to dense connective tissue. Under the microscope, collagen fibers are the most prevalent structures seen.

4. Bone

Bone is the hardest of the connective tissues. The extracellular matrix is organized in layers called **lamellae** and consists of collagen fibers, ground substance, and inorganic salts. Inorganic salts, especially calcium salts, give bone its hardness; collagen fibers provide strength and flexibility. **Osteocytes** (mature bone cells) are trapped in spaces called **lacunae.** Small canals called **canaliculi** connect lacunae to each other and to larger canals that contain blood vessels. Nutrients diffuse from a blood vessel through the canaliculi to the osteocytes, and waste materials diffuse back to the blood vessel for removal.

The two types of bone tissue are **compact (cortical) bone** and **spongy (cancellous or trabecular) bone.** These two types of osseous tissue differ in the amount and size of spaces present. The extracellular matrix of compact bone is organized into repeating structural units called **osteons.** In the center of each osteon is a large central canal. Spongy bone has large spaces compared to compact bone and does not have osteons. Instead, the extracellular matrix of spongy bone is arranged in **trabeculae** (little beams)—flat plates with a lattice-like network of thin, bony columns.

5. Blood

Blood is composed of red blood cells, white blood cells, platelets, and plasma. **Plasma** is the extracellular matrix, and its fibers are produced and observed only during blood clotting. Red blood cells obtain their red color from hemoglobin, and the mature cells do not contain a nucleus (anucleate). White blood cells, which appear white or clear on unstained slides, are nucleated. Platelets, anucleate fragments of cells that are much smaller than red blood cells, participate in blood clotting.

SAFETY NOTE: *Use disposable gloves when handling fresh tissue specimens.*

Before Going to Lab

1 Label and examine the photomicrographs of the connective tissue types in Figures 6.11–6.22.
 • Observe the extracellular matrix in each tissue.
 • Compare the amount of fibers and ground substance.
 • Identify the connective tissue cells.

2 Review the location and function for each connective tissue type in Tables 6.7–6.11.

LAB ACTIVITY 2 **Microscopic Examination of Connective Tissue Types**

1 Examine prepared microscope slides or use **Real Anatomy** (Histology) to observe the different connective tissue types. Use Figures 6.11–6.22 to help you locate the major structures in each tissue. Draw each tissue in the space provided and label the major structures.

2 Identify the type of fibers in each connective tissue type. Collagen fibers are the thickest fibers. Elastic fibers are long, thin, dark fibers. Reticular fibers are short, thin, branching fibers.

3 Draw each tissue in the space provided and label the major structures. ■

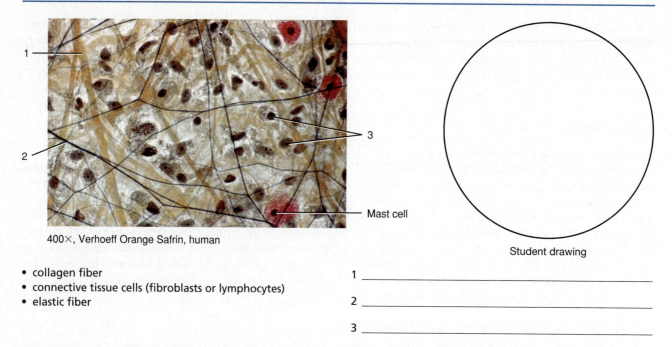

400×, Verhoeff Orange Safrin, human

- collagen fiber
- connective tissue cells (fibroblasts or lymphocytes)
- elastic fiber

Student drawing

1 _____

2 _____

3 _____

FIGURE 6.11 **Areolar connective tissue spread of mesentery.**

400×, human

- reticulocyte
- reticular fiber

Student drawing

1 _____

2 _____

FIGURE 6.12 **Sectional view of reticular tissue of the lymph node.**

400×, H&E, human

Student drawing

• lipid storage area
• nucleus of adipocyte

1 _____

2 _____

FIGURE 6.13 **Sectional view of adipose tissue.**

TABLE 6.7 **Location and Function of Loose Connective Tissues**

LOOSE CONNECTIVE TISSUE TYPE	LOCATION	FUNCTION
Areolar	Beneath all epithelial tissues	Binds epithelium to underlying tissues and allows nutrients to diffuse to epithelial cells.
Reticular	Liver, spleen, lymph nodes	Forms support (framework) of these soft organs.
Adipose	Under skin and surrounding organs	Stores lipids for fuel and thermal insulation; cushioning organs.

Skeletal muscle fibers

1

2

400×, H&E, human

• collagen fiber bundle
• fibroblast

Student drawing

1 _____

2 _____

FIGURE 6.14 **Sectional view of dense, regular connective tissue-forming tendons.**

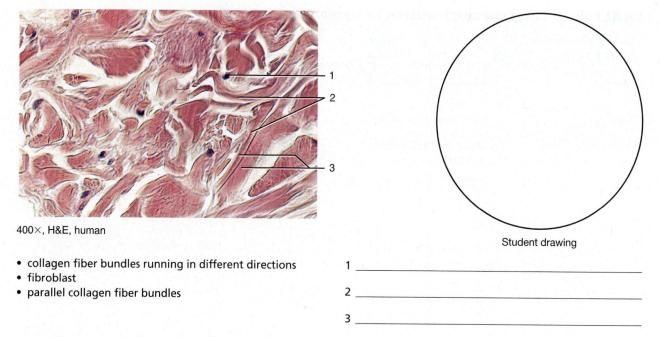

1

2

3

400×, H&E, human

• collagen fiber bundles running in different directions
• fibroblast
• parallel collagen fiber bundles

Student drawing

1 _____

2 _____

3 _____

FIGURE 6.15 **Sectional view of dense, irregular connective tissue in skin.**

400×, human

- bundle of elastic fibers
- fibroblast
- individual elastic fiber

Student drawing

1 _____

2 _____

3 _____

FIGURE 6.16 **Sectional view of elastic connective tissue in the wall of the aorta.**

TABLE 6.8 Location and Function of Dense Connective Tissues

DENSE CONNECTIVE TISSUE TYPE	LOCATION	FUNCTION
Dense regular	Forms ligaments (connects bone to bone), tendons (connects muscle to bone), and aponeuroses (sheet-like tendons that connect muscle to muscle or muscle to bone)	Resists pulling forces at attachment points.
Dense irregular	Skin	Resists pulling forces from many different directions that would tear skin when skin is stretched.
Elastic	Lungs, trachea, and bronchi Aorta and other elastic arteries	Allows respiratory organs to recoil after inhalation. Recoil of elastic tissue helps push blood through cardiovascular system.

400×, H&E, human

- extracellular matrix
- lacuna (la-KOO-na)
- nucleus of chondrocyte

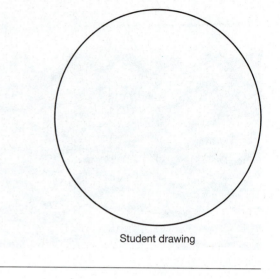

Student drawing

1 _____

2 _____

3 _____

FIGURE 6.17 **Sectional view of hyaline cartilage in the wall of the trachea.**

400×

- elastic fibers
- lacuna
- nucleus of chondrocyte

Student drawing

1 _____

2 _____

3 _____

FIGURE 6.18 **Sectional view of elastic cartilage of the ear.**

400×

- chondrocytes
- collagen fibers
- lacuna

Student drawing

1 _____

2 _____

3 _____

FIGURE 6.19 **Sectional view of fibrocartilage of tendon.**

TABLE 6.9 Location and Function of Cartilage

CARTILAGE TYPE	LOCATION	FUNCTION
Hyaline cartilage	Ends of long bones (articular cartilage)	Smooth surface that is resilient and reduces friction at joint.
	Trachea and bronchi	Provide support and flexibility to ensure an open airway.
	Anterior ends of ribs (costal cartilage)	Connect ribs to sternum with flexible joint (breastplate).
	Embryonic skeleton	Provides template for bone formation.
Fibrocartilage	Intervertebral discs	Provide strength to discs that form joints between vertebrae and act as shock absorbers.
	Cartilage pads in knee	Provide cushioning for bones forming knee joints.
	Pubic symphysis	Forms strong, flexible joint between hip bones.
Elastic cartilage	External ear	Provides support and maintains shape of external ear.
	Auditory tube	Provides support and elasticity to auditory tube as it changes diameter to equalize pressure in middle ear (ear pops).
	Epiglottis of larynx	Provides support and elasticity to epiglottis as it folds to block entrance to trachea while swallowing food and liquid.

400×, unstained, human

- canaliculus
- central canal
- lamella
- lacuna

Student drawing

1 _____

2 _____

3 _____

4 _____

FIGURE 6.20 **Sectional view of dried compact bone.**

Red marrow cell

400×

- osteocyte
- trabecula

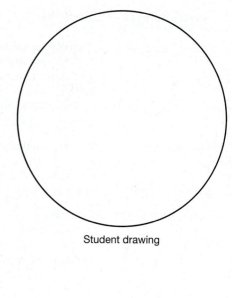

Student drawing

1 _____

2 _____

FIGURE 6.21 **Sectional view of spongy bone.**

TABLE 6.10 Location and Function of Osseous Tissue

TYPE OF OSSEOUS TISSUE	LOCATION	FUNCTION
Compact	Exterior of bones	Support, protection, storage of minerals.
Spongy	Interior of bones	Support, protection, storage of minerals.

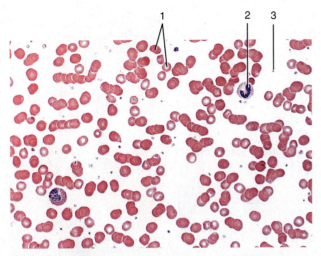

400×, H&E, human

- red blood cells
- nucleus of white blood cell
- platelet

Student drawing

1 _____

2 _____

3 _____

FIGURE 6.22 Blood smear.

TABLE 6.11 Location and Function of Blood Components

BLOOD COMPONENT	LOCATION	FUNCTION
Plasma	Liquid part of blood in arteries, capillaries, veins	Transport of nutrients, blood gases, wastes, chemical messengers, blood cells and platelets.
Red blood cells	Formed element of blood in arteries, capillaries, veins	Transport of blood gases.
White blood cells	Formed element of blood in arteries, capillaries, veins. WBCs can also leave blood vessels and enter infected tissues.	Attack pathogens and other substances that invade the body.
Platelets	Formed element of blood in arteries, capillaries, veins	Participate in blood clotting.

C. Muscle Tissue Overview

Muscle tissue is very cellular, with most of the tissue consisting of muscle cells (Figure 6.23). All muscle tissues are highly vascularized and are innervated. Muscle cells are elongated cells called **fibers** that shorten (contract) when stimulated, causing movement. There are three types of muscle tissue—skeletal, cardiac, and smooth—with each type having a distinct appearance. **Skeletal muscle** cells are large, multinucleated, cylindrical cells. **Cardiac muscle** cells are smaller, branching cells with one or two centrally located nuclei per cell. **Intercalated discs** are dark bands where cardiac muscle cells connect end to end. Skeletal and cardiac muscle cells exhibit **striations** (light and dark bands). **Smooth muscle** cells are small, spindle-shaped (tapered at both ends) cells that are uninucleated and nonstriated.

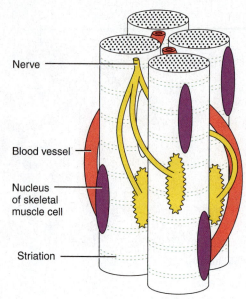

Nerve

Blood vessel

Nucleus
of skeletal
muscle cell

Striation

FIGURE 6.23 **Skeletal muscle tissue.**

Before Going to Lab

1 Label and examine the photomicrographs of muscle tissue in Figures 6.24–6.26.
 • Compare the size of the three types of muscle fibers, their shape, and the number of nuclei.
 • Look for striations in skeletal and cardiac muscle fibers.
2 Review the location and function for each muscle tissue type in Table 6.12.

LAB ACTIVITY 3 **Muscle Tissue**

1 Examine prepared microscope slides or use **Real Anatomy** (Histology) to observe the different muscle tissue types. Use Figures 6.24–6.26 to help you locate the major structures in each tissue. Draw each tissue in the space provided and label the major structures.

2 Draw each tissue in the space provided and label the major structures.

TABLE 6.12 Location and Function of Muscle Tissue Types

MUSCLE TISSUE TYPE	LOCATION	FUNCTION
Skeletal muscle	Attached to bones and skin	Movement of bones and skin. Contraction generates heat.
Cardiac muscle	Wall of heart	Movement of blood through cardiovascular system.
Smooth muscle	Walls of digestive tract organs	Movement of food through digestive tract.
	Walls of arteries and veins	Contraction and relaxation controls blood flow and blood pressure.
	Walls of ureters, urinary bladder, and urethra	Movement of urine through urinary tract.
	Intrinsic muscles of eye	Contraction and relaxation controls pupil size.

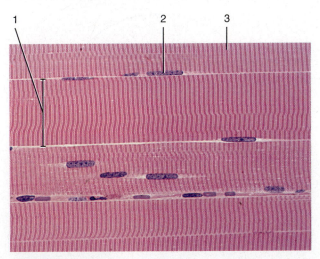

400×, H&E, human

- nucleus
- striation (stry-AY-tion)
- width of individual muscle fiber

FIGURE 6.24 Sectional view of skeletal muscle fibers.

Student drawing

1 _____

2 _____

3 _____

400×, iron H&E

- branches of cardiac muscle fiber
- intercalated discs
- nucleus
- width of cardiac muscle fiber

FIGURE 6.25 Longitudinal view of cardiac muscle fibers.

Student drawing

1 _____

2 _____

3 _____

4 _____

Smooth muscle layers
Connective tissue
Epithelial lining
Lumen

Student drawing

60×
(a) Survey photomicrograph

1 2

Student drawing

400×
(b) Smooth muscle fibers in cross-section
 and longitudinal section

• nucleus of smooth muscle fiber in cross-section
• nucleus of smooth muscle fiber in longitudinal section

1 _____

2 _____

FIGURE 6.26 Sectional view of smooth muscle in wall of the ureter.

D. Nervous Tissue Overview

Nervous tissue, which forms the brain, spinal cord, and nerves, is very cellular (Figure 6.27). Two basic categories of nervous tissue cells are **neurons** and **neuroglia** (*neuro* = nerve; *glia* = glue). Neurons receive and send information, whereas neuroglia support the neurons and help them to function. Neurons have one or more processes (cellular extensions) that receive or send information as nerve impulses. **Dendrites** (*dendro* = tree) are processes that receive signals from sensory receptors or other neurons. An **axon** is a process that sends signals to other neurons, muscles, or glands. Neurons have one axon but may have many dendrites. The main part of the cell where the nucleus is located is called the **cell body.** In the following activity you will observe only **multipolar neurons,** which are neurons with many processes.

FIGURE 6.27 **Nervous tissue.**

Before Going to Lab

1 Label and examine the photomicrograph of nervous tissue in Figure 6.28. Compare the sizes of the neuron and neuroglia.

LAB ACTIVITY 4 **Nervous Tissue**

1 Examine prepared microscope slides or use **Real Anatomy** (Histology) to observe nervous tissue containing multipolar neurons. Use Figure 6.28 to help you locate the major structures.

2 Draw each tissue in the space provided and label the major structures. ■

LM 260×

- cell body of multipolar neuron
- nucleus
- processes

FIGURE 6.28 **Photomicrograph of nervous tissue.**

Student drawing

1 _____

2 _____

3 _____

SAFETY NOTE: *Wear safety glasses and gloves when using preserved or fresh tissue. Wash hands thoroughly with soap and water when you are finished.*

LAB ACTIVITY 5 Observation of Fresh Tissue in Chicken Leg

1 Obtain a fresh chicken leg, dissecting pan, forceps, pointed dissecting probes, blunt dissecting probe, scalpel, dissecting microscope, compound microscope, and lens paper.

2 Observe the appearance of epidermis.
 - Epidermis of chicken skin is a waterproof layer composed of stratified epithelial tissue.
 - Notice the "goosebumps" on the surface of the epidermis. These goosebumps are formed when the chicken feathers are plucked out.
 - The goosebumps contain a follicle that produces a feather.

3 Observe the appearance of areolar connective tissue
 - Using the forceps, slowly pull the skin away from the muscle. Notice the very thin areolar connective tissue that holds the skin to the muscle tissue.

4 Observe the ability of dense, irregular connective tissue to resist pulling from different directions.
 - The dermis of the skin (layer of skin deep to the epidermis) contains dense irregular connective tissue.
 - Stretch the skin in different directions and observe how difficult it is to pull the skin apart.

5 Observe the appearance of adipose tissue.
 - Use a scalpel to cut a very small piece of adipose tissue from the proximal end of the leg bone or from under the skin, place it on a clean slide, and observe it under a dissecting microscope.
 - Using two pointed probes, tease small droplets of fat apart from the edge of the tissue.
 - If teased enough, you can observe that the adipose tissue is made of many tiny fat globules adhering together.
 - Using a compound microscope, observe the same tiny fat droplets with the scanning objective lens (4×) and then the low-power objective lens (10×). Compare their appearance with what you observed using the dissecting microscope.

6 Observe the appearance of the dense, regular connective tissue forming a tendon.
 - Use a scalpel to cut a very thin piece of tendon from the end of a muscle and place it on a clean slide.

 - Under the dissecting microscope, first notice the regular arrangement of collagen fibers that appear as horizontal "wrinkles" perpendicular to the length of the tendon.
 - As you tease the fibers apart with two dissecting probes, you will notice the deeper fibers run longitudinally in the tendon.
 - Using the compound microscope observe the collagen fibers with the scanning objective lens (4×).

7 Observe the appearance of hyaline cartilage covering the ends of the leg bone.
 - Use a scalpel to cut a very thin piece of cartilage from the proximal or distal end of the bone and place it on a clean slide.
 - Tease small pieces of cartilage with two probes using a dissecting microscope. Notice that this tissue looks very different from the others and the surface of the cartilage looks like the surface of the moon.
 - Using the compound microscope examine the cartilage slide with the scanning objective lens (4×).

8 Observe skeletal muscles, blood vessels, and nerves.
 - Observe the thin layer of connective tissue covering skeletal muscles. This layer is composed of dense regular connective tissue and is called fascia.
 - Remove fascia from muscle and, using a blunt probe, separate the individual muscles.
 - Look for blood vessels and nerves between muscles. Blood vessels are brownish-red hollow tubes, while nerves are white, thread-like structures. Cut the blood vessel to observe it in cross-section. Coagulated blood may be found within the blood vessel.

9 Observe the appearance of skeletal muscle tissue.
 - Use your dissecting probe to separate a small piece of skeletal muscle and place it on a microscope slide.
 - Add a drop of methylene blue stain.
 - Cover the microscope slide with a cover slip.
 - Use the scanning objective lens (4×) of the compound microscope to center the muscle in the field of view and to bring the muscle in focus. Switch to the low-power objective (10×) to find muscle cells. Switch to the high-power objective (100×) to observe the muscle cells.
 - Observe the shape of the muscle cells, nuclei, and striations.

10 Compare the appearance of each of these tissues with their prepared microscope slides.

11 Clean up as directed by your instructor. ■

Reviewing Your Knowledge

A. Primary Tissue Structure and Function

Name the primary tissue type (epithelial, connective, muscle, or nervous) that is described.

_____ **1.** Cells contain processes that receive and generate electrical signals to communicate with other cells.

_____ **2.** Tissue has elongated cells that shorten and cause movement.

_____ **3.** Tissue contains more extracellular matrix than cells.

_____ **4.** Cells either form a barrier that controls passage of molecules or form glands.

_____ **5.**

_____ **6.** } Primary tissue types that exhibit cellularity.

_____ **7.**

_____ **8.**

_____ **9.** } Cells determine function of these primary tissue types.

_____ **10.**

_____ **11.** The extracellular matrix determines function of this primary tissue.

Identify the primary tissue types in Figure 6.29.

12. _____ 13. _____ 14. _____ 15. _____

FIGURE 6.29 Photomicrographs of primary tissues.

B. Epithelial Tissues

Write the name of the epithelial tissue type that matches each description. An epithelial tissue type may be used more than once.

_____ 1. Lines the mouth and protects underlying tissues in areas subject to abrasion.

_____ 2. Located in the alveoli (the air sacs of the lung) and provides a short distance for the diffusion of oxygen and carbon dioxide.

_____ 3. Forms kidney tubules and is involved in absorption and secretion.

_____ 4. Lines the nasal cavities and moves substances over the epithelial surfaces.

_____ 5. Forms the mesothelium of the peritoneum and secretes serous fluid into the peritoneal cavity.

_____ 6. Lines the stomach and its microvilli; increases surface area for absorption and secretion.

_____ 7. Lines the bladder and ureter and is distensible.

C. Connective Tissues

Write the name of the connective tissue type that matches each description. A connective tissue can be used more than once.

_____ 1. Contains elastic fibers and is found in the lungs. This tissue allows the lungs to inflate during inhalation and return to their original shape after exhaling.

_____ 2. Packed with parallel bundles of collagen fibers and found in tendons. This tissue resists pulling forces applied by muscles.

_____ 3. Has a firm, gelatinous ground substance containing collagen fibers. This tissue is found in the tracheal wall to support and prevent the trachea from collapsing.

_____ 4. Found under covering and lining epithelium. Its extracellular matrix contains a loose arrangement of fibers, and its viscous ground substance facilitates the flow of interstitial fluid containing nutrients to epithelial tissues. It also cushions and supports epithelia.

_____ 5. Contains many elastic fibers in a firm gelatinous ground substance. Located in external ear, auditory tube, and epiglottis.

_____ 6. Hard extracellular matrix forms trabeculae.

_____ 7. Forms a framework in the spleen, bone marrow, and lymph nodes. It contains fine, branching fibers.

_____ 8. Is packed with bundles of collagen fibers running in different directions. It is found in skin and allows skin to resist pulling forces from many different directions.

_____ 9. Fluid extracellular matrix used to transport substances throughout the body.

_____ 10. Contains a large number of lipid-storing cells. It is found throughout the body, cushions and insulates organs, and stores lipids for future energy needs.

_____ **11.** Hard extracellular matrix containing osteons; involved in protection and support.

_____ **12.** Firm gelatinous ground substance packed with bundles of collagen fibers. This tissue is found in intervertebral discs, pubic symphysis, and knee meniscus.

_____ **13.**

_____ **14.**

_____ **15.**

Fibroblasts produce extracellular matrix for these connective tissue types.

_____ **16.**

_____ **17.**

_____ **18.**

_____ **19.**

Osteoblasts produce extracellular matrix.

_____ **20.**

_____ **21.**

_____ **22.** Chondroblasts produce extracellular matrix.

_____ **23.**

_____ **24.** Fibers not present unless injury occurs. Extracellular matrix not produced by cells present in this tissue.

D. Muscle and Nervous Tissue

Write the name of the tissue that matches each function. For muscle tissue, write the name of the muscle tissue type.

_____ **1.** Movement of urine through the urinary tract

_____ **2.** Movement of bone and/or skin

_____ **3.** Movement of blood through the heart and into arteries

_____ **4.** Receives and sends information

_____ **5.** Movement of food through the digestive tract

_____ **6.** Controls blood flow through arteries and veins and controls blood pressure

E. Epithelial and Connective Tissue Identification

Identify the epithelia and connective tissues types in Figure 6.30.

1. _____

2. _____

3. _____

4. _____

5. _____

6. _____

7. _____

8. _____

9. _____

10. _____

11. _____

12. _____

13. _____

14. _____

15. _____

16. _____

17. _____

18. _____

FIGURE 6.30 Identification of epithelial and connective tissue types.

FIGURE 6.30 Identification of epithelial and connective tissue types, *continued.*

F. Muscle and Nervous Tissue Identification

Identify the muscle tissue types and the nervous tissue in Figure 6.31.

1. _____

2. _____

3. _____

4. _____

1

2

3

4

FIGURE 6.31 **Muscle and nervous tissue identification.**

Using Your Knowledge

A. Genetic Diseases and Tissue Function

Each of the following diseases is due to a gene mutation and the production of an abnormal protein. Look up the diseases in your textbook or other resource. For each disease identify the abnormal protein and state how the abnormal protein affects the function of the tissue.

Cystic Fibrosis and Epithelial Tissue

1. Abnormal protein:

2. The effect of the abnormal protein on tissue function:

Duchenne Muscular Dystrophy and Skeletal Muscle Tissue

3. Abnormal protein:

4. The effect of the abnormal protein on tissue function:

B. Nutritional Deficiencies and Tissue Function

Nutritional deficiencies interfere with cell metabolism and tissue structure. Look up scurvy in your textbook or other resource and name the nutritional deficiency that causes scurvy and describe the change in extracellular matrix and how this causes the symptoms of scurvy.

Scurvy and Connective Tissue

5. Nutritional deficiency:

6. Extracellular matrix changes and symptoms:

C. Tissues Forming Organs

Name the tissue types that form the wall of each organ in questions 7–10. Use your textbook as a reference and list the tissues in order from lumen to outer surface.

7. esophagus: _____

8. small intestine (within peritoneal cavity): _____

9. trachea: _____

10. ureter: _____

The Integumentary System Structure and Function

OBJECTIVES

1 Describe the structure and function of the integumentary system

2 Identify the layers of the epidermis, dermis, and hypodermis

3 Identify the accessory structures of the skin

4 Describe the function of the epidermal accessory structures

5 Compare eccrine sweat gland density on the forehead, forearm, palm, and anterior leg

6 Identify the main types of fingerprint patterns: arch, loop, and whorl

MATERIALS

- integumentary system model or chart
- compound microscopes and lens paper
- prepared microscope slides or use **Real Anatomy** (Histology): thick skin, thin skin with hair (scalp)
- **Eccrine Gland Density:** cornstarch, vegetable oil, Betadine (contains iodine) or 2% tincture of iodine, cotton swabs, two 100-ml beakers or small bowls
- **Fingerprinting:** ink pad, 3" × 5" index cards (2 per student), towelettes, magnifying glasses, data collection sheet

The integumentary system (*inte-* = whole; *-gument* = body covering) consists of organs including the skin, hair, nails, sweat and sebaceous glands, and associated muscle and nervous tissue. This system provides a protective barrier for the body, contains cutaneous sensory receptors, aids in the production of vitamin D, is important in regulating body temperature, and plays a minor role in excretion and absorption. The **skin** is classified as an organ because it has many types of tissues that work together to perform specific functions. The skin is also known as the **integument** or **cutaneous membrane.**

A. Major Divisions of the Skin

The skin has two major divisions: the superficial epidermis and the deep dermis. The **epidermis** (*epi-* = above), the outer layer of the skin, is composed of epithelial tissue. The **dermis** is the connective tissue layer that is firmly attached to the epidermis by a basement membrane, provides the avascular epidermis with nutrients, and connects the epidermis to the underlying hypodermis. Although it is not a part of the integumentary system, the **hypodermis** (*hypo-* = below) or **subcutaneous layer** (subQ) is located below the skin and is usually discussed with the skin. The hypodermis is a major storage site for adipose tissue.

1. Epidermis

The epidermis is **keratinized stratified squamous epithelium.** This is a very thick epithelial layer in comparison with other epithelial layers of the body.

Four types of cells are found in the epidermis: keratinocytes, melanocytes, Langerhans cells, and Merkel cells. **Keratinocytes** (*keratino-* = horn-like; *-cytes* = cells) comprise 90% of the cells of the epidermis and produce keratin, a tough fibrous protein that protects the skin and deeper tissues from chemicals, microbes, and heat. These cells also produce granules that secrete a lipid-rich product that helps to waterproof the skin. **Melanocytes** (*melano-* = black) comprise 8% of the epidermal cells and produce and secrete the pigment melanin. **Langerhans cells** are immune system cells that attack pathogens that enter the skin. **Merkel cells** are the least abundant cell type and are found only in the deepest layer of the epidermis. These cells function as touch receptors and are associated with sensory neurons.

The layers of the epidermis are in order from deepest to most superficial: stratum basale, stratum spinosum, stratum granulosum, stratum lucidum, and stratum corneum. The **stratum basale** (*strata* = layers; *basa-* = base) or stratum germinativum (*germ-* = sprout), a single row of cells attached to the basement membrane, contains stem cells that divide to form new keratinocytes. As new cells are formed, the older cells are pushed toward the surface and undergo a process called keratinization, which produces tough, dead cells in the superficial layer. Melanocytes, Langerhans cells, and Merkel cells are also found within the stratum basale.

The **stratum spinosum** (*spinos-* = thorn-like) contains 8 to 10 rows of cells, mainly keratinocytes. In prepared slides, cells in this layer have thorn-like projections caused by the tissue preparation process. Granules are observed within keratinocytes in the **stratum granulosum** (*granulos-* = little grains). This layer contains 3 to 5 rows of flattened keratinocytes that are beginning to die. No dividing cells are present in this layer or in more superficial layers.

The **stratum lucidum** (*lucid-* = clear) contains 3 to 5 rows of flat, dead keratinocytes. This layer is translucent in specimens of fresh skin, but in prepared slides it may be clear or stained.

The outermost layer, the **stratum corneum** (*corn-* = hard or hoof-like) is a very thick layer containing 25 to 30 or more rows of dead, squamous-shaped keratinocytes. This layer is tough and water-repellent. These cells continually slough off and are replaced by cells in the adjacent layer.

The epidermis differs in thickness in thin and thick skin. **Thick skin** is found on the palms of the hands and the soles of the feet and has all five strata. **Thin skin,** which covers the rest of the body, does not have a visible stratum lucidum and has a thinner stratum corneum than thick skin.

2. Dermis

The dermis consists of two regions: the papillary (*papilla* = nipples) region and the reticular (*reticul-* = net-like) region. The **papillary region** is a thin layer of areolar connective tissue that is deep to the stratum basale of the epidermis and the basement membrane. **Dermal papillae** are finger-like projections of the papillary region that extend into the epidermis. In the palms, fingers, soles, and toes, the dermal papillae cause genetically determined whorls in the epidermis called **epidermal ridges** that increase surface area, friction, and grip. Sweat glands deposit their secretions onto these ridges, resulting in fingerprints when these ridges touch surfaces. The **reticular layer** is the deeper and much thicker region of the dermis. It is composed mainly of dense, irregular connective tissue whose collagen fibers provide the skin with strength and whose elastic fibers provide elasticity. Some adipose tissue is also found in the reticular region. The dermis is highly vascularized, allowing nutrients to diffuse from the dermis into the avascular epidermis. Hair follicles, sweat glands, and sebaceous glands are all derived from epithelial tissue and extend into the dermis. Nerves and lymphatic vessels are also found in this layer.

Before Going to Lab

1 Label the skin structures in Figures 7.1 and 7.2. Observe where the basement membrane separates the epidermis and dermis.

2 Label the layers of the epidermis in thick skin in Figure 7.3.

LAB ACTIVITY 1 Major Divisions of the Skin

1 Identify the skin structures in Figures 7.1, 7.2, and 7.3 on a model or chart.

2 Examine a prepared microscope slide of thick skin or use **Real Anatomy** (Histology).
- Using the scanning or low-power objective, identify the structures in Figure 7.2.
- Using the high-power objective, identify the layers of the epidermis in Figure 7.3. Observe the different cell shapes in the various layers of the epidermis.
- Observe the areolar connective tissue in the papillary layer of the dermis, the dense irregular connective tissue in the reticular layer of dermis, and the adipose tissue in the hypodermis.

FIGURE 7.1 Diagram of the skin.

- dermal papillae
- epidermis
- hypodermis
- papillary layer of dermis
- reticular layer of dermis

1 epidermis
2 papillary of dermis
3 reticular of dermis
4 hypodermipillae
5 dermal papillae

LM 240x

30×

- epidermis (EPI-derm-is)
- dermal papillae (puh-PILL-ee)
- hypodermis (HY-poh-der-mis)
- papillary (PAP-il-lary) layer of dermis
- reticular layer of dermis

1 dermal papillae
2 epidermis
3 papillary layer
4 reticular layer
5 hypodermis

FIGURE 7.2 Photomicrograph of the skin.

- dermis
- stratum basale (bay-SAL)
- stratum corneum (kor-NEE-um)
- stratum granulosum (gran-you-LOW-sum)
- stratum lucidum (LOU-sih-dum)
- stratum spinosum (spy-NO-sum)

6 1 dermis
1 2 stratum corneum
2 3 lucidum
3 4 granulosum
4 5 spinosum
5 6 basale

FIGURE 7.3 Photomicrograph of the epidermal layers.

B. Accessory Structures of the Skin

The accessory structures of the skin include hairs, hair follicles, nails, sweat glands, sebaceous glands, ceruminous glands, and mammary glands. All of these are derived from epidermal tissue and extend into the dermis.

1. Sudoriferous and Sebaceous Glands

Sudoriferous glands (*sudori-* = sweat; *-ferous* = bearing), or sweat glands, secrete a watery substance that is important in excretion and body temperature regulation. There are two types of sudoriferous glands: eccrine glands and apocrine glands. **Eccrine glands** (*eccrine* = sweating outwardly) are the most common type of sudoriferous gland and are found on most areas of the body. Ducts from the eccrine glands deposit their secretions, called sweat, on the epithelial surface. **Apocrine glands** are found only in the axilla, genital area, and pigmented area around the nipples (areolae). Apocrine glands produce a secretion similar in composition to sweat but more viscous. This secretion, which is deposited on the distal end of the hair root, is odorless until broken down by bacteria. Ceruminous glands and mammary glands are modified sudoriferous glands. **Ceruminous glands** (*ceri-* = wax) are found in the ear canal and secrete a waxy substance called cerumen that prevents foreign substances (including insects) from entering the auditory canal. **Mammary glands** are found in the breasts and synthesize and secrete milk after appropriate hormonal stimulation. **Sebaceous glands** (*sebace-* = greasy), or oil glands, are found surrounding hair follicles and deposit sebum, an oily substance that lubricates the skin and hair, into the neck of the follicle.

2. Hair and Hair Follicles

Hairs are found all over the body with the exception of the palms, soles, lips, and parts of the external genitalia. Hair consists of dead, keratinized epithelial cells and has two main sections: the **shaft,** which projects from the skin surface, and the **root,** which extends into the dermis of the skin and sometimes the hypodermis. The **hair follicle,** which surrounds the hair root, is formed from epidermal layers that project into the dermis. The expanded base of the hair follicle, the **hair bulb,** contains the papilla of the hair and the matrix. The **papilla** of the hair is a projection of connective tissue into the hair follicle and contains blood vessels that provide nutrients to the dividing cells of the matrix. The **matrix,** which is derived from the stratum basale of the epidermis, forms new hair cells that are added to the base of the hair root. Surrounding the hair follicle is a **connective tissue sheath** comprised of dermal tissue.

The **arrector pili** (*arrect-* = to raise; *pili* = hair) is a bundle of smooth muscle cells attached to the connective tissue sheath around the hair follicle. Contraction of the arrector pili muscle moves the hair from its normal angle to a 90° angle (perpendicular) with the skin surface, elevating the skin surrounding the hair shaft and causing goose bumps. Arrector pili muscles contract in response to stress (including cold temperature).

Examination of the hair in cross-section shows three layers: the cuticle (outer layer), the cortex (middle layer), and the medulla (inner layer). The **cuticle,** the layer of hair that we see, is a thin layer of dead, flattened, keratinized cells that overlap each other like shingles on a roof. Split ends occur when the free ends of these cells are pulled away from each other. Cells in the **cortex and medulla** contain pigment granules that give hair its color. The shape of hair in cross-section indicates whether the hair is curly, wavy, or straight. Curly hair is flat in cross-section, wavy hair is oval, and straight hair is round.

Before Going to Lab

1 Label the diagram of the skin and accessory structures in Figure 7.4.

2 Label the photomicrograph in Figure 7.5.

LAB ACTIVITY 2 **Hair, Hair Follicles, Sebaceous Glands, and Sudoriferous Glands**

1 Point to the structures in Figure 7.4 on a model or chart of the integumentary system.

2 Examine a prepared microscope slide of hairy skin or use **Real Anatomy** (Histology) to identify the structures in Figure 7.5.

3 Examine a strand of hair with the compound microscope.
 - Pull out one strand of scalp hair and lay the hair horizontally on the stage over the light source.
 - Observe the dark cuticle and lighter medulla of the hair shaft with the 4× objective lens. Switch to the low- and high-power lenses and look for the cells of the cuticle overlapping each other. Observe the root end of the hair. What differences do you see between the root end and the shaft area?
 - If someone is willing to donate hair with "split ends," compare the cuticle of a hair with split ends with undamaged hair.

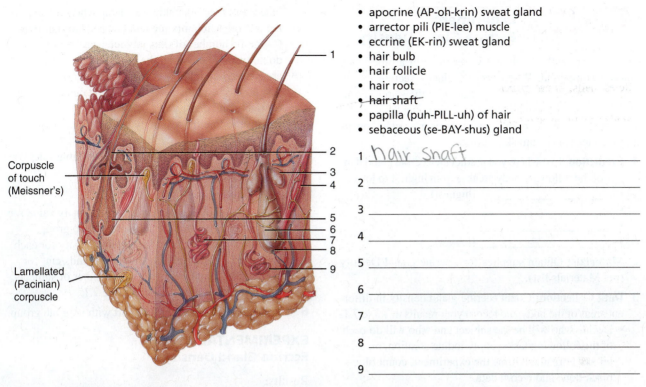

- apocrine (AP-oh-krin) sweat gland
- arrector pili (PIE-lee) muscle
- eccrine (EK-rin) sweat gland
- hair bulb
- hair follicle
- hair root
- hair shaft
- papilla (puh-PILL-uh) of hair
- sebaceous (se-BAY-shus) gland

Corpuscle of touch (Meissner's)

Lamellated (Pacinian) corpuscle

1 hair shaft _____

2 _____

3 _____

4 _____

5 _____

6 _____

7 _____

8 _____

9 _____

FIGURE 7.4 **Diagram of the skin and accessory structures.**

- hair bulbs
- hair follicle
- hair root
- papilla of hair
- sebaceous gland

1 _____

2 _____

3 _____

4 _____

5 _____

10×

FIGURE 7.5 **Photomicrograph of the skin and accessory structures.**

LAB ACTIVITY 3 Experiment: Eccrine Gland Density

Iodine reacts with starch and water to form a blue-black colored compound. When eccrine glands secrete sweat, the water enables the iodine to react with the starch (cornstarch) to produce a small blue-black dot about the size of a straight pin point where the duct of the eccrine gland opens and secretes water onto skin surface.

1 Prediction: List where you predict the eccrine gland density of the following body areas is from highest to lowest.
- _____ (highest)
- _____
- _____
- _____ (lowest)

2 Materials: Obtain supplies for Eccrine Gland Density (see Materials list).

3 Data Collection: Count eccrine gland density in different areas of the body and record your results in Table 7.1.
- Decide who will be the subject and who will do each of the following tasks: apply iodine solution, mix oil and cornstarch, time the experiment, count blue-black dots, and record data.
- Mix equal parts of vegetable oil and cornstarch. Approximately 1 tablespoon of each should be enough.
- If the room is cool and students are not producing enough sweat, have them exercise a little so the experiment will work.
- Find areas of your body that do not have large crease lines: forehead, anterior forearm, palm, and anterior leg.
- Clean and thoroughly dry these areas. If the skin is still damp, the procedure will not work and areas may be splotchy.
- For each location, use a cotton swab to apply iodine solution to a 1-cm^2 area of skin. Allow Betadine to *thoroughly* dry.
- Apply the oil/cornstarch solution to each area of skin. Make sure the oil/cornstarch solution exceeds the iodine-covered area. Iodine reacts with starch and water to form a blue-black colored compound. After about 10 minutes you should see a few blue-black pinprick-size dots. Each of these dots represents a eccrine duct gland opening. If you see large splotchy

blue areas, then your skin was damp when you applied the oil/cornstarch mixture and you need to start over.
- Count the number of dots in your 1-cm^2 area. If you do not want to count all the dots, you can rank the areas based on relative eccrine gland density (list them from greatest eccrine gland density to least eccrine gland density).
- Remove iodine from skin by washing the area with soap and water.
- Record the number of dots per cm^2 in Table 7.1.

4 Clean up as directed by your instructor.

5 Data Analysis:
- Collect the individual eccrine gland density value for each body area from the subject for each group.
- Calculate the average eccrine gland density for each body area by adding the number of glands/cm^2 for each subject and dividing by the number of subjects.
- Record the average values in Table 7.1.

6 Complete the Experimental Report with your lab group.

EXPERIMENTAL REPORT
Eccrine Gland Density

Results:

1. Using the class averages in Table 7.1, state which body area has the highest average eccrine gland density and which has the lowest.

2. State whether variation in eccrine gland density per body area was observed among the subjects.

Discussion:

1. Discuss how eccrine gland density correlates with the amount of sweat that appears on these body areas during exercise or nervousness. Base this on your own observations and experiences.

2. State whether eccrine gland secretion is continuous and discuss the role of insensible water loss.

Conclusion:

1. Write a sentence that postulates a correlation between average eccrine gland density and the amount of sweat produced by an area.

2. Write a sentence that postulates why eccrine gland density per area varied or did not vary among individuals.

TABLE 7.1 Eccrine Sweat Gland Density

BODY AREA	ECCRINE SWEAT GLAND DENSITY (no. of glands/cm^2)	
	Individual Value	Class Average
Forehead		
Right anterior forearm		
Right palm		
Right anterior leg		

3. Nails

Nails, which are found on the distal ends of the digits, assist in grabbing and manipulating objects, and protect the digits. The nail consists of a nail body, a free edge, and a root. The **nail body** is the part of the nail that is visible, the **free edge** extends beyond the digit, and the **root** is within the fold of skin at the proximal end of the nail body. The **lunula** (*lunula* = little moon) is the crescent-shaped area of the nail body distal to the nail root. The **cuticle** or **eponychium** (*epi-* = above; *-onych* = nail) is the thickened epithelial tissue along the proximal border of the nail body. The **hyponychium** (*hypo-* = below) or nail bed is deep to the free edge and attaches the nail to the fingertip. The **nail matrix** is the epithelial tissue deep to the nail root that divides to produce new cells that are added to the nail body as it grows.

Before Going to Lab

1 Identify the nail structures on the diagram in Figure 7.6.

2 Identify nail structures on your own nails.

4. Fingerprints

Fingerprints are useful for identifying unknown bodies, missing children, adults with amnesia or Alzheimer's, or criminals. The practice of using fingerprints to identify someone is called **dactyloscopy.** Fingerprints develop from dermal papillae that form epidermal ridges in the 3rd and 4th fetal month of development. The ridge arrangement on every finger of every human being is unique and does not alter with growth or age. Superficial injuries do not affect the ridge structure or alter the dermal papillae, and the original pattern is duplicated in any new skin. An

(a) Dorsal view (b) Sagittal section through finger

(a) Dorsal view
• eponychium (eh-poh-NICK-ee-um)
• free edge
• lunula
• nail body

(b) Sagittal section through finger
• eponychium
• free edge
• hyponychium (hypo-NICK-ee-um)
• lunula (LOON-you-luh)
• nail root
• nail body
• nail matrix

1 _____

2 _____

3 _____

4 _____

5 _____

6 _____

7 _____

8 _____

9 _____

10 _____

11 _____

FIGURE 7.6 Nail structures.

(a) Arch (b) Loop

(c) Plain whorl

FIGURE 7.7 Main types of fingerprint patterns.

injury that destroys the dermal papillae, however, will permanently destroy the ridges. The epidermal ridges of thick skin in the distal pads of phalanges (fingers and toes) are dotted with openings for oil and perspiration glands that transfer the image of the ridges onto a surface. The fatty acid and amino acid secretions leave invisible fingerprints that can be on a nonporous or a porous surface, and those same prints can be visible if there is any dirt or additional material on the fingertips.

Fingerprints are categorized into 3 generally recognized patterns, as shown in Figure 7.7: loops, whorls, and arches. Loops constitute about 65% of the total fingerprint patterns, whorls make up about 30%, and arches account for the other 5%. Forensic detectives acquire more details about the pattern area, including bifurcations, ridge endings, and islands. They also use the size of the patterns and the position of the patterns on the finger for further identification. The United States FBI further divides the basic 3 fingerprint classifications into 8 subclassifications.

LAB ACTIVITY 4 Fingerprinting Lab

1 For each person in your group, obtain an ink pad, two 3″ × 5″ index cards, ¾″ transparent tape, magnifying glass, and towelettes. Have the subject wash and dry the hands.

2 Using the lined side of the index card, write the subject's name and R or L hand.

3 The subject should stand, extending the arm, and *relax* while the lab partner stands to the side of the subject and

with both hands holds onto the index finger. Roll the finger, from the *pad* to at least the first joint, from side to side in the ink pad and then carefully roll the inked finger the same way onto the index card. If the fingerprint smears, or is too light or dark, repeat the procedure.

4 Repeat the same process with the middle finger, but when applying the print to the same card, roll the finger in the opposite direction so the index print doesn't get smeared. Clean your fingers with a towelette.

5 Repeat the procedure with the other hand and use the 2nd card.

6 Switch subjects and repeat with your lab partner.

7 Classify your prints into one of the recognized pattern groups of loops, whorls, or arches and mark your classification on the back of the cards.

8 Exchange fingerprint record cards with another group and compare classifications of your prints as well as identifying which two fingerprint cards belong to the same individual. If the other group agrees with your classification, give your instructor your fingerprint classification.

9 Your instructor will collect all fingerprint classification data from the class and write the results on the board, giving the class percentages of loops, whorls, and arches.

10 Make a bar graph of the class classification data and compare with the norms for the 3 recognized patterns: 65% loops (look like hairpins), 30% whorls (circles), 5% arches (hills).

11 Answer the Discussion Questions with your lab group.

DISCUSSION QUESTIONS
Fingerprints

1 Are the prints from your index and middle fingers on the same hand similar? Are those fingerprints similar compared to your other hand?

2 What classification pattern do you have?

3 How does your class compare with the total percentages of the 3 recognized patterns?

Reviewing Your Knowledge

A. Skin Layers and Structures

Write the name of the skin layer or structure that fits the description.

_____ 1. Layer of epidermis where there is the most rapid cell division.

_____ 2. Tough, water-repellent epidermal layer; contains dead squamous-shaped cells.

_____ 3. Areolar connective tissue layer beneath basement membrane.

_____ 4. Projections of dermis that cause epidermal ridges.

_____ 5. Translucent layer found in thick skin, absent in thin skin.

_____ 6. Appears to have thorn-like projections in prepared slides.

_____ 7. Thick dermal layer containing dense, irregular connective tissue.

_____ 8. Epidermal layer containing visible granules.

_____ 9. Epidermal layer with stem cells.

_____ 10. Dermal layer with hair follicles and glands.

B. Cells of the Epidermis

Write the name of the epidermal cell type that fits the description.

_____ 1. Defends skin against microbes.

_____ 2. Produces keratin.

_____ 3. Produces the pigment melanin which shields cell nuclei from ultraviolet (UV) radiation.

_____ 4. Associated with sensory neurons and functions in sensation of touch.

_____ 5. Compose about 8% of epidermal cells.

_____ 6. Compose about 90% of epidermal cells.

_____ 7.⎫
 ⎬ Compose about 2% of epidermal cells.
_____ 8.⎭

C. Accessory Structures of the Skin

Write the name of the accessory structure that best fits the description.

_____ **1.** Sudoriferous glands located in axillary and genital areas; become active after puberty.

_____ **2.** Connective tissue projection; provides blood supply for hair matrix.

_____ **3.** Secures nail to digit.

_____ **4.** Epithelial layer that surrounds hair root.

_____ **5.** Secretes sebum onto hair and skin.

_____ **6.** Part of nail that is visible.

_____ **7.** Part of nail that extends beyond digit.

_____ **8.** Part of hair that contains the matrix.

_____ **9.** Sudoriferous glands that deposit sweat onto epidermal ridges causing fingerprints.

_____ **10.** Part of hair in epidermis and extending beyond skin surface.

_____ **11.** Moves hair shaft perpendicular to skin; causes goose bumps.

_____ **12.** Thickened epithelial tissue at proximal end of nail body.

_____ **13.** Crescent-shaped area of nail body near cuticle.

_____ **14.** Part of hair within dermis.

_____ **15.** Part of nail within skin.

_____ **16.** Part of nail deep to nail root that produces new cells, causing the nail to grow.

Using Your Knowledge

A. Identification of Skin from Different Body Locations

Observe diagrams of skin from different body locations in Figure 7.8. Based on number of epithelial layers and skin accessory structures, determine if skin is from the axillary area, forearm, or sole of foot.

1. _____

2. _____

3. _____

1　　　　　　　　　　　　2　　　　　　　　　　　　3

FIGURE 7.8 **Identification of skin from different body locations.**

B. Application

Answer the following questions in the space provided.

4. Cortisone is a steroid that is applied to the skin to reduce inflammation. Cortisone acts on cells within the dermis and can travel through unbroken epidermis to reach cells in the dermis. If the epidermis is such a good barrier, how can cortisone easily travel through it?

5. Explain how dandruff is formed.

6. Joey has a splinter in his finger. His mother pulled the skin away from the splinter with a sterile needle and removed the splinter. Joey did not feel pain and did not bleed. Explain.

7. Identify the epidermal layer(s) in which the following cancers arise: (a) basal cell carcinoma; (b) malignant melanoma; (c) squamous cell carcinoma.

8. Name the part(s) of the skin that "peel(s)" off after a minor sunburn.

9. As we age, our skin wrinkles due to changes in collagen and elastic fibers in the dermis and a decrease in their production. Explain why topical application of collagen and elastic fibers would not eliminate wrinkles.

10. Permanent tattoos are made by injecting pigment into the skin. Into which part of the skin is the pigment injected, the epidermis or dermis? (You can answer this without research; just use your knowledge.)

11. If you had a third-degree burn on your scalp, would you expect for your hair to grow back in that area? Explain.

12. Pulling a hair out of your head hurts, but cutting your hair doesn't. Explain the difference.

13. How do vitiligo and albinism differ? How are they similar?

14. What structure of skin helps you grip a wet glass?

15. Describe the mechanism for acquiring a suntan.

Bone Structure and Function

Bones are organs composed of a complex arrangement of several tissues. A typical bone has compact and spongy osseous tissues, connective tissues, cartilage, and adipose tissue. In addition, bones contain blood vessels and nerves.

A. Classification of Bones

Human bones have different shapes and distinct gross anatomical features. Bones are placed in five classifications according to their shapes: long, short, flat, irregular, and sesamoid. **Long bones** are longer than they are wide, with a thick compact bone exterior. Distribution of spongy bone in long bones is covered in detail later in this exercise. **Short bones** are almost equal in length and width and contain a thick interior of spongy bone covered by a thin veneer of compact bone. **Flat bones** are relatively flat, but may be curved, and contain a thin, spongy bone interior covered by a thin veneer of compact bone. **Irregular bones** are self-explanatory and do not easily fit into any of these categories. **Sesamoid bones** (*sesame* = seed) are small bones that develop in tendons (e.g., patella) for protection against wear and tear.

LAB ACTIVITY 1 — Classification of Bones According to Shape

1 Classify each of the bones your instructor has displayed according to shape.

2 Discuss with the rest of the class which bones are difficult to classify and why. ■

B. Gross Features of Long Bones

The enlarged proximal and distal ends of long bones are called **epiphyses** (*epi-* = above, over; *-physis* = growing), and the middle shaft area is called the **diaphysis** (*dia-* = through). Compact bone forms the exterior (or cortex) of long bones and most of the diaphysis. A small layer of spongy bone lines the interior of the diaphysis. Spongy bone also forms the interior of the epiphyses. The **metaphyses** are the areas in an adult bone where the epiphyses and diaphysis join. In a growing bone, the metaphyses contain a layer of hyaline cartilage called the **epiphyseal plate.** Division of cartilage allows the bone to grow in length. Bone growth stops when the epiphyseal plate cartilage becomes ossified and forms a bony structure called the **epiphyseal line.**

Articular cartilage (*articul-* = joint), composed of hyaline cartilage, covers both epiphyses; and the rest of the bone exterior is covered with a tough, connective tissue membrane, the **periosteum** (*peri-* = around; *osteo-* = bone).

The hollow center of the bony diaphysis is called the **medullary cavity** (*medulla* = marrow, pith), and a small amount of spongy bone is found in this cavity. The medullary cavity is lined with a connective tissue membrane called the **endosteum** (*endo-* = within). The endosteum also lines the cavities within the spongy bone of the epiphyses. Both the periosteum and the endosteum contain **osteoblasts** (*-blast* = builder) and **osteoclasts** (*-klasis* = breaking) for bone formation, bone tissue repair, and bone remodeling.

Yellow marrow is a fatty substance found within the medullary cavity. **Red marrow** is found within the cavities of spongy bone and produces blood cells.

The **nutrient artery** is a large artery that enters compact bone near the middle of the diaphysis. The nutrient artery immediately branches into proximal and distal portions which supply blood to the inner layer of compact bone, spongy bone, and red marrow. The **nutrient foramen** is the foramen through which the nutrient artery enters.

Before Going to Lab

1 Label Figure 8.1.

Longitudinal section

- articular cartilage
- compact bone
- diaphysis (die-AF-ih-sis)
- distal epiphysis (e-PIF-ih-sis)
- endosteum (en-DOS-tee-um)
- epiphyseal (ep-i-PHY-zee-al or ee-PIF-ih-seal) line
- medullary (MED-yoo-lar-y) cavity
- nutrient artery
- periosteum (peri-OS-tee-um)
- proximal epiphysis
- spongy bone with red marrow
- yellow marrow

1 Proximal epiphysis
2 diaphysis
3 distal epiphysis
4 articular cartilage
5 epiphyseal line
6 compact bone
7 medullary cavity
8 yellow marrow
9 periosteum
10 nutrient artery
11 endosteum
12 spongy bone

FIGURE 8.1 Features of an adult long bone.

LAB ACTIVITY 2 Gross Features of a Human Long Bone

1 Identify as many items as you can from Figure 8.1 on a human long bone that is partially sectioned longitudinally and transversely.

2 Use a dissecting microscope to observe spongy bone in the epiphyses and lining the medullary cavity. ■

SAFETY NOTE: *Wear gloves when using fresh tissue!*

LAB ACTIVITY 3 Grow Features of a Fresh Long Bone

1 Using gloves, examine a fresh chicken tibia (leg bone) or femur (thigh bone) and identify as many structures as you can from Figure 8.1.
- Note the shiny appearance of the articular cartilage.
- Use a blunt probe and forceps to loosen the periosteum, ligaments, or tendons.
- Locate the center of the diaphysis and look for the nutrient artery passing through a hole (nutrient foramen) in the compact bone.
- Examine a cross-section of the bone to observe marrow in the medullary cavity, spongy bone, and compact bone.

2 Examine a longitudinal section of fresh beef bone.
- Find the epiphyseal line or the epiphyseal plate.
- Use a dissecting microscope to observe spongy bone in the epiphyses and lining the medullary cavity.
- Note the difference between the yellow and red bone marrow.

3 Clean up as directed by your instructor. ■

C. Microscopic Structure of Compact and Spongy Bone Tissue

Compact (cortical) bone is composed of repeating units of **osteons,** with each unit having a **central (Haversian) canal** running longitudinally. The central canal contains blood vessels, lymphatic vessels, and nerves that serve compact bone tissue. The blood vessels, lymphatic vessels, and nerves travel from the **periosteum,** dense regular connective tissue covering the bone surface, to the central canal through **perforating (Volkmann) canals.** These canals run horizontally in compact bone and connect with the central canal. The main feature of each osteon is the concentric rings, or **concentric lamellae** (*lamella* = small plate or ring), which look similar to the rings of a tree trunk

cut in cross-section. When viewed on a stained slide of compact bone, there are dark areas with thin lines extending between the lamellae. The dark areas are **lacunae** (*lacuna* = little lake) that are found between concentric lamellae, and the thin lines are **canaliculi** (small channels) that connect the lacunae. **Osteocytes** are mature bone cells that reside in the lacunae, and osteocyte processes extend through the canaliculi. Canaliculi allow nutrients from the blood vessels in the central canal to diffuse to the osteocytes embedded in the solid bone material. The canaliculi are also the route by which waste materials are removed from these cells. **Interstitial lamellae** (*inter-* = between; *-stitial* = to stand) fill in the spaces between the osteons.

Spongy (cancellous or trabecular) bone does not contain osteons but instead has **trabeculae** (little beams)—flat plates with a lattice-like network of thin, bony columns lined with endosteum. The trabeculae have **lamellae, lacunae, osteocytes,** and **canaliculi.** Spongy bone has many spaces filled with red marrow. Blood vessels within the red marrow provide the osteocytes with nutrients. This fragile spongy bone needs the protection of an outer layer of compact bone. Spongy bone is found in the epiphyses of long bones and in the interior of short, flat, and irregular bones.

Before Going to Lab

1 Label the structures in Figures 8.2(a) and (b), and Figure 8.3(a) and (b).

LAB ACTIVITY 4 Microscopic Structure of Ground Compact Bone and Spongy Bone

1 Examine a prepared slide or use **Real Anatomy** (Histology) to observe ground compact bone. Blood vessels, osteocytes, and periosteum cannot be observed on ground bone slides.
- Using the low-power objective lens, identify an osteon, a central canal, concentric lamellae, and interstitial lamellae. Count the number of osteons in one field of view.
- Using the high-power objective lens, identify canaliculi and lacunae.

2 Examine a prepared microscope slide or use **Real Anatomy** (Histology) to observe spongy bone. Note that on slides of spongy bone the preparations are generally demineralized (mineral salts removed) and do not show the lamellae and canaliculi as depicted in the drawing in Figure 8.2(b).
- Using the low-power objective lens, identify the trabeculae.
- Using the high-power objective lens, identify the lacunae, which appear as darker areas in the trabeculae. ■

Compact bone
Spongy bone
Periosteum
Medullary cavity

Medullary cavity

3
4

Osteon

5
6
7

8

9
10

1
11

(a) Osteons (Haversian systems) in compact bone and trabeculae in spongy bone

13
14

Space for red bone marrow

12

Osteoblasts aligned along trabeculae of new bone

(b) Enlarged aspect of spongy bone trabeculae

(a)
• blood vessels
• canaliculus (can-a-LIK-yoo-lus)
• central canal
• compact bone
• concentric lamellae (la-MEL-lee)
• lacuna (la-COO-na)
• osteocyte (OS-tee-o-site)
• perforating canal

1 _Spongy bone_
2 _trabeculae of Spongy bone_
3 _concentric lamellae_
4 _blood vessels_
5 _canaliculus_
6 _lacuna_
7 _osteocyte_
8 _periosteum_

• periosteum (per-ee-OS-tee-um)
• spongy bone
• trabeculae (trah-BEK-yoo-lee) of spongy bone covered with endosteum

(b)
• interstitial lamellae _14_
• osteocyte in lacuna _13_
• trabeculae covered with endosteum _12_

9 _central canal_
10 _perforating canal_
11 _compact bone_

12 _____
13 _____
14 _____

FIGURE 8.2 **Microscopic features of bone.**

(a) Longitudinally sectioned femur (thigh bone)

(a)
- compact bone
- medullary cavity
- metaphysis
- proximal epiphysis
- spongy bone

1 _____

2 _____

3 _____

4 _____

5 _____

LM 285×

(b) Cross-section of ground compact bone

(b)
- canaliculi
- central canal
- concentric lamella
- lacuna

6 _____

7 _____

8 _____

9 _____

Red marrow cell
10

11

LM 200×

(c) Photomicrograph of spongy bone

(c)
- osteocyte
- trabecula

10 _____

11 _____

FIGURE 8.3 **Gross and microscopic features of compact and spongy bone.**

D. Role of Collagen and Mineral Salts in Osseous Tissue

The properties of osseous tissue are determined by its extracellular matrix, which contains approximately 25% water, 25% collagen fibers, and 50% mineral salts. Collagen fibers are a fibrous protein that provide tensile strength and flexibility so that bone does not break with normal stress. The mineral salts consist mainly of calcium phosphate and calcium carbonate salts, giving the "backbone" or hardness to bone. As we age, the collagen content of osseous tissue decreases, causing bones to become brittle and break more easily. Decreased mineral content of bone, as occurs with rickets, causes bones to be soft and to bend due to body weight.

LAB ACTIVITY 5 **Role of Collagen and Mineral Salts in Osseous Tissue**

1 With your group, examine a chicken thigh or leg bone baked at about 350°F for a minimum of 2 hours or until brittle. Compare what happens when you try to bend a baked and fresh bone.

2 Examine a chicken thigh or leg bone that has been soaked in an acidic solution (vinegar or nitric acid) for 5 to 7 days. Try bending the bone soaked in the acidic solution and compare this with the fresh bone and the baked bone.

3 Clean up as directed by your instructor.

4 Answer Discussion Questions with your lab group.

DISCUSSION QUESTIONS
Collagen and Mineral Salts

1 Compare the flexibility of the untreated, baked, and acid-treated bones.

2 Which bone is brittle?

3 What substance has been damaged in the brittle bone? _____ Why would this increase bone breakage?

4 Which bone is the softest?

5 What substance has been leached from the soft bone? _____ What clinical disorder is this bone simulating?

Reviewing Your Knowledge

A. Gross Features of Long Bones

Write the answer in the space provided for questions 1–5.

1. What area of the long bone is covered with cartilage? _____

2. What type of cartilage is articular cartilage? _____

3. What area (epiphysis or diaphysis) is made up of a *thin* layer of compact bone and a *thick* spongy bone?

4. What area (epiphysis or diaphysis) is made up of a *thick* layer of compact bone and a *very thin* layer of spongy bone?

5. The long bone in Figure 8.1 is from an adult. If this figure were from a child, what structure would be present instead

of the epiphyseal line? _____

B. Microscopic Features of Long Bones

Identify the term that describes the phrase about long bones.

_____ 1. bone cell found in lacunae

_____ 2. vertical canal in an osteon

_____ 3. cavity that contains yellow marrow in adults

_____ 4. bone shaft

_____ 5. spaces where the osteocytes are located

_____ 6. horizontal canal in an osteon

_____ **7.** small canal connecting lacunae

_____ **8.** membrane lining medullary cavity

_____ **9.** membrane covering surface of bone

_____ **10.** thin bony columns in spongy bone

C. Comparison of Compact and Spongy Bone

Identify whether the statements below describe compact bone, spongy bone, or both.

_____ **1.** composed of osteons

_____ **2.** contains osteocytes and lacunae

_____ **3.** has lamellae

_____ **4.** has trabeculae

_____ **5.** has perforating canals

_____ **6.** located in the epiphyses

_____ **7.** located in the diaphysis

_____ **8.** has a central canal

_____ **9.** spaces filled with red marrow

_____ **10.** has canaliculi

D. Chemical Composition of Bone

Fill in the blanks with the correct term.

_____ **1.** The hardness of bone is due to _____.

_____ **2.** The flexibility and tensile strength of bone are due to _____.

_____ **3.** What type of a macromolecule (_carbohydrate, lipid, protein_) is collagen?

_____ **4.** A bone that has the collagen removed is _flexible_ or _inflexible_?

_____ **5.** A bone that has calcium removed is _flexible_ or _inflexible_?

Using Your Knowledge

A. Bone Tissue

Fill in the blanks with the correct term.

1. Two bone cells located in the periosteum and endosteum are _____ and _____.

2. Which type of bone tissue, compact bone or spongy bone, significantly degenerates first in osteoporosis?

3. As we age, the amount of collagen in the extracellular matrix of bone decreases and bones become more brittle. Identify the osseous tissue cell that secretes collagen. _____

4. Explain the importance of the integumentary system to bone formation.

B. Bone Images

Identify the structures indicated on Figure 8.4 and the structures and corresponding name of bone on Figure 8.5.

5 _____

6 _____

FIGURE 8.4 **Longitudinal section of a long bone.**

7 _____

8 _____

FIGURE 8.5 **X-ray of a child's knee joint.**

9. Observe the bones in Figure 8.6(a) and (b). Identify the child's hand and the adult's hand.

(a) _____ (b) _____

(a)

(b)

FIGURE 8.6 **X-rays of the child and adult hand.**

10. Observe Figure 8.7(a) and (b). Identify the normal bone and osteoporotic bone.

(a) _____ (b) _____

SEM 30×

SEM 30×

(a)

(b)

FIGURE 8.7 **Normal and osteoporotic bone tissue.**

Axial Skeleton

OBJECTIVES

1 Identify the 3 main parts of the axial skeleton

2 Identify the major bones of the axial skeleton

3 Identify selected bone surface markings and sutures

4 Identify paranasal sinuses and bones of the nasal septum, hard palate, and orbit of eye

5 Identify the fontanels in the fetal skull

6 Identify the parts of a typical vertebra

7 Compare the 3 individual types of vertebrae and the identifying features of each

8 Compare and contrast the normal and abnormal curvatures of the vertebral column

9 Identify the bones of the thoracic cage and selected surface markings

MATERIALS

• articulated skeleton or use **Real Anatomy** (Skeletal)

• disarticulated skeleton

• articulated skulls, Beauchene (disarticulated) skull, and fetal skull

• articulated vertebral column with intervertebral discs

• small plastic straws (coffee stirrers) or pipe cleaners for pointers

There are **206 named bones in** the adult skeleton, which can be separated into the **axial** and the **appendicular** divisions. The **axial skeleton** is composed of 80 bones located along a vertical line, the longitudinal axis of the body. Its bones support and protect the organs of the head, neck, and torso. The **appendicular skeleton** (*appendere* = to hang upon) is composed of 126 bones that make up the upper limbs (or extremities), lower limbs, and the bones of the girdles that attach the limbs to the axial skeleton. The appendicular skeleton will be studied in the next exercise.

The axial skeleton includes the **skull, hyoid bone, vertebral column,** and **thoracic cage** (rib cage). Smaller bones included in this division are the **ear ossicles** (bones).

A. The Skull

The major bones of the skull include the cranial and facial bones. The **cranial bones** form a bony cavity that harbors and protects the brain and houses organs of hearing and equilibrium. **Facial bones** provide the shape of the face, house the teeth, and provide attachments for all the muscles of

facial expression. Other major features of the skull include **sutures, orbit of eye, bone markings, paranasal sinuses, nasal septum, hard palate,** and **fontanels** in the fetal skull.

1. Cranial Bones

There are a total of 8 bones in the cranium, 2 paired and 4 single bones.

- **(2) parietal bones** (*paries* = wall)—superior lateral walls of cranial cavity
- **(2) temporal bones**—inferior lateral walls of cranial cavity; house organs of inner ear
- **(1) frontal bone**—anterior portion of cranial cavity
- **(1) occipital bone** (*occipit-* = atlas)—posterior wall of cranial cavity
- **(1) sphenoid bone** (*sphen-* = wedge; *-eidos* = form)—floor of cranial cavity posterior to ethmoid
- **(1) ethmoid bone** (*ethmos* = sieve)—floor and anterior wall of cranial cavity

2. Sutures

Sutures are immovable joints between bones of the skull. The 4 main sutures in the skull are the:

- **coronal** (*corona* = crown)—joins frontal and parietal bones
- **sagittal** (*sagitta-* = arrow)—joins parietal bones
- **lambdoid** (shape of the Greek letter *lambda*)—joins both parietal bones with occipital bone
- **(2) squamous** (*squama* = scale)—join temporal and parietal bones

3. Facial Bones

There are a total of 14 facial bones, 6 paired and 2 single bones.

- **(2) maxillae** (*mala* = jaw)—fused upper jaw bones
- **(2) zygomatic bones** (*zygoma* = yoke or bar)—cheek bones
- **(2) lacrimal bones** (*lacri-* = tear)—portion of orbit of eyes near nasal bones
- **(2) nasal bones**—bridge of nose
- **(2) inferior nasal conchae** (*concha* = shell or scroll-shaped) or turbinate—forms lateral walls of nasal cavity
- **(2) palatine bones** (*palatum* = palate)—fused bones that form posterior part of hard palate
- **(1) mandible** (*mandere* = to chew)—lower jaw bone
- **(1) vomer** (*vomer* = plowshare)—inferior portion of nasal septum

NOTE: *BE ACCOUNTABLE FOR THE CARE OF SKULLS AND BONES.*
Please be careful not to use pencils, pens, or markers as pointers while you are studying the skull and other bones. Thin bones can be broken, and all bones can be permanently marked. Your instructor will provide a small plastic straw, pipe cleaner, or other suitable pointer.

Before Going to Lab

1 Label the cranial bones, facial bones, and sutures in Figures 9.1–9.5.

LAB ACTIVITY 1 **Cranial Bones, Facial Bones, and Sutures**

1 Identify the bones and sutures in Figures 9.1–9.5 on an articulated skull, disarticulated skull, or use the search text box to locate these structures in **Real Anatomy** (Skeletal).

2 Palpate as many of the cranial and facial bones as you can on your own skull and say their names.

3 Palpate the joint between the temporal and mandibular bone (temporomandibular joint, or TMJ) by putting your fingers on either side of your jawbone just anterior to the ears. Open and close your mouth to feel this joint. ◼

CLINICAL NOTE: *A dislocated or defective TMJ has a history of causing clinical problems characterized by pain, tenderness, dysfunction, and clicking noises.*

- coronal (cor-RONE-al) suture
- ethmoid bone
- frontal (FRON-tal) bone
- lacrimal (LAK-rih-mal) bone
- lambdoid (LAMB-doid) suture
- mandible
- maxilla (ma-XIL-la)
- nasal bone
- occipital (oc-CI-pi-tal) bone
- parietal (pa-RYE-e-tal) bone
- sphenoid (SFEE-noid) bone
- squamous (SQUAY-mus) suture
- temporal (TEM-por-ul) bone
- zygomatic (zy-go-MA-tic) bone

1 _____

2 _____

3 _____

4 _____

5 _____

6 _____

7 _____

8 _____

9 _____

10 _____

11 _____

12 _____

13 _____

14 _____

FIGURE 9.1 Lateral view of skull.

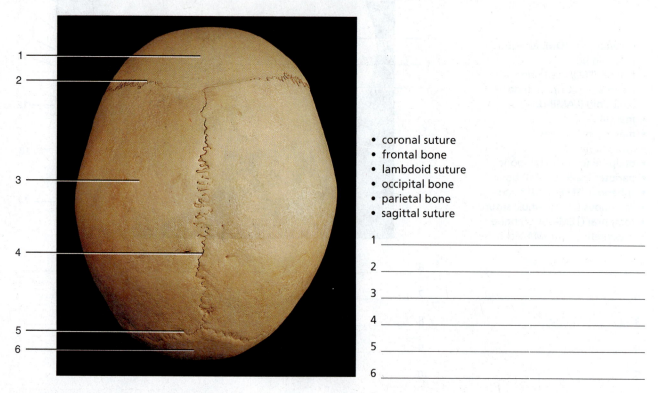

- coronal suture
- frontal bone
- lambdoid suture
- occipital bone
- parietal bone
- sagittal suture

1 _____

2 _____

3 _____

4 _____

5 _____

6 _____

FIGURE 9.2 Superior view of skull.

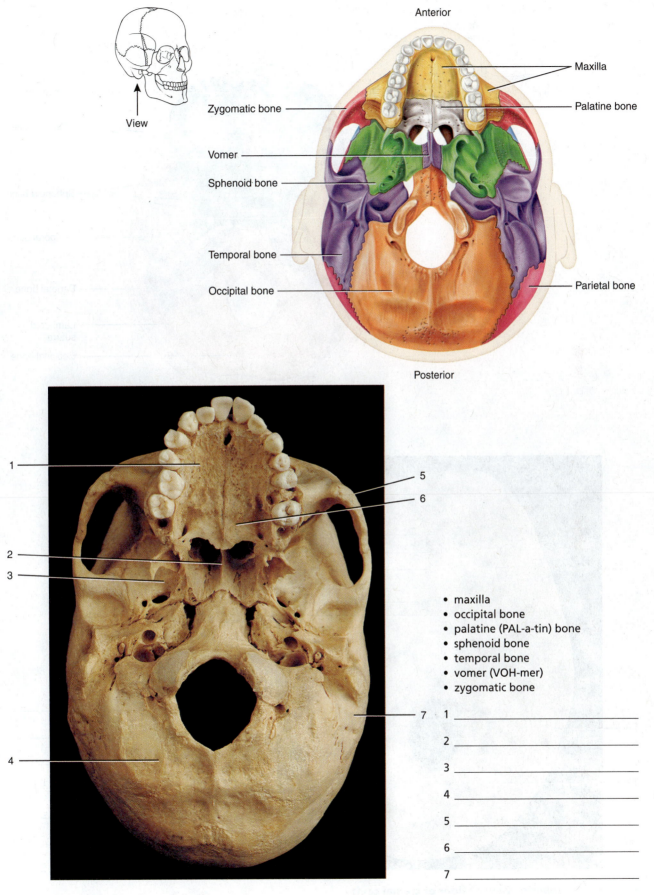

View

Anterior

Maxilla

Zygomatic bone

Palatine bone

Vomer

Sphenoid bone

Temporal bone

Parietal bone

Occipital bone

Posterior

1
2
3
4
5
6
7

- maxilla
- occipital bone
- palatine (PAL-a-tin) bone
- sphenoid bone
- temporal bone
- vomer (VOH-mer)
- zygomatic bone

1 _____

2 _____

3 _____

4 _____

5 _____

6 _____

7 _____

FIGURE 9.3 **Inferior view of skull.**

View
Transverse plane

Anterior

Frontal bone

Ethmoid bone

Sphenoid bone

Temporal bone

Parietal bone

Lambdoid suture

Occipital bone

Posterior

• ethmoid bone
• frontal bone
• lambdoid suture
• occipital bone
• parietal bone
• sphenoid bone
• temporal bone

1 _____
2 _____
3 _____
4 _____
5 _____
6 _____
7 _____

FIGURE 9.4 **Superior view of floor of cranial cavity.**

Parietal bone

Frontal bone

Sphenoid bone

Ethmoid bone

Temporal bone

Nasal bone

Lacrimal bone

Zygomatic bone

Inferior nasal concha

Maxilla

Vomer

Mandible

Sphenoid bone

Ethmoid bone

- ethmoid bone
- frontal bone
- inferior nasal concha (CON-cha) or turbinate
- lacrimal (LAC-ri-mal) bone
- mandible (MAN-di-ble)
- maxilla
- nasal bone
- parietal bone
- sphenoid bone
- temporal bone
- vomer
- zygomatic bone

1 _____

2 _____

3 _____

4 _____

5 _____

6 _____

7 _____

8 _____

9 _____

10 _____

11 _____

12 _____

FIGURE 9.5 Anterior view of skull.

4. Orbit of the Eye

The orbit of the eye is a mosaic of 7 different bones, 3 cranial bones (**frontal, sphenoid,** and **ethmoid**), and 4 facial bones (**maxilla, zygomatic, lacrimal,** and **palatine**). The piece of the palatine bone that is part of the orbit is small and not readily observable in Figure 9.6.

Before Going to Lab

1 Label the bones forming the orbit of the eye in Figure 9.6.

LAB ACTIVITY 2 | Orbit of the Eye

1 On a skull or using the search text box in **Real Anatomy** (Skeletal), locate the 6 bones that constitute the orbit of the eye. Palatine bone is difficult to see. ■

SUPERIOR

INFERIOR

- ethmoid bone
- frontal bone
- lacrimal bone
- maxilla
- sphenoid bone
- zygomatic bone

1 _____

2 _____

3 _____

4 _____

5 _____

6 _____

FIGURE 9.6 Orbit of the eye.

5. Selected Bone Surface Markings of the Skull

The bones of the skull are not totally smooth but have depressions, openings, and projections. These bone surface markings have specific functions. Depressions and openings form joints or passageways for blood vessels and nerves. Processes form joints or points of attachment for ligaments or tendons. The major types of bone markings found on the axial skeleton are listed in Table 9.1 along with their descriptions and functions. Selected bone surface markings are listed below.

Cranial Bones: Selected Bone Surface Markings

1. *Frontal bone* (1)
 - **supraorbital foramina** (*supra-* = above; *orbit* = wheel rut; *foram-* = opening)—1 opening located above the orbit of each eye for supraorbital nerve and artery
 - **supraorbital ridges or margins**—thickening of frontal bone superior to orbit of each eye

2. *Temporal bone* (2)—Each temporal bone has one of each structure listed below:
 - **external auditory meatus** (*meatus* = passageway)—tube-like opening for the ear canal
 - **mastoid process** (*mastoid* = breast-like)—rounded projection posterior to external auditory meatus; attachment for muscles
 - **styloid process** (*stylo-* = point)—long, thin projection on inferior skull surface; attachment for muscles and ligaments of tongue and neck
 - **zygomatic process**—projection that articulates with the zygomatic bone
 - **mandibular fossa**—depression in mandible for articulation with condylar process (mandibular condyle)
 - **foramen lacerum** (*lacerare* = torn or lacerated)—jagged opening filled with cartilage in a living person
 - **carotid foramen (canal)**—foramen for internal carotid artery
 - **jugular foramen** (*jugula* = throat)—foramen for jugular vein and cranial nerves IX, X, and XI
 - **stylomastoid foramen**—opening for an artery and cranial nerve VII
 - **internal auditory meatus**—opening for cranial nerve VIII

TABLE 9.1 Bone Markings

MARKING	DESCRIPTION	PURPOSE
Depressions or Openings		
1. Fissure (FISH-er)	Narrow slit or cleft in a bone	Opening for blood vessels and nerves
2. Foramen (for-AY-men)	Opening or hole (foramina, pl.)	Opening for blood vessels and nerves
3. Fossa (FOS-sa)	Shallow depression (fossae, pl.)	Muscle attachment or articulation (joint)
4. Meatus (me-AY-tus)	Tube-like passageway or opening (meati, pl.)	Passageway or canal for blood vessels and nerves
Processes		
5. Condyle (CON-dile)	Smooth, rounded articular process	Articulation
6. Ramus (RAY-mus)	A small branch (rami, pl.)	Articulation
7. Spine	Pointed process	Articulation

3. *Occipital bone* (1)
 - **foramen magnum** (*magnum* = large)—opening through which spinal cord connects to lower brain
 - **hypoglossal foramina**—openings for cranial nerves XII
 - **occipital condyles**—rounded processes that articulate with the atlas (C1)
 - **external occipital protuberance**—projection at base of skull posterior to foramen magnum

4. *Ethmoid bone* (1)
 - **cribriform plates** (*cribr-* = sieve)—one on either side of crista galli; form roof of nasal cavity
 - **crista galli** (*crist-* = crest; *galli* = rooster)—projection for attachment of membranes covering brain
 - **olfactory foramina** (*olfact-* = smell)—tiny holes in cribriform plates for cranial nerve I
 - **perpendicular plate**—forms superior part of nasal septum
 - **middle nasal conchae**—scroll-like projections on each lateral wall of nasal cavity
 - **superior nasal conchae**—scroll-like projections on each lateral wall of nasal cavity

5. *Sphenoid bone* (1)
 - **foramina ovale** (*ovale* = oval)—openings for mandibular branch of cranial nerve V
 - **foramina rotundum**—openings for maxillary branch of cranial nerve V
 - **sella turcica** (*sella* = saddle; *turcica* = Turkish)—bony projection that surrounds and protects pituitary gland
 - **greater and lesser wings**—form anterior and lateral floor of cranial cavity
 - **optic foramina**—openings for cranial nerve II
 - **inferior orbital fissures**—openings for blood vessels and nerves
 - **superior orbital fissures** (*fissure* = cleft)—openings for blood vessels and cranial nerves III, IV, V, and VI (ophthalmic branch)
 - **pterygoid processes (medial and lateral)** (*pterygoid* = wing-shaped)—wing-like projections on the base of the skull in the middle section of the sphenoid bone

Facial Bones: Selected Bone Surface Markings

1. *Maxillae* (2)
 - **alveoli** (*alve-* = socket)—tooth sockets (alveolus, sing.)
 - **palatine process**—fused processes that form the anterior part of hard palate

2. *Mandible* (1)
 - **alveoli**—tooth sockets
 - **body**—curved, anterior portion of mandible
 - **mental foramina** (*menta* = chin)—openings in chin for nerves and blood vessels
 - **rami**—posterior branches, one on either side of the body of mandible
 - **condylar processes (mandibular condyles)**—rounded processes on rami that articulate with temporal bone at the mandibular fossa to form the TMJ
 - **coronoid processes**—triangular projections of rami anterior to the condylar processes

3. *Lacrimal bone* (2)
 - **lacrimal fossa**—canal that houses lacrimal sac; formed from the maxilla and lacrimal bone

4. *Zygomatic bone* (2)
 - **temporal process**—projects posteriorly; temporal process of zygomatic bone and zygomatic process of temporal bone form the **zygomatic arch**

1 Label the bone markings listed in Figures 9.7–9.10.

LAB ACTIVITY 3 Bone Surface Markings

1 Identify the bone markings in Figures 9.7–9.10 on an articulated skull, unarticulated skull, or use the search text box to locate these structures in **Real Anatomy** (Skeletal).

2 Using two small plastic straws, poke each straw into an orbit of the eye and through an optic foramen. Observe the straws in the superior view of the cranial floor and notice that the straws cross above the sella turcica. This simulates the crossing of the optic nerves, called the optic chiasma.

3 Palpate the supraorbital ridges of the frontal bone and the body, ramus, and condylar process of the mandible.

4 Palpate the mastoid process of the temporal bone by placing your fingers on the projection inferior and posterior to the external auditory meatus.

5 Palpate the condylar process by placing fingers anterior to the external auditory meatus and opening and closing your mouth. ■

- inferior orbital fissure
- mental foramen
- middle nasal concha
- orbit of eye
- perpendicular plate of ethmoid
- superior orbital fissure
- supraorbital (supra-OR-bi-tal) foramen
- supraorbital margin

1 _____

2 _____

3 _____

4 _____

5 _____

6 _____

7 _____

8 _____

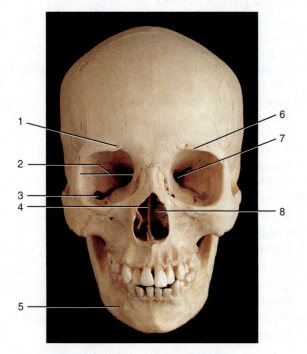

FIGURE 9.7 **Anterior view of skull.**

Zygomatic process
of temporal bone

External auditory
meatus

Mastoid process

Styloid process

Lacrimal fossa

Coronoid process

Condylar process
(mandibular condyle)

Body of mandible

Ramus of mandible

Hyoid bone

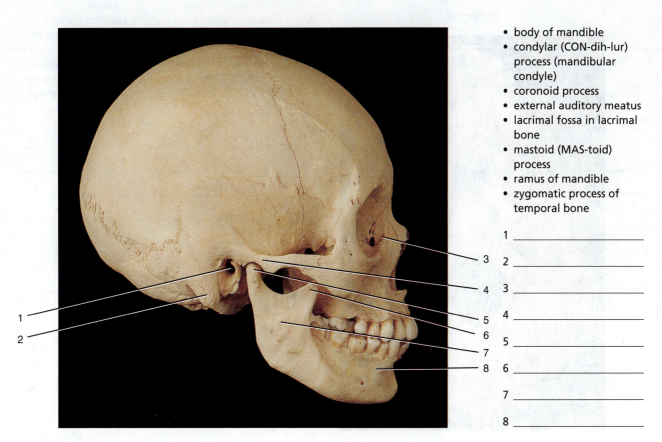

- body of mandible
- condylar (CON-dih-lur) process (mandibular condyle)
- coronoid process
- external auditory meatus
- lacrimal fossa in lacrimal bone
- mastoid (MAS-toid) process
- ramus of mandible
- zygomatic process of temporal bone

1 _____

2 _____

3 _____

4 _____

5 _____

6 _____

7 _____

8 _____

FIGURE 9.8 Lateral view of skull.

Anterior

Palatine process
of maxilla

Hard palate

Palatine bone

Pterygoid processes

Foramen ovale

Foramen lacerum

Mandibular fossa

Styloid process

Carotid foramen

External auditory meatus

Jugular foramen

Stylomastoid foramen

Occipital condyle

Mastoid process

Hypoglossal
foramen

Foramen magnum

External occipital
protuberance

Posterior

- carotid foramen
 (canal)
- external occipital
 protuberance
- foramen lacerum
 (LA-sir-um)
- foramen magnum
- foramen ovale
 (o-VAL-ee)
- hard palate
- hypoglossal foramen
- jugular foramen
- mandibular fossa
- mastoid process
- occipital (ox-CI-pi-tal)
 condyle
- palatine bone
- palatine process of
 maxilla
- pterygoid processes
- stylomastoid (sty-lo-
 MAS-toid) foramen

ANTERIOR

POSTERIOR

1 _____
2 _____
3 _____
4 _____
5 _____
6 _____
7 _____
8 _____
9 _____
10 _____
11 _____
12 _____
13 _____
14 _____
15 _____

FIGURE 9.9 **Inferior view of skull.**

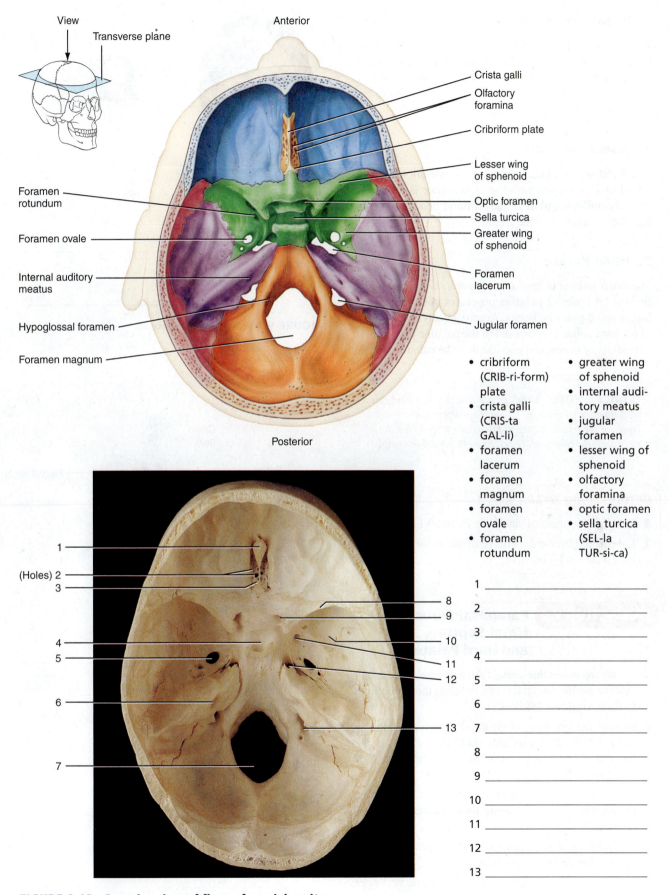

View
Transverse plane

Anterior

Crista galli
Olfactory foramina
Cribriform plate
Lesser wing of sphenoid
Optic foramen
Sella turcica
Greater wing of sphenoid
Foramen lacerum
Jugular foramen

Foramen rotundum
Foramen ovale
Internal auditory meatus
Hypoglossal foramen
Foramen magnum

Posterior

- cribriform (CRIB-ri-form) plate
- crista galli (CRIS-ta GAL-li)
- foramen lacerum
- foramen magnum
- foramen ovale
- foramen rotundum
- greater wing of sphenoid
- internal auditory meatus
- jugular foramen
- lesser wing of sphenoid
- olfactory foramina
- optic foramen
- sella turcica (SEL-la TUR-si-ca)

1
(Holes) 2
3
8
9
4
10
5
11
12
6
13
7

1 _____
2 _____
3 _____
4 _____
5 _____
6 _____
7 _____
8 _____
9 _____
10 _____
11 _____
12 _____
13 _____

FIGURE 9.10 **Superior view of floor of cranial cavity.**

6. Paranasal Sinuses

Paranasal (*para-* = next to; *nasal* = nose) **sinuses** (*sinu-* = hollow) are cavities lined with mucous membranes that are located near and have openings into the nasal cavities. The **ethmoid, frontal, maxillary,** and **sphenoid** bones contain paranasal sinuses.

7. Nasal Septum

The **nasal septum** (*saeptum* = wall or partition) is comprised of 2 bones and cartilage. The **vomer** is inferior to the **perpendicular plate** of the **ethmoid** bone. The **septal cartilage** is anterior to the 2 bones.

8. Hard Palate

The **hard palate,** or roof of the mouth, is formed by the fusion of 4 bones: 2 **palatine processes of the maxillary bones** and 2 **palatine bones.** Approximately three-fourths of the hard palate is composed of the palatine processes of the maxillary bones, and one-fourth is the palatine bones.

CLINICAL NOTE: *A cleft palate is a common congenital malformation of the hard palate. This defect is the result of bones of the palate failing to fuse along the midline during embryonic development. The cleft can be partial or complete.*

Before Going to Lab

1 Label the 4 paranasal sinuses in Figure 9.11.
2 Label the bones and cartilage of the nasal septum and hard palate in Figure 9.12.

LAB ACTIVITY 4 Paranasal Sinuses, Nasal Septum, and Hard Palate

1 Locate the bones that contain the paranasal sinuses on a skull or use the search text box to locate these structures in **Real Anatomy** (Skeletal).

2 Identify the structures of the nasal septum on an articulated skull (and a Beauchene skull, if available) or use the search text box to locate these structures in **Real Anatomy** (Skeletal).

3 Locate the bones of the hard palate on a skull or use the search text box to locate these structures in **Real Anatomy** (Skeletal).

4 Palpate your nasal septum to find the junction of the nasal bones and nasal cartilage.

5 Palpate your own palate with your tongue or finger, noting the junction of the hard and soft palates. ■

(a) (b)

- ethmoidal sinus 1 _____
- frontal sinus
- maxillary sinus 2 _____
- sphenoidal sinus
 3 _____

 4 _____

FIGURE 9.11 Paranasal sinuses.

Crista galli
Sphenoidal sinus
Sella turcica
Frontal sinus
Nasal bone

1
2
3

- perpendicular plate of ethmoid 1 _____
- septal cartilage 2 _____
- vomer
 3 _____

FIGURE 9.12 Nasal septum.

9. Fetal and Newborn Skull

The newborn skull is not completely ossified but has membranous sections composed of fibrous connective tissue called **fontanels** (*fontaine* = fountain) or "soft spots." Fontanels allow the cranial bones to compress during the journey through the birth canal, and they also permit rapid growth of the brain and skull during the early years. At birth, the physician or midwife can tell how the baby is presenting itself in the birth canal by palpating the fontanel. If the fontanel is a diamond shape, the baby is presenting the frontal bone first and is face down. If the fontanel is triangular in shape, the baby is presenting the occipital bone and is face up. These membrane templates that are present at birth eventually ossify.

Before Going to Lab

1 Label the fontanels in Figures 9.13(a)–(c).

LAB ACTIVITY 5 Fetal Skull and Major Fontanels

1 Identify the fontanels on a fetal skull (if available) or use the search text box to locate these structures in **Real Anatomy** (Skeletal).

2 Compare the location of the fontanels and sutures in the fetal skull with adult sutures.

3 Compare the size of the facial bones in the fetal skull with the adult skull. ■

Parietal bone
Future squamous suture
Posterior fontanel
Posterolateral fontanel
Occipital bone
Anterior fontanel
Future coronal suture
Frontal bone
Anterolateral fontanel
Sphenoid bone
Temporal bone

(a) Right lateral view

Parietal bone
Future coronal suture
Future squamous suture
Sphenoid bone
Temporal bone
1
2
3
Frontal suture
Frontal bone
Perpendicular plate
Maxilla
Mandible

(b) Right anterolateral view

Parietal bone
Future sagittal suture
4
Future lambdoid suture
Occipital bone

(c) Posterior view

- anterior (frontal) fontanel (fon-ta-NEL)
- anterolateral (sphenoidal) fontanel
- posterior (occipital) fontanel
- posterolateral (mastoid) fontanel

1 _____

2 _____

3 _____

4 _____

FIGURE 9.13 Major fontanels of the fetal skull.

B. Hyoid Bone

The hyoid bone is not attached to the axial skeleton but is included with the axial skeleton because of its mid-line location and proximity to the mandible and vertebral column. This bone is U-shaped and has the distinction of not articulating with any other bones. It is secured in place by ligaments and muscles, including many muscles of the tongue and neck. It is located in the anterior neck region between the mandible and larynx. (This is the bone that is crushed during strangulation.)

Before Going to Lab

1 Locate the hyoid bone in Figure 9.8.

LAB ACTIVITY 6 Hyoid Bone

1 Identify the hyoid bone on an articulated skeleton and on a disarticulated skeleton or use the search text box to locate these structures in **Real Anatomy** (Skeletal).

2 Palpate your hyoid bone by placing the thumb and middle finger of one hand on either side of the neck about 1 inch inferior to the mandible. This bone can be moved laterally (from side to side). ■

C. Vertebral Column

The vertebral column protects the spinal cord and provides attachment points for back and abdominal muscles. The curved **vertebral column** (backbone) is a flexible structure that can be bent, twisted, and rotated, especially in the cervical region. The vertebral column consists of **cervical** (*cervic-* = neck), **thoracic** (*thorac-* = chest), and **lumbar** (*lumb-* = loin) **vertebrae,** the **sacrum** (*sacer* = sacred), and **coccyx** (*coccyx* = cuckoo's beak). The sacrum consists of 5 sacral vertebrae that are fused in the adult. The coccyx (tailbone) is usually composed of 4 small, fused coccygeal vertebrae. An infant has 33 individual vertebrae, and the adult has 26 due to fusion of the sacral and coccygeal vertebrae. The vertebral column articulates with the skull, the ribs, and the pelvis.

1. Regions and Normal Curvatures of the Vertebral Column

The 5 regions of the vertebral column and the number of vertebrae in the adult, from superior to inferior, are: **cervical** (7), **thoracic** (12), **lumbar** (5), **sacral** (1), and **coccygeal** (1). To remember the number of vertebrae in the first 3 groups, students say they eat "breakfast at 7, lunch at 12, and dinner at 5." There are 4 normal curvatures that correspond with the regions of the vertebral column: **cervical, thoracic, lumbar,** and **sacral** (pelvic) **curvatures.** A newborn has a single anteriorly concave primary curve that will become the thoracic curve and sacral curve (Fig. 9.14). Two anteriorly convex secondary curves—the cervical and lumbar—develop several months later. The cervical curve develops when the baby can hold its head erect, while the lumbar curve develops when the baby can stand.

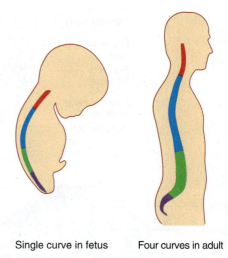

Single curve in fetus Four curves in adult

FIGURE 9.14 Fetal and adult curves of the vertebral column.

Before Going to Lab

1 Label the 5 regions of the vertebral column in Figure 9.15.
2 Label the 4 normal curvatures in Figure 9.15.

LAB ACTIVITY 7 The Vertebral Column

1 Identify the 5 regions on an articulated skeleton or vertebral column model.

2 Identify the 4 curvatures on an articulated skeleton or vertebral column. ■

POSTERIOR ANTERIOR

Intervertebral disc

Intervertebral disc

Intervertebral foramen

- cervical curve
- cervical vertebrae (VER-te-bray)
- coccyx (COCK-six)
- lumbar curve
- lumbar vertebrae
- sacral curve
- sacrum (SAY-crum)
- thoracic curve
- thoracic vertebrae

1 _____

2 _____

3 _____

4 _____

5 _____

6 _____

7 _____

8 _____

9 _____

FIGURE 9.15 **Normal spinal curvatures of the vertebral column.**

2. Parts of a Typical Vertebra

The cervical, thoracic, and lumbar vertebrae have similar as well as different structures. The similarities are studied in this activity, and the differences will be studied in the next activity.

- **body**—located anteriorly; is the largest part of the vertebra
- **pedicle** (*pediculus* = little foot)—attached to and extends posteriorly on either side of the body
- **transverse process**—extends laterally from each pedicle
- **lamina** (*lamin-* = plate)—connects transverse processes to the spinous process
- **spinous process**—projects posteriorly from fused lamina
- **vertebral arch**—formed by the fusion of pedicles and laminae

- **vertebral foramen**—large opening formed by the vertebral arch that protects the spinal cord
- **superior** and **inferior articular processes** (with **facets**)—extend from the vertebra at the junction of the pedicle and lamina to articulate with a superior and inferior vertebra, respectively

Before Going to Lab

1 Locate the parts of a typical vertebra shown in Figure 9.16.

LAB ACTIVITY 8 **Parts of a Typical Vertebra**

1 Identify the parts of a typical vertebra on a disarticulated thoracic vertebra or use the search text box to locate these structures in **Real Anatomy** (Skeletal).

2 Palpate the spinous processes of your vertebral column.

POSTERIOR

Facet of superior articular process
Spinal cord
Spinous process
Transverse process
Vertebral arch:
Lamina
Vertebral foramen
Pedicle
Facet for head of rib
Body

ANTERIOR

FIGURE 9.16 **Parts of a typical vertebra.**

3. Comparison of Cervical, Thoracic, and Lumbar Vertebrae

The first two of the seven cervical vertebrae look different from the others as seen in Figure 9.17(a). The **atlas** is the first cervical vertebra (C1) and is named for the mythical Greek god who held up the world on his shoulders. The large **superior articular facets** of the atlas articulate with the **occipital condyles** of the skull and allow the head to tilt up and down in a nodding motion. The **axis,** C2, has a superior tooth-like protuberance called the **dens** (*dens* = tooth in Latin) or **odontoid process** (*odous* = tooth in Greek). The dens extends superiorly into the vertebral foramen of the atlas and allows the atlas to pivot laterally in a "no" type motion. Some head traumas push the head onto the vertebral column, and the dens is shoved into

the medulla oblongata (brain stem). This is the cause of death in such head injuries. C3 through C6 have bifurcated (forked) spinous processes. In Figure 9.17(a), observe the **vertebra prominens** (C7) which has a prominent single spinous process that protrudes at the base of the neck and can be seen and palpated. A **whiplash,** which occurs in some rear-ended auto accidents, causes partial or complete dislocation of the cervical vertebrae.

The easiest way to distinguish any **cervical vertebra** is by its three foramina (pl.; foramen, sing.)—a **vertebral foramen** plus two **transverse foramina.** The transverse foramina, found only in cervical vertebrae, allow passage of the vertebral arteries, veins, and sympathetic nerves to and from the brain. Since the cervical vertebrae do not carry much weight, they are lightweight and have small bodies.

A **thoracic vertebra** has a medium-size body and usually a long, narrow spinous process that commonly slants inferiorly at a sharp angle. Thoracic vertebrae also have facets (*facette* = little face) on the transverse processes and demifacets (*demi-* = half) on the bodies, both of which articulate with ribs.

Lumbar vertebrae have the largest bodies to support more weight and thick, hatchet-shaped spinous processes that extend horizontally.

Intervertebral discs can be found between the bodies of vertebrae from C2 to the sacrum. As you recall, they are made of fibrocartilage to give strength, permit movement, and absorb shock. The outer portion of the disc is called the **annulus fibrosus** (*annulus* = ring-like) and consists of tough fibrocartilage. The inner part, the **nucleus pulposus** (*pulposus* = pulp-like), is soft and pulpy. A *herniated* or *slipped disc* occurs when the fibrocartilage of the annulus is stressed and torn or cracked. In this situation, the inner pulp-like center protrudes outward and causes pressure on the spinal cord or a spinal nerve. A *laminectomy* (removal of the lamina and sometimes spinous process of the vertebra) relieves the pressure.

Intervertebral foramina are formed between adjacent vertebrae when vertebrae are stacked on one another. Spinal nerves exit the vertebral column through these foramina.

Before Going to Lab

1 Label the parts of an atlas, axis, and cervical vertebra in Figure 9.17(b), (c), and (d).

2 Label the parts of a thoracic vertebra in Figure 9.18(a) and (b) and a lumbar vertebra in Figure 9.19(a) and (b).

3 Label the parts of articulated vertebrae in Figure 9.20(a) and a herniated disc in (b).

LAB ACTIVITY 9 Comparison of Cervical, Thoracic, and Lumbar Vertebrae

1 Place all three types of disarticulated vertebrae side by side and note their differences. Mix them up and identify each type of vertebra with a partner.

2 Palpate your own spinous process of C7 (vertebra prominens).

3 Identify the structures in Figures 9.17–9.19 on disarticulated vertebrae or use the search text box to locate these structures in **Real Anatomy** (Skeletal).

4 Using articulated vertebrae, identify the intervertebral disc and intervertebral foramen in Figure 9.20. ■

Location of cervical vertebrae

(a) Posterior view of articulated cervical vertebrae

FIGURE 9.17 Cervical vertebrae.

Posterior

Transverse
foramen

Anterior

(b) Superior view of the atlas (C₁)

Posterior

Bifurcated
spinous process

Transverse
process

Dens

Anterior

(c) Superior view of the axis (C₂)

Posterior

Bifurcated
spinous process

Lamina

Superior
articular facet

Pedicle

DANK

Anterior

(d) Superior view of a typical cervical vertebra (C₃)

(b) Atlas
• superior articular facet
• transverse foramen
• transverse process

(c) Axis (*Note:* The transverse foramen cannot be seen in this photo.)
• dens (odontoid process)
• lamina
• spinous process

(d) Typical cervical vertebra
• bifurcated spinous process
• body
• pedicle
• transverse process

1 _____

2 _____

3 _____

4 _____

5 _____

6 _____

7 _____

8 _____

9 _____

10 _____

FIGURE 9.17 **Cervical vertebrae (*continued*).**

POSTERIOR

Facet for articular part of tubercle of rib

Superior demifacet

(a) Superior view

(a) Superior view

ANTERIOR

Superior articular facet

Superior demifacet

Inferior demifacet

(b) Right lateral view

SUPERIOR

POSTERIOR

ANTERIOR

(b) Right lateral view

(a) Superior view
• facet for articular part of tubercle of rib
• superior articular facet
• superior demifacet
• transverse process

(b) Right lateral view
• facet for articular part of tubercle of rib
• inferior demifacet
• slanted spinous process
• superior articular facet
• superior demifacet

1 _____

2 _____

3 _____

4 _____

5 _____

6 _____

7 _____

8 _____

9 _____

FIGURE 9.18 **Thoracic vertebrae.**

POSTERIOR

1

2

3

4

5

ANTERIOR

(a) Superior view

SUPERIOR

6

7

POSTERIOR ANTERIOR

(b) Right lateral view

(a) Superior view
- body
- pedicle
- superior articular process
- transverse process
- vertebral foramen

1 _____

2 _____

3 _____

4 _____

5 _____

(b) Right lateral view
- hatchet-shaped spinous process
- inferior articular facet

6 _____

7 _____

FIGURE 9.19 Lumbar vertebrae.

Posterior

1

Posterior Anterior

2

3

4

(a) Right lateral view of articulated vertebrae
 with intervertebral disc

Posterior

Spinal cord

Spinal nerve

5

6

7

Anterior

(b) Superior view of herniated disc

(a) Articulated vertebrae
- annulus fibrosus
- intervertebral disc
- intervertebral foramen
- nucleus pulposus

1 _____

2 _____

3 _____

4 _____

(b) Herniated disc
- annulus fibrosus
- herniation
- nucleus pulposus

5 _____

6 _____

7 _____

FIGURE 9.20 Intervertebral discs.

4. Sacrum and Coccyx

The **sacrum** is formed by the fusion of 5 sacral vertebrae and has a slightly curved, triangular shape. The broad superior portion is called the **base** and the two lateral wing-like projections are called **alae** (ala, sing.). The **sacral promontory** (*promontory* = a projecting part), which protrudes anteriorly from the base, is an important landmark in females during labor and delivery. The sacrum has **sacral foramina** that provide exits for spinal nerves.

From the posterior view, the opening of the **sacral canal,** a continuation of the vertebral canal, is located just posterior to the body of the sacrum. The inferior opening of the sacral canal, the **sacral hiatus,** is due to failure of the lamina of the 5th sacral vertebrae (and sometimes the 4th) to fuse. (*Note:* On many plastic models, the sacral canal is closed.) Two **superior articular processes** with facets that articulate with the fifth lumbar vertebra are located on either side of the opening of the sacral canal. The lateral surfaces of the sacrum, the **auricular surfaces,** are roughened to articulate with the iliac portion of the os coxa on each side, forming **sacroiliac joints.** The vertebral column ends with a small **coccyx,** or tailbone. The tiny, fused vertebrae of the coccyx are attached to the sacrum with ligaments.

Before Going to Lab

1 Label the parts of the sacrum and coccyx in Figure 9.21.

LAB ACTIVITY 10 Sacrum and Coccyx

1 Identify the structures in Figure 9.21(a) and (b) on a model of the sacrum and coccyx or use the search text box to locate these structures in **Real Anatomy** (Skeletal). ■

(a) Anterior view

(b) Posterior view

(a)
- base
- coccyx
- sacral ala
- sacral foramen
- sacral promontory

(b)
- auricular surface (for sacroiliac joint)
- sacral canal
- sacral hiatus
- superior articular facet

1 _____ 6 _____

2 _____ 7 _____

3 _____ 8 _____

4 _____ 9 _____

5 _____

FIGURE 9.21 Sacrum and coccyx.

5. Abnormal Vertebral Curvatures

Three abnormal curves of the vertebral column are scoliosis, kyphosis, and lordosis. The vertebral column bends laterally in **scoliosis** (*scolio-* = crooked). **Kyphosis** is an exaggerated thoracic curve that results in a hunched back with rounded shoulders. **Lordosis** is an exaggerated lumbar curve that appears as a swayback with the abdomen protruding anteriorly.

Before Going to Lab

1 Label the 3 different types of abnormal curves in Figure 9.22(a), (b), and (c).

(a) (b) (c)

- kyphosis
- lordosis
- scoliosis

a _____

b _____

c _____

FIGURE 9.22 Abnormal curves of the vertebral column.

D. Thoracic Cage (Rib Cage)

The bony cage that encircles the chest is called the **thoracic cage,** or rib cage, and is composed of the sternum, ribs, costal cartilages, and thoracic vertebrae.

The **sternum** is a narrow flat bone that is composed of three fused bones: the manubrium, the body of the sternum, and the xiphoid process. The **manubrium** (handle), the superior portion of the sternum, has a concave superior surface called the **suprasternal notch** or **jugular notch.** Between the **body** of the sternum and the manubrium is the **sternal angle,** an important clinical landmark indicating the attachment of the second rib, below which is the second intercostal space. The manubrium and body articulate (*articul-* = pertaining to a joint) with the **costal cartilages** of the **ribs.** The **xiphoid process** (sword-like) is the inferior portion of the sternum that is shaped like a small sword. There are 12 pairs of **ribs** in both males and females. The first 7 pairs are called the **true ribs** or **vertebrosternal ribs** because their **costal (rib) cartilages** have a direct attachment to the sternum. The last 5 rib pairs (8–12) are called **false ribs.** Rib pairs 8 through 10 are **vertebrochondral ribs** (*vertebro* = ribs; *chondral* = cartilage) because their costal cartilages do not have a direct attachment to the sternum but attach to the costal cartilage of the seventh rib instead. Rib pairs 11 and 12 do not have any attachment to the sternum or cartilage and are called **floating ribs** or **vertebral ribs.** The space between the ribs is called the **intercostal space.**

The main parts of a rib are the head, neck, tubercle, and body. The **head** projects from the posterior part of the rib and articulates with **demifacets** on the bodies of thoracic vertebrae. The **neck** is the constricted part lateral to the head. The **tubercle** is a small, knob-like projection close to the neck that articulates with the facet of a transverse process. The **body** is the main part of the rib.

Before Going to Lab

1 Label the parts of the thorax in Figure 9.23.

2 Label the parts of a rib and its articulation with a vertebra in Figure 9.24(a) and (b).

LAB ACTIVITY 11 Thoracic Cage and Rib Articulations

1 Identify the structures in Figure 9.23 on an articulated thorax or use the search text box to locate these structures in **Real Anatomy** (Skeletal).

2 Identify the parts of a rib and its articulation with a vertebra on an articulated thorax or use the search text box to locate these structures in **Real Anatomy** (Skeletal).

3 Identify the four parts of a rib and its articulations on a skeleton.

4 Palpate the **suprasternal notch** and then move your fingers inferiorly until a ridge (**sternal angle**) is reached. Move your fingers laterally until the **2ⁿᵈ rib** can be palpated. Palpate the **body of the sternum** inferior to the sternal angle. Palpate down the sternum until you reach the inferior point of the **xiphoid process.** ■

- body of sternum
- costal cartilage
- false ribs
- floating ribs
- manubrium
- sternal angle
- sternum
- suprasternal notch (jugular notch)
- true ribs
- xiphoid process

1 _____ 6 _____

2 _____ 7 _____

3 _____ 8 _____

4 _____ 9 _____

5 _____ 10 _____

FIGURE 9.23 Thoracic cage.

Head
Neck
Tubercle
Articular facet
Costal angle
Costal groove
Superior facet
Inferior facet
Body
Facet for costal cartilage

(a) Posterior view of left rib

POSTERIOR

1
2
3
4

ANTERIOR

(a) Superior view

(a) Superior view
• articular part of tubercle
• facet for articular part of tubercle
• head of rib
• tubercle

1 _____
2 _____
3 _____
4 _____

(b) Right lateral view
• body of rib
• head of rib
• inferior demifacet of vertebra
• intervertebral foramen
• superior demifacet of vertebra
• tubercle

5 _____
6 _____
7 _____
8 _____
9 _____
10 _____

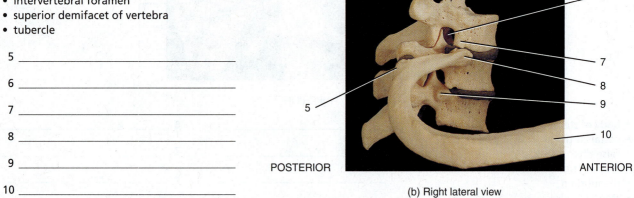

SUPERIOR

6
7
8
9
10

5

POSTERIOR ANTERIOR

(b) Right lateral view

FIGURE 9.24 Rib and its articulation with thoracic vertebrae.

Reviewing Your Knowledge

A. Cranial Bones and Bone Surface Markings

Identify the cranial bones that have the following bone markings.

_____ **1.** Middle nasal concha

_____ **2.** Foramen magnum

_____ **3.**

⎫ Alveoli (2 bones)

_____ **4.** ⎭

_____ **5.** Sella turcica

_____ **6.** External auditory meatus

_____ **7.** Crista galli

_____ **8.** Greater wing

_____ **9.** Mastoid process

_____ **10.** Occipital condyle

_____ **11.** Optic foramen

_____ **12.** Mandibular fossa

_____ **13.** Olfactory foramina

_____ **14.** Supraorbital ridges

_____ **15.** Lacrimal fossa

_____ **16.** Zygomatic process

_____ **17.** Styloid process

_____ **18.** Palatine process

_____ **19.** Mental foramina

_____ **20.** Carotid foramen (canal)

_____ **21.** Condylar processes (mandibular condyle)

Identify the bone surface marking(s) that have the following function:

_____ 22.⎫
 ⎬ Form the TMJ
_____ 23.⎭

_____ 24. Opening for carotid artery

_____ 25. Opening for the ear canal

_____ 26. Holes for cranial nerve I (olfactory nerves)

_____ 27. Holes for cranial nerve II (optic nerve)

_____ 28. Opening through which the spinal cord connects to lower brain

_____ 29. Anterior part of hard palate

_____ 30. Rounded processes on skull that articulate with the atlas

_____ 31. Tooth sockets

_____ 32.⎫
 ⎬ Form zygomatic arch
_____ 33.⎭

_____ 34. Protects the pituitary gland

B. Vertebral Column and Thorax

Write the name of the structure that is described.

_____ 1. Softens jolts to the vertebral column

_____ 2. Opening for spinal nerve exit

_____ 3. Bony protection for thoracic organs

_____ 4. First cervical vertebra

_____ 5. Second cervical vertebra

_____ 6. Articulates with rib posteriorly

_____ 7. Tailbone

_____ 8. Encloses and protects the spinal cord

_____ 9.⎫
 ⎬ Primary curvatures
_____ 10.⎭

_____ 11.⎫
 ⎬ Secondary curvatures
_____ 12.⎭

_____ 13. Formed by the fusion of pedicles and lamina

_____ **14.** Passageway for vertebral arteries in cervical vertebrae

_____ **15.**

_____ **16.**
 } Name the 4 parts of the thoracic cage
_____ **17.**

_____ **18.**

_____ **19.**

_____ **20.** } Name the 3 bones of the sternum

_____ **21.**

_____ **22.** Concave depression in superior surface of manubrium

_____ **23.** False ribs

_____ **24.** True ribs

_____ **25.** Floating ribs

_____ **26.** Space between ribs

_____ **27.** A special feature present on the transverse process of cervical vertebrae

_____ **28.**

_____ **29.** } Special features present on the thoracic vertebrae for rib articulations

_____ **30.**

_____ **31.** A part of the rib that articulates with the body of the thoracic vertebrae

_____ **32.** A part of the rib that articulates with the transverse process of the thoracic vertebrae

_____ **33.** The odontoid process is a part of which bone?

C. Other Major Features of the Axial Skeleton

Identify the major skull feature or structure that is described.

_____ **1.** The suture that joins the occipital bone to the parietal bones.

_____ **2.** Paired facial bones that contain a paranasal sinus.

_____ **3.** The articulation between the mandible and skull.

_____ **4.** Neck and tongue muscles are attached to this bone.

_____ **5.** These structures allow the baby's skull to compress during childbirth.

_____ **6.** Name the bone marking that forms the superior part of the nasal septum.

_____ **7.** Name the inferior bone of the nasal septum.

D. Skull Bones and Bone Markings

Identify the bones and bone markings in Figure 9.25(a) and (b).

(a) Anterior view

(b) Inferior view

FIGURE 9.25 **Skull bones and bone markings.**

(a)

1. _____
2. _____
3. _____
4. _____
5. _____
6. _____
7. _____
8. _____
9. _____
10. _____
11. _____
12. _____
13. _____
14. _____
15. _____

(b)

1. _____
2. _____
3. _____
4. _____
5. _____
6. _____
7. _____
8. _____
9. _____
10. _____
11. _____
12. _____
13. _____
14. _____
15. _____
16. _____
17. _____
18. _____
19. _____

Using Your Knowledge

A. Identifying Bones and Bone Markings on Radiographs of Skull

Identify the structures in Figure 9.26(a).

1 _____ (bone marking)

2 _____ (bone)

3 _____ (bone)

4 _____ (bone)

5 _____ (bone)

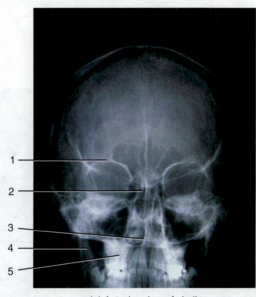

(a) Anterior view of skull.

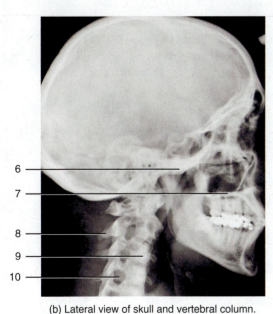

(b) Lateral view of skull and vertebral column.

FIGURE 9.26 **Radiographs of skull.**

Identify the structures in Figure 9.26(b).

6 _____ (bone marking)

7 _____ (bone marking)

8 _____ (bone marking)

9 _____ (bone marking)

10 _____ (space between bones)

B. Identifying Bones and Bone Markings on CT Scans

Identify the structures in Figure 9.27(a).

11 _____ (bone)

12 _____ (bone)

13 _____ (sinus)

14 _____ (bone marking in sphenoid)

15 _____ (opening)

(a) Lateral view of skull.

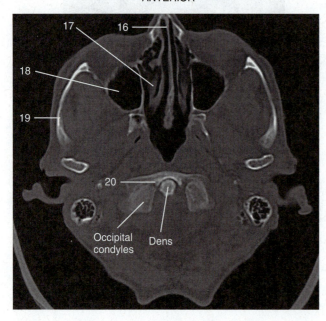

ANTERIOR

POSTERIOR

(b) Thansverse section through skull at the level of the atlas.

FIGURE 9.27 CT scans of skull.

Identify the structures in Figure 9.27(b).

16 _____ (bone marking)

17 _____ (bone marking)

18 _____ (bone sinus)

19 _____ (bone marking)

20 _____ (bone)

Appendicular Skeleton

OBJECTIVES

1 Identify the bones of the appendicular skeleton and be able to identify selected bones as right or left

2 Identify the principal bone markings of the pectoral (shoulder) girdle and the upper limb

3 Describe the articulation of the pectoral girdle with the humerus

4 Describe the articulation of the humerus with the forearm bones

5 Identify the bone markings of the pelvic girdle and the lower limbs

6 Describe the articulation of the bones of the pelvic girdle with the femur

7 Describe the articulation of the femur with the leg bones

8 Determine gender and height using femur measurements

9 Determine height using the length of the radius or the humerus

MATERIALS

- articulated skeletons or use **Real Anatomy** (Skeletal)
- disarticulated skeletons
- small plastic straws (coffee stirrers) or pipe cleaners for pointers
- tape measure

The appendicular skeleton (*appendere* = to hang upon) has larger bones than the axial skeleton and bears more weight. The bones of this division are separated into four main areas: the **pectoral girdles** (*pectus* = breast; *girdle* = encircles), the **upper limbs** (**extremities**), the **pelvic girdle** (*pelvis* = basin), and the **lower limbs.** The bone markings include sites of muscle attachment and articulations with other bones to form a joint. Bone surface markings that pertain to the appendicular skeleton are given in Table 10.1.

TABLE 10.1 **Selected Bone Surface Markings of the Appendicular Skeleton**

MARKING	DESCRIPTION	PURPOSE
1. Foramen	Opening or hole	Opening for blood vessels and nerves
2. Fossa	Shallow depression	Muscle attachment or articulation
3. Crest	Prominent ridge	Muscle attachment
4. Condyle	Smooth, rounded articular process	Articulation
5. Epicondyle	Projection above a condyle	Muscle attachment
6. Head	Rounded articular projection supported on the neck of a bone	Articulation
7. Line	Long, narrow ridge (less prominent than a crest)	Muscle attachment
8. Trochanter	Very large projection	Muscle attachment
9. Tubercle	Small, rounded projection	Muscle attachment
10. Tuberosity	Large, roughened projection	Muscle attachment

A. The Pectoral Girdle

There are 2 pectoral girdles, and each attaches an upper limb to the axial skeleton. Each **pectoral** (or shoulder) **girdle** is composed of a **scapula** and a **clavicle** (*clavicle* = key) (Figure 10.1).

Clavicle (collar bone)
- **sternal end**—blunt, medial end
- **acromial end** (*acrom-* = topmost)—broader, flat, roughened, lateral end

Scapula (shoulder blade)
- **spine**—sharp ridge located on posterior side
- **acromion**—flattened process at lateral end of spine
- **glenoid cavity** (*glene* = joint socket) or **fossa**—depression inferior to acromion
- **coracoid process** (*coracoid* = crow's beak)—superior and medial to glenoid cavity; projects anteriorly
- **supraspinous fossa** (*supra-* = above; *spinous* = spine)—depression superior to spine
- **infraspinous fossa** (*infra-* = below)—depression inferior to spine
- **subscapular fossa** (*sub-* = under)—depression on anterior surface of scapula
- **lateral** or **axillary border**—margin near axilla
- **medial** or **vertebral border**—margin near vertebral column

To form the pectoral girdle, the acromial end of the clavicle articulates with the acromion (acromial process) of the scapula laterally. The pectoral girdle is attached to the axial skeleton by the articulation of the sternal end of the clavicle with the manubrium of the sternum. The scapula does not articulate directly with the axial skeleton but is attached to it with muscles.

> **Before Going to Lab**
>
> **1** Label the parts of the clavicle and scapula in Figure 10.2(a), (b), and (c).

LAB ACTIVITY 1 The Pectoral Girdle

1 Identify the bones and bone markings in Figure 10.2 on disarticulated bones or use the search text box to locate these structures in **Real Anatomy** (Skeletal).

2 Palpate these structures on your own body: clavicle, acromion (process), spine of scapula, and muscles located in the supraspinous and infraspinous fossae.

3 Distinguish between right and left scapula.
- Spine of scapula is superior and posterior.
- Glenoid cavity is laterally located.

Pectoral girdle:
Clavicle
Scapula

(a) Anterior view

Clavicle
Scapula

(b) Posterior view

FIGURE 10.1 Pectoral girdle.

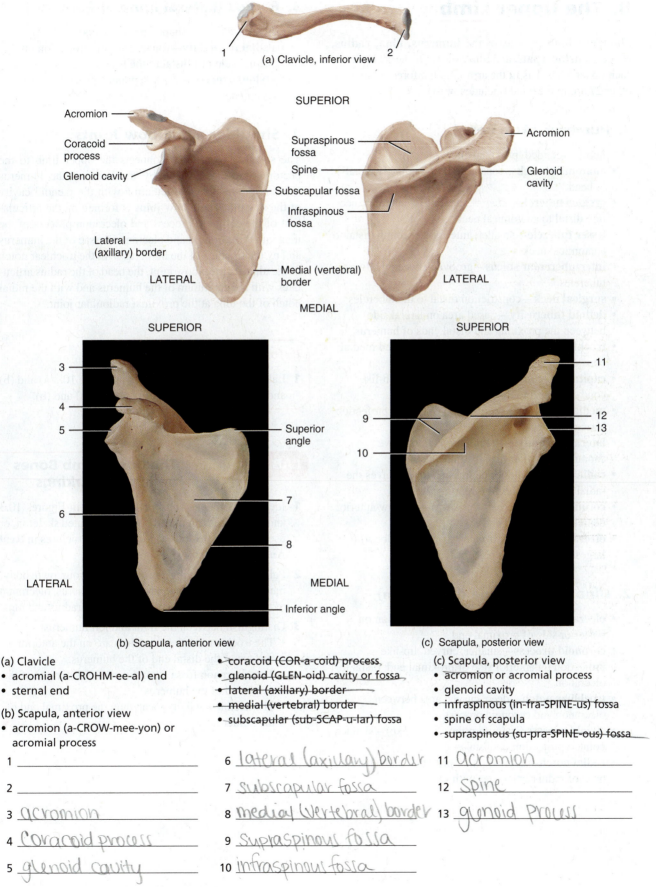

(a) Clavicle, inferior view

SUPERIOR

Acromion

Coracoid process

Glenoid cavity

Lateral (axillary) border

LATERAL

Supraspinous fossa

Spine

Subscapular fossa

Infraspinous fossa

Medial (vertebral) border

MEDIAL

Acromion

Glenoid cavity

LATERAL

SUPERIOR

Superior angle

LATERAL

MEDIAL

Inferior angle

(b) Scapula, anterior view

SUPERIOR

(c) Scapula, posterior view

(a) Clavicle
• acromial (a-CROHM-ee-al) end
• sternal end

(b) Scapula, anterior view
• acromion (a-CROW-mee-yon) or acromial process

• coracoid (COR-a-coid) process
• glenoid (GLEN-oid) cavity or fossa
• lateral (axillary) border
• medial (vertebral) border
• subscapular (sub-SCAP-u-lar) fossa

(c) Scapula, posterior view
• acromion or acromial process
• glenoid cavity
• infraspinous (in-fra-SPINE-us) fossa
• spine of scapula
• supraspinous (su-pra-SPINE-ous) fossa

1 _____

2 _____

3 _acromion_____

4 _Coracoid process_____

5 _glenoid cavity_____

6 _lateral (axillary) border_

7 _subscapular fossa_____

8 _medial (vertebral) border_

9 _supraspinous fossa_____

10 _infraspinous fossa_____

11 _acromion_____

12 _Spine_____

13 _glenoid process_____

FIGURE 10.2 The right pectoral girdle.

B. The Upper Limb

The upper limb consists of the **humerus, ulna, radius, carpals, metacarpals,** and **phalanges.** Of the 30 bones in each upper limb, 1 is in the arm, 2 in the forearm, and the other 27 are in the hand (includes wrist).

1. Humerus (Arm Bone)

- **head**—rounded, proximal end
- **anatomical neck**—constriction immediately distal to head
- **greater tubercle** (*tuber* = swelling)—lateral projection distal to anatomical neck
- **lesser tubercle**—smaller, anterior projection distal to anatomical neck
- **intertubercular sulcus**—groove between the two tubercles
- **surgical neck**—constriction distal to the tubercles
- **deltoid tuberosity**—raised area on lateral side between the proximal and distal ends of humerus
- **trochlea** (*trochlea* = pulley)—spool-shaped medial condyle on distal end
- **capitulum** (*caput* = head)—rounded, knob-like condyle lateral to trochlea
- **medial epicondyle** (*epi-* = upon)—rough projection above trochlea
- **lateral epicondyle**—rough projection above capitulum; smaller than medial epicondyle
- **radial fossa**—anterior depression that receives the radial head with flexed forearm
- **coronoid fossa** (*corona* = crown)—shallow anterior depression on distal end
- **olecranon fossa** (*olekranon* = tip of the elbow)—largest depression on posterior, distal end

2. Ulna (Medial Bone of Forearm)

- **olecranon**—large, curved, lip-like projection on posterior side of proximal end
- **coronoid process**—smaller, curved, lip-like projection on anterior side of proximal end; distal to olecranon
- **trochlear notch**—deep, curved area between olecranon and coronoid process
- **styloid process** (*stylos* = pole; *-oid* = like)—slender, pointed projection on distal end
- **radial notch**—depression on proximal end where head of radius articulates with ulna

3. Radius (Lateral Bone of Forearm)

- **head**—flat, disc-shaped proximal end
- **radial tuberosity**—rough, anterior projection on medial side just distal to the head
- **styloid process**—slender, pointed projection; distal end

4. Shoulder and Elbow Joints

The **shoulder joint** that connects the upper limb to the pectoral girdle is formed by the head of the humerus (*humeri* = shoulder), articulating with the glenoid cavity of the scapula. The **elbow joint** is formed by the articulation of the coronoid process and olecranon process of the ulna into the coronoid and olecranon fossae of the humerus and by the trochlea of the humerus with the trochlear notch of the ulna. At the elbow joint, the head of the radius articulates with the capitulum of the humerus and with the radial notch of the ulna at the proximal radioulnar joint.

Before Going to Lab

1 Label the parts of the humerus in Figure 10.3(a) and (b) and the ulna and radius in Figure 10.4(a) and (b).

LAB ACTIVITY 2 **The Upper Limb Bones and Bone Markings**

1 Identify the bones and bone markings in Figures 10.3 and 10.4 on disarticulated bones, articulated skeleton, or use the search text box to locate these structures in **Real Anatomy** (Skeletal).

2 Palpate the following bone parts on your own body: medial and lateral epicondyle of the humerus, olecranon of the ulna, and styloid processes of the radius and ulna.

3 Distinguish between the right and left humerus.
- The trochlea and the capitulum are on the anterior surface of the distal end of the humerus.
- The olecranon fossa is on the posterior surface of the distal end of the humerus.
- The head is medially located on the proximal end of the humerus.

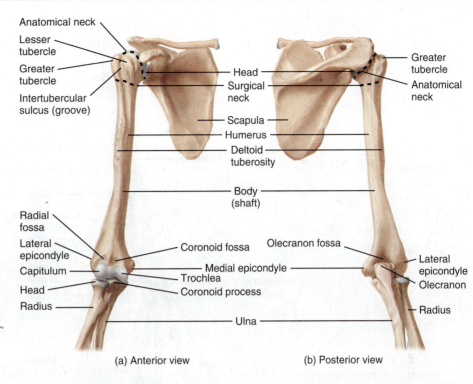

(a) Anterior view
(b) Posterior view

(a) Anterior view
- anatomical neck
- capitulum (ca-PIT-u-lum)
- coronoid (COR-a-noid) fossa
- deltoid tuberosity
 (tu-ber-OS-ity)
- greater tubercle
 (TU-ber-cul)
- head
- intertubercular (in-ter-tu-BER-
 cue-lar) sulcus
- lateral epicondyle
 (epi-CON-dile)
- lesser tubercle
- medial epicondyle
- trochlea

1 greater tubercle
2 intertubercular sulcus
3 lesser tubercle
4 lateral epicondyle
5 capitulum
6 head
7 anatomical neck
8 deltoid tuberosity
9 coronoid fossa
10 medial epicondyle
11 trochlea

(b) Posterior view
- lateral epicondyle
- medial epicondyle
- olecranon (o-LEH-cra-non)
 fossa

12 olecranon fossa
13 medial epicondyle
14 lateral epicondyle

SUPERIOR SUPERIOR

Surgical neck

Radial fossa

LATERAL MEDIAL LATERAL

(a) Anterior view (b) Posterior view

FIGURE 10.3 **Right humerus.**

Humerus

Olecranon fossa

Capitulum
Coronoid fossa
Trochlea
Olecranon
Head of radius
Coronoid process
Head of radius
Neck of radius
Ulnar tuberosity
Neck of radius
Radial tuberosity
Radius
Radius

Ulna

Interosseous membrane

Head of ulna

Styloid process of ulna

Styloid process of radius
Carpals
Styloid process of radius

Lateral Medial Lateral

(a) Anterior view (b) Posterior view

(a) Anterior view
- ~~coronoid process~~
- ~~head of radius~~
- ~~olecranon (process)~~
- radial notch
- ~~radial tuberosity~~
- ~~styloid (STY-loid) process of radius~~
- ~~styloid process of ulna~~
- ~~trochlear notch (semilunar)~~

1 head of radius
2 radial tuberosity
3 styloid process of radius
4 olecranon
5 trochlear notch
6 coronoid process
7 radial notch
8 styloid process of ulna

(b) Posterior view
- radius
- ulna

9 ulna
10 radius

SUPERIOR SUPERIOR

4
5
6
1
7
2

9 10

LATERAL MEDIAL LATERAL

3 8

(a) Anterior view (b) Posterior view

FIGURE 10.4 Right ulna and radius.

5. Carpus (Wrist)

The carpus is composed of 8 short bones of the wrist, the **carpal** bones, which are lined up to form a proximal and a distal row of bones. Two of the carpal bones articulate with the radius, but there is no articulation of the carpal bones with the ulna.

6. Metacarpus (Palm of Hand)

The metacarpus is composed of 5 **metacarpal** bones that make up the palm of the hand. They are numbered as Roman numerals I to V from the metacarpal of the thumb (lateral side) to the little finger side. The metacarpals articulate with the carpals proximally and with the phalanges distally.

7. Phalanges (Fingers)

The phalanges (phalanx, sing.) make up the fingers or digits. The fingers are also numbered I to V from the thumb (*pollex*)

to the little finger. The thumb has 2 phalanges, proximal and distal. Digits II through V each have **proximal, middle,** and **distal phalanges** (*phalanx* = closely knit row).

Before Going to Lab

1 Label the bones of the hand in Figure 10.5. For each phalanx, include the Roman numeral.

LAB ACTIVITY 3 The Hand

1 Identify the bones of the hand in Figure 10.5 on an articulated hand, articulated skeleton, or use the search text box to locate these structures in **Real Anatomy** (Skeletal).

2 Identify and palpate the bones in Figure 10.5 on yourself.

- carpals
- distal phalanx (FAY-lanx) V
- metacarpals (meta-CAR-puls)
- middle phalanx V
- proximal phalanx V

1 _____

2 _____

3 _____

4 _____

5 _____

FIGURE 10.5 **Right hand and wrist.**

C. Bones and Selected Bone Markings of the Pelvic Girdle

The **pelvic girdle** (*pelvis* = basin) is composed of 2 hip (coxal) bones called the **ossa coxae** (*os-* = bone; *cox-* = hip) that attach the lower limb to the axial skeleton. Each **os coxa** is formed by the fusion of 3 separate bones: the **ilium** (*ilia* = flank), **ischium** (*ischion* = hip joint), and **pubis** (*pub-* = grown up or adult) bones. These 3 bones are identifiable as separate bones in children.

1. Os Coxa

ilium—largest and most superior of the three components of os coxa
- **iliac crest**—superior border of ilium
- **anterior superior iliac spine**—anterior end of iliac crest
- **anterior inferior iliac spine**—below the anterior superior iliac spine
- **posterior superior iliac spine**—posterior end of iliac crest
- **posterior inferior iliac spine**—below the posterior superior iliac spine
- **greater sciatic notch**—large notch on posterior side
- **iliac fossa**—depression on anterior surface

ischium—inferior, posterior portion of os coxa
- **ischial tuberosity**—large, roughened projection on posterior and inferior edge
- **ischial spine**—posterior projection between greater and lesser sciatic notches
- **lesser sciatic notch**—smaller indentation between ischial spine and ischial tuberosity

pubis—anterior inferior portion of os coxa
- **pubic symphysis** (*symphysis* = growing together)—joint where the two pubic bones join anteriorly

bone markings formed by ilium, ischium, and pubis
- **acetabulum** (*acetabulum* = little saucer)—deep indentation, or cup, for head of the femur
- **obturator foramen**—largest foramen in the skeleton

2. Pelvis

- **pelvic brim**—divides the false pelvis from the true pelvis; begins at the sacral promontory and extends laterally and inferiorly to end at the pubic symphysis
- **false pelvis**—portion of pelvis superior to pelvic brim; wide area extending to top of iliac crest
- **true pelvis**—portion of pelvis inferior to pelvic brim; surrounds the pelvic cavity
- **pelvic inlet**—superior opening of true pelvis; bordered by pelvic brim
- **pelvic outlet**—inferior opening of true pelvis; bordered by the coccyx, ischial spines, and ischial tuberosities

There are definite anatomical differences between the male and female pelves (pl.). The bones of the male are typically heavier and rougher, with larger bone markings than the female. The **male pelvis** is more vertical and narrower, has a pelvic inlet that is heart-shaped, and has a 90° or less pubic arch angle. The **female pelvis** generally has more space in the true pelvis for childbirth and is tilted backward and flared. The pelvic inlet is round or oval, the angle of the pubic arch is generally greater than 90°, and the angle of the sciatic notch is wider.

The pelvic girdle articulates with the axial skeleton at the **sacroiliac joints** (*sacro-* = sacrum; *iliac* = ilium). The sacroiliac joints are located where the articulated ossa coxae unite posteriorly with the sacrum.

Before Going to Lab

1 Label the bones and bone markings of an os coxa in Figure 10.6.
2 Label the bones and bone markings of the pelvis in Figure 10.7.

LAB ACTIVITY 4 Os Coxa and Pelvis

1 Identify the bones and bone markings in Figures 10.6 and 10.7 on a disarticulated os coxa, an articulated pelvis, or use the search text box to locate these structures in **Real Anatomy** (Skeletal).

2 Distinguish between the right and left os coxa.
- The iliac crest is on the superior part of the os coxa.
- The ischium is the inferior and posterior part of the os coxa.
- The pubic bone is on the anterior and medial part of the os coxa.
- The acetabulum is on the lateral part of the os coxa.

3 Palpate the iliac crest, the anterior superior iliac spine, and the pubic symphysis on your own body.

4 Locate these markings on an articulated pelvis or skeleton.

5 Identify the differences between male and female pelves on models or articulated skeletons.

Iliac crest

Ilium

Anterior superior
iliac spine

Posterior
superior
iliac spine

Anterior inferior
iliac spine

Posterior
inferior
iliac spine

Acetabulum

Greater sciatic notch

Ischial spine

Pubis

Lesser sciatic notch

Obturator foramen

Ischial tuberosity

Ischium

SUPERIOR

POSTERIOR

- acetabulum (asa-TAB-u-lum)
- anterior inferior iliac spine
- anterior superior iliac spine
- greater sciatic (sigh-A-tic) notch
- iliac crest
- ilium (IL-lee-um)
- ischial (ISH-ee-ul) spine
- ischial tuberosity (tu-ber-OS-ity)
- ischium (ISH-ee-um)
- lesser sciatic notch
- obturator (OB-tur-a-tur) foramen
- posterior inferior iliac spine
- posterior superior iliac spine
- pubis (PYU-bis)

1 (Bone)

2

3

4

5 (Bone)

6

7

8

9

10

11

12

13

14
(Bone)

Lateral view

1 _Ilium_

2 _Posterior superior iliac spine_

3 _____

4 _____

5 _____

6 _____

7 _____

8 _____

9 _____

10 _____

11 _____

12 _____

13 _____

14 _____

FIGURE 10.6 Right os coxa.

Ilium

Iliac crest

Iliac fossa

False pelvis

Anterior superior iliac spine

Sacrum

Pelvic brim

Ischial spine

Coccyx

Acetabulum

True pelvis

Obturator foramen

Ischium

Pubic symphysis

Pubis

Pubic arch

ANTERIOR

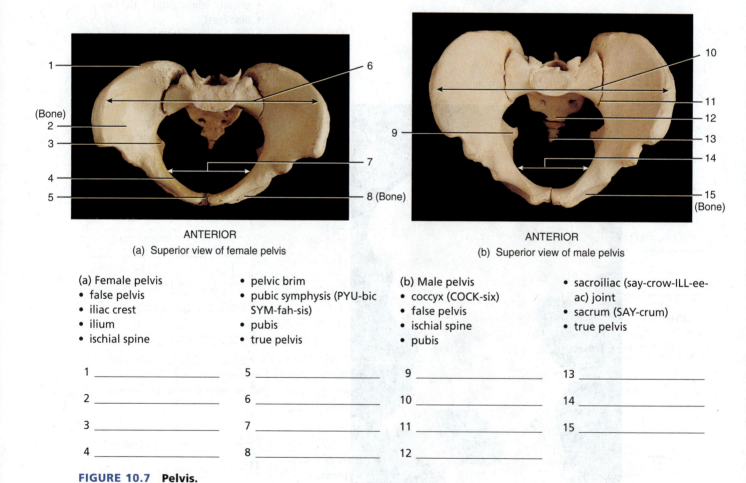

(Bone)

ANTERIOR

(a) Superior view of female pelvis

ANTERIOR

(b) Superior view of male pelvis

(a) Female pelvis
• false pelvis
• iliac crest
• ilium
• ischial spine

• pelvic brim
• pubic symphysis (PYU-bic SYM-fah-sis)
• pubis
• true pelvis

(b) Male pelvis
• coccyx (COCK-six)
• false pelvis
• ischial spine
• pubis

• sacroiliac (say-crow-ILL-ee-ac) joint
• sacrum (SAY-crum)
• true pelvis

1 _____

2 _____

3 _____

4 _____

5 _____

6 _____

7 _____

8 _____

9 _____

10 _____

11 _____

12 _____

13 _____

14 _____

15 _____

FIGURE 10.7 Pelvis.

D. Bones and Selected Bone Markings of the Lower Limb

The lower limb consists of bones of the **femur, patella, tibia, fibula, tarsals, metatarsals,** and the **phalanges.** Of the 30 bones in each lower limb, 4 are in the thigh and leg, and the other 26 are in the foot (including ankle).

1. Femur (Thigh Bone)

The femur is the largest and strongest bone in the human skeleton. This bone is bowed anteriorly in a slight curve.

- **head**—large, rounded, knob-like proximal end
- **neck**—narrower, constriction distal to head
- **greater trochanter** (*trochanter* = runner)—large and roughened superior projection; lateral to neck
- **lesser trochanter**—smaller, posterior-medial prominence distal to greater trochanter
- **medial condyle**—rounded, medial process on posterior side of distal end
- **lateral condyle**—similar to medial condyle on lateral side
- **intercondylar fossa**—deep fossa between medial and lateral condyles
- **medial epicondyle**—bump-like projection superior to medial condyle
- **lateral epicondyle**—bump-like projection superior to lateral condyle; a little smaller
- **gluteal tuberosity**—posterior surface of body of femur; roughened projection inferior to lesser trochanter
- **linea aspera**—vertical ridge on posterior surface

2. Patella (Kneecap)

This small, triangular bone has an anterior surface that is smoother than the posterior surface. Shallow, irregular-shaped articular facets are on the posterior surface that articulate with the condyles of the femur.

3. Tibia (Leg Bone)

The tibia is the weight-bearing bone of the 2 leg bones and is medially located.

- **medial condyle**—flattened, expanded medial projection on proximal end
- **lateral condyle**—similar to medial condyle on lateral side
- **tibial tuberosity**—large, roughened projection on anterior surface, inferior to condyles
- **anterior border (crest)**—slender ridge on anterior surface; shin

- **medial malleolus**—medial process on distal end, forms medial bump of ankle

4. Fibula (Leg Bone)

The slender fibula is the lateral leg bone that is important for muscle attachment but not for bearing weight. The **lateral malleolus** is distal and articulates with the talus laterally.

- **head**—proximal end
- **lateral malleolus**—distal end, forms lateral bump of ankle

5. Hip and Knee Joints

The **hip** or **coxal joint** is formed by the acetabulum articulating with the head of the femur to form a ball-and-socket joint. There is a strong ligament that connects these two structures deep inside the joint itself.

The **knee joint** is formed by the articulation of the medial and lateral condyles of the femur with the medial and lateral condyles of the tibia. The patella articulates with the condyles of the femur. The fibula does not form part of the knee joint; however, the head of the fibula articulates with the tibia but not the femur.

Before Going to Lab

1 Label the bones and bone markings of the femur in Figure 10.8(a) and (b) and the patella, tibia, and fibula in Figure 10.9(a) and (b).

LAB ACTIVITY 5 The Thigh and Leg

1 Identify the bones and bone markings in Figures 10.8 and 10.9 on disarticulated bones, an articulated skeleton, or use the search text box to locate these structures in **Real Anatomy** (Skeletal).

2 Distinguish between the right and left femur.
- The head is on the proximal end and faces medially.
- The medial and lateral condyles are on the anterior surface of the distal end of the femur (the medial condyle is larger).

3 Distinguish between the right and left tibia.
- The tibial tuberosity is anterior and superior to the anterior crest.
- The medial malleolus is on the medial surface of the distal end of the tibia.

4 Palpate these bone markings on your own body: greater trochanter, medial and lateral epicondyles, patella, head of the fibula, tibial tuberosity, anterior crest (shin) of the tibia, medial malleolus of the tibia, and lateral malleolus of the fibula.

• greater trochanter (tro-CAN-ter)
• head of femur
• lateral condyle (CON-dile)
• lateral epicondyle (epi-CON-dile)
• lesser trochanter
• linea aspera (LIN-ee-uh ASP-er-uh)
• medial condyle
• medial epicondyle
• neck

(a) Anterior view

1 _____

2 _____

3 _____

4 _____

5 _____

6 _____

(b) Posterior view

7 _____

8 _____

9 _____

FIGURE 10.8 Right femur.

(a) Anterior view (b) Posterior view

SUPERIOR

Intercondylar
eminence

- fibula (FIB-u-la)
- head of fibula
- lateral condyle
- lateral malleolus (mal-LAY-e-lus)
- medial condyle
- medial malleolus
- tibia
- tibial tuberosity

1 _____

2 _____

3 _____

4 _____

5 _____

6 _____

7 _____

8 _____

Anterior
border

MEDIAL

(a) Anterior view (b) Posterior view

FIGURE 10.9 **Right tibia, fibula, and patella.**

6. Tarsus (Ankle)

The tarsus is composed of 7 tarsal bones of the foot, with 2 of them being larger than the rest. The largest tarsal bone is the **calcaneus** (*calcaneum* = heel), also known as the heel bone. The other large tarsal bone is the **talus** (*talus* = ankle), which articulates with the medial malleolus of the tibia and lateral malleolus of the fibula.

7. Metatarsus

The metatarsus is composed of 5 metatarsal bones (*meta-* = after or next) that are analogous to the metacarpals in the hand. They are numbered the same way, I to V, from the great toe to the little toe.

8. Phalanges (Toes)

The phalanges (toes or digits) are similar to the phalanges in the hand. The toes are numbered I to V from the great toe (*hallux*) to the little toe. The great toe is made of 2 phalanges (**proximal** and **distal**), and digits II to V have 3 bones each—**proximal, middle,** and **distal phalanges.**

Before Going to Lab

1 Label Figure 10.10(a) and (b). For each phalanx, include the Roman numeral.

LAB ACTIVITY 6 The Foot

1 Identify the bones of the foot in Figure 10.10 on an articulated foot or use the search text box to locate these structures in **Real Anatomy** (Skeletal).

2 Palpate these parts on your own body: lateral malleolus, calcaneus, and talus.

(a) Superior view
- calcaneus (cal-CANE-ee-us)
- distal phalanx II (FAY-lanx)
- middle phalanx II
- metatarsals (meta-TAR-suls)
- phalanges
- proximal phalanx II
- talus (TA-lus)
- tarsals (TAR-suls)

1 _____ 5 _____

2 _____ 6 _____

3 _____ 7 _____

4 _____ 8 _____

(a) Superior view

FIGURE 10.10 **Bones of the right foot.**

(b) Lateral view
- calcaneus
- lateral malleolus of fibula
- tibia

- metatarsals
- phalanges
- talus
- tarsals

9 _____ 13 _____

10 _____ 14 _____

11 _____ 15 _____

12 _____

(b) Lateral view

FIGURE 10.10 **Bones of the right foot,** *continued.*

LAB ACTIVITY 7 **Assembly of a Complete Disarticulated Skeleton**

1 Obtain a disarticulated skeleton from your instructor.

2 With your lab partners, assemble the skeleton.

3 Have your instructor check it to see that you have assembled it properly. ■

LAB ACTIVITY 8 **Determining Gender and Height Using Femur Measurements**

1 **Determining gender:** The diameter of the head of a femur is a very accurate way of determining gender.
- Use a femur from a disarticulated or articulated skeleton.
- Use a tape measure to measure in cm the circumference of the head of the femur. Record your value in Table 10.2.

- Calculate the diameter of the femur using the following equation.

$$\text{Diameter (cm)} = \frac{\text{Circumference (cm)}}{3.14}$$

Record your value in Table 10.2.
- Males have a diameter greater than 4.5 cm (45 mm) and females have a diameter less than 4.3 cm (43 mm).[1] Record gender in Table 10.2.

2 **Estimating height:**
- Measure the longest possible length of the femur in cm. Record your values in Table 10.2.
- Use the appropriate equation for each bone (male or female) to estimate height in cm.[2] Record your value in Table 10.2.

Male: Height (cm) =
\qquad (1.88 × length of femur in cm) + 81.306
Female: Height (cm) =
\qquad (1.945 × length of femur in cm) + 72.844

- To convert height in cm to height in inches, divide height in cm by 2.54. Record your value in Table 10.2. ■

TABLE 10.2 **Determining Gender and Height from a Femur**

ESTIMATING GENDER		ESTIMATING HEIGHT	
Femur circumference (cm)		Femur length (cm)	
Femur diameter (cm)		Height (cm)	
Gender		Height (in)	

[1]Institute for Algorithmic Medicine. *The Medical Algorithms Project*. Chapter 38: Forensic Medicine; Determination of Gender from Physical Remains; Determination of Gender of Measurement of the Femur. www.medal.org (accessed September 27, 2007).

[2]Institute for Algorithmic Medicine. *The Medical Algorithms Project*. Chapter 38: Forensic Medicine; Estimation of Body Height from Physical Remains; Pearson's Formulas for Estimating Adult Body Height from Length of Long Bones. www.medal.org (accessed September 27, 2007).

LAB ACTIVITY 9 Estimating Your Height from Bone Length

1 Estimate your height from length of radius:

- With the hand in anatomical position, palpate the lateral epicondyle of your humerus and then move your hand just a little distally to the head of the radius. Slowly pronate and supinate your forearm to feel the rotation of the disc-shaped head of the radius. To find the styloid process of the radius, palpate the lateral side of your wrist.
- Measure the length of the radius in inches from the head to the styloid process. Record the value in Table 10.3.
- Use the appropriate formulas to calculate your height in inches.[3] Record the value in Table 10.3.

 Males: Height (in) =

 (length of radius in inches \times 3.3) + 34

 Females: Height (in) =

 (length of radius in inches \times 3.3) + 32

2 Estimate your height from length of humerus:

- Palpate the head of your humerus as you rotate your arm in the shoulder socket. Then locate the medial epicondyle by palpating this "bump" just above the elbow hinge.
- Measure the length of the humerus in inches from the head of the humerus to the medial epicondyle.

- Use the appropriate formulas to calculate your height in inches.[3] Record your value in Table 10.3.

 Males: Height (in) =

 (length of humerus in inches \times 2.9) + 27.8

 Females: Height (in) =

 (length of humerus in inches \times 2.8) + 28.1

3 Measure your height in inches and record the value in Table 10.3.

4 Compare the calculated and measured values. ◼

TABLE 10.3 Estimating Your Height from Bone Length

RADIUS	
Length of radius (in)	
Calculated height (in)	
Measured height (in)	
HUMERUS	
Length of humerus (in)	
Calculated height (in)	
Measured height (in)	

[3]Scientific American Frontiers, "Science Safari Teaching Guide: The First People," *Scientific American Frontiers Archives* (Fall 1990 to Spring 2000). www.pbs.org/safarchive/4_class/45_pguides/pguide_702/4572_firstpeople.html (accessed September 27, 2007).

Reviewing Your Knowledge

A. Pectoral Girdle (Shoulder)

Fill in the blank with the correct term.

_____ 1. The acromion (process) articulates with what bone?

_____ 2. Is the clavicle anterior or posterior compared to the scapula?

_____ 3. The clavicle articulates medially with which bone?

_____ 4. The humerus articulates with what bone marking of the scapula?

_____ 5. Is the subscapular fossa located anterior or posterior to the supraspinatus and infraspinatus fossae?

_____ 6. Name the two bones that make up the pectoral girdle.

B. Upper Limb

Fill in the blank with the correct term.

_____ 1. What part of the radius articulates with the humerus?

_____ 2. What part of the ulna fits into the olecranon fossa of the humerus?

_____ 3. The coronoid process articulates with what depression on the distal end of the humerus?

_____ 4. Is the ulna medial or lateral compared with the radius?

_____ 5. What are the bones called that make up the fingers?

_____ 6. What are the bones called that make up the palm of the hand?

_____ 7. What is the name of the lateral condyle on the humerus?

_____ 8. What is the name of the medial condyle on the humerus?

_____ 9. What is the name of the slender, pointed projection on the distal end of the radius?

_____ 10. What is the name of the slender, pointed projection on the distal end of the ulna?

C. Pelvic Girdle and Pelvis

Fill in the blank with the correct term.

_____ **1.** When you put your hands on your hips, which bone marking of each os coxa are you touching?

_____ **2.** With your hands on your hips, you can feel a point of the pelvis protruding out anteriorly just above your thigh. Name this bone marking.

_____ **3.** Name the 2 bones that form the pelvic girdle.

_____ **4.** What prominent bone marking on each os coxa do you sit on?

_____ **5.** Name the bones of the ossa coxae that articulate anteriorly.

_____ **6.** The female pelvis has smoother bone markings than the male. True or False?

_____ **7.** Name the anterior joint between pubic bones.

_____ **8.** Deep indentation formed by fusion of ilium, ischium, and pubis.

_____ **9.** Largest foramen in the skeleton.

_____ **10.** The pelvic outlet is larger in females. True or False?

D. Lower Limb

Fill in the blank with the correct term.

_____ **1.** Is the fibula medial or lateral to the tibia?

_____ **2.** What is the correct term for the process at the distal end of the tibia that forms the medial bump of the ankle?

_____ **3.** What is the heaviest and strongest bone of the leg (not thigh)?

_____ **4.** Name the thigh bone.

_____ **5.** The fibula is a weight-bearing bone. True or False?

_____ **6.** Name the tarsal bone that articulates with the tibia.

_____ **7.** Name the heel bone.

_____ **8.** Name the bones of the distal part of the instep.

_____ **9.** Bone marking that is commonly called the shin.

_____ **10.** Bone that is commonly called the kneecap.

_____ **11.** Processes on the femur and tibia that form the knee joint.

_____ **12.** Name of the bone marking of the femur that articulates with the pelvic girdle.

_____ **13.** Does the fibula form part of the knee joint?

_____ **14.** Name the phalanges in the great toe.

_____ **15.** The number of metatarsal bones.

E. Bones and Bone Markings of the Pectoral Girdle and Upper Limb

Identify the bones and bone markings in Figure 10.11.

Anterior view

FIGURE 10.11

1. _____
2. _____
3. _____
4. _____
5. _____
6. _____
7. _____
8. _____
9. _____
10. _____
11. _____
12. _____
13. _____
14. _____
15. _____
16. _____
17. _____
18. _____
19. _____
20. _____

F. Bones and Bone Markings of the Pelvic Girdle and Lower Limb

Identify the bones and bone markings in Figure 10.12.

Anterior view

FIGURE 10.12 Pelvic girdle and lower limb.

1. _____
2. _____
3. _____
4. _____
5. _____
6. _____
7. _____
8. _____
9. _____
10. _____
11. _____
12. _____
13. _____
14. _____
15. _____
16. _____
17. _____
18. _____
19. _____
20. _____
21. _____
22. _____

Using Your Knowledge

A. Upper Limb

Identify the bones and bone markings of the upper limb as shown in Figure 10.13(a), (b), and (c).

1 _____ (bone marking)

2 _____ (bone marking)

3 _____ (bone marking)

4 _____ (bone)

FIGURE 10.13a Radiograph of the shoulder joint, posterior view.

5 _____ (bone)

6 _____ (bone)

7 _____ (bone)

8 _____ (bone marking)

9 _____ (bone marking)

FIGURE 10.13b Radiograph of the elbow joint.

10 _____ (bone marking)

11 _____ (bone)

12 _____

Draw a band on the specific bone where the wedding ring is worn.

13 _____

Name the bone the ring is on.

FIGURE 10.13c Radiograph of the left hand.

B. Lower Limb

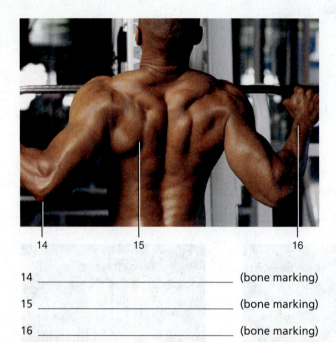

14 _____ (bone marking)

15 _____ (bone marking)

16 _____ (bone marking)

FIGURE 10.14a **Surface anatomy of the upper limb and pectoral girdle.**

17 _____ (bone marking)

18 _____ (bone)

19 _____ (bone marking)

20 _____ (bone marking)

FIGURE 10.14b **Surface anatomy of the hip and lower limb.**

Joints and Synovial Joint Movements

A **joint or articulation** (*articulare* = to divide into joints) connects a bone with another bone, cartilage, or tooth. Joints are commonly classified according to their structure and function.

A. Structural and Functional Classification of Joints

The structural classification of joints depends on the type of connective tissue forming the joint and on whether or not there is a space or synovial cavity between the bones.

Fibrous joints have dense fibrous connective tissue with strong collagen fibers that hold the joints firmly together with no synovial cavity. This type of joint permits little to no movement. Examples are the skull joints, teeth sockets, and the distal joint between the tibia and fibula.

Cartilaginous joints have either hyaline cartilage or fibrocartilage connecting the bones with no synovial cavity. Usually, there is a small degree of movement with this type of joint. Examples are the intervertebral joints, the pubic symphysis, and the joint between the manubrium and body of the sternum. **Synovial joints** (*syn-* = together; *ovum* = egg) have a small synovial cavity (space) between the two bones that permits a greater amount of movement than

fibrous or cartilaginous joints. The term *synovial* comes from the synovial fluid present in the synovial cavity that resembles the albumin of an uncooked egg, only more viscous. Dense fibrous connective tissue on the exterior of the joint holds the bones together. The majority of the joints in the human body are synovial joints—for example, the shoulder, elbow, hip, and knee joints.

The functional classification of joints is made on the basis of the amount of movement the joint allows. Immovable joints or **synarthroses** (*syn-* = union; *arthro-* = joint) include the sutures between the skull bones and the teeth sockets. Intervertebral joints, the tibiofibular joint, (the joint between the manubrium and the body of the sternum), and the pubic symphysis are examples of slightly movable joints or **amphiarthroses** (*amphi-* = on both sides). Most of the joints in the body, about 90%, are freely movable joints, or **diarthroses** (*di-* = apart; away from).

As you can see from the description of structural and functional joints, there are similarities in the amount of movement and certain types of structural joints. Most of

the fibrous joints, such as sutures and teeth sockets, are immovable joints. However, the fibrous tibiofibular joint is a slightly movable joint. Most of the cartilaginous joints are slightly movable joints, such as the intervertebral discs and the pubic symphysis. The cartilaginous epiphyseal plates of long bones, however, are immovable joints. All synovial joints are diarthroses.

Before Going to Lab

1 Label the structural and functional category of each joint in Figure 11.1 (two answers for each joint).

LAB ACTIVITY 1 Structural Classification of Joints

1 Identify the joints in Figure 11.1 on an articulated skeleton.

2 Point out these joints and palpate the ones you can on your own body. ■

(a) Anterior view (b) Tooth in alveolus

FIGURE 11.1 Structural classification of joints.

- amphiarthrosis (amphi-ar-THROW-sis)
- cartilaginous (car-tih-LA-jih-nous) joint
- diarthrosis (die-ar-THROW-sis)
- fibrous joint
- synarthrosis (syn-ar-THROW-sis)
- synovial (sih-NO-vee-ul) joint

1 _____

2 _____

3 _____

4 _____

5 _____

6 _____

7 _____

8 _____

9 _____

10 _____

11 _____

B. Basic Structure of Synovial Joints

Although synovial joints vary in structure, they have several common features as seen in Figure 11.2(a).

- **synovial cavity**—small space between the two articulating bones
- **articular cartilage**—hyaline cartilage covering the ends of the bones in the synovial cavity
- **articular capsule**—structure that encloses the synovial joint and synovial cavity; has two layers: the fibrous membrane and synovial membrane
- **fibrous membrane**—outer dense fibrous connective tissue layer of the articular capsule that is continuous with the periosteum of the bone; also forms ligaments when fibrous bundles are parallel
- **synovial membrane**—inner layer of the articular capsule; composed of areolar connective tissue containing elastic fibers and adipocytes
- **synovial fluid**—secreted by the synovial membrane; lubricates the articular cartilages to reduce friction

Before Going to Lab

1 Label the simple synovial joint in Figure 11.2b.

- articular (ar-TIH-ku-lar) bone
- articular capsule
- articular cartilage
- fibrous membrane
- synovial cavity (contains synovial fluid)
- synovial membrane

1 _____

2 _____

3 _____

4 _____

5 _____

6 _____

(a) Frontal section

(b) Frontal section

FIGURE 11.2 Typical synovial joint.

SAFETY NOTE: *If you are using fresh or frozen joints, wear protective gloves. Wash your hands thoroughly with soap and water after the dissection.*

LAB ACTIVITY 2 Typical Synovial Joint and Dissection

1 Dissect a whole synovial joint following the dissection instructions.
 - After placing your **synovial joint** in a dissecting tray, observe the external white **articular capsule** that holds the bones together. Note that the articular capsule is continuous with the periosteum of the bone.
 - Using a scalpel, cut open the joint capsule. Feel the slippery **synovial fluid** between your fingers.
 - Try to detect the difference between the outer **fibrous membrane** layer and the inner **synovial membrane** layer of the articular capsule.
 - Note the **synovial cavity** or space between the **articular bones.**
 - Observe the ends of the joint bones for the **articular cartilage.** Cut out a small piece of hyaline cartilage to observe its thickness.

2 Clean up as directed by your instructor. ■

C. The Knee Joint

The knee joint is specifically studied here because it has the distinction of being the most complex and highly stressed joint, as well as being the location of many joint injuries. The knee joint is classified as a hinge joint, but when flexed it also demonstrates gliding and rotation movements.

In addition to the common joint features, the knee joint also has the following accessory structures.

Accessory structures of knee joint:
 - **medial meniscus**—inside of joint cavity; cushions knee joint
 - **lateral meniscus**—inside of joint cavity; cushions knee joint
 - **medial** (tibial) **collateral ligament**—extracapsular ligament; adds strength to joint medially

 - **lateral** (fibular) **collateral ligament**—extracapsular ligament; adds strength to joint laterally
 - **anterior cruciate ligament** (*cruci-* = cross) or **ACL**—intracapsular ligament; attaches the femur and tibia anteriorly
 - **posterior cruciate ligament** or **PCL**—intracapsular ligament; stabilizes joint posteriorly
 - **patellar ligament**—extension of tendon from quadriceps muscle; connects patella to tibial tuberosity and stabilizes the joint anteriorly
 - **infrapatellar fat pad** (*infra-* = beneath)—cushion between patellar ligament and synovial capsule
 - **bursa**—reduces friction; 13 bursae in knee

Before Going to Lab

1 Label the parts of the knee joint in Figure 11.3(a).

LAB ACTIVITY 3 The Knee Joint

1 Identify knee joint structures in Figures 11.3(a)–(c) on a model or chart of the knee or use the search text box in **Real Anatomy** (Arthrology) to find these structures.

2 In Figure 11.3(d), locate the suprapatella bursa, prepatellar bursa, infrapatellar bursa, and infrapatellar fat pad. ■

 - articular cartilage of femur 1
 - anterior cruciate (KRU-she-ate) ligament 5
 - lateral (fibular) collateral ligament 2
 - lateral meniscus (meh-NIS-cus) 3
 - medial (tibial) collateral ligament 7
 - medial meniscus 6
 - posterior cruciate ligament 4

1 _____

2 _____

3 _____

4 _____

5 _____

6 _____

7 _____

1
2
3

4
5
6
7
Transverse
ligament of
the knee

Fibula

Tibia

LATERAL

MEDIAL

(a) Anterior deep view

Articular
cartilage

Posterior
cruciate
ligament (PCL)

Tibial
collateral
ligament

Medial
meniscus

Oblique
popliteal
ligament (cut)

Tibia

Femur

Anterior
cruciate
ligament (ACL)

Fibular
collateral
ligament

Lateral
meniscus

Posterior ligament
of head of fibula

Fibula

MEDIAL

LATERAL

(b) Posterior deep view

Femur

Articular
cartilage

Lateral
meniscus

Fibular
collateral
ligament

Fibula

POSTERIOR

Tendon of
quadriceps
femoris
muscle

Synovial
membrane

Patella

Articular
capsule (cut)

Patellar
ligament

Periosteum

ANTERIOR

Sagittal
plane

(c) Lateral deep view

Femur
Suprapatellar
bursa
Patella
Articular
cartilage
Prepatellar
bursa
Infrapatellar
fat pad
Patellar
ligament
Infrapatellar
bursa
Medial meniscus
Tibia

POSTERIOR

ANTERIOR

(d) Sagittal section

FIGURE 11.3 **Knee joint.**

D. Types of Movement at Synovial Joints

Skeletal muscle contraction causes bone movement at synovial joints. Professionals in kinesiology and physical therapy use particular terms to describe the movements of these joints. Many movements are grouped to demonstrate opposite movements. The four main categories of synovial joint movements are gliding, angular, rotation, and special movements (Table 11.1).

Before Going to Lab

1 Label the movements illustrated in Figures 11.4–11.8.

LAB ACTIVITY 4 Types of Movement at Synovial Joints

1 Demonstrate each movement in Figures 11.4–11.8 with a partner(s).

2 Use an articulated skeleton to demonstrate each movement. During pronation, observe the radius crossing over the ulna.

3 Pronate forearm to feel the radius crossing over the ulna. ■

TABLE 11.1 Types of Movement at Synovial Joints

MOVEMENT	DESCRIPTION
A. GLIDING	Nearly flat bone surfaces slide or glide over each other.
B. ANGULAR	
Flexion (*flex-* = to bend)	Decrease in the angle between bones of a joint; usually occurs on a sagittal plane.
Extension (*exten-* = to stretch out	Increase in the angle between bones of a joint; restore to anatomical position.
Hyperextension (*hyper-* = excessive)	Excessive extension movement beyond normal anatomical position.
Abduction (*ab-* = away; *duct-* = to lead)	Move appendage away from the midline.
Adduction (*ad-* = toward)	Move appendage toward midline.
Circumduction (*circ-* = circle)	Move a distal part of an appendage in a circle.
C. ROTATION (*rota-* = revolve)	Turn on a pivot with a circle.
D. SPECIAL JOINT MOVEMENTS	
Elevation	Upward movement raising body part vertically.
Depression	Downward movement lowering body part vertically.
Protraction (*pro-* = in front of; *trahere* = to draw)	Move a body part forward or anterior on a horizontal plane.
Retraction (*retractare* = to draw back)	Move a body part backward or posterior.
Supination (*supine* = lying on the back)	Turn palm of the hand to face forward, or, if arm is outstretched, turn palm upward.
Pronation (*pronate* = lying face downward)	Turn palm of the hand to face backward, or, if arm is outstretched, turn palm downward.
Inversion	Turn the sole of the foot inward.
Eversion	Turn the sole of the foot outward.
Dorsiflexion	Point your toes upward; stand on your heels.
Plantar flexion	Point your toes downward; raise your heels.

(a) Atlanto-occipital and cervical intervertebral joints

(b) Shoulder joint

(c) Elbow joint

(d) Wrist joint

(e) Hip joint

(f) Knee joint

- extension
- flexion
- hyperextension

(a)

1 _____

2 _____

3 _____

(b)

4 _____

5 _____

6 _____

(c)

7 _____

8 _____

(d)

9 _____

10 _____

11 _____

(e)

12 _____

13 _____

14 _____

(f)

15 _____

16 _____

FIGURE 11.4 **Angular joint movements of flexion, extension, and hyperextension.**

(a) Shoulder joint

(b) Hip joint

Lateral Medial

(c) Right wrist joint

(d) Shoulder joint

(e) Hip joint

(f) Metacarpophalangeal joints of the fingers (not the thumb)

- abduction (ab-DUK-shun)
- adduction (ad-DUK-shun)
- circumduction (sir-cum-DUC-shun)

(a)

1 _____

2 _____

(b)

3 _____

(c)

4 _____

5 _____

(d)

6 _____

(e)

7 _____

8 _____

(f)

9 _____

10 _____

FIGURE 11.5 Angular joint movements of abduction, adduction, and circumduction.

(a) Atlanto-axial joint (b) Shoulder joint

- lateral rotation
- medial rotation
- rotation

1 _____

2 _____

3 _____

FIGURE 11.6 Joint movements of rotation.

(a) Temporomandibular joint (b) (c) Temporomandibular joint (d)

(e) Radioulnar joint

- depression
- elevation
- pronation (pro-NAY-shun)
- protraction (pro-TRAC-shun)
- retraction (re-TRAC-shun)
- supination (soup-in-NAY-shun)

1 _____

2 _____

3 _____

4 _____

5 _____

6 _____

FIGURE 11.7 Special joint movements.

(a) Intertarsal joint (b) (c) Ankle joint

- dorsiflexion
- eversion
- inversion
- plantar flexion

1 _____ 3 _____

2 _____ 4 _____

FIGURE 11.8 Special joint movements of the foot.

TABLE 11.2 **Types of Synovial Joints**

NAME OF JOINT	ARTICULAR SURFACE DESCRIPTION	MOVEMENT
Planar (gliding)	Flat or slightly curved plane	Gliding motion back and forth and/or side to side
Hinge	Convex bone surface articulates with a concave bone surface	Flexion and extension
Pivot	Rounded or pointed projection articulates with ring formed by bone and ligament	Rotation
Condyloid	Oval convex projection articulates with oval concave depression	Flexion and extension, abduction and adduction, circumduction
Saddle	Saddle-shaped depression articulates with projection that fits into the saddle	Same as condyloid joint, except more exaggerated
Ball-and-socket	Ball-shaped head articulates with cup-shaped socket	Freely movable joint; flexion and extension; abduction and adduction; circumduction; rotation

E. Six Types of Synovial Joints

There are 6 types of synovial joints based on the structure of the articulating bone surfaces at the joints (Table 11.2). The joint structure determines the movement of the joint.

LAB ACTIVITY 5 **Palpation of Synovial Joint Movements**

1 Palpate the following synovial joints on your body and use an articulated skeleton to observe the structure of the joint.
 - **Planar joint (gliding)**—flex your arm up and feel the gliding movement at the acromioclavicular joint.
 - **Hinge joint**—flex and extend your elbow and your knee as you feel this joint type. Note how this joint type only moves by flexion and extension.
 - **Pivot joint**—feel the proximal part of your forearm until you locate the head of the radius. Rotate the radioulnar joint as you palpate this rotation movement.
 - **Condyloid joint**—feel the joint between the 2nd metacarpal and the 2nd proximal phalanx. Extend and flex, abduct and adduct, and circumduct this joint.
 - **Saddle joint**—the thumb joint is the only saddle joint in the body. Feel the movement of trapezium bone (carpal bone) with the 1st metacarpal as you move this joint on two different axes.
 - **Ball-and-socket joint**—feel your shoulder joint as you move your arm in different motions. Make a full circle with your shoulder joint.

2 Identify the joints listed above on an articulated skeleton and note the shape of the articulating surfaces. Refer to Table 11.2.

3 Identify the bones involved in these joints with your lab group. ■

Reviewing Your Knowledge

A. Structural and Functional Classification of Joints

Fill in the blank with the most appropriate term: **cartilaginous, fibrous, synovial, amphiarthrosis, diarthrosis,** or **synarthrosis.**

_____ **1.** Functional category that has the greatest amount of movement.

_____ **2.** Functional category that has the least amount of movement.

_____ **3.** Functional category that is slightly movable.

_____ **4.** Structural category that has an articular (joint) capsule.

_____ **5.** Structural category that has cartilage joining the ends of the articulating bones.

_____ **6.** Structural category with a joint cavity.

_____ **7.** Structural category that is tightly held together by fibrous connective tissue.

B. Basic Structure of Synovial Joints

Fill in the blank with the correct answer.

_____ **1.** Name the type of cartilage that covers the articular ends of bones.

_____ **2.** Name the fluid that lubricates, reduces friction, and gives nutrition to a joint.

_____ **3.** Name the tissue at the end of a bone that reduces friction in a joint.

_____ **4.** Identify the sac-like structure in a synovial joint that is sometimes present to reduce friction.

_____ **5.** Identify the structure that secretes synovial fluid.

C. Types of Synovial Joints

There are six different types of synovial joints. Fill in the blank with the correct type.

_____ **1.** The type of synovial joint between the atlas and axis.

_____ **2.** The type of synovial joint between the humerus head and the glenoid cavity at the shoulder (between scapula and humerus).

_____ **3.** The type of synovial joint at the knee.

_____ **4.** The type of synovial joint between the scapula and clavicle.

_____ **5.** The type of synovial joint between the trapezium (carpal bone) and the 1st metacarpal.

_____ **6.** The type of synovial joint between the metacarpal and proximal phalanx.

D. Movement at Synovial Joints

Fill in the blank with the correct term for the type of movement of the synovial joint.

_____ **1.** Moves appendage toward the midline

_____ **2.** Increases the angle of a joint

_____ **3.** Downward movement, lowering the body part vertically

_____ **4.** Palm of hand faces forward or upward

_____ **5.** Pointing toes downward; raising the heel, on your tiptoes

_____ **6.** Decreases the angle of a joint

_____ **7.** Moves an appendage away from the midline

_____ **8.** Turns on a pivot with a circular motion

_____ **9.** Movement that raises mandible

_____ **10.** Move body part forward along horizontal plane

_____ **11.** Turn sole of foot inward

Using Your Knowledge

A. Synovial Joints

Identify the synovial joint structures of the synovial joint in Figure 11.9(a), (b), and (c).

SUPERIOR

LATERAL

Frontal plane

Acromion of scapula

1

(space) 2

3

MEDIAL

Supraspinatus muscle

Scapula

Subscapularis muscle

1 _____

2 _____

3 _____

INFERIOR

(a) Right shoulder joint, frontal section

SUPERIOR

Articular (joint) capsule

Sagittal plane

Subcutaneous bursa

4

Ulna

5

Trochlea of humerus

Articular capsule

Head of radius

4 _____

5 _____

POSTERIOR

ANTERIOR

INFERIOR

(b) Right elbow joint, frontal section

FIGURE 11.9 Synovial joints.

Frontal
plane

(bone marking) 6

9 (bone)

10 (bone marking)

(space) 7

Acetabular labrum

Ligament of the head
of the femur

(bone marking) 8

Articular capsule

Subcutaneous fat

LATERAL

MEDIAL

(c) Frontal section

6 _____

9 _____

7 _____

10 _____

8 _____

FIGURE 11.9 **Synovial joints,** *continued.*

B. Synovial Joint Movements

Identify the joint movements involved in walking that are described in the sentences below. Use the following terms **dorsiflex, plantar flex, flex,** and **extend** to describe the movement of the right leg.

First you flex the thigh, (11) the leg at the knee, and (12) the foot. Then you (13) the leg at the knee, (14) the thigh, and (15) the foot.

11. _____

12. _____

13. _____

14. _____

15. _____

Skeletal Muscle Structure

OBJECTIVES

1 Describe the structure of skeletal tissue and skeletal muscle fibers
2 Identify the connective tissue structures in skeletal muscle
3 Describe the structure of the sarcomere
4 Describe the structure of the neuromuscular junction

MATERIALS

- compound microscope, lens paper, prepared microscope slides of skeletal muscle tissue and neuromuscular junction
- model of 3-D skeletal muscle fiber(s)

Skeletal muscles are organs composed of skeletal muscle tissue and connective tissue. These organs also contain nerves and blood vessels. The skeletal muscle fibers within skeletal muscles contract (shorten) and cause movement of our skeleton or skin. The signal for contraction is carried by neurons that innervate each skeletal muscle fiber. We consciously control contraction of skeletal muscles, so the contraction is called *voluntary*.

A. Skeletal Muscle Tissue and Connective Tissue Coverings

Skeletal muscle fibers (cells) are striated and multinucleated. The striations are light and dark stripes along the muscle cell. A skeletal muscle fiber is actually many embryonic cells that have fused together to form one large cell with multiple nuclei.

Individual skeletal muscle fibers are surrounded by a layer of areolar connective tissue called **endomysium** (*endo-* = within; *mys* = muscle). Skeletal muscle fibers are grouped into bundles called **fascicles** that are surrounded by a layer of dense regular connective tissue called **perimysium** (*peri-* = around). A muscle is formed from a number of fascicles that are surrounded by a dense regular connective tissue layer called **epimysium** (*epi-* = on; upon). **Tendons,** connective tissues that attach the muscle to bone, are formed from endomysium, perimysium, and epimysium that extend beyond each skeletal muscle fiber.

The connective tissues surrounding skeletal muscle fibers separate and electrically insulate these cells. Therefore, electrical impulses that initiate contraction in one skeletal muscle fiber are not spread to another fiber. Recall that two or more tissues working together form an organ. Because skeletal muscle tissue and connective tissues form a skeletal muscle, each skeletal muscle can be considered an organ.

- Using the low-power objective, locate a cross-section through the muscle showing a **fascicle.** Identify the **perimysium** surrounding the fascicle and an **endomysium** surrounding a **skeletal muscle fiber.**
- Locate a section of skeletal muscle fiber in longitudinal section.
- Using the high-power objective, identify **nuclei** and **striations.**

LAB ACTIVITY 1 Skeletal Muscle Tissue and Connective Tissue Coverings

1 Examine a prepared microscope slide of skeletal muscle.

(a) Skeletal Muscle Components

(b) Fascicle Components

(a) Skeletal muscle components
- endomysium (endo-MY-zee-um) 4
- epimysium (epi-MY-zee-um) 2
- fascicle (FAS-i-kul) 3
- muscle fiber 5
- myofibril 6
- perimysium (peri-MY-zee-um) 1

1 _____
2 _____
3 _____
4 _____
5 _____
6 _____

(b) Fascicle components
- endomysium
- filament
- muscle fiber
- myofibril
- perimysium
- sarcolemma

7 _____
8 _____
9 _____
10 _____
11 _____
12 _____

FIGURE 12.1 Skeletal muscle.

(a) Cross-section of a fascicle

400×, H&E, human

- endomysium
- fascicle
- perimysium
- skeletal muscle fiber in cross-section

1 _____

2 _____

3 _____

4 _____

(b) Longitudinal section of muscle fibers

400×, H&E, human

- nucleus
- striation (stry-AY-tion)
- width of skeletal muscle fiber

5 _____

6 _____

7 _____

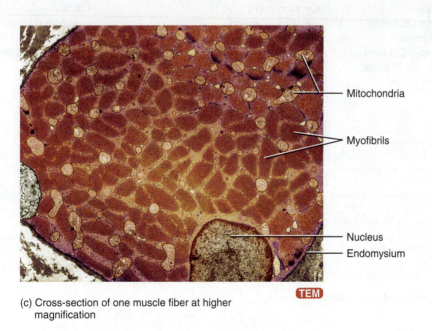

Mitochondria

Myofibrils

Nucleus

Endomysium

TEM

(c) Cross-section of one muscle fiber at higher magnification

FIGURE 12.2 **Sectional views of a fascicle and skeletal muscle fibers.**

B. Skeletal Muscle Fibers

Skeletal muscle tissue contains elongated cells called **muscle fibers.** Muscle fibers have specialized features important to their function. The plasma membrane, called the **sarcolemma** (*sarco* = flesh; *lemma* = sheath) in muscle fibers, can conduct an electrical signal called an *action potential*. The **sarcoplasmic reticulum** is endoplasmic reticulum that is specialized to store calcium ions. Muscle cells contain **thick** and **thin filaments** or **myofilaments** (*myo* = muscle) that contain different contractile proteins. Movement of the myofilaments causes muscle shortening or contraction.

The sarcolemma has tube-like invaginations called **T tubules** (transversetubules) that transmit the action potential deep inside the fiber to widened areas of the sarcoplasmic reticulum called **terminal cisternae.** The arrangement of two terminal cisternae with a T tubule between them is called a **triad.** An action potential traveling down the T tubules causes calcium to be released from the terminal cisternae into the cytoplasm of the muscle fiber.

Thick and **thin filaments,** which are bundled into **myofibrils,** occupy 80% of the volume of the fiber. Each myofibril is a chain of **sarcomeres,** the smallest structural and contractile unit of a muscle myofibril. Sarcomeres are composed of an orderly arrangement of thin and thick filaments and shorten when the muscle fiber is stimulated to contract.

Thin filaments contain actin, tropomyosin, and troponin molecules. **Actin** molecules make up the majority of the thin filament and contain a **binding site** for the myosin molecules of the thick filament. A strand of **tropomyosin** molecules is nestled between the actin molecules. Several **troponin** molecules bind at precise intervals along the tropomyosin strand.

Thick filaments are composed of **myosin** molecules. A myosin molecule resembles two golf clubs twisted together. The tails of many myosin molecules form the thick filament, whereas the **heads** project toward the thin filaments. When the myosin-binding sites on the actin molecules are uncovered, myosin heads attach and pull the thin filaments past the thick filaments. When myosin heads attach to actin, they are called **crossbridges.**

> *Before Going to Lab*
>
> **1** Label the skeletal muscle fiber in Figure 12.3.

(a) Skeletal muscle fiber
- myofibril (myo-FY-bril)
- sarcolemma (sar-co-LEM-ma)
- sarcoplasmic reticulum (sar-co-PLAZ-mic re-TIC-u-lum)
- terminal cisternae (cis-TER-nee) of sarcoplasmic reticulum
- triad
- T tubule

1 _____

2 _____

3 _____

4 _____

5 _____

6 _____

(b) Thick and thin filaments
- sarcomere
- thick filament
- thin filament

7 _____

8 _____

9 _____

(c) Contractile proteins
- actin
- myosin (MY-oh-sin) heads
- myosin tails
- tropomyosin (tro-poh-MY-o-sin)
- troponin

10 _____

11 _____

12 _____

13 _____

14 _____

FIGURE 12.3 **Skeletal muscle fiber.**

(a) Muscle fiber

Mitochondrion

Z disc 8 9 M line

(b) Thick and thin filaments

(c) Contractile proteins 10 11 12

Myosin tail 13 14 Myosin heads

FIGURE 12.3 **Skeletal muscle fiber,** *continued.*

C. The Sarcomere

The regular arrangement of thin and thick filaments in the sarcomeres gives the myofibrils and the skeletal muscle fibers light and dark bands (Figure 12.4). Lighter color bands with a dark stripe down the middle are the **I bands.** The I bands contain only thin filaments, and the stripes are **Z discs** which secure the thin filaments. **A bands** are dark-colored bands that extend the length of the thick filaments. In the A band of relaxed muscle, thin filaments overlap the thick filaments except in the middle of the A band, the **H zone,** which appears lighter. The **M line** is found in the middle of the H zone and secures the thick filaments.

Muscle fibers shorten because myosin heads on the thick filaments attach to the thin filaments and pull the thin filaments past the thick filaments toward the M line. This action increases the overlap of thin and thick filaments (Figure 12.4). As these filaments slide past each other in each sarcomere within each myofibril, the ends of the muscle fiber are brought closer together, the I bands decrease in length, and the H zone (lighter area) disappears. The increase in overlap of thin and thick filaments can be readily observed in sarcomeres of skeletal muscle.

FIGURE 12.4 **Sarcomere structure.**

Before Going to Lab

1 Label the electron micrograph of the sarcomere in Figure 12.5.

Sarcomere **TEM** 21,600×

- A band
- H zone
- I band
- M line
- Z disc

1 _____

2 _____

3 _____

4 _____

5 _____

FIGURE 12.5 **Transmission electron micrograph illustrating sarcomere structure.**

LAB ACTIVITY 2 The Sarcomere

1 Using Figure 12.6, answer Discussion Questions with your lab group.

DISCUSSION QUESTIONS
Sarcomere Structure

1 Do the lengths of the thin or thick filaments change when a muscle contracts? Explain.

2 Does the A band length change when a muscle contracts? Explain.

3 Does the I band length change when a muscle contracts? Explain.

4 Does the H zone length change when a muscle contracts? Explain.

5 Is the sarcomere in Figure 12.5 fully contracted? Explain.

(a) Relaxed muscle

(b) Partially contracted muscle

(c) Maximally contracted muscle

FIGURE 12.6 Sliding filament mechanism of muscle contraction.

D. The Neuromuscular Junction

Each skeletal muscle fiber is stimulated to contract by a **motor neuron.** A motor neuron and all the skeletal muscle fibers it innervates are a **motor unit.** Within the muscle, the axon of a motor neuron divides into many branches or **axon terminals,** each of which forms a **neuromuscular junction** with a skeletal muscle fiber. At the neuromuscular junction, each axon terminal divides into **synaptic end bulbs.** Within the synaptic end bulbs are **synaptic vesicles** filled with acetylcholine (neurotransmitter molecules). When a nerve impulse reaches the synaptic end bulb, **acetylcholine** is released and diffuses across the **synaptic cleft** (the space between the synaptic end bulb and the sarcolemma). **Acetylcholine** molecules bind to receptors in the **motor end plate,** the region of the sarcolemma directly across from the synaptic end bulb. If enough acetylcholine binds, an action potential is generated, stimulating the skeletal muscle fiber to contract.

Before Going to Lab

1 Label the neuromuscular junction in Figure 12.7.

2 Label the photomicrograph in Figure 12.8.

LAB ACTIVITY 3 Neuromuscular Junction

1 Examine a prepared microscope slide of the neuro-muscular junction and identify the structures listed in Figure 12.8.

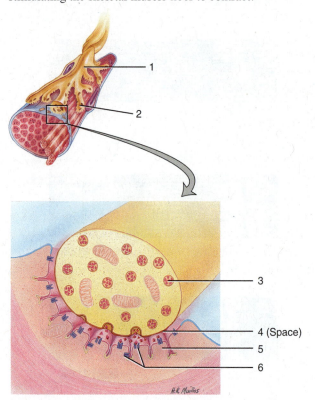

- axon terminal
- motor end plate
- receptors
- synaptic cleft
- synaptic end bulb
- synaptic vesicle with acetylcholine

1 _____

2 _____

3 _____

4 _____

5 _____

6 _____

FIGURE 12.7 Structure of the neuromuscular junction.

365×

- skeletal muscle fibers
- axon terminal with synaptic end bulbs
- motor neuron axon

1 _____

2 _____

3 _____

FIGURE 12.8 Section of skeletal muscle showing axon terminals and synaptic bulbs.

Reviewing Your Knowledge

A. Skeletal Muscle Structure

Number the following nestled, cylindrical structures in order from largest (1) to smallest (5).

____ myofibril

____ muscle fiber

____ fascicle

____ muscle

____ filaments

B. Skeletal Muscle Structural Terms

Identify the structure that matches the description.

_____ 1. Connective tissue covering surrounding a fascicle

_____ 2. Finger-like invaginations of plasma membrane; extend into interior of fiber and surround myofibrils

_____ 3. Plasma membrane of skeletal muscle fiber

_____ 4. Connective tissue covering surrounding the muscle

_____ 5. Smallest contractile unit within individual muscle fibers

_____ 6. Stores calcium within muscle fiber

_____ 7. Connective tissue covering surrounding individual muscle fibers

_____ 8. Two terminal cisternae and a T tubule

_____ 9. Rod-like structures within skeletal muscle fiber that contain thin and thick filaments organized into sarcomeres

C. Sarcomere Structure and the Sliding Filament Mechanism of Contraction

Write the name of the sarcomere structure defined. A structure may be used more than once, and a definition may apply to more than one structure.

_____ **1.** Length does not change when sarcomere shortens.

_____ **2.** This area is the length of thick filaments.

_____ **3.** Center point of attachment for thick filaments.

_____ **4.** Length decreases when sarcomere shortens.

_____ **5.** This area contains only thin filaments.

_____ **6.** Point of attachment for thin filaments.

_____ **7.** This area contains only thick filaments.

_____ **8.** This area contains overlapping thin and thick filaments.

_____ **9.** The area from Z disc to Z disc.

_____ **10.** This area disappears in a fully contracted muscle.

D. The Neuromuscular Junction

Write the name of the structure(s) described.

_____ **1.** Found in synaptic end bulbs of axon terminal; contains neurotransmitter molecules

_____ **2.** Area of sarcolemma across from synaptic end bulbs of axon terminal; contains neurotransmitter receptors

_____ **3.** Space between synaptic end bulbs of axon terminal and sarcolemma

_____ **4.** Divides into synaptic end bulbs at neuromuscular junction

_____ **5.** Parts of axon terminal that form neuromuscular junction

Using Your Knowledge

1. With age, the collagen-containing connective tissue coverings of skeletal muscles increase and the number of muscle fibers decreases. Is meat from an older or a younger animal more tender? Explain.

2. Weight training increases muscle fiber size by increasing the number of myofibrils. Explain.

3. The diaphragm is a muscle that controls inspiration. Is control of the diaphragm voluntary, involuntary, or both? Explain.

Using your textbook or another reference, for each condition below indicate which part of the neuromuscular junction is affected: the motor end plate or the axon terminal.

_____ **4.** Myasthenia gravis

_____ **5.** Curare poisoning

_____ **6.** Botulinum toxin poisoning

Put a check mark next to the muscles that are skeletal muscles. **Hint:** Consider location and whether muscle contraction is voluntary or involuntary.

_____ **7.** Arrector pili

_____ **8.** Tongue

_____ **9.** Muscle in gallbladder

_____ **10.** Muscles that control movement of the eyeball

11–13. In Figure 12.9, rank the TEMs from 1 (least) to 3 (greatest) according to number of cross-bridges formed.

TEM 21,600×

11. _____

TEM 21,600×

12. _____

TEM 21,600×

13. _____

FIGURE 12.9 **Skeletal muscle contraction.**

Contraction of Skeletal Muscle

OBJECTIVES

1 Describe the role of ATP in skeletal muscle contraction

2 Describe the three muscle fiber types and their influence on contraction

3 Identify and describe the three phases of a twitch contraction

4 Describe how skeletal muscles achieve a smooth, sustained contraction

5 Describe how skeletal muscles vary the force of contraction

6 Describe and compare isotonic and isometric contractions

7 Define threshold stimulus, maximal stimulus, motor unit recruitment, wave summation, unfused tetanus, fused tetanus, and fatigue, and explain how to observe them

MATERIALS

- **Role of ATP in Contraction of Skeletal Muscle Fibers:** ATP-glycerinated muscle kit from biological supply house and the following items per group: 1 Petri dish, 3 test tubes, marker, 2 teasing needles or straight pins, watchmaker forceps, sharp scissors, 3 microscope slides, 1 millimeter ruler, and 3 plastic transfer pipettes. Dissecting microscope, compound microscope, and coverslips are optional.

- **Control of Muscle Tension:** rulers and ankle weights (students may bring them from home).

- **PowerPhys Experiments:**
 - Twitch Contractions and Summation
 - Recruitment and Isometric and Isotonic Contractions

- **BIOPAC**® Systems, Inc. WWW.BIOPAC.COM **Biopac Laboratory Guide Experiments:**
 - Recruitment and Fatigue
 - Isometric and Isotonic Contractions

- **Record Frog Gastrocnemius Muscle Contractions:** frogs, needle probes, dissecting equipment, Ringer's solution in squeeze bottle, femur clamp, recorder, stimulator, force transducer.

Muscle cells have the ability to convert the chemical energy of ATP into mechanical energy of contraction. All muscles (skeletal, cardiac, and smooth), in turn, exert force and produce movement. The following activities focus on skeletal muscles at the molecular and cellular level of contraction.

A. Contraction of Skeletal Muscle Fibers

Three main events occur in contraction of a skeletal muscle fiber—electrical excitation of a muscle fiber, excitation-contraction coupling, and muscle fiber contraction due to the sliding filament mechanism.

- **Electrical excitation of a muscle fiber.** Skeletal muscle fibers (cells) can be stimulated either by a motor neuron in the body or by a voltage stimulator in the lab. Stimulation given by either method results in a depolarization of the sarcolemma. If the depolarization reaches threshold, an action potential (electrical signal) is initiated.

- **Excitation-contraction coupling.** The action potential is transmitted along the sarcolemma and down the T tubules (transverse tubules). This action causes calcium ions to be released from the terminal cisternae of the sarcoplasmic reticulum. Calcium ions couple electrical excitation to muscle fiber contraction by binding to troponin, which is attached to the actin filament and tropomyosin. Troponin changes shape and pulls tropomyosin away from the myosin-binding sites on the actin filament.

- **Muscle fiber contraction.** A muscle fiber contracts (shortens) when thin filaments (actin) slide past the thick filaments (myosin). Each contraction cycle shortens each muscle fiber about 1% of its resting length. The 4 steps of the contraction cycle are:

 1. **ATP hydrolysis.** Myosin heads contain an ATP binding site and an ATPase. When ATP binds to the ATP-binding site, the ATPase hydrolyzes ATP to form ADP and inorganic phosphate. Hydrolysis of ATP energizes the myosin head.

 2. **Attachment of myosin to actin to form crossbridges.** Energized myosin heads bind to the unblocked myosin-binding site on actin and the inorganic phosphate is released from the myosin head.

 3. **Power stroke.** The release of the inorganic phosphate starts the power stroke, which is the rotation of the myosin head that pulls the thin filament toward the center of the sarcomere. During the power stroke, ADP is released from the myosin head, but the myosin head remains attached.

 4. **Detachment of myosin from actin.** Another ATP molecule binds to the myosin ATP-binding pocket, releasing the myosin head from actin, and the contraction cycle begins again.

The contraction cycle continues as long as intracellular calcium levels remain high. As the intracellular calcium levels drop, tropomyosin blocking of the myosin-binding sites on actin returns, energized myosin is prevented from binding, and the muscle fiber relaxes.

The amount of shortening that occurs when a muscle is stimulated to contract depends on how long the contraction cycle continues. With each contraction cycle, sarcomeres shorten a little more until maximal contraction of a sarcomere is reached. Muscle fibers can shorten up to 40% of their resting length.

Exposure of skeletal muscle fibers to glycerin creates holes in the sarcolemma allowing ions and ATP to diffuse into and out of the muscle fiber. **Glycerination** also disrupts the troponin-tropomyosin complex so that *calcium is not needed* to bind to troponin and pull tropomyosin to unblock the myosin-binding sites on the actin molecules. Although calcium is not needed for contraction of glycerinated muscle fibers, other ions are needed to ensure the proper functioning of enzymes. In this activity, you will observe contraction in glycerinated skeletal muscle fibers.

LAB ACTIVITY 1 Experiment: Role of ATP in Contraction of Skeletal Muscle Fibers

1 **Prediction:** With your lab group, predict which solution will cause muscle fiber shortening by selecting one of the choices below.
 - 0.25% ATP in distilled water
 - 0.25% ATP, 0.05 M KCl, and 0.001 M $MgCl_2$ in distilled water
 - 0.05 M KCl and 0.001 M $MgCl_2$ in distilled water, no ATP

2 **Materials:** Obtain materials for Role of ATP in Contraction of Skeletal Muscle Fibers.

3 **Data Collection:** Measure muscle length of glycerinated skeletal muscle fibers before and after exposure to the different solutions.
 - Decide who will mix the solutions; prepare the muscle strands; apply the solutions to the muscle strands; and time, measure, and record.
 - Label test tubes and microscope slides 1, 2, and 3.
 - Place the following in the appropriate test tube:

 Test tube 1 (0.25% ATP in distilled water)— 5 drops from dropper bottle labeled 0.25% ATP in distilled water
 Test tube 2 (0.25% ATP in salt solution)— 5 drops from dropper bottle labeled 0.25% ATP, 0.05 M KCl, and 0.001 M $MgCl_2$ in distilled water
 Test tube 3 (salt solution, no ATP)—5 drops from dropper bottle labeled 0.05 M KCl and 0.001 M $MgCl_2$ in distilled water

 - Obtain a Petri dish containing a 2-cm-long piece of glycerinated skeletal muscle in a small amount of glycerin from your instructor.
 - Using teasing needles or straight pins, separate the skeletal muscle (in the Petri dish) into at least 9 strands not more than 0.2 mm in diameter (2–4 muscle fibers per strand).
 - Using forceps, place 3 (of the 9) thin strands of muscle fibers on a microscope slide. Arrange the strands so they are straight and parallel. Do not cover the strands with a cover slip. *Note:* The amount of glycerol that is transferred with the strands should be enough to keep them moist. Add a small drop of

glycerol only if the strands are exposed to heat from the microscope lamp for a sustained period.

- Measure the length of each muscle strand with a millimeter ruler and record the value in Table 13.1. Measurements can be made using a dissecting microscope, if desired.
- Using a clean transfer pipette, transfer all the solution from test tube 1 to the microscope slide and measure the length of the strands after 40 seconds. Record the results in Table 13.1.
- Repeat the above steps for the solutions in test tubes 2 and 3 using 3 new muscle strands, transfer pipettes, and microscope slides for each solution. Record the results in Table 13.1.
- *Optional—Save all 3 slides to observe striations. See instructions below.*

4 Observation of Striations in Skeletal Muscle Fibers (optional):
- Place a coverslip over each microscope slide.
- Using a compound microscope, observe one microscopic slide with the low-power objective lens and note the striations on each strand.

- Switch to the high-dry objective lens. Identify the light-colored I bands and the dark-colored A bands.
- Compare the distance between the bands.
- Repeat for each slide.

5 Clean up as directed by your instructor.

6 Data Analysis:
- Calculate the percentage of contraction by dividing the length after exposure to the solution by the resting length.
- Average the values for each solution and record in Table 13.1.

7 Complete the Experimental Report with your lab group.

TABLE 13.1 Role of ATP in Contraction of Skeletal Muscle Fibers

	TEST TUBE 1 0.25% ATP DISTILLED WATER	TEST TUBE 2 0.25% ATP 0.05 M KCl 0.001 M MgCl$_2$	TEST TUBE 3 0 ATP 0.05 M KCl 0.001 M MgCl$_2$
Muscle Strand 1			
Length before solution			
Length after solution			
% Contraction			
Muscle Strand 2			
Length before solution			
Length after solution			
% Contraction			
Muscle Strand 3			
Length before solution			
Length after solution			
% Contraction			
Average % contraction			

EXPERIMENTAL REPORT
Role of ATP in Contraction of Skeletal Muscle Fibers

Results:
- Name the solution(s) that caused muscle strand contraction.

- State the average percentage contraction for each solution.

Discussion:
- Estimate the number of contraction cycles that occurred in the contracted muscle fibers.

- Discuss whether the sarcomere length increased, decreased, or stayed the same in muscle fibers from each solution.

- Explain what limits the amount of shortening that is observed in the skeletal muscle fibers.

- Discuss why calcium was not needed to cause muscle contraction in glycerinated muscle fibers.

- Discuss the role of KCl and $MgCl_2$ in the experiment.

Conclusion:
- State how adding ATP to the glycerinated muscle fibers affected muscle fiber and sarcomere length.

B. Influence of Muscle Fiber Type on Skeletal Muscle Contraction

There are 3 types of skeletal muscle fibers: **slow oxidative (SO), fast oxidative-glycolytic (FOG),** and **fast glycolytic (FG).** These muscle fiber types differ in diameter, myosin ATPase, and methods of ATP production.

1. Muscle Diameter and Force of Contraction

Fast glycolytic fibers have the largest diameter of the muscle fiber types and develop more force than slow oxidative fibers that have the smallest diameter of the three muscle types. Muscle fibers with a larger diameter develop more force because they have more myofibrils, and therefore, more myosin heads that can attach to actin. The amount of force a muscle fiber develops is dependent on the number of myosin heads attached to actin.

2. Myosin ATPase and Speed of Contraction

Fast glycolytic fibers and fast oxidative-glycolytic fibers have a myosin ATPase that breaks down ATP faster than the myosin ATPase in slow oxidative fibers. The speed at which the contraction cycle can occur is determined by how fast myosin ATPase can break down ATP. The faster the myosin ATPase, the greater the speed at which the contraction cycle can occur, and the faster the muscle can contract.

3. Methods of ATP Generation and Fatigability

Fibers with fast myosin ATPase need ATP to be produced quickly to achieve a fast contraction cycle. Fast glycolytic fibers produce the majority of ATP by glycolysis, while slow oxidative fibers use aerobic cellular respiration to produce ATP. Fast oxidative-glycolytic fibers use both ATP production methods.

Glycolysis produces ATP quickly but cannot produce large amounts of ATP over time. Therefore, muscle fibers that primarily use glycolysis fatigue quickly. Glycolysis requires large glycogen (glucose source) stores, and muscle fibers that primarily use glycolysis to make ATP are paler due to lower myoglobin content and less blood supply.

Aerobic cellular respiration produces ATP slowly but can produce large amounts of ATP over time if there is a sufficient blood supply and myoglobin (red pigment that stores oxygen) stores. Therefore, muscle fibers that use aerobic respiration take longer to fatigue. These muscle

fibers have lower glycogen stores, are darker due to higher myoglobin content, and have a greater blood supply. These fibers also have greater numbers of mitochondria because the enzymes for aerobic cellular respiration are located in the mitochondria.

4. Recruitment of Muscle Fiber Types

Skeletal muscles contain all 3 fiber types. The proportion of muscle fiber types in a muscle type differs depending on muscle action. Slow oxidative fibers are stimulated first. If more force is needed, then the fast oxidative-glycolytic fibers are stimulated and finally the fast glycolytic fibers.

Before Going to Lab

1 Label the muscle fiber types in Figure 13.1. This slide has been stained so that the different muscle fiber types can be easily identified.

LAB ACTIVITY 2 **Influence of Muscle Fiber Type on Skeletal Muscle Contraction**

1 Observe a cross-section of skeletal muscle. Are all the muscle fibers the same diameter?

2 Answer the Discussion Questions with your lab group.

LM 440×

- fast glycolytic fiber
- fast oxidative-glycolytic fiber
- slow oxidative fiber

1 _____

2 _____

3 _____

FIGURE 13.1 **Cross-section of skeletal muscle showing all three fiber types.**

DISCUSSION QUESTIONS
Influence of Muscle Fiber Type on Skeletal Muscle Contraction

1 Discuss why fast glycolytic fibers develop more force than slow oxidative fibers.

2 Discuss why fast glycolytic fibers have a greater contraction velocity than slow oxidative fibers.

3 Discuss why slow oxidative fibers are more fatigue-resistant than fast glycolytic fibers.

4 Discuss why slow oxidative fibers have more capillaries and mitochondria than fast glycolytic fibers.

5 Discuss why fast glycolytic fibers have more glycogen stores than slow oxidative fibers.

6 Weight training causes fast glycolytic fibers to hypertrophy (increase in diameter). Explain why this increases strength.

7 Identify which muscle fiber types would be recruited for each of the following functions. Explain your answer.
a. walking

b. standing

c. lifting a heavy object and immediately putting it down

C. Control of Muscle Tension

Muscle contraction results in development of tension or force, usually measured in grams. The muscle tension developed by a contracting muscle is determined by the frequency of stimulation of motor units and the number of motor units stimulated.

1. Frequency of Stimulation and Muscle Tension

The minimal stimulus that results in a muscle twitch is called the **threshold stimulus.** A threshold stimulus applied to a motor nerve or directly to the muscle results in a single contraction called a **twitch contraction.** The amount of force developed by this threshold stimulus is dependent on the number of motor units that are stimulated to contract.

The 3 phases of a twitch contraction are the latent period, the contraction period, and the relaxation period. The **latent period** lasts about 2–10 msec (milliseconds) and is the time between stimulation of muscle cells and force generation. The **contraction period** lasts about 10–100 msec and is the period during which force (measured in grams) is increasing, whereas the **relaxation period,** which lasts 10–100 msec, is the period when force is decreasing. Normal muscle contractions are not twitch contractions but are sustained contractions of varying force.

If the contracting motor unit(s) is (are) stimulated again before the relaxation phase of a muscle twitch is complete, then the next contraction will produce a greater force or tension. This is called **wave summation.** Increasing the frequency of muscle stimulation produces sustained force generation. **Unfused tetanus** (*tetan* = rigid) occurs when there is a partial relaxation between muscle twitches. **Fused tetanus** is a sustained contraction with no relaxation observed between twitches. Most sustained voluntary skeletal muscle contractions are unfused tetanic contractions with different motor units stimulated at different times **(asynchronous contractions).** The asynchronous contractions delay **muscle fatigue,** which is an inability to contract caused by long periods of muscle contraction.

2. Number of Motor Units Stimulated and Muscle Tension

Increasing the number of motor units contracting at the same time, **motor unit recruitment,** also increases force generated. Lifting a feather requires fewer motor units than lifting your anatomy and physiology textbook. Slow oxidative motor units are recruited first, and if more force is needed, fast oxidative-glycolytic motor units are also recruited. If maximal force is required, fast glycolytic motor units are recruited.

Maximal force development occurs when all motor units of a muscle are stimulated and all muscle fibers are contracting. In the lab, the stimulus that produces maximal force is called the **maximal stimulus.** Therefore, a stimulus to the muscle greater than maximal does not produce a greater force.

LAB ACTIVITY 3 Control of Muscle Tension

1 Draw a twitch contraction that has a latent period of 2 msec, a contractile period of 10 msec with a maximum force of 2 g, and a relaxation period of 10 msec. Use the graph in Figure 13.2. Add an arrow for the stimulus.

2 Observe unfused tetanus, motor unit recruitment, and fatigue in the muscles that cause bending of the knee.
- In your lab group, decide who will be the subject, observer, and recorder.
- Have the subject stand while holding onto the lab bench for support. Have subject bend the left leg to a 90-degree angle and hold this position. The hamstring muscles on the posterior thigh are used to bend the leg.
- Observe the unfused tetanic contraction. Start timing.
- Time how long it takes for the muscle to fatigue, which is demonstrated by any vertical movement in the leg.
- Time until fatigue: _____ min.
- Place a 5-lb weight on the subject's right ankle and repeat the steps.
- Time until fatigue: _____ min.

3 Answer the Discussion Questions with your lab group.

4 Complete **PowerPhys** Experiment: Twitch Contractions and Summation.

5 Complete the **Biopac Laboratory Guide** Experiment: Recruitment and Fatigue.

FIGURE 13.2 **Student drawing of twitch contraction.**

DISCUSSION QUESTIONS
Control of Muscle Tension

1 Explain why you can maintain contraction of the hamstring muscles over time.

2 Explain why you can sustain the same contraction with a 5-lb weight attached to the ankle.

3 Explain why the hamstring muscles fatigue faster with the 5-lb ankle weight.

4 State the order of recruitment of muscle fiber types.

D. Isotonic and Isometric Contractions

Muscle contraction results in development of tension or force usually measured in grams. Muscles will shorten if they develop more force than the force that is opposing them. For example, contracting muscles in our arm will shorten and allow us to lift a book if the force developed by the muscles is greater than the weight (force) of the book. The weight of the book is also called *load* because it is a force against which the muscle is contracting.

If a muscle is generating a constant force to move a load, that contraction is called an **isotonic contraction.** There are two types of isotonic contraction: concentric and eccentric isotonic contractions. In **concentric isotonic contraction,** the muscle is shortening while it is contracting. An example of this is using the biceps brachii (large anterior muscle of arm) to lift a book off the table. In **eccentric isotonic contraction,** the muscle is lengthening while it is contracting. This occurs when you slowly lower your arm to return the book to the table. The biceps brachii is still contracting, but it is lengthening while it is contracting. This enables you to lower the book in a controlled manner.

Isometric contractions are contractions in which the muscle is developing force but not shortening and no visible movement is seen. In this case, the force developed by the muscle equals the force (load) it is contracting against. An example of this would be holding a book in the same position. Isometric muscle contractions are maintaining the position against the weight (load) of the book, but the biceps

brachii muscle is not moving the book because it is not generating a force greater than the weight (force) of the book.

LAB ACTIVITY 4 **Isotonic and Isometric Contractions**

1 Recalling Lab Activity 3, in which the subject bent a knee and held it at 90 degrees, state which action was an isotonic contraction and which was an isometric contraction.

2 Is the isotonic contraction observed when the hamstrings contracted a concentric isotonic contraction or an eccentric isotonic contraction? What would you ask the subject to do to observe the other type of isotonic contraction?

3 Complete **PowerPhys** Experiment: Recruitment and Isometric and Isotonic Contractions

4 Complete the **Biopac Laboratory Guide** Experiment: Isometric and Isotonic Contractions.

E. Recording Force Generated by Skeletal Muscle

Polygraphs amplify and record physiological changes that are detected by various sensors. Examples of **sensors** include those that detect changes in pH, force, temperature, and pressure. A **myograph** is a type of polygraph used to record the force generated by muscle contractions along with the following equipment:

1. A **stimulator** stimulates skeletal muscle fibers to contract, and the stimulus strength is measured in volts. The **frequency of stimulation** is the number of stimuli per second. Increasing the frequency of stimulation increases the frequency of contractions.

2. A **transducer,** which is a sensor that converts the force generated by a contracting muscle into an electric signal (current), is attached to the muscle.

3. An **amplifier** increases the amplitude of the transducer signal and transmits it to a recorder channel.

4. The **recorder** uses the electric signal from the transducer to move a pen across a sheet of paper that is moving at preset speed. This allows students to observe if the force changes over time. Computers can also act as recorders and display the pen deflections as lines moving across the computer screen. The visual displays can be saved and printed.

Recorders can have more than one **channel,** allowing several different measurements to be recorded at one time. For example, an electrode can be attached to the muscle and channel 1 of a recorder to record the electric signal or action potential of a stimulated muscle while a force transducer connected to channel 2 records the force of contraction (myogram). Time and event markers of recorders are pens that mark the time when events occur, such as stimulation of a muscle.

To calculate the force recorded from a contracting muscle, the transducer must be calibrated before it is attached to the muscle. Different gram weights are attached to the transducer to observe the electric signal or pen deflection for each weight (or force). Pen deflection is adjusted so that each gram of force will result in a pen deflection of a known length.

Figure 13.3 illustrates the use of equipment to record a **myogram** (recording of force generated by a frog gastrocnemius muscle).

FIGURE 13.3 **Apparatus used to measure force generated in frog gastrocnemius muscle.**

LAB ACTIVITY 5 Record Frog Gastrocnemius Muscle Contraction

1 Measure the force generated in grams and the length of the latent, contraction, and relaxation periods in milliseconds on the myogram in Figure 13.4 and record your results on the figure.

2 With your group, discuss how you would measure threshold stimulus, maximal stimulus, and maximal force in the frog gastrocnemius preparation.

3 Record frog gastrocnemius muscle contractions. Follow the instructions supplied by your instructor to prepare and conduct the experiment.

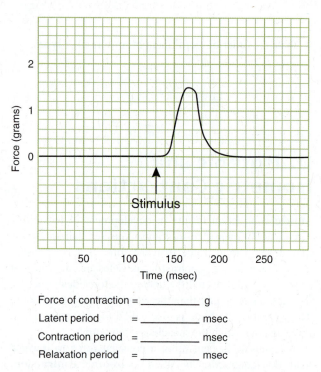

Force of contraction = _____ g

Latent period = _____ msec

Contraction period = _____ msec

Relaxation period = _____ msec

FIGURE 13.4 **Myogram of frog gastrocnemius muscle contraction.**

Reviewing Your Knowledge

A. Role of ATP in Contraction of Skeletal Muscle Fibers

Write T for true or F for false for the following statements.

_____ 1. ATP causes the detachment of myosin from actin.

_____ 2. Glycerinated skeletal muscle fibers need calcium in order to contract.

_____ 3. Glycerinated skeletal muscle fibers need ATP in order to contract.

B. Influence of Muscle Fiber Type on Skeletal Muscle Contraction

Name the fiber type—slow oxidative (SO), fast oxidative-glycolytic (FOG), or fast glycolytic (FG)—that matches each fiber characteristic. Questions may have more than one answer.

_____ 1. largest diameter

_____ 2. is pink in color

_____ 3. myoglobin content is low

_____ 4. has the fastest contraction velocity

_____ 5. has many capillaries

_____ 6. is red in color

_____ 7. generates ATP by aerobic respiration

_____ 8. has the slowest myosin ATPase

_____ 9. has glycogen stores

_____ 10. smallest diameter

_____ 11. recruited first

_____ 12. recruited last

C. Control of Muscle Tension

Name the term that matches each definition.

_____ 1. Inability of muscle fibers to contract after a long period of contraction

_____ 2. Time between muscle fiber stimulation and measurement of force generation

_____ 3. Motor neuron and all the muscle fibers it innervates

_____ 4. Period of force generation

_____ 5. Type of wave summation with partial relaxation observed between twitches

_____ 6. Increasing the number of motor units that are stimulated to contract

_____ 7. Stimulus that results in maximal force generation

_____ 8. Lowest stimulus that results in force generation

_____ 9. Single contractile event in response to single action potential

_____10. Period during which more crossbridges detach than reattach to thin filaments

_____11. Type of wave summation with no observable relaxation between twitches

D. Isotonic and Isometric Contractions

Identify the following contractions of the biceps brachii muscle (large anterior muscle of forearm) as concentric isotonic, eccentric isotonic, or isometric.

_____ 1. Lifting a glass off the table

_____ 2. Holding a glass in the same position

_____ 3. Lowering a glass to the table

E. Recording Force Generated by Skeletal Muscle Contractions

Name the equipment described.

_____ 1. Increases the amplitude of the transducer signal and transmits it to a recorder channel

_____ 2. Makes skeletal muscle fibers contract

_____ 3. Machine that amplifies and records physiological changes detected by sensors

_____ 4. Uses the electric signal from a transducer to make a tracing on moving paper

_____ 5. Machine that records force generated by muscle contractions

_____ 6. Sensor that converts force generated by a contracting muscle into an electric signal

Using Your Knowledge

Answer the following questions with a short answer.

1. Explain why muscles are stiff (contracted) when rigor mortis occurs.

2. Large muscles, such as the muscles of the leg, have more muscle fibers than small muscles, such as the muscles of the finger. Explain why the muscles of the finger cannot develop as much force as the muscles of the leg.

3. Which type of muscle would fatigue faster, one that has many blood vessels or one that has fewer blood vessels? Explain.

Using your textbook or another reference, for each contraction below indicate whether it is an example of a muscle twitch, unfused tetanus, or fused tetanus.

_____ **4.** muscle spasm

_____ **5.** facial muscle tic

_____ **6.** cardiac fibrillation

_____ **7.** smiling

The percentage of each muscle fiber type in any given muscle of your body is genetically determined. Also, physical activity can cause slight changes in muscle fiber type. Discuss the relative amount of each muscle type in the thigh muscles of Olympic athletes participating in the following events:

8. marathon

9. weight lifting

10. 100-meter dash

Skeletal Muscles and Their Actions

There are approximately 700 skeletal muscles in the human body, and most of these originate and/or insert on an area of the skeleton. Because most muscles or their tendons cross a joint, muscle contractions cause skeletal movement at the joint.

A. Criteria for Naming Skeletal Muscle

Learning the criteria used to name skeletal muscles will assist you in identifying and learning the function of these muscles. Skeletal muscles are named for their orientation relative to the midline of the body, size, shape, action, number of origins, location, or origin and insertion. The muscle attachment points, the origin and insertion, are usually skin or bone. The **origin** is the nonmoving point of attachment when a muscle contracts, and the **insertion** point moves toward the origin. The larger region of the muscle between the origin and the insertion is called the **belly.** For characteristics used to name skeletal muscles, refer to your textbook.

B. Skeletal Muscle Identification and Action

When many skeletal muscles contract, they pull their point of insertion on a bone toward their point of origin. For the muscles of facial expression, this action moves the skin on the face and allows us to express emotion. In the appendages, muscles that perform similar functions are surrounded by fascia, forming **compartments** that are served by the same major blood vessels and nerves. Also, within a location some muscles are grouped according to whether they are **superficial** or **deep.**

Sometimes muscles will work in groups when they are involved in multifaceted movements. Smaller muscles assist or enhance the action of larger muscles. Based on their function in a group, these muscles can be categorized as: an **agonist** (prime mover), an **antagonist** (working against the agonist), or a **synergist** (helps the agonist).

1. Muscles of the Head and Neck

Muscles of Facial Expression and Mastication

Muscles located in the head include those involved in facial expression, mastication, blinking, and eye movements. Their origins are found in the superficial fascia beneath the skin and attached to skull bones, while their insertions are in the skin of the face. Muscles that insert on the exterior of the eyeball and move the eye will be studied in the exercise on Special Senses—The Eye.

Muscles That Move the Head and Neck

Muscles that flex the head and neck are found in pairs on the anterior and lateral surface of the neck, while muscles that extend the head and neck are found on the posterior surface of the neck. However, the levator scapulae, a posterolateral neck muscle, elevates the scapula but does not move the neck.

The **sternocleidomastoid muscle,** a muscle that runs diagonally from the sternum and the clavicle to the mastoid process of the temporal bone, divides each side of the neck into an anterior and posterior triangle. The **anterior triangle** includes the carotid artery and the internal jugular vein, while the **posterior triangle** includes the external jugular vein and part of the subclavian artery.

Muscles That Move the Hyoid Bone and Larynx

Muscles that move the hyoid bone and larynx include the suprahyoid muscles and the infrahyoid muscles. The **suprahyoid muscles** are located superior to the hyoid bone and elevate the hyoid bone, floor of the oral cavity, and tongue while swallowing. The **infrahyoid muscles** are located inferior to the hyoid bone and depress the hyoid bone and larynx while swallowing and talking.

Before Going to Lab

1 Label Figure 14.1(a), (b), and (c) and Figure 14.2(a), (b), and (c).

LAB ACTIVITY 1 Muscles of the Head and Neck

1 Use Figures 14.1 and 14.2 and Table 14.1 to identify the muscles on models, diagrams, cadaver photographs or use the search text box in **Real Anatomy** (Muscular) to find these structures.

2 Locate the superficial muscles on yourself and perform the action indicated in Table 14.1. Palpate each muscle while performing the action.

3 Use Appendix B to locate the origin and insertion on an articulated skeleton. Pull the insertion toward the origin to simulate muscle action. ■

TABLE 14.1 Muscles of the Head and Neck

MUSCLE	LOCATION AND *ACTION*
MUSCLES OF FACIAL EXPRESSION	
Occipitofrontalis muscle	Combination of frontalis and occipitalis muscles connected by an aponeurosis (a flat, broad tendon).
Frontal belly	Lies over frontal bone. _Raises eyebrows and wrinkles forehead._
Occipital belly	Lies over occipital bone. _Pulls scalp posteriorly._
Orbicularis oculi (*orbicularis* = little circle; *oculi* = eye)	Circular muscle that encircles the eye. _Closes eye._
Zygomaticus major (*zygoma* = bar)	Between zygomatic bone and corner of mouth; inferior to zygomatic minor. _Raises corners of mouth (smiling)._
Zygomaticus minor	Between zygomatic bone and corner of mouth. _Raises upper lips, exposing upper teeth._
Orbicularis oris (*oris* = mouth)	Circular muscle that encircles the mouth. _Closes and purses lips._
Platysma (*platy* = flat)	Wide, flat muscle that covers lower mandible and entire anterior neck, and ends on chest. _Tenses neck skin; depresses mandible (pouting)._

- frontalis (fron-TA-lis)
- orbicularis oculi (or-bi-kyoo-LAR-is OC-yoo-lie)
- orbicularis oris (OR-is)
- platysma (pla-TIZ-ma)
- zygomaticus (zy-go-MAH-tu-kus) major
- zygomaticus minor

1 _____

2 _____

3 _____

4 _____

5 _____

6 _____

(a) Anterior superficial view

FIGURE 14.1 **Selected muscles of the head.**

TABLE 14.1 Muscles of the Head and Neck (*continued*)

MUSCLE	LOCATION AND *ACTION*
MUSCLES OF MASTICATION	
Superficial Muscles	
Temporalis (*tempora* = temples)	Lies over temporal bone. *Elevates and retracts mandible.*
Masseter (*maseter* = chewer)	Anterior to ear between zygomatic arch and posterior portion of mandible. *Elevates and retracts mandible.*
Deep Muscles	
Buccinator (*bucca* = cheek)	Deep to masseter. Fibers run transversely and form fleshy part of cheek. *Presses cheeks inward to whistle, blow, and suck. Helps hold food between teeth while chewing.*
Lateral pterygoid (*pterygoid* = like a wing)	Deep to masseter and superior to medial pterygoid. Transverse muscle fibers between pterygoid process of sphenoid and mandible. *Protracts mandible, depresses mandible (opening mouth), and moves it sideways.*
Medial pterygoid	Deep to masseter and inferior to lateral pterygoid. Vertical muscle fibers between pterygoid process of sphenoid and mandible. *Elevates and protrudes mandible and moves it sideways.*

(b) Right lateral superficial view

- buccinator
- frontalis
- masseter

- occipitalis (ok-si-pi-TA-lis)
- orbicularis oculi
- orbicularis oris
- zygomaticus major

7 _____

8 _____

9 _____

10 _____

11 _____

12 _____

13 _____

FIGURE 14.1 Selected muscles of the head, *continued.*

Temporalis

Lateral pterygoid

Medial pterygoid

Buccinator

Orbicularis oris

DANK

14

Mandible (cut)

15

16

17

18

- buccinator
- lateral pterygoid (TER-ih-goid)
- medial pterygoid
- orbicularis oris
- temporalis (tem-por-A-lis)

14 _____

15 _____

16 _____

17 _____

18 _____

(c) Right lateral deep view

FIGURE 14.1 Selected muscles of the head, *continued.*

TABLE 14.1 **Muscles of the Head and Neck (*continued*)**

MUSCLE	LOCATION AND *ACTION*
MUSCLES THAT MOVE THE HEAD AND NECK	
Anterior and Lateral Muscles	
Sternocleidomastoid (*sterno-* = breastbone; *cleido* = clavicle; *mastoid* = mastoid process)	Strap-like muscle on anterior and lateral neck. Fibers run diagonally across the neck from the sternum and clavicle to the mastoid process. *Bilateral contraction flexes head (prayer muscle). Unilateral contraction laterally flexes head and rotates head to opposite side (as in saying no).*
Scalenes (*scalenos* = uneven)	Three muscles—anterior, middle, and posterior scalenes—located in lateral neck deep to sternocleidomastoid muscle. *Bilateral contraction flexes head and elevates 1st and 2nd ribs during deep inspiration. Unilateral contraction laterally flexes head and rotates head to opposite side (as in saying no).*
Posterior Muscles	
Splenius capitis	Posterior neck deep to trapezius. *Extends head when both muscles contract. Laterally flexes and rotates head to same side as contracting muscle when only one muscle contracts.*
Trapezius (superior portion)	Superficial muscle of the upper back and posterior neck. *Extends head and elevates scapula.*
MUSCLE THAT MOVES ONLY THE SCAPULA	
Levator scapulae (levare = to raise)	Posterior to sternocleidomastoid; runs diagonally from posterior head to scapula. *Elevates scapula and rotates it downward; stabilizes scapula.*
MUSCLES THAT MOVE HYOID BONE	
Suprahyoid Muscles	
Digastric (*di-* = two; *gastric* = belly)	
Anterior belly	Strap-like muscle under chin that is parallel to midline of chin and extends from posterior belly. *Elevates hyoid bone and depresses mandible (as in opening mouth).*
Posterior belly	Runs along posterior border of chin. *Elevates hyoid bone and depresses mandible (as in opening mouth).*
Stylohyoid	Chin muscle, medial to posterior belly of digastric. *Elevates hyoid bone and moves it posteriorly.*
Mylohyoid	Deep muscle of chin extending from right side of mandible to left side. *Elevates hyoid bone and floor of oral cavity and depresses mandible.*
Infrahyoid Muscles	
Omohyoid	Lateral to sternohyoid. *Depresses hyoid bone.*
Sternohyoid (*sterno* = sternum)	Runs along midline of anterior neck. *Depresses hyoid bone.*

Sternocleidomastoid

Splenius capitis

Trapezius

Levator scapulae

Platysma

Scalene

Epicranial aponeurosis

Occipitofrontalis
(occipital belly)

Occipitofrontalis
(frontal belly)

Orbicularis oculi

Nasalis

Levator labii superioris
Zygomaticus major

Buccinator
Orbicularis oris

Thyroid cartilage

Sternohyoid

1
2

3
4
5

• levator scapulae
 (lee-VAY-tor SKA-
 pyoo-lee)
• scalene
• splenius capitis
• sternocleidomastoid
 (ster-no-kli-do-MAS-toid)
• trapezius (tra-PEE-
 zee-us)

1 _____

2 _____

3 _____

4 _____

5 _____

(a) Lateral superficial view

FIGURE 14.2 **Selected muscles of the chin and neck.**

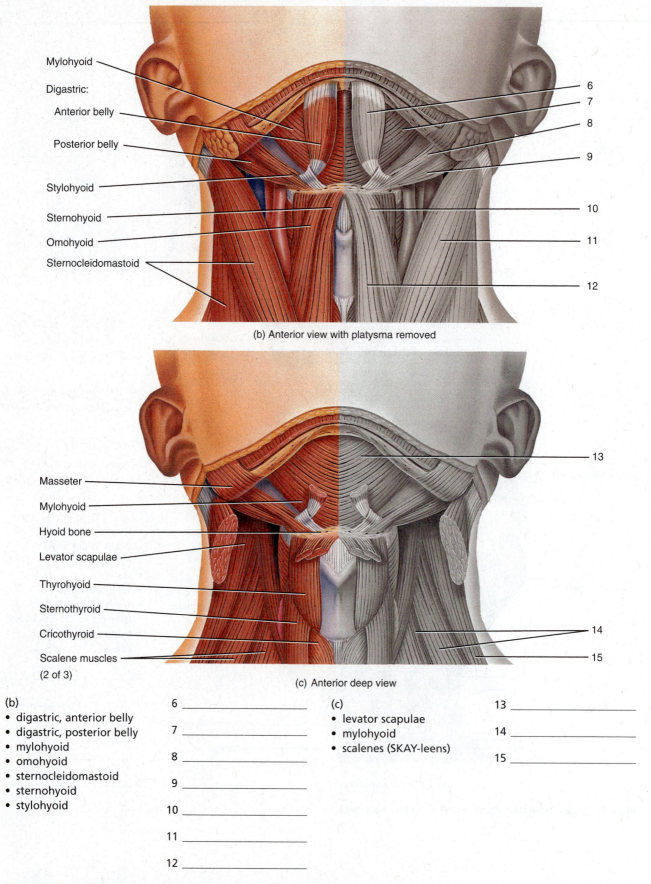

Mylohyoid

Digastric:

Anterior belly

Posterior belly

Stylohyoid

Sternohyoid

Omohyoid

Sternocleidomastoid

6

7

8

9

10

11

12

(b) Anterior view with platysma removed

Masseter

Mylohyoid

Hyoid bone

Levator scapulae

Thyrohyoid

Sternothyroid

Cricothyroid

Scalene muscles
(2 of 3)

13

14

15

(c) Anterior deep view

(b)
- digastric, anterior belly
- digastric, posterior belly
- mylohyoid
- omohyoid
- sternocleidomastoid
- sternohyoid
- stylohyoid

6 _____

7 _____

8 _____

9 _____

10 _____

11 _____

12 _____

(c)
- levator scapulae
- mylohyoid
- scalenes (SKAY-leens)

13 _____

14 _____

15 _____

FIGURE 14.2 **Selected muscles of the chin and neck,** *continued.*

2. Muscles of the Trunk

Muscles That Move the Arm

These muscles originate on the anterior or posterior trunk and insert on the humerus. Muscle contraction pulls the humerus toward the insertion, resulting in movement of the humerus at the shoulder joint. Because of the structure of the shoulder joint and the way these muscles surround the humerus, contraction can cause a full range of motion at a synovial joint—flexion, extension, abduction, adduction, circumduction, and rotation.

The rotator cuff muscles are deep muscles that move the arm and surround the shoulder joint like a sleeve cuff and help strengthen and stabilize the shoulder joint. They include the **subscapularis, supraspinatus, infraspinatus,** and **teres minor.**

Muscles That Move the Scapula

Movements of the scapula include elevation, depression, abduction (protraction), adduction (retraction), and rotation. Shrugging the shoulders or lifting an object over the head elevates the scapula. Depression of the scapula occurs when doing a pullup. Performing a "pushup" abducts the scapula, and standing at attention adducts the scapula.

Muscles That Move the Abdominal Wall and Vertebral Column

Muscles on the anterior surface of the abdomen (abdominal muscles) flex the vertebral column, while the muscles of the back extend it. However, because the transversus abdominis muscle runs horizontally it cannot flex the vertebral column.

Muscles Used in Breathing

Muscles of inspiration increase the diameter or length of the thoracic cavity when they contract, while muscles of expiration decrease the diameter or length of the thoracic cavity. In this activity we will be examining the diaphragm, external intercostals, and internal intercostals. All muscles involved in breathing will be examined together when we study the respiratory system.

Before Going to Lab

1 Label Figure 14.3(a), (b), (c), (d), and (e) and Figure 14.4(a) and (b).

LAB ACTIVITY 2 Muscles of the Trunk

1 Use Figures 14.3 and 14.4 and Tables 14.2 and 14.3 to identify the muscles on models, charts, cadaver photographs or use the search text box in **Real Anatomy** (Muscular) to find these structures.

2 Locate each muscle on yourself and perform the action indicated in Tables 14.2 and 14.3. Palpate each muscle while performing the action.

3 Identify the rotator cuff muscles on Figures 14.3 and 14.4.

4 Use Appendix B to locate the origin and insertion on an articulated skeleton. Pull the insertion toward the origin to stimulate muscle action. ■

TABLE 14.2 Muscles of the Anterior Trunk

MUSCLE	LOCATION AND *ACTION*
MUSCLES THAT MOVE THE ARM AT THE SHOULDER JOINT	
Superficial Muscles	
Deltoid (*delta* = triangle)	Large, rounded, triangular shoulder muscle. **Anterior fibers:** *flex and medially rotate arm.* **Lateral fibers:** *abduct arm.* **Posterior fibers:** *extend and laterally rotate arm.*
Pectoralis major (*pectus* = chest)	Large muscles on superior, anterior chest between sternum and arm. *Whole muscle adducts and medially rotates arm, clavicular head only flexes arm, and sternocostal head only extends arm.*
Deep Muscle	
Subscapularis	On ventral side of the scapula in the subscapular fossa. *Medially rotates arm.*
MUSCLES THAT MOVE THE SCAPULA	
Serratus anterior (*serratus* = saw-toothed)	Serrated-looking muscles on the lateral trunk inferior to arms and rib. *Abducts scapula and rotates it upward. This occurs when throwing a punch and is often called the boxer's muscle.*
Pectoralis minor	Deep to pectoralis major attaching to 3rd through 5th ribs. *Abducts scapula and rotates it downward; elevates 3rd through 5th ribs during forced inspiration.*
MUSCLES THAT MOVE THE ABDOMINAL WALL	
Superficial Muscles	
Rectus abdominis (*rectus* = parallel fibers)	Midline abdominal muscles located between sternum and inguinal ligament. *Flexes vertebral column and compresses abdomen. Does not laterally flex vertebral column.*
External oblique	Lateral and anterior sheet-like abdominal muscles whose fibers run obliquely down toward the midline (linea alba). *Bilateral contraction flexes vertebral column and compresses abdomen. Unilateral contraction laterally flexes and rotates vertebral column to the opposite side.*
Deep Muscles	
Internal oblique	Abdominal muscles deep to external oblique whose fibers run obliquely up toward linea alba. *Bilateral contraction flexes vertebral column and compresses abdomen. Unilateral contraction laterally flexes and rotates vertebral column to the same side.*
Transversus abdominis	Abdominal muscles deep to internal oblique whose fibers run transversely. *Compresses abdomen and stabilizes trunk.*
MUSCLES USED IN BREATHING	
Deep Muscles	
External intercostals (*inter-* = between; *costal* = ribs)	Oblique muscles located between ribs superficial to internal intercostals. Fibers run obliquely down toward the midline. *Increase thoracic diameter by elevating ribs during inspiration.*
Internal intercostals	Oblique muscles located between ribs deep to external obliques. Fibers run obliquely up toward the midline. *Decrease thoracic diameter by depressing ribs during forced expiration.*
Diaphragm	Dome-shaped muscle that separates the thoracic and abdominal cavities. Contains openings for the esophagus, aorta, and vena cava. *Flattens, increasing length of thoracic cage, during normal inspiration.*

Platysma

Clavicle

Deltoid

Pectoralis major

Latissimus dorsi

Serratus anterior

Linea alba

External oblique

Rectus sheath

Inguinal ligament

Inguinal ring (superficial)

Sternocleidomastoid

Trapezius

Pectoralis minor

Internal intercostals

External intercostals

Internal oblique

Transverse abdominis

Rectus abdominis

Anterior superficial view Anterior deep view

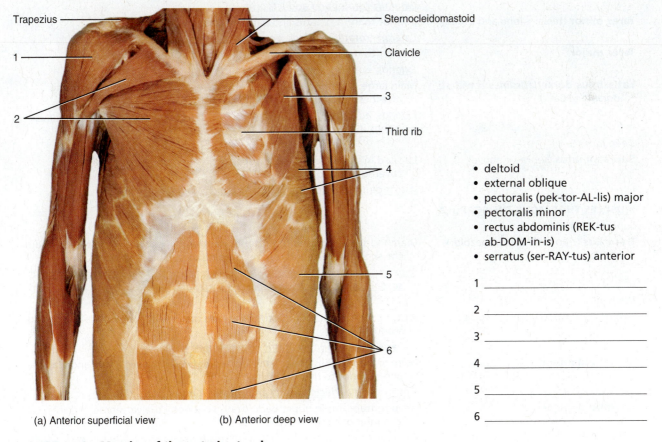

Trapezius

Sternocleidomastoid

Clavicle

1

3

2

Third rib

4

5

6

(a) Anterior superficial view (b) Anterior deep view

- deltoid
- external oblique
- pectoralis (pek-tor-AL-lis) major
- pectoralis minor
- rectus abdominis (REK-tus ab-DOM-in-is)
- serratus (ser-RAY-tus) anterior

1 _____

2 _____

3 _____

4 _____

5 _____

6 _____

FIGURE 14.3 **Muscles of the anterior trunk.**

(c) Right anterolateral
superficial view

(d) Right anterolateral
deep view

(e) Right anterolateral
deeper view

FIGURE 14.3 **Muscles of the anterior trunk,** *continued.*

TABLE 14.3 Muscles of the Posterior Trunk

MUSCLE	LOCATION AND *ACTION*
MUSCLES THAT MOVE THE ARM AT THE SHOULDER JOINT	
Superficial muscles	
Deltoid (*delta* = triangle)	Large, rounded, triangular shoulder muscle. *Posterior portion of deltoid extends and laterally rotates arm.*
Infraspinatus (*infra-* = below; *spinatus* = spine of scapula	Located in infraspinous fossa (inferior to spine of scapula) and partially covered by deltoid. *Laterally rotates and adducts arm.*
Teres minor (*teres* = long and round)	Inferior to infraspinatus and partially covered by deltoid. *Laterally rotates, extends, and adducts arm.*
Teres major	Inferior to teres minor and partially covered by deltoid. *Extends, adducts, and medially rotates arm.*
Latissimus dorsi (*latissimus* = widest; *dorsum* = back)	Main large, flat muscle of middle and lower back that runs laterally and attaches to superior portion of humerus. *Extends, adducts, and medially rotates arm (if arm is elevated over head, it brings it down).*
Deep muscles	
Supraspinatus (*supra-* = above)	Located in supraspinous fossa (superior to spine of scapula) and totally covered by deltoid. *Assists deltoid muscle in abducting arm.*
MUSCLES THAT MOVE THE SCAPULA	
Superficial muscle	
Trapezius (*trapezoides* = trapezoid shape)	Diamond-shaped muscle of posterior neck and upper back that extends from the skull and vertebral column to spine of scapula and lateral clavicle. *Superior portion elevates scapula and extends head, middle portion adducts scapula, inferior portion depresses scapula.*
Deep muscles	
Levator scapulae (*levare* = to raise)	Posterior to sternocleidomastoid and runs diagonally from posterior head to scapula. *Elevates scapula and rotates it downward.*
Rhomboid minor	Deep to trapezius in upper back located between the vertebrae and scapula, and superior to rhomboid major. *Elevates and adducts scapula and rotates it downward, stabilizes scapula.*
Rhomboid major	Deep to trapezius in upper back located between the vertebrae and scapula, and inferior to rhomboid minor. *Elevates and adducts scapula and rotates it downward, stabilizes scapula.*

Sternocleidomastoid

Trapezius

Deltoid

Infraspinatus

Teres minor

Teres major

Latissimus dorsi

External oblique

Splenius capitis

Levator scapulae

Rhomboid minor

Rhomboid major

Supraspinatus

Infraspinatus

Teres minor

Teres major

Anterior superficial view Anterior deep view

1

2

Teres major

3

4
5
6
Acromion of scapula
Greater tubercle of humerus
7
8
9
10
Triceps brachii

(a) Posterior superficial view (b) Posterior deep view

- deltoid
- infraspinatus (in-fra-spi-NAY-tus)
- latissimus dorsi (la-TIS-i-mus DOR-sigh)
- levator scapulae
- rhomboid (ROM-boid) major
- rhomboid minor
- supraspinatus (soo-pra-spi-NAY-tus)
- teres (TEH-rez) major
- teres minor
- trapezius (tra-PEE-zee-us)

1 _____

2 _____

3 _____

4 _____

5 _____

6 _____

7 _____

8 _____

9 _____

10 _____

FIGURE 14.4 Muscles of the posterior trunk.

TABLE 14.3 Muscles of the Posterior Trunk (*continued*)

MUSCLE	LOCATION AND *ACTION*
MUSCLES THAT MOVE THE VERTEBRAL COLUMN	
Erector spinae (*erector* = erect posture; *spinae* = spine) consists of iliocostalis, longissimus, and spinalis groups	Group of muscles next to the vertebral column deep to the trapezius, latissimus dorsi, and scapula. *Extend vertebral column and maintain erect posture when both muscles contract. Laterally flexes vertebral column when only one muscle.*

Longissimus capitis

Spinalis cervicis

Longissimus cervicis

Iliocostalis thoracis

Spinalis thoracis

Iliocostalis lumborum

Semispinalis capitis

Spinalis capitis

Splenius capitis

Semispinalis cervicis

Longissimus thoracis

Semispinalis thoracis

(c) Deep erector spinal muscles of the back

FIGURE 14.4 Muscles of the posterior trunk, *continued*.

3. Muscles of the Arm

These muscles have their origin on the humerus or scapula, cross the elbow joint, and insert on the bones of the forearm. When these muscles contract, they flex or extend the forearm at the elbow joint or cause movement at the radioulnar joints. The muscles that have origins on the scapula can also move the arm at the shoulder joint. Muscles of the limbs that perform similar functions often are surrounded by fascia-forming compartments. Muscles in the **anterior compartment** of the arm, such as the biceps brachii and brachialis muscles, flex the forearm, whereas those in the **posterior compartment,** the triceps brachii, extend the forearm.

Muscles of the Forearm

Muscles located in the forearm flex or extend the hand or cause movement at the radioulnar joint. The muscles that move the hand cross the wrist and insert on the carpals, metacarpals, or phalanges. These muscles are found in the anterior (flexor) compartment and posterior (extensor) compartment of the forearm. The muscles of the superficial anterior compartment of the forearm from lateral to medial are: flexor carpi radialis, palmaris longus, flexor carpi ulnaris, and flexor digitorum superficialis. The flexor digitorum superficialis is actually deep to the other three muscles. These muscles flex the hand or digits and originate on the medial epicondyle of the humerus. Muscles of the superficial posterior compartment of the forearm from medial to lateral are: extensor carpi ulnaris, extensor digitorum

minimi, extensor digitorum, extensor carpi radialis brevis, and extensor carpi radialis longus. These muscles extend the hand and digits and originate on the lateral epicondyle of the humerus. Flexor and extensor muscles near the lateral surface of the forearm may also abduct the wrist, whereas those near the medial surface may adduct it. Other muscles located in the forearm include the brachioradialis, pronator teres, and supinator, which move the forearm at the radioulnar joint.

Before Going to Lab

1 Label Figure 14.5(a) and (b) and Figure 14.6(a), (b), (c), and (d).

LAB ACTIVITY 3 Muscles of the Arm and Forearm

1 Use Figures 14.5 and 14.6 and Tables 14.4 and 14.5A and B to identify the muscles on models, charts, cadaver photographs or use the search text box in **Real Anatomy** (Muscular) to find these structures.

2 Locate each muscle on yourself and perform the action indicated in Tables 14.4 and 14.5. Palpate each muscle while performing the action.

3 Use Appendix B to locate the origin and insertion on an articulated skeleton. Pull the insertion toward the origin to stimulate muscle action. ■

TABLE 14.4 Muscles of the Arm

MUSCLE	LOCATION AND *ACTION*
ANTERIOR SURFACE	
Biceps brachii (*biceps* = 2 heads; *brachii* = arm)	Large muscle with 2 heads. *Flexes and supinates forearm, and flexes arm.*
Brachialis	Deep to biceps brachii and anterior to humerus. *Flexes forearm.*
POSTERIOR SURFACE	
Triceps brachii	Large muscle with three heads. **Long head:** Superficial muscle on medial side. **Medial head:** Deep muscle on medial side. **Lateral head:** Lateral side. *All three heads extend forearm and help extend arm.*

Subscapularis

Teres major

Biceps brachii

Brachialis

DANK

Coracoid process
of scapula

Supraspinatus

Greater tubercle
of humerus

1

2

3

4

(a) Anterior view

- biceps brachii (BI-ceps BRAY-key-eye)
- subscapularis
- teres major
- brachialis (BRAY-key-AL-is)

1 _____

2 _____

3 _____

4 _____

FIGURE 14.5 **Muscles of the arm.**

Triceps brachii
long head

Triceps brachii
lateral head

Triceps brachii
medial head

Spine of scapula

Infraspinatus

Teres major

Greater tubercle
of humerus

Teres minor

5

6

7

Medial epicondyle
of humerus

Olecranon of ulna

(b) Posterior view

- triceps brachii, lateral head
- triceps brachii, long head
- triceps brachii, medial head

5 _____

6 _____

7 _____

FIGURE 14.5 **Muscles of the arm,** *continued.*

TABLE 14.5A Muscles of the Anterior Forearm

MUSCLE	LOCATION AND *ACTION*
MUSCLES THAT MOVE THE RADIUS AND ULNA	
Superficial muscles	
Brachioradialis (*brachi-* = arm; *radialis* = radius)	Large, superficial lateral muscle. *Flexes forearm and pronates and supinates forearm to neutral position.*
Pronator teres (*pronare* = to bend forward; *teres* = long, cylindrical)	Lateral to brachioradialis with oblique muscle fibers that cross antecubital fossa. *Pronates and weakly flexes forearm.*
Deep muscle	
Supinator (*supinate* = supine; palm upward)	Anterior portion is deep to brachioradialis. *Supinates forearm.*
MUSCLES THAT MOVE THE HAND	
Superficial muscles of anterior (flexor) compartment (Lateral to medial)	
Flexor carpi radialis (*carpi-* = wrist)	Medial and inferior to pronator teres and medial to brachioradialis in forearm. *Flexes and abducts hand.*
Palmaris longus (*palma* = palm)	Medial to flexor carpi radialis. *Weakly flexes wrist.*
Flexor carpi ulnaris	Small part of muscle is seen anteriorly; most medial muscle of posterior forearm. *Flexes and adducts hand.*
Flexor digitorum superficialis	Deep to tendons of flexor carpi radialis, palmaris longus, and flexor carpi ulnaris. *Flexes hand and proximal and middle phalanx of each finger.*
Deep muscle of anterior (flexor) compartment	
Flexor pollicis longus	Inferior and lateral to flexor digitorum superficialis. *Flexes distal phalanx of thumb.*

Biceps brachii

Brachialis

Medial epicondyle of humerus

Pronator teres

Brachioradialis

Flexor carpi radialis

Palmaris longus

Flexor carpi ulnaris

Flexor digitorum superficialis

Flexor pollicis longus

Abductor pollicis brevis

Flexor pollicis brevis

Adductor pollicis

Pisiform bone

Abductor digiti minimi

Flexor digiti minimi brevis

1
2
3
4
5

(a) Anterior superficial view

Supinator

Flexor digitorum profundus

Flexor pollicis longus

Carpal tunnel

Opponens digiti minimi

Dorsal interossei

6

7

8

(b) Anterior deep view

(a)
- brachioradialis
- flexor carpi (KAR-pee) radialis
- flexor carpi ulnaris
- palmaris (pall-MARE-is) longus
- pronator (PRO-nay-tor) teres

1 _____

2 _____

3 _____

4 _____

5 _____

(b)
- flexor digitorum profundus
- flexor pollicis (POL-ih-kis) longus
- supinator

6 _____

7 _____

8 _____

FIGURE 14.6 **Muscles of the forearm, wrist, and hands.**

TABLE 14.5B Muscles of the Posterior Forearm

MUSCLE	LOCATION AND *ACTION*
MUSCLE THAT MOVES THE RADIUS AND ULNA	
Supinator	Posterior portion is deep to lateral muscles of posterior forearm. *Supinates forearm.*
MUSCLES THAT MOVE THE HAND	
Superficial muscles of posterior (extensor) compartment (Medial to lateral)	
Extensor carpi ulnaris	Lateral to flexor carpi ulnaris. *Extends and adducts hand.*
Extensor digitorum minimi	Lateral to extensor carpi ulnaris. *Extends hand and proximal phalanx of 5th digit.*
Extensor digitorum	Lateral to extensor digitorum minimi. *Extends hand and proximal, middle, and distal phalanges.*
Extensor carpi radialis brevis	Lateral to extensor digitorum. *Extends and abducts hand.*
Extensor carpi radialis longus	Lateral to extensor radialis brevis. *Extends and abducts hand.*
Deep muscle of posterior (extensor) compartment	
Extensor pollicis longus	Deep to lateral extensor muscles on posterior forearm. *Extends distal phalanx of thumb and first metacarpal of thumb and abducts hand.*

Medial epicondyle

Extensor carpi
radialis longus

Extensor carpi ulnaris

Extensor carpi radialis
brevis

Extensor digitorum

Extensor digiti minimi

Abductor pollicis longus

Extensor pollicis brevis

Flexor carpi
ulnaris

Extensor retinaculum

9

10

11

12

13

14

(c) Posterior superficial view

(c)
• extensor carpi radialis brevis
• extensor carpi radialis longus
• extensor carpi ulnaris
• extensor digitorum
• extensor digiti minimi
• flexor carpi ulnaris

9 _____

10 _____

11 _____

12 _____

13 _____

14 _____

Supinator

Abductor pollicis longus

Extensor
pollicis longus

Extensor pollicis brevis

15

16

(d)
• extensor pollicis longus
• supinator (SOUP-ih-nay-tor)

15 _____

16 _____

(d) Posterior deep view

FIGURE 14.6 **Muscles of the forearm, wrist, and hands,** *continued.*

4. Muscles of the Thigh

Muscles of the thigh flex, extend, abduct, adduct, circumduct, and rotate the thigh. Muscles that move the thigh have origins on the pelvis or vertebrae and insertions on the femur or tibia. Those with origins on the pelvis move the thigh at the hip joint and the leg at the knee joint, whereas those with origins on the femur move only the leg at the knee joint. Muscles of the **anterior compartment** of the thigh, the quadriceps femoris group, extend the leg. The rectus femoris muscle of this group also flexes the thigh. The sartorius, which runs diagonally from the hip to the medial tibia, marks the boundary between the anterior and medial compartments. Muscles of the **medial compartment** are pectineus, adductor magnus, adductor longus, adductor brevis, and gracilis. These muscles adduct and flex the thigh. Muscles of the **posterior compartment** are the hamstring muscle group, which flex the leg and extend the thigh.

> **Before Going to Lab**
>
> **1** Label Figure 14.7(a), (b), and (e).

LAB ACTIVITY 4 Muscles of the Thigh

1 Use Figure 14.7 and Table 14.6 to identify the muscles on models, diagrams, cadaver photographs or use the search text box in **Real Anatomy** (Muscular) to find these structures.

2 Locate the muscles on yourself and perform the actions indicated in Table 14.6. Palpate the muscles while performing the action.

3 Use Appendix B to locate the origin and insertion on an articulated skeleton. Pull the insertion toward the origin to simulate muscle action. ■

TABLE 14.6 Muscles of the Thigh

MUSCLE	LOCATION AND *ACTION*
ANTERIOR SURFACE (*Lateral to medial*)	
Tensor fasciae latae (*tensor* = to stretch; *fascia* = band; *latus* = wide)	Small lateral hip muscle with fascia lata (deep fascia) that extends down thigh. *Flexes and abducts thigh.*
Quadriceps femoris (*quad-* = four; *cep-* = head or origins)	Group of 4 muscles on anterior thigh.
Rectus femoris	Located along midline of thigh. *Extends leg and flexes thigh.*
Vastus lateralis	Lateral to rectus femoris. *Extends leg.*
Vastus medialis	Medial to rectus femoris. *Extends leg.*
Vastus intermedius	Deep to rectus femoris and intermediate to vastus laterali and vastus medialis. *Extends leg.*
Sartorius (*sartor* = tailor)	Diagonal muscle. Extends from anterior superior iliac spine to medial surface of tibia. *Flexes leg and flexes, abducts, and laterally rotates thigh (allows us to flex and cross our legs).*
Iliopsoas	Combination of psoas major and iliacus.
Psoas major (*psoas* = loin) **Iliacus** (*iliacus* = ilium)	A small portion of these muscles are observed in anterior/medial compartment of thigh inferior to inguinal ligament. *Together flex thigh, rotate thigh laterally, and flex trunk.*
Pectineus (*pecten* = comb)	Medial to iliopsoas. *Adducts and flexes thigh.*
Adductor longus	Medial to pectineus. *Adducts and flexes thigh and medially rotates thigh.*
Adductor magnus	Observed medial and inferior to adductor longus. A larger portion of this muscle is deep to the adductor longus. *Adducts and medially rotates thigh. Anterior portion flexes thigh and posterior portion extends thigh.*
Adductor brevis	Deep to adductor longus, it is totally covered by superficial muscles in Figure 14.7. *Adducts and flexes thigh and medially rotates thigh.*
Gracilis	Medial to adductor magnus; straight muscle on inside of thigh. *Adducts thigh and medially rotates thigh. Flexes leg.*

(a) Anterior superficial view

(a)
• adductor longus
• gracilis (grah-SIL-us)
• iliacus (il-ee-AK-us)
• pectineus (pek-tin-EE-us)
• psoas (SO-us) major
• rectus femoris (FEM-or-is)
• sartorius (sar-TOR-ee-us)
• tensor fasciae latae (FA-schee LA-tee)
• vastus intermedius
• vastus lateralis
• vastus medialis

1 _____ 7 _____

2 _____ 8 _____

3 _____ 9 _____

4 _____ 10 _____

5 _____ 11 _____

6 _____

(b)
• adductor brevis
• adductor magnus
• gracilis

12 _____

13 _____

14 _____

(b) Anterior deep view

FIGURE 14.7 Muscles of the thigh.

TABLE 14.6 **Muscles of the Thigh (*continued*)**

MUSCLE	LOCATION AND *ACTION*
POSTERIOR SURFACE	
Gluteus maximus	Largest buttocks muscle. *Extends and laterally rotates thigh.*
Gluteus medius	Observed superior to gluteus maximus but is partially covered by it. *Abducts and medially rotates thigh.*
Gluteus minimus	Deep muscle covered by gluteus medius. *Abducts and medially rotates thigh.*
Piriformis (*pirum* = pear; *-forma* = shape)	Deep muscle inferior to gluteus minimus; important landmark for sciatic nerve, which passes deep to piriformis. *Laterally rotates and abducts thigh.*
Hamstrings	Three muscles on posterior thigh.
Biceps femoris	Most lateral hamstring. *Flexes leg and extends thigh.*
Semitendinosus (*semi-* = half or part; tendinosus = long tendon)	Medial to biceps femoris and superficial to semimembranosus. *Flexes leg and extends thigh.*
Semimembranosus (*membranosus* = partly membrane)	Most medial of the hamstrings; (remember 'm' for medial). *Flexes leg and extends thigh.*

(c) Posterior superficial view

(d) Posterior deep view

FIGURE 14.7 **Muscles of the thigh, *continued*.**

15
16
Superior
gemellus
Obturator
internus
Inferior
gemellus
Quadratus
femoris
Adductor
minimus
Adductor
magnus
17
Popliteal fossa

18

Vastus
lateralis
(covered by
iliotibial tract)

19

20

Deep Superficial

(e) Posterior view

- biceps femoris
- gluteus maximus
- gluteus medius
- piriformis
- semimembranosus
- semitendinosus

15 _____

16 _____

17 _____

18 _____

19 _____

20 _____

FIGURE 14.7 Muscles of the thigh, *continued.*

5. Muscles of the Leg and Foot

Palpate the tibia of your leg to find the anterior crest and the lateral and medial borders of the tibia. Observe that the muscles on the anterior surface of the tibia are located lateral to the anterior crest of the tibia. The anterior and lateral superficial muscles are in order starting at the anterior crest just below the knee and moving laterally: tibialis anterior, extensor digitorum longus, and the fibularis (peroneus) longus. The tibialis anterior and extensor digitorum are muscles of the **anterior compartment** of the leg. These muscles dorsiflex the foot and also extend the toes if they insert on the phalanges.

Muscles of the **posterior compartment** of the leg include the gastrocnemius, soleus, and flexor digitorum longus. These muscles plantar flex the foot and also flex the toes if they insert on the phalanges. Eversion of the foot is due to contraction of muscles near the lateral surface of the leg, whereas inversion is due to contraction of muscles near the medial surface of the leg.

Before Going to Lab

1 Label Figure 14.8(a), (b), (c), and (d).

LAB ACTIVITY 5 Muscles of the Leg and Foot

1 Use Figure 14.8 and Table 14.7 to identify the muscles on models, diagrams, cadaver photographs or use the search text box in **Real Anatomy** (Muscular) to find these structures.

2 Locate each muscle on yourself and perform the action indicated in Table 14.7. Palpate each muscle while performing the action.

3 Use Appendix B to locate the origin and insertion on an articulated skeleton. Pull the insertion toward the origin to simulate muscle action. ■

TABLE 14.7 Muscles of the Leg and Foot

MUSCLE	LOCATION AND *ACTION*
LATERAL AND ANTERIOR SURFACES *(Superficial muscles)*	
Tibialis anterior	Lateral to the anterior crest of the tibia. *Dorsiflexes and inverts foot.*
Extensor digitorum longus	Lateral to tibialis anterior. *Dorsiflexes foot and extends toes.*
Fibularis (peroneus) longus	Lateral to the extensor digitorum longus. *Plantar flexes and everts foot.*
POSTERIOR SURFACE *(Superficial muscles)*	
Gastrocnemius	Large superficial muscle on posterior leg. *Plantar flexes foot and flexes leg.*
Soleus	Deep to gastrocnemius and posterior to fibularis (peroneus) longus. *Plantar flexes foot only.*
(Deep muscles)	
Flexor digitorum longus	Deep to soleus, medial muscle (not shown on Figure 14.8). *Plantar flexes foot and flexes toes.*
Flexor hallucis longus	Deep to soleus, lateral muscle. *Plantar flexes foot and flexes great toe.*

Tibialis anterior

Extensor digitorum longus

Fibularis (peroneus) longus

Fibularis (peroneus) brevis

Gastrocnemius

Soleus

Flexor digitorum longus

Tendon of extensor hallucis longus

Patella

Patellar ligament

Tibia

1

2

3

4

Fibularis tertius

(a) Anterior superficial view

(a)
• extensor digitorum longus
• fibularis (peroneus) brevis
• fibularis (peroneus) longus
• tibialis anterior

1 _____

2 _____

3 _____

4 _____

Plantaris

Gastrocnemius

Soleus

Fibularis (peroneus) longus

Extensor digitorum longus

Tibialis anterior

Patella

Lateral condyle of femur

Patellar ligament

5

6

7

8

9

Fibularis brevis

Fibularis tertius

Calcaneal (Achilles) tendon

Extensor hallucis longus

Lateral malleolus of fibula

(b) Lateral superficial view

(b)
• extensor digitorum longus
• fibularis (peroneus) longus
• gastrocnemius (gas-trok-NEE-mee-us)
• soleus
• tibialis anterior

5 _____

6 _____

7 _____

8 _____

9 _____

FIGURE 14.8 **Muscles of the leg, ankle, and foot.**

(c) Posterior superficial view

(d) Posterior deep view

FIGURE 14.8 **Muscles of the leg, ankle, and foot,** *continued.*

(c)
• gastrocnemius
• soleus

10 _____

11 _____

(d)
• flexor digitorum longus
• flexor hallucis (HAL-a-kis) longus

12 _____

13 _____

C. Dissection of Skeletal Muscles

If you are dissecting a cat to observe skeletal muscles, refer to the appropriate accompanying dissection manual. Many skeletal muscles of the cat are similar to human muscles.

This dissection will reinforce your knowledge of human skeletal muscles and allow you to observe the fascia that surrounds, protects, and compartmentalizes these muscles.

Real Anatomy (Dissection), a cadaver dissection, can be used to complement or substitute for animal dissection of skeletal muscles.

Reviewing Your Knowledge

A. Anatomy Review

For Figures 14.9 (a) and 14.9 (b), write the name of the skeletal muscle next to the number. Color with colored pencils.

(a) Anterior view

FIGURE 14.9 Superficial skeletal muscles.

(b) Posterior view

FIGURE 14.9 **Superficial skeletal muscles,** *continued.*

B. Muscle Function

Identify the muscles that perform each function.

Muscles of the Head and Neck

a. masseter
b. orbicularis oculi
c. orbicularis oris
d. sternocleidomastoid
e. temporalis
f. trapezius
g. zygomaticus major

_____ **1.** Smiling muscle

_____ **2.** Kissing muscle

_____ **3.** Closes eyelid

_____ **4.**
_____ **5.** } Two muscles that close mouth

_____ **6.** Paired muscle that flexes head and rotates head to side

_____ **7.** Extends head

Muscles of the Anterior Trunk

a. deltoid
b. external oblique
c. internal oblique
d. pectoralis major
e. rectus abdominis
f. serratus anterior
g. transversus abdominis

_____ **8.** Adducts and flexes arm

_____ **9.** Anterior portion flexes arm; lateral portion abducts arm

_____ **10.** Abducts scapula and rotates it upward (boxer's muscle)

_____ **11.** Flexes vertebral column and compresses abdomen

_____ **12.**
_____ **13.** } Two muscle pairs that flex vertebral column, compress abdomen, and laterally flex vertebral column

_____ **14.** Only compresses abdomen

Muscles of the Posterior Trunk

a. deltoid
b. erector spinae
c. latissimus dorsi
d. teres major
e. trapezius

____ 15.
} Extends, adducts, and medially rotates arm
____ 16.

____ 17. Posterior portion extends arm; lateral portion abducts arm

____ 18. Superior portion elevates scapula, middle portion adducts scapula, inferior portion depresses scapula

____ 19. Paired muscle that extends vertebral column, maintains erect posture, and laterally flexes vertebral column

Muscles of the Arm and Forearm

a. biceps brachii
b. brachialis
c. brachioradialis
d. extensor carpi radialis
e. extensor carpi ulnaris
f. extensor digitorum
g. flexor carpi radialis
h. flexor carpi ulnaris
i. palmaris longus
j. triceps brachii

____ 20. Extends forearm at elbow and extends arm

____ 21. Flexes forearm at elbow and flexes arm

____ 22. Flexes forearm

____ 23. Flexes forearm and pronates and supinates forearm

____ 24. Flexes and abducts hand

____ 25. Flexes and adducts hand

____ 26. Weakly flexes hand

____ 27. Extends hand and extends phalanges

____ 28. Extends and adducts hand

____ 29. Extends and abducts hand

Muscles of the Thigh

a. adductor magnus, longus, brevis
b. biceps femoris
c. gluteus maximus
d. gluteus medius

e. gracilis
f. rectus femoris
g. sartorius
h. semimembranosus
i. semitendinosus
j. tensor fasciae latae
k. vastus intermedius
l. vastus lateralis
m. vastus medialis

____ **30.** Extends leg at knee and flexes thigh at hip

____ **31.**

____ **32.** } Three muscles that extend leg only

____ **33.**

____ **34.** Flexes leg and flexes, abducts, and laterally rotates thigh (allows us to flex and cross our legs)

____ **35.** Adducts thigh and flexes leg

____ **36.** Group of muscles that adducts and flexes thigh

____ **37.** Abducts thigh

____ **38.** Flexes and abducts thigh

____ **39.** Extends thigh

____ **40.**

____ **41.** } Three muscles that flex leg and extend thigh

____ **42.**

Muscles of the Leg
a. extensor digitorum longus
b. fibularis (peroneus) longus
c. flexor digitorum longus
d. gastrocnemius
e. soleus
f. tibialis anterior

____ **43.** Plantar flexes and everts foot

____ **44.** Plantar flexes foot and flexes toes

____ **45.** Plantar flexes foot only

____ **46.** Plantar flexes foot and flexes leg

____ **47.** Dorsiflexes foot and extends toes

____ **48.** Dorsiflexes and inverts foot

Using Your Knowledge

Identify the *muscles* and/or *actions* in the following questions.

1. Identify the *muscles of the head* that you use when you *whistle*.

2. Identify the *muscles of the head* used to *blink*.

3. Identify the major *thigh* and *leg* muscles you use when you *kick* a ball.

4. Give the action of each muscle listed in question 3.

5. Identify the *forearm* muscles you use when you *turn* a key in a lock.

6. Give the action of each muscle listed in question 5.

7. Identify the major *arm* muscles you use when you *lift* a glass to drink from it.

8. Give the action of each muscle listed in question 7.

9. Identify the *neck* muscles you use when you *look both* ways to cross a street.

10. Give the action of each muscle listed in question 9.

11. Identify the *abdominal* muscles you use when you do the dance called the "*twist*."

12. Give the action of each muscle listed in question 11.

13. Identify the muscles used to move the *upper limb* when you do *jumping jacks*.

14. Give the action of each muscle listed in question 13.

15. Identify the muscles that move the *lower limb* when you do *jumping jacks*.

16. Give the action of each muscle listed in question 15.

17. Identify the muscles used to bend the *trunk* when you *touch your toes*.

18. Give the action of each muscle listed in question 17.

19. Identify the *upper limb* muscles used to *play* the cymbals in a band.

20. Give the action of each muscle listed in question 19.

Surface Anatomy

OBJECTIVES

1 Define and describe the importance of surface anatomy

2 Locate by palpation and identify important anatomical landmarks on the external surface of the body

3 Identify the borders of the anterior and posterior triangles of the neck and the femoral triangle using anatomical landmarks on the external surface of the body

4 Identify organ location using anatomical landmarks on the external surface of the body

MATERIALS

- hand mirror
- articulated skeleton
- skeletal muscle models or charts

S urface anatomy is the study of anatomical landmarks observed on the external surface of the body. These landmarks can be used to locate underlying structures by palpation. **Palpation** (*palpare* = to touch gently) is a technique that uses the hands or fingers to locate internal body structures and to determine the size and texture of the structures.

This exercise is a review and can be completed at home or in the lab.

LAB ACTIVITY 1 Surface Anatomy of the Head and Neck

1 Read the location and description of each structure.

2 Locate and label each structure in Figure 15.1(a), (b), (c), and (d).

3 Palpate the designated structures on your body.

A. Anterior Surface of the Head

Structures located in Figure 15.1(a).

Bones and Bone Surface Markings

- **supraorbital margins**—Superior borders of the frontal bone that border the eye orbits. Palpate the supraorbital margins.
- **nasal bone**—Forms the bridge of the nose. Place fingers along the bridge of the nose to feel nasal bones. Find the anterior border of the nasal bones and palpate the **nasal cartilage** inferior to nasal bones.
- **body of mandible**—Horizontal portion of the lower jawbone. Palpate this main part of the mandible that includes the chin.
- **mental protuberance**—Palpate this anterior tip of the chin.

Skeletal Muscles

- **frontalis muscle**—Lies over the forehead. Palpate the frontalis muscle with fingers while raising eyebrows.
- **zygomaticus major muscle**—Originates on the zygomatic bone and inserts on the corner of the mouth. Palpate this muscle near the zygomatic bone while smiling.

B. Lateral Surface of the Head

Structures located in Figure 15.1(b).

Bones and Bone Surface Markings

- **temporomandibular joint** (TMJ)—Located anterior to the external auditory meatus of the ear. Palpate the joint while opening and closing the mouth.

- **condylar process** (mandibular condyle) *[not shown on Figure 15.1(a)]*—Rounded process at the posterior portion of the ramus. It articulates with the mandibular fossa of the temporal bone to form the TMJ.
- **ramus of mandible**—Vertical process of mandible. Palpate ramus inferior to the TMJ.
- **mastoid process**—Rounded projection on the inferior portion of the temporal bone posterior to the ear. Palpate area of the skull inferior and posterior to external auditory meatus of ear.
- **external occipital protuberance**—Rounded projection superior to foramen magnum of the occipital bone. Palpate the base of the skull near the midline. Lateral to the external occipital protuberance are two curved ridges called the **superior nuchal lines,** which mark the boundary between the head and neck.

Skeletal Muscles

- **temporalis muscle**—Located superior to the ear. Palpate the temporalis muscle while closing the mouth and clenching your teeth.
- **masseter muscle**—Located anterior to ramus of the mandible. Palpate the masseter muscle while closing the mouth and clenching your teeth.
- **occipitalis muscle**—Lies over the inferior portion of the occipital bone. Firmly palpate the posterior surface of the skull immediately above the neck as you raise and lower eyebrows.

Other Structures

- **parotid gland**—Located anterior and inferior to the ears. The superior border is at the level of the zygomatic arch and the inferior border is at the angle formed by ramus and body of the mandible. The parotid gland covers the masseter muscle and is approximately 2 fingers wide.

(a) Anterior view of head
- body of mandible
- frontalis muscle
- mental protuberance
- nasal bone
- supraorbital margin
- zygomaticus major muscle

1 _____

2 _____

3 _____

4 _____

5 _____

6 _____

(b) Right lateral view of head
- external occipital protuberance
- masseter muscle covered by parotid gland
- mastoid process
- occipitalis muscle
- ramus of mandible
- temporalis muscle
- temporomandibular joint

7 _____

8 _____

9 _____

10 _____

11 _____

12 _____

13 _____

1

Orbicularis oculi
muscle

4

5

2

Orbicularis oris
muscle

Depressor labii
inferioris muscle

6

3

(a) Anterior view of the head

7

8

9

10

11

12

13

Hyoid
bone

(b) Right lateral view of the head

FIGURE 15.1 Surface anatomy of the head and neck.

C. Anterior Surface of the Neck

Structures located in Figure 15.1(c).

Bones and Bone Surface Markings

- **hyoid bone**—Located in the anterior neck between the mandible and larynx. With the head in anatomical position, palpate the hyoid bone by placing the thumb and middle finger of one hand on either side of the neck about 1 inch inferior to the mandible. This bone can be moved laterally from side to side.

- **clavicle**—Starting at the manubrium of the sternum, trace the clavicle's S-shaped curvature laterally to its acromial end. Palpate the acromioclavicular joint just posterior to the acromial end of the clavicle as you thrust your shoulder joint anteriorly.

- **suprasternal (jugular) notch**—Located at the base of the neck between the sternal heads of the sternocleidomastoid muscles and just superior to the sternum. Palpate the sternum and move the fingers superiorly to feel the suprasternal notch.

Skeletal Muscles

- **sternocleidomastoid muscle**—Originates on the sternum and clavicle and inserts on the mastoid process of the temporal bone.

Other Structures

- **thyroid cartilage of larynx**—Largest cartilage of the larynx that has a prominence called the Adam's apple. Located in anterior neck inferior to hyoid bone. Move fingers 1 inch inferiorly from the hyoid bone until a firm structure is reached. Swallow to feel this cartilage move superiorly.

- **cricoid cartilage of larynx**—Located inferior to the thyroid cartilage and used as a landmark when locating the trachea during a tracheostomy. Move your fingers down the thyroid cartilage. There is a depression between the thyroid cartilage and the cricoid cartilage. Palpate the cricoid cartilage inferior to this depression.

- **thyroid gland**—Located inferior to the larynx on either side of trachea. Palpate by placing fingers on neck inferior and lateral to the thyroid cartilage and feeling for a soft mass.

- **common carotid arteries**—Located in the lateral neck between the trachea and the sternocleidomastoid muscle. They branch into the external and internal carotid arteries at the superior border of the larynx. Palpate the right common carotid artery by placing your fingers on the right side of the neck just lateral to the trachea.

- **external jugular vein**—Located lateral to the sternocleidomastoid muscle. To observe this vein, look in the mirror while clenching your teeth as in anger or placing fingers on the skin above the clavicle and pressing firmly to prevent blood from draining the external jugular vein.

- **internal jugular vein**—Located between the common carotid artery and external jugular vein *[not shown on Figure 15.1(a)]*.

D. Lateral Surface of the Neck

Structures located in Figure15.1(d).

Skeletal Muscles

- **sternocleidomastoid muscle**— A portion of this muscle is look into a mirror while turning your head to the side. Observe the sternal and clavicular heads of the sternocleidomastoid muscle and palpate the muscle.

- **scalenes**—Located posterior-lateral to the sternocleidomastoid muscle, just superior to the clavicle.

- **levator scapulae**—Located superior to the scalenes.

- **trapezius muscle**— A portion of this muscle is located in the posterior and lateral neck. Inflammation of trapezius muscle may result in "stiff neck." Place fingers on the posterior portion of the lateral neck and palpate while flexing and extending the neck.

Anterior and Posterior Triangles

- **anterior triangle** *[not shown in Figure 15.1(d)]*— Superiorly bounded by the inferior margin of the mandible, anteriorly by the midline of the neck, and posteriorly by the sternocleidomastoid. The **common carotid artery** and **internal jugular vein** are located in this triangle. Palpate to identify the triangle's borders and palpate your carotid pulse in the anterior triangle.

- **posterior triangle** *[not shown in Figure 15.1(d)]*— Anteriorly bordered by the sternocleidomastoid, posteriorly by the trapezius and inferiorly by the clavicle. Palpate to identify the triangle's borders. The **brachial plexus** and **external jugular vein** are located in this triangle.

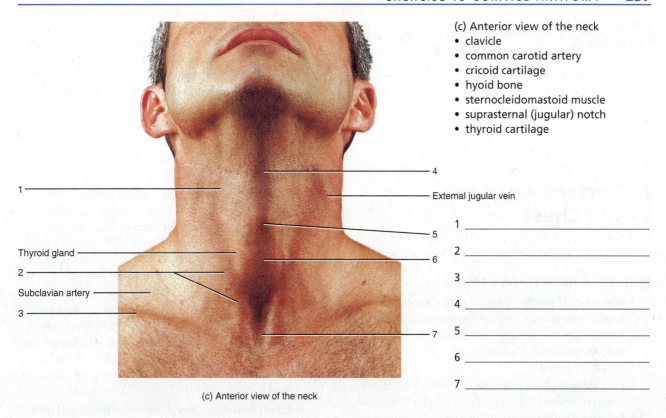

(c) Anterior view of the neck
- clavicle
- common carotid artery
- cricoid cartilage
- hyoid bone
- sternocleidomastoid muscle
- suprasternal (jugular) notch
- thyroid cartilage

4

External jugular vein

5

Thyroid gland

6

2

Subclavian artery

3

7

1 _____

2 _____

3 _____

4 _____

5 _____

6 _____

7 _____

(c) Anterior view of the neck

(d) Lateral view of the neck
- levator scapulae
- scalenes
- sternocleidomastoid
- trapezius

1 _____

2 _____

3 _____

4 _____

1

2

3

4

(d) Lateral view of the neck

FIGURE 15.1 Surface anatomy of the head and neck, *continued*.

LAB ACTIVITY 2 **Surface Anatomy of the Trunk**

1 Read the location and description of each structure.

2 Locate and label each structure in Figure 15.2(a), (b), (c), (d), (e), (f) and (g).

3 Palpate the designated structures on your body. ■

E. Surface Anatomy of the Chest

Structures located in Figure 15.2(a).

Bones and Bone Surface Markings

- **suprasternal notch**—Located at the base of the neck between the sternal heads of the sternocleidomastoid muscles and just superior to the sternum. Palpate the sternum and move the fingers superiorly to feel the suprasternal notch.

- **manubrium of the sternum**—Superior part of the sternum between the suprasternal notch and body of the sternum. Palpate by placing fingers at suprasternal notch and moving them inferiorly until a ridge (sternal angle) is reached.

- **body of sternum**—Palpate area of the sternum inferior to the sternal angle.

- **sternal angle**— Slightly raised area that can be felt at border of manubrium and body of sternum.

- **ribs**—The ribs can be palpated lateral to the sternum. The **second rib** is located at the level of the sternal angle. Palpate the sternal angle and move fingers laterally until the second rib can be felt. Move the fingers inferiorly, counting each rib and each intercostal space between the ribs.

- **costal margin**—Anterior edge of costal cartilage of ribs 7–10 that begins at the xiphisternal joint.

- **xiphisternal joint**—Joint between body of the sternum and xiphoid process of the sternum. Palpate the costal margin. Move fingers anteriorly along the costal margin until they reach the superior edge of the costal margin. Medial to this point is the xiphisternal joint.

- **xiphoid process of sternum**—Inferior portion of the sternum. Palpate the xiphoid process inferior to the xiphisternal joint.

Skeletal Muscles

- **pectoralis major muscle**—Major muscle of the chest. In males the inferior border can be observed as a curved line under the breasts. This line is at the level of the fifth rib. Bend over your lab bench and push yourself up with one limb. Use your opposite hand to palpate the pectoralis major muscle.

- **serratus anterior muscle**—Located on the lateral chest wall, extending from the ribs under the arm to the scapula. Flex your forearm and abduct your elbow. Use your opposite hand to palpate the serratus anterior muscle as it abducts the scapula. If you move too far posteriorly, you will be palpating the latissimus dorsi instead.

- **diaphragm**—Located between the fourth and fifth intercostal space. Position changes during inhalation and exhalation.

Other Structures *[not shown in Figure 15.2(a)]*

- **trachea**—Located between cricoid cartilage of larynx and sternal angle. The trachea is approximately 2 fingers in diameter.

- **primary bronchi**—The trachea divides into the right and left bronchi at the level of the sternal angle.

- **lungs**—Apex of lungs are slightly above the clavicle. The base of the lungs rests on the diaphragm.

- **heart**—Located between the second and sixth ribs. The heart rests on the diaphragm deep to the xiphisternal joint. The heart is about the size of your fist, and 2/3 of the heart lies to the left of the midline.

- **aortic arch**—Superior border of aortic arch is posterior to the sternal angle and anterior to the trachea.

Clavicle

Suprasternal notch

6

7

8 (muscle)

9

10

1

2

3

4

(muscle) 5

(a) Anterior view of the chest

(a) Anterior view of chest
- body of sternum
- costal margin
- manubrium of sternum
- pectoralis major muscle
- ribs (deep to muscle)
- second rib (deep to muscle)
- serratus anterior muscle
- sternal angle
- xiphisternal joint
- xiphoid process of sternum

1 _____

2 _____

3 _____

4 _____

5 _____

6 _____

7 _____

8 _____

9 _____

10 _____

FIGURE 15.2 Surface anatomy of the trunk.

F. Surface Anatomy of the Abdomen

Structures located in Figure 15.2(b) and (c).

Bone and Bone Surface Structures

- **anterior superior iliac spine**—Located at the anterior end of the iliac crest. Palpate the iliac crest with your fingers and move your fingers medially until you feel the "bump" at the anterior end of the iliac crest.
- **iliac crest**—Marks the inferior border of the abdomen. Palpate the iliac crest by placing your hands on your hips.
- **pubic symphysis** [*not shown in Figure* 15.2(*b*)]—Anterior joint between os coxae (hip bones). Palpate midline of the inferior pelvic area.

Skeletal Muscles

- **serratus anterior muscle**—Located on the lateral chest wall, extending from the ribs under the arm to the scapula. Flex your forearm and abduct your elbow. Use your opposite hand to palpate the serratus anterior muscle as it abducts the scapula. If you move too far posteriorly, you will be palpating the latissimus dorsi instead.
- **external oblique muscle**—Inferior to serratus anterior muscles. Palpate while twisting trunk toward the opposite side.
- **rectus abdominis muscle**—Longitudinal muscles lateral to the linea alba. Lie back in your chair and palpate these muscles as you sit up.

Other Structures

- **linea alba**—Tendinous raphe between the xiphoid process and pubic symphysis forming a vertical groove along the midline. This is a common incision site because there is little damage to muscles and little bleeding.
- **linea semilunaris**—Lateral margin of rectus abdominis that is observable in lean people as a groove.
- **tendinous intersection**—Transverse grooves across rectus abdominis muscles that are observable in muscular individuals.
- **umbilicus**—Most notable feature of the abdomen and a common incision site. The umbilicus is located between **L3** (3rd lumbar vertebra) and **L4**.
- **liver** [*not shown on Figure 15.1(b) and (c)*]—Inferior to the diaphragm. The position of the liver varies with body position, respiration, and degree of distension of the stomach and intestines.
- **gallbladder** [*not shown on Figure 15.1(b) and (c)*]—Located deep to the lateral margin of the rectus abdominis.
- **appendix**—Deep to **McBurney's point**, which is located along the line between the umbilicus and the right anterior superior iliac spine, about 1 to 2 inches away from the latter. During appendicitis, pressure on McBurney's point results in tenderness. This is the most common site of incision for an appendectomy.
- **common iliac arteries** [*not shown on Figure 15.1(b) and (c)*]—The abdominal aorta bifurcates into the common iliac arteries at the level of the anterior superior iliac spines.

(b) Anterior view of abdomen
- anterior superior iliac spine
- external oblique muscle
- iliac crest
- linea alba
- linea semilunaris
- McBurney's point
- rectus abdominis muscle
- tendinous intersection
- umbilicus

1 _____

2 _____

3 _____

4 _____

5 _____

6 _____

7 _____

8 _____

9 _____

(c) Anteriolateral view of abdomen
- anterior superior iliac spine
- external oblique muscle
- iliac crest
- McBurney's point
- rectus abdominis muscle
- serratus anterior muscle
- tendinous intersection

1 _____

2 _____

3 _____

4 _____

5 _____

6 _____

7 _____

(b) Anterior view of abdomen

(c) Anteriolateral view of abdomen

FIGURE 15.2 Surface anatomy of the trunk, *continued.*

G. Surface Anatomy of the Back

Structures located in Figure 15.2(d) and (e).

Bones and Bone Surface Markings

- **acromion**—The flattened lateral end of the spine of the scapula located at the peak of the shoulder. Palpate this prominent projection.

- **spine of scapula** [*not shown in Figure 15.2(d) and (e)*]—The ridge across the posterior surface of the scapula that extends from the acromion to the medial border of the scapula. Palpate this bone with your fingers. The spine is easier to palpate on a lean body.

- **vertebral border of scapula**—Draw left shoulder back and run the fingers of your right hand just lateral to the the midline to feel the vertebral border of the scapula.

- **spinous processes of vertebrae**—Located along the midline of the back. Run your fingers along midline to feel the spinous processes.

- **vertebra prominens**—Located at base of neck. Prominent single spinous process of **C7** (7th cervical vertebra). The vertebra prominens, that can be seen as a bump at the base of the neck and palpated.

- **T3**—Tip of spinous process of 3rd thoracic vertebra is at the level of medial end of spine of scapula.

- **L4**—Tip of spinous process of 4th lumbar vertebra is at same level as highest point of iliac crest. Lumbar punctures to obtain CSF are usually done between **L3** and **L4**.

Skeletal Muscles

- **deltoid muscle**—This large, triangular muscle forms the rounded protrusion of the shoulder. It is used as a site for intramuscular injections. Palpate this muscle as you flex, abduct, and extend the arm to locate the deltoid's anterior, lateral, and posterior portions.

- **trapezius muscle**—Large muscle of the posterior neck and middle of back.

- **supraspinatus muscle**—Located superior to the spine of the scapula. Palpate this muscle.

- **infraspinatus muscle**—Located inferior to the spine of the scapula. Palpate area inferior to the spine of the scapula to feel this muscle.

- **teres major muscle**—Located inferior to the infraspinatus muscle.

- **latissimus dorsi muscle**—Broad muscle located between the lumbar region and axillary region. Together with teres major muscle forms the posterior axillary fold.

- **erector spinae muscle**—Large muscle of lower back. Bend over and palpate this muscle while extending the vertebral column.

Other

- **triangle of auscultation**—Triangular region of the back formed by the latissimus dorsi, trapezius, and vertebral border of scapula. This region is not covered by superficial muscles enabling respiratory sounds to be heard with a stethoscope.

(d) Posterior view of back
- acromion
- erector spinae muscle
- infraspinatus muscle
- latissimus dorsi muscle
- spinous process of thoracic vertebrae
- supraspinatus
- teres major muscle
- trapezius muscle
- vertebral border of scapula

1 _____

2 _____

3 _____

4 _____

5 _____

6 _____

7 _____

8 _____

9 _____

(e) Posterior view of back with arms flexed
- deltoid muscle , lateral fibers
- deltoid muscle posterior fibers
- latissimus dorsi muscle
- spinous process of lumbar vertebra
- spinous process of thoracic vertebra
- teres major muscle
- trapezius muscle
- triangle of auscultation
- vertebral border of scapula

1 _____

2 _____

3 _____

4 _____

5 _____

6 _____

7 _____

8 _____

9 _____

(d) Posterior view of the back

(e) Posterior view of the back with arms flexed

FIGURE 15.2 **Surface anatomy of the trunk,** *continued.*

H. Surface Anatomy of the Posterior Pelvis and Gluteal Region

Structures located in Figure 15.2(f) and (g).

Bones and Bone Surface Markings

- **iliac crest**—Superior boundary of the ileum. The iliac crest can be palpated when you "put your hands on your hips."

- **posterior superior iliac spine**—Located on posterior end of iliac crest. Locate the dimple in the skin located just lateral to sacrum.

- **sacroiliac joint** [*not shown on Figure 15.2(f) and (g)*]—Joint between sacrum and ilium. Middle of sacroiliac joint is deep to posterior superior iliac spine.

- **sacrum**—Superior to the gluteal cleft, palpate the medial sacral crest (fused spinous processes) portion of the buttocks.

- **coccyx**—The tip of the coccyx can be palpated in the superior portion of the gluteal cleft.

- **ischial tuberosity**—Sit down to palpate this structure that is near the gluteal folds of the inferior buttocks. This is the bony prominence of the hip bone that you sit upon.

- **greater trochanter**—Palpate this bony landmark on the lateral side of your hip. Flex and extend your thigh to feel this structure move with the joint action.

Skeletal Muscles

- **gluteus maximus muscle**—Most students are familiar with these muscles that make up the largest portion of the buttocks. Palpate this prominent muscle as you extend your thigh.

- **gluteus medius muscle**—These muscles are inferior to the iliac crests in the upper, outer quadrant of the buttocks. Palpate this muscle as you shift your weight onto the palpated leg, causing the gluteus medius to contract. These muscles are the site of intramuscular (IM) injections.

Other Structures

- **gluteal (natal) cleft**—A midline crevice between the buttocks.

- **kidneys** [*not shown in Figure 15.2(f) and (g)*]—Place your hands on the iliac crest. Your thumbs land at the approximate location of the kidneys.

(f) Surface anatomy of the posterior male pelvis and gluteal region
- coccyx
- gluteus maximus muscle
- gluteus medius muscle
- greater trochanter
- iliac crest
- ischial tuberosity
- posterior superior iliac spine
- sacrum

1 _____

2 _____

3 _____

4 _____

5 _____

6 _____

7 _____

8 _____

(g) Surface anatomy of the posterior female pelvis and gluteal region
- sacrum
- greater trochanter of the femur
- coccyx
- gluteal cleft
- iliac crest
- posterior superior iliac spine
- gluteus maximus muscle

1 _____

2 _____

3 _____

4 _____

5 _____

6 _____

7 _____

(f) Surface anatomy of the posterior male pelvis and gluteal region

(g) Surface anatomy of posterior female pelvis and gluteal region

FIGURE 15.2 Surface anatomy of the trunk, *continued.*

LAB ACTIVITY 3 **Surface Anatomy of the Upper Limb**

1 Read the location and description of each structure.

2 Locate and label each structure in Figure 15.3(a), (b), (c), and (d).

3 Palpate the designated structures on your body. ■

I. Surface Anatomy of the Shoulder

Structures located in Figure 15.3(a).

Bones and Bone Surface Markings

- **clavicle**—Starting at the manubrium of the sternum, trace the clavicle's S-shaped curvature laterally to its acromial end.

- **acromion**—The flattened lateral end of the spine of the scapula located at the peak of the shoulder. Palpate this prominent projection.

- **acromioclavicular joint**—Palpate the acromioclavicular joint just posterior to the acromial end of the clavicle as you thrust your shoulder joint anteriorly.

- **spine of scapula**—The ridge across the posterior surface of the scapula that extends from the acromion to the medial border of the scapula. Palpate this bone with your fingers. The spine is easier to palpate on a lean body.

- **greater tubercle of humerus**—This bony landmark can be palpated through the deltoid muscle on the

lateral part of the shoulder. As you bend your forearm at the elbow and abduct your arm, move your fingers a little anteriorly to feel the intertubercular groove where the proximal tendon of the long head of the biceps traverses. The greater tubercle is just posterior to this groove.

Skeletal Muscles

- **deltoid muscle**—This large, triangular muscle forms the rounded protrusion of the shoulder. It is used as a site for intramuscular injections. Palpate this muscle as you flex, abduct, and extend the arm to locate the deltoid's anterior, lateral, and posterior portions.

J. Lateral Surface of Arm

Structures located in Figure 15.3(b).

Bones and Bone Surface Markings

- **lateral epicondyle of humerus**—Move your fingers lateral of the olecranon to palpate the lateral epicondyle of the humerus.

Skeletal Muscles

- **biceps brachii muscle**—Flex your forearm and tighten the biceps brachii as you palpate this muscle on the anterior surface of your arm.

- **triceps brachii muscle**—This muscle can be palpated on the posterior surface of the arm when the forearm is extended against resistance. If a person has a muscular arm with definition, all three heads of the triceps can be distinguished.

(a) Right lateral view of shoulder
- acromion
- acromioclavicular joint
- clavicle
- deltoid muscle (lateral fibers)
- greater tubercle of humerus
- spine of scapula

1 _____

2 _____

3 _____

4 _____

5 _____

6 _____

(b) Lateral surface of the arm
- acromion
- deltoid muscle (lateral fibers)
- biceps brachii muscle
- lateral epicondyle of humerus
- triceps brachii muscle

1 _____

2 _____

3 _____

4 _____

5 _____

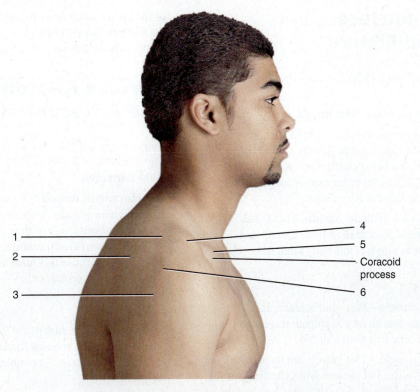

(a) Right lateral view of shoulder

(b) Lateral surface of the arm and elbow

FIGURE 15.3 Surface anatomy of the upper limb.

K. Medial Surface of Arm and Elbow

Structures located in Figure 15.3(c).

Bones and Bone Surface Markings

- **medial epicondyle of humerus**—Move your fingers medially from the olecranon to feel the medial epicondyle. The groove between the olecranon and the medial epicondyle has the **ulnar nerve** that you can palpate when the forearm is extended.

- **olecranon of ulna**—With the forearm flexed, palpate this prominent point at the elbow. Extend and flex the forearm as you feel this bony landmark.

Skeletal Muscles

- **biceps brachii muscle**—Flex your forearm and tighten the biceps brachii as you palpate this muscle on the anterior surface of your arm.

- **triceps brachii muscle**—This muscle can be palpated on the posterior surface of the arm when the forearm is flexed against resistance. If a person has a muscular arm with definition, all three heads of the triceps can be distinguished.

Other Structures

- **groove for brachial artery**—Palpate the medial border of the flexed biceps brachii muscle. You can feel your **brachial pulse** if you press into this groove with your fingers.

- **tendon of biceps brachii muscle**—The distal tendon of the biceps brachii can best be palpated with the forearm flexed. The tendon is located in the middle of the anterior arm and runs from just proximal of the **cubital fossa,** through the cubital fossa, to insert on the radial tuberosity.

L. Surface Anatomy of the Antecubital Region

Structures located in Figure 15.3(d).

Skeletal Muscles

- **brachioradialis muscle**—As its name indicates, this muscle originates in the distal arm (brachium), borders the lateral side of the cubital fossa, and extends to the proximal forearm to insert on the radius. Observe and palpate this muscle as you flex your forearm against resistance.

Other Structures

- **basilic vein**—This superficial vein is seen on the medial side of the anterior forearm and ascends to the arm. This vein is discernible under the skin of lean people.

- **cephalic vein**—This superficial vein can be viewed on the lateral side of the anterior forearm and ascends to the arm. This vein is discernible under the skin of lean people.

- **cubital fossa** (antecubital)—This triangular-shaped concavity is on the anterior surface of the elbow joint.

- **median cubital vein**—This superficial vein diagonally crosses the anterior surface of the elbow in the cubital (antecubital) fossa. This vein connects the lateral cephalic vein with the medial basilica vein. The median cubital vein is a typical site for drawing blood or for inserting an intravenous (IV) catheter.

(c) Medial view of the arm and elbow
- biceps brachii muscle
- groove for brachial artery
- medial epicondyle of humerus
- olecranon of ulna
- tendon of biceps brachii muscle
- triceps brachii muscle

1 _____

2 _____

3 _____

4 _____

5 _____

6 _____

(d) Anterior view of antecubital region
- basilic vein
- biceps brachii muscle
- brachioradialis muscle
- cephalic vein
- cubital fossa
- median cubital vein

1 _____

2 _____

3 _____

4 _____

5 _____

6 _____

(c) Medial view of the arm and elbow

(d) Anterior view of cubital region

FIGURE 15.3 **Surface anatomy of the upper limb,** *continued*.

M. Anterior Surface of the Wrist and Hand

Structures are located in Figure 15.3(e).

The muscles of the forearm are difficult to differentiate from surface anatomy, but four of the muscle tendons are readily discernible in the anterior forearm.

Tendons

- **flexor carpi radialis tendon**—Make a fist to observe and palpate this laterally located tendon.

- **flexor carpi ulnaris tendon**—Make a fist to palpate this most medially located tendon. Tendon is not readily apparent.

- **flexor digitorum superficialis tendon**—Make a fist to observe and palpate this tendon located medial to the palmaris longus tendon.

- **palmaris longus tendon**—Make a fist to observe and palpate this center-most tendon that is located just medial to the flexor radialis tendon.

Bone and Bone Surface Markings

- **pisiform bone**—Located on the medial side just distal to the wrist crease. Palpate this small bone that feels like a bump.

Other Structures

- **radial artery**—Note the location of the radial artery just lateral to the flexor carpi radialis tendon. With three fingers, press down on the radial artery and feel your pulse.

- **hypothenar eminence**—Palpate this small hand pad located medially just distal to the pisiform bone.

- **thenar eminence**—Palpate this larger elevation located between the wrist crease and the thumb.

N. Dorsum of Hand and Wrist

Structures located in Figure 15.3(f).

Bones and Bone Surface Markings

- **head of ulna**—Palpate this distal end of the ulna on the medial side of the wrist.

- **styloid process of ulna**—Palpate just distal to the head of the ulna.

- **styloid process of radius**—Palpate this distal end of the radius on the lateral side of wrist.

Other Structures

- **cephalic vein and dorsal venous arch**—The dorsal venous arch on the dorsum of the hand drains into the cephalic vein. The plexus of veins in the dorsal venous arch is also a site for drawing blood and inserting an IV catheter.

- **anatomical snuffbox**—Palpate this depression located on the dorsum of the hand between the tendon of the extensor pollicis longus muscle and the tendon of the extensor pollicis brevis muscle.

- **tendons of extensor muscles**—Extend the fingers and palpate the tendons of the extensor digitorum muscles.

Tendon of palmaris longus muscle

Tendon of flexor carpi radialis muscle

Radial artery

Wrist crease

Thenar eminence

Tendon of flexor digitorum superficialis muscle

Tendon of flexor carpi ulnaris muscle

Pisiform bone

Hypothenar eminence

(e) Anterior right wrist and hand

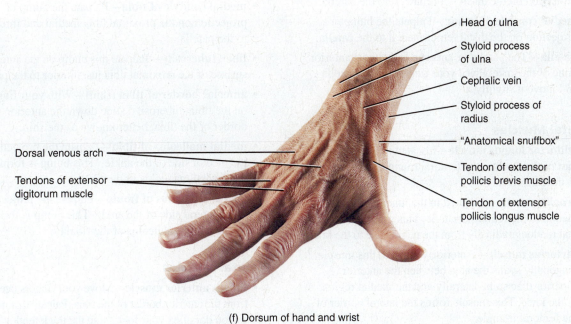

Head of ulna

Styloid process of ulna

Cephalic vein

Styloid process of radius

"Anatomical snuffbox"

Tendon of extensor pollicis brevis muscle

Tendon of extensor pollicis longus muscle

Dorsal venous arch

Tendons of extensor digitorum muscle

(f) Dorsum of hand and wrist

FIGURE 15.3 **Surface anatomy of the upper limb,** *continued.*

LAB ACTIVITY 4 **Surface Anatomy of the Lower Limb**

1 Read the location and description of each structure.

2 Locate and label each structure in Figure 15.4(a) and (b).

3 Palpate the designated structures on your body. ■

O. Anterior Surface of the Lower Limb

Structures located in Figure 15.4(a).

1. Anterior Thigh

Bones and Bone Surface Markings

- **lateral condyle of femur**—Palpate the bump or projection on the distal femur lateral to the patella.
- **medial condyle of femur**—Palpate the bump or projection on the distal femur medial to the patella.
- **patella**—The patella or kneecap is the most anterior bone of the knee. Place your fingers on the patella and move it slightly.

Skeletal Muscles

- **adductor longus muscle**—Located in the superior part of the thigh in the femoral triangle, just medial to the sartorius.
- **gracilis muscle**—Located in the inner thigh region, medial to the femoral triangle. This muscle has fibers that run longitudinally from the pubic bone to the knee.
- **sartorius muscle**—Anteriorly located, this muscle diagonally spans the area between the anterior superior iliac spine laterally and the medial region of the knee. This muscle forms the lateral border of the femoral triangle.
- **rectus femoris muscle**—Located in the anterior compartment of the thigh, lateral to the sartorius. This muscle is in the midline and is the most superficial of the quadriceps femoris muscle group.
- **vastus lateralis muscle**—Large muscle of the quadriceps femoris group located on the lateral surface of the thigh.
- **vastus medialis muscle**—Large muscle of the quadriceps femoris group located on the medial surface of the thigh.

Other Structures

- **patellar ligament**—The quadriceps femoris tendon extends beyond the patella as the patellar ligament and inserts on the tibial tuberosity. Press your fingers on the tendon between the patella and the tibial tuberosity to palpate the patellar ligament as you extend and flex the leg at the knee.
- **femoral triangle** [*not shown on Figure 15.4(a)*]—Triangle of the medial thigh bordered laterally by the sartorius and medially by the gracilis. The **femoral artery, vein,** and **nerve** traverse through this triangle. Palpate your femoral artery to feel your pulse.

2. Leg, Ankle, and Dorsum of Foot

Bones and Bone Surface Markings

- **lateral condyle of tibia**—Palpate the bump or projection on the proximal tibia lateral and inferior to the patella.
- **medial condyle of tibia**—Palpate the bump or projection on the proximal tibia medial and inferior to the patella.
- **tibial tuberosity**—Palpate this bump on the anterior surface of the proximal tibia just inferior to the patella.
- **anterior border of tibia** (shin)—With your fingers on the tibial tuberosity, slide down the anterior border of the tibia, better known as the shin.
- **medial malleolus of tibia**—Palpate the projection at the medial side of the ankle. This bump is formed by the medial malleolus of the tibia.
- **lateral malleolus of fibula**—Palpate the projection at the lateral side of the ankle. This bump is formed by the lateral malleolus of the fibula.

Skeletal Muscles

- **tibialis anterior muscle**—Move your fingers laterally from the anterior border of the tibia. Palpate this muscle as you dorsiflex your foot. Note the thick tendon of this muscle as it crosses medially at the ankle to insert on the medial cuneiform bone and the first metatarsal.
- **extensor digitorum longus muscle**—Palpate the tendon as you dorsiflex your second through fifth toes.
- **fibularis (peroneus) longus muscle**—This lateral muscle of the leg is superficial to the fibula and can be palpated as you plantar flex the foot. Its tendon inserts on the plantar surface of the foot.
- **extensor hallucis longus muscle**—Palpate the tendon as you dorsiflex your great toes.

(a) Anterior surface of the lower limb

FIGURE 15.4 Surface anatomy of the lower limb.

(a) Anterior surface of the lower limb
- adductor longus muscle
- gracilis muscle
- rectus femoris muscle
- sartorius muscle
- vastus lateralis muscle
- vastus medialis muscle

1 _____

2 _____

3 _____

4 _____

5 _____

6 _____

- lateral condyle of femur
- lateral condyle of tibia
- medial condyle of femur
- medial condyle of tibia
- patella
- patellar ligament
- tibial tuberosity

7 _____

8 _____

9 _____

10 _____

11 _____

12 _____

13 _____

- anterior border of tibia (shin)
- lateral malleolus of fibula
- fibularis (peroneus) longus muscle
- medial malleolus of tibia
- tendons of extensor digitorum longus muscle
- tendon of extensor hallucis longus muscle
- tibialis anterior muscle

14 _____

15 _____

16 _____

17 _____

18 _____

19 _____

20 _____

P. Posterior Surface of Lower Limb

Structures located in Figure 15.4(b).

1. Posterior Thigh

Skeletal Muscles

- **biceps femoris muscle**—Lateral hamstring muscle. Palpate muscle while flexing the knee.
- **biceps femoris tendon**—Move your fingers inferiorly until you feel the rope-like that crosses the knee joint.
- **semitendinosus and semimembranosus muscles**—Medial hamstring muscles. Palpate these muscles while flexing the knee.
- **tendon of semitendinosus muscle**—Move your fingers inferiorly until you feel the tendons that crosses the knee joint.

2. Posterior Leg and Ankle

Bone and Bone Surface Markings

- **calcaneus**—Heel bone of the foot.

Skeletal Muscles

- **gastrocnemius muscle** (medial and lateral heads)—Two large bellies of the calf muscles that form most of the superior portion of the calf. To palpate these two heads, extend the knee and place your fingers over the medial and lateral heads of the gastrocnemius. Plantar flex the foot.
- **soleus muscle**—The flatter calf muscle that lies deep to the gastrocnemius. This muscle extends laterally from the gastrocnemius, and its tendon is in the mid to-distal portion of the calf. Palpate the soleus as you plantar flex the foot.

Other Structures

- **popliteal fossa**—A depression located on the posterior of the knee.
- **calcaneal (Achilles) tendon**—The tendons of both the gastrocnemius and the soleus merge to insert on the calcaneus bone of the foot. Palpate this thick tendon as you alternately plantar flex and dorsiflex the foot.

(b) Posterior surface of the lower limb
- biceps femoris muscle
- biceps femoris tendon
- calcaneus
- gastrocnemius muscle (medial and lateral heads)
- semitendinosus and semimembranosus muscles
- soleus muscle
- tendon of semitendinosus

(b) Posterior surface of the lower limb

FIGURE 15.4 **Surface anatomy of the lower limb, continued.**

1 _____

2 _____

3 _____

4 _____

5 _____

6 _____

7 _____

Nervous Tissue

OBJECTIVES

1 Describe the functions of the nervous system

2 Name the 2 major divisions of the nervous system and their organs

3 Explain the difference in function between neurons and neuroglia

4 Identify neuron structures and describe their functions

5 Describe how neurons are classified structurally and functionally

6 Identify unipolar, bipolar, and multipolar neurons

7 Describe the difference between myelinated and unmyelinated axons

8 Identify where the gray and white matter are located in the brain and spinal cord

MATERIALS

- compound microscope, lens paper, prepared slides of astrocytes, motor neurons, dorsal root ganglia, cerebral cortex, and teased myelinated nerve fibers or use **Real Anatomy** (Histology)
- Simulation of Schwann Cell and Axon: zippered 1-quart plastic bag (1 per group), long pencil (1 per group), small smooth pebble or dry bean (1 per group)
- section of brain (human or animal)
- section of spinal cord (human or animal)
- **PowerPhys** Experiment: Graded and Action Potentials

Nervous tissue is found in the organs of the nervous system—the nerves, brain, and spinal cord—and contains cells that enable the nervous system to generate and transmit electrical signals called nerve impulses or action potentials.

A. Overview of Nervous System

The nervous system senses changes in our internal and external environments, coordinates and integrates data, and initiates and transmits action potentials. It is organized into two basic components: the central nervous system (CNS), which consists of the brain and spinal cord, and the peripheral nervous system (PNS). The PNS contains an afferent division composed of sensory receptors and sensory neurons, and an efferent division composed of motor neurons.

Sensory receptors detect changes in the environment and transmit this information along sensory or afferent nerves to the CNS. The CNS coordinates and integrates information received from sensory receptors and initiates responses that are transmitted by neurons to effectors (neurons, muscle cells, or glands). Motor nerves transmit impulses from the CNS to effectors in the PNS. The nervous system is streamlined to send rapid signals from cell to cell to maintain homeostasis and coordinate body organs and functions.

B. Structure of Nervous Tissue

Nervous tissue is located in the brain, spinal cord, ganglia, and nerves, and is composed of 2 types of cells: neurons and neuroglia. **Neurons** conduct action potentials and are the structural and functional units of nervous tissue. **Neuroglia** (*neuro-* = nerve; *-glia* = glue) are cells that support, protect, and furnish nutrients to neurons, and augment the speed of neuron transmission.

1. Neuroglia

Neuroglial cells are generally smaller and more abundant than neurons. Although they do not create action potentials, neuroglial cells have important roles in the nervous system. Of the 6 types of neuroglial cells, 4 are in the CNS and 2 are in the PNS. The 4 neuroglial cells in the CNS are astrocytes, oligodendrocytes, microglia, and ependymal cells. **Astrocytes** (*astro-* = star; *-cyte* = cell) have many processes that make them look star-shaped. Their **perivascular** (*peri-* = around; *vascular* = vessel) **feet** wrap around and cover neurons and blood vessels to keep neurons in place. Astrocytes also guide neurons during development and control the composition of the chemical environment of the neurons by forming a **blood–brain barrier.** This barrier allows only certain substances to enter the nervous tissue at the blood vessel sites. **Oligodendrocytes** (*oligo-* = few; *dendro-* = tree) support the CNS neurons and have processes that form **myelin sheaths** around axons to increase the speed of nerve impulses. **Microglia** (*micro-* = small) are the phagocytes

(*phago-* = to eat) of the CNS that engulf debris, necrotic tissue, and invading bacteria or viruses. **Ependymal cells** (*epen-* = above; *dym-* = garment) line all 4 ventricles (spaces or cavities) of the brain, as well as the central canal of the spinal cord. These cells form cerebrospinal fluid (CSF), and their cilia move the CSF through the ventricles. The 2 neuroglial cells in the PNS are Schwann cells and satellite cells. **Schwann cells** are flattened cells that wrap around the axons in the PNS. Many Schwann cells form the myelin sheath around one axon. The myelin sheath increases nerve impulse speed and aids in the regeneration of PNS axons. **Satellite cells** have processes that are flattened and surround the sensory neuron cell bodies located in ganglia in the PNS. They give support to these neurons and regulate their chemical environment.

Before Going to Lab

1 Complete Table 16.1 using the following list of cells. Under the "Location" heading, circle CNS or PNS.
 - astrocyte
 - ependymal cell
 - microglia
 - oligodendrocyte
 - satellite cell
 - Schwann cell

2 Label the neuroglial cells in Figure 16.1(a) and (b).

TABLE 16.1 Neuroglia

CELL TYPE	LOCATION	FUNCTION
1. _____	CNS or PNS	Entire cell forms myelin sheath around a segment of an axon; helps regeneration of axons.
2. _____	CNS or PNS	Lines four brain ventricles; forms and circulates CSF.
3. _____	CNS or PNS	Engulfs invading microbes; clears debris; migrates to injured nerves.
4. _____	CNS or PNS	Maintains environment around neurons; forms blood–brain barrier.
5. _____	CNS or PNS	Covers sensory neuron cell bodies; maintains neuron environment.
6. _____	CNS or PNS	Processes from cell form myelin sheaths around axons of neurons.

Cells of pia mater

1

2

Neuron

Blood capillary

3

4

Microvillus

Cilium

Node of Ranvier

Myelin sheath

Axon

Neurons

(a) CNS

Neuron cell body in a ganglion

5

6

Axon

(b) PNS

- astrocytes (AS-troh-cytes)
- ependymal (ee-PIN-dih-mahl) cell
- microglial (my-CROG-lee-al) cell
- oligodendrocyte (OL-ih-go-DEN-droh-site)
- satellite cell
- Schwann (shh-WAN) cell

(a) CNS

1 _____

2 _____

3 _____

4 _____

(b) PNS

5 _____

6 _____

FIGURE 16.1 **Neuroglia of the CNS and PNS.**

2. Structure of a Neuron

The longest cells in the body are **neurons,** which can be over 3 feet long. Think about one neuron being long enough to reach from your spinal cord to the tips of your fingers or toes. There are 3 basic parts to any neuron: dendrites, a cell body, and an axon. Both the dendrites and the single axon are **processes** (extensions) of the neuron cell body. **Dendrites** receive information from receptors or other neurons and send it as a change in membrane potential to the neuron **cell body** or **soma** (*soma* = body). Neuron cell bodies have most of the organelles that are present in other types of cells. There is usually a triangular or cone-shaped area of the cell body called the **axon hillock** (*hillock* = small hill). The **axon** (*axon* = axis), a longer process than the dendrites, extends from the axon hillock. Changes in membrane potential travel to the axon hillock where they are integrated to determine whether an action potential will be initiated in the axon. The first part of the axon is known as the **trigger area** (initial segment), where the action potential begins. The axon may be a single process, or it may have side branches called **axon collaterals.** Axons and axon collaterals conduct action potentials along their full lengths to end at fine processes called **axon terminals.** Neurotransmitter molecules are released from axon terminals and transmit signals across a synapse to other neurons or to effectors such as muscles or glands.

Before Going to Lab

1 Label the structures in Figure 16.2(a) and (b). Note the magnification of the cells in these figures.

2 Label the structures in Figure 16.3.

LAB ACTIVITY 1 **Comparison of a Neuron and an Astrocyte**

1 Examine a prepared slide of **astrocytes.**
- Using the low-power objective, find an astrocyte and center it in the field of view.
- Using the high-power objective, identify the **cell body** and **processes.**
- Note the size of the astrocytes.

2 Examine a prepared slide of **motor neurons** or use **Real Anatomy** (Histology).
- Using the low-power objective, find a large motor neuron and center it in the field of view.
- Using the high-power objective, identify the **cell body, nucleus,** and **dendrites.** The axon is difficult to distinguish.
- Note the size of the motor neuron and the numerous small, dark-stained **neuroglial cells** near the neuron.

3 With your lab group compare the size of astrocytes and motor neurons. ■

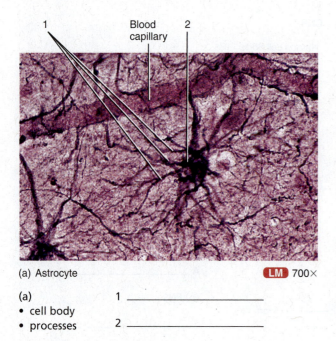

(a) Astrocyte **LM** 700×

(a)
- cell body
- processes

1 _____
2 _____

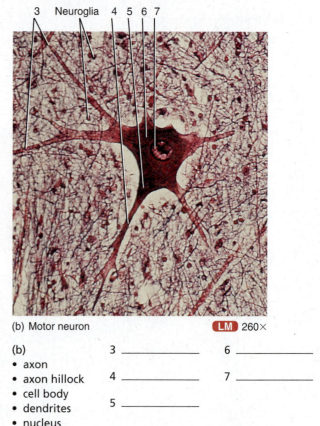

3 Neuroglia 4 5 6 7

(b) Motor neuron **LM** 260×

(b)
- axon
- axon hillock
- cell body
- dendrites
- nucleus

3 _____ 6 _____
4 _____ 7 _____
5 _____

FIGURE 16.2 **Photomicrographs of nervous tissue cells.**

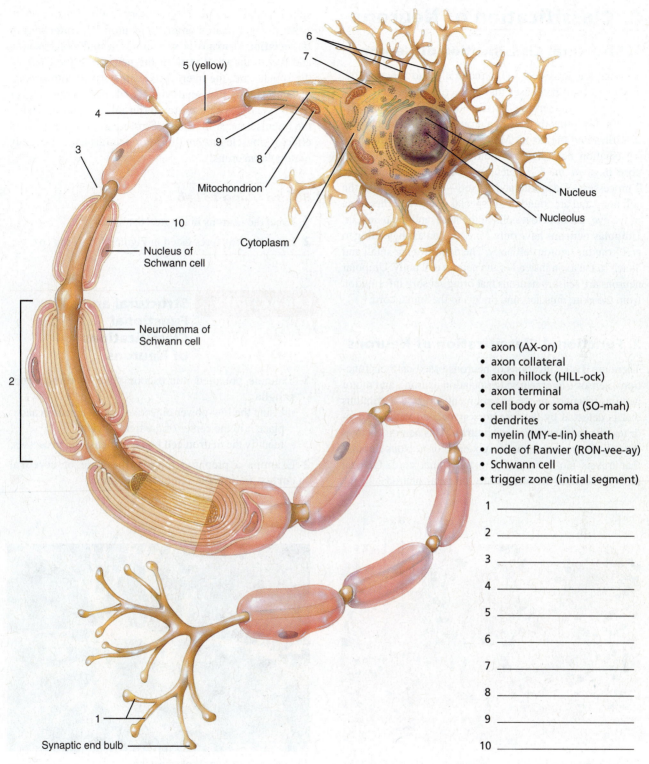

Nucleus

Nucleolus

Mitochondrion

Cytoplasm

5 (yellow)

4

3

9

8

7

6

10

Nucleus of
Schwann cell

Neurolemma of
Schwann cell

2

1

Synaptic end bulb

- axon (AX-on)
- axon collateral
- axon hillock (HILL-ock)
- axon terminal
- cell body or soma (SO-mah)
- dendrites
- myelin (MY-e-lin) sheath
- node of Ranvier (RON-vee-ay)
- Schwann cell
- trigger zone (initial segment)

1 _____

2 _____

3 _____

4 _____

5 _____

6 _____

7 _____

8 _____

9 _____

10 _____

FIGURE 16.3 Parts of a motor neuron.

C. Classification of Neurons

1. Structural Classification of Neurons

Neurons are classified both structurally and functionally. The number of processes that project from the cell body of the neuron determine its structural classification. The **multipolar neurons** have numerous processes, with many dendrites and one axon. Motor neurons and interneurons (association neurons) are multipolar neurons and compose most of the CNS neurons. **Bipolar neurons** have 2 processes—1 dendrite and 1 axon—on either side of the cell body and are found in the special senses like the retina of the eye, the olfactory cells of the nose, and the inner ear. **Unipolar neurons** have only 1 process, an axon, leading to and from the neuron cell body. The dendrites are small and attach to the axon instead of the neuron cell body. Unipolar neurons are sensory neurons that bring sensory information from the skin, muscles, and organs to the spinal cord.

2. Functional Classification of Neurons

There are 3 classifications of neurons based on their functions: sensory, interneuron (association neuron), and motor neuron. Changes in the environment produce a stimulus that is detected by the receptors associated with the dendrites of a **sensory (afferent) neuron.** This neuron changes the stimulation into an action potential or nervous impulse that travels along the axon to the spinal cord. General sensory neurons are structurally unipolar neurons. In the spinal cord, the axon of the sensory neuron synapses with either a motor neuron or an interneuron. The **interneuron (association neuron)** is structurally a multipolar neuron and makes up about 90% of the neurons in the CNS. In the spinal cord, the interneuron can synapse with a chain of interneurons that sends the signal to the brain, and/or it can synapse with a **motor (efferent) neuron** that takes the impulse out of the spinal cord via a spinal nerve to an effector (muscle or gland). Motor neurons are structurally multipolar neurons.

Before Going to Lab

1 Label the neurons in Figure 16.4(a) and (b).

2 Label the structures listed in Figure 16.5(a) and (b).

LAB ACTIVITY 2 · Structural and Functional Classifications of Neurons

1 Examine prepared microscope slide of **dorsal root ganglia.**
- Using the low-power objective, locate a neuron and place it in the center of the field of view.
- Identify the **neuron cell body, nucleus,** and **processes.**

2 Examine a prepared microscope slide of **cerebral cortex.** Follow the same steps as above. ■

(a) Dorsal root ganglia, unipolar neurons **LM** 300×

(b) Cerebral cortex, multipolar neurons

(a)
- neuron cell body
- nucleus
- process
- satellite cells

1 _____

2 _____

3 _____

4 _____

(b)
- axon
- dendrites
- neuron cell body

5 _____

6 _____

7 _____

FIGURE 16.4 **Sectional views of the dorsal root ganglion and cerebral cortex.**

(a) Structural classification of neurons

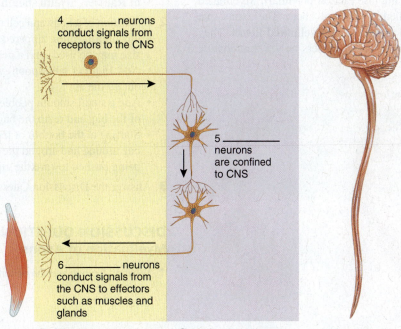

4 _____ neurons conduct signals from receptors to the CNS

5 _____ neurons are confined to CNS

6 _____ neurons conduct signals from the CNS to effectors such as muscles and glands

Peripheral nervous system Central nervous system

(b) Functional classification of neurons

(a) Structural classifications
- bipolar neuron
- multipolar neuron
- unipolar neuron

(b) Functional classifications
- interneuron (association neuron)
- motor neuron (efferent)
- sensory neuron (afferent)

1 _____

2 _____

3 _____

4 _____

5 _____

6 _____

FIGURE 16.5 **Structural and functional classifications of neurons.**

D. Myelination of Axons

Two neuroglial cells, the **oligodendrocytes** (CNS) and the **Schwann cells** (PNS), form insulated wrappings called **myelin sheaths** around axons. As Schwann cells wrap around axons, their cytoplasm is pushed to the periphery and is called **neurolemma.** The multiple layers of myelin that surround the axons are composed of lipoprotein (about 80% lipids and 20% protein), similar to the makeup of plasma membranes. The high amount of lipid in the myelin sheath gives the axons a whitish appearance. Because babies and children do not have the complete myelin coverings encircling the axons, their need for dietary fat is different from that of an adult and is necessary for proper nervous system development. There is not a continuous myelin sheath around the axon, and gaps do exist between the cells. Gaps in the myelin sheaths are called **nodes of Ranvier** and are more numerous in the PNS than in the CNS. The myelin sheath provides protection and insulation for the axon, and also increases the speed of conductivity of the nervous impulse (action potential). Myelinated axons are also called myelinated fibers. The axons that are not myelinated are called **unmyelinated fibers.** They still have neuroglial cells but possess only a thin coating of the Schwann cell or oligodendrocyte plasma membrane covering the axons. Unmyelinated fibers conduct impulses slower than myelinated fibers.

Before Going to Lab

1 Label the structures on Figure 16.6(a) and (b) and identify (a) and (b) as myelinated or unmyelinated axons.

2 Label the structures in the photomicrograph of a teased myelinated nerve fiber in Figure 16.7.

LAB ACTIVITY 3 Myelination of Axons

1 Examine a microscope slide of a teased myelinated nerve fiber.
- Using the low-power objective, locate a nerve fiber and place it in the center of the field of view.
- Using the high-power objective, identify the **nodes of Ranvier, myelin sheath, axon,** and **neurolemma.**

2 Simulation of a Schwann cell and axon.
- Obtain a 1-gallon zippered plastic bag (sandwich size may be used) and a new pencil.
- Add 15 ml (1 tablespoon) water to the bag and push out *all* the air.
- Add a small smooth pebble or dry bean to the bottom of the bag and rezip the bag.
- Starting at the bottom of the bag, wrap the plastic bag around and around the pencil, watching the bean being pushed toward the zippered end.

3 Answer the Discussion Questions with your lab group.

DISCUSSION QUESTIONS
Myelination of Axons

1 What do the bag and the pencil represent?

2 What do the water and bean represent?

3 Where do the water and the bean end up after the bag is totally wrapped around the pencil?

4 Why is it preferable to use a 1-gallon bag versus a sandwich-size plastic bag?

(a) _____ (b) _____

(a)
- axon
- myelin sheath
- node of Ranvier
- Schwann cell cytoplasm

1 _____

2 _____

3 _____

4 _____

(b)
- axons
- Schwann cell cytoplasm

5 _____

6 _____

FIGURE 16.6 Myelinated and unmyelinated axons.

FIGURE 16.7 **Teased myelinated nerve fiber.**

- axon
- myelin sheath (surrounding axon)
- neurolemma (ner-oh-LEM-ma)
- node of Ranvier

1 _____

2 _____

3 _____

4 _____

E. Gray and White Matter in the CNS

Myelin sheaths are white in color, giving nervous tissue with many myelinated axons a white color. Nervous tissue with few myelinated axons appears gray, the color of nervous tissue cells. Groups of myelinated axons in the CNS form **tracts** and are called **white matter,** whereas unmyelinated areas comprised of neuron cell bodies, dendrites, axon terminals, and neuroglia are called **gray matter.**

In the brain, there is an outer area or cortex of gray matter, with an inner layer of white matter. There are also deeper areas within the brain that have isolated areas of gray matter; these are called **nuclei** of the brain and also contain neuron cell bodies and their dendrites. Unlike the brain, the spinal cord has an outer layer of white matter and a central H-shaped area of gray matter.

Before Going to Lab

1 Label gray and white matter in Figure 16.8(a) and (b).

SAFETY NOTE: *Wear safety glasses and gloves when handling preserved or fresh tissue. Always wash your hands thoroughly with soap and water when you are done.*

LAB ACTIVITY 4 **Gray and White Matter**

1 Examine a section of brain and spinal cord, and identify the gray and white matter in each. ■

(a) Sectional view of brain

(b) Transverse view of spinal cord

- gray matter in brain and spinal cord
- white matter in brain and spinal cord

1 _____

2 _____

FIGURE 16.8 **Gray and white matter of brain and spinal cord.**

F. Synapses Between Neurons, Graded Potentials, and Action Potentials

We have previously discussed a type of chemical synapse at the neuromuscular junction where the synaptic end bulb released a neurotransmitter across the synaptic cleft to the receptors in the sarcolemma. Now we will study communication between neurons at a **chemical synapse** that is similar, except the signal transmission is between two neurons. The neuron sending the neurotransmitter is the **presynaptic neuron,** and the neuron receiving the chemical is the **postsynaptic neuron.** The postsynaptic neuron produces a type of **graded potential** called a **postsynaptic potential.** The postsynaptic neuron converts the postsynaptic potential into an electrical signal.

There are three types of neuron-to-neuron synapses that are named according to where the presynaptic neuron anatomically forms a synapse with the postsynaptic neuron. Most synapses are either **axoaxonic** (from axon to axon), **axodendritic** (from axon to dendrite), or **axosomatic** (from axon to the cell body).

Before Going to Lab

1 Identify the type of synapse (#1 through 3) in Figure 16.9, using the bulleted list.
- axosomatic
- axoaxonic
- axodendritic

1. _____
2. _____
3. _____

2 Write the numbers 4 through 8 shown in Figure 16.9 next to the phrase that describes the action that is occurring.

_____ action potential started on postsynaptic neuron

_____ graded potential sent toward trigger zone

_____ action potential reaches axon terminal

_____ action potential travels down axon

_____ dendrites on postsynaptic neurons receive neurotransmitter at synapse

LAB ACTIVITY 5 Graded and Action Potentials

1 Complete the **PowerPhys** Experiment: Graded and Action Potentials. ■

FIGURE 16.9 Types of neuron-to-neuron synapses.

Reviewing Your Knowledge

A. Organization of Nervous System

Fill in the blanks with the correct term chosen from the following:

afferent motor
effectors peripheral
efferent receptors

If you touch a hot stove with your hand, the sensory **(1)** _____ in your hand send

a signal of pain to the CNS through the **(2)** _____ nerve fibers of the

(3) _____ nervous system. When the information

reaches the CNS and is processed, a(n) **(4)** _____ response

is sent through the **(5)** _____ nerve fibers of the

PNS system to skeletal muscles that are **(6)** _____.

B. Nervous Tissue Cells

Write the name of the nervous tissue cell described.

_____ **1.**
_____ **2.** } Supporting cells of PNS

_____ **3.**
_____ **4.**
_____ **5.** } Supporting cells of CNS
_____ **6.**

_____ **7.**
_____ **8.** } Form myelin sheaths

_____ **9.**
_____ **10.** } Regulate chemical environment of neurons

_____ **11.** Generate and transmit nerve impulses

_____ **12.** Line cavities of brain and spinal cord; form and move CSF

_____ **13.** Phagocytes that destroy debris, dead tissue, and pathogens

C. Structural Classification of Neurons

Identify the neuron type described.

_____ **1.** Many processes associated with the cell body

_____ **2.** Has two cell processes

_____ **3.** One short process extends from the cell body and divides

_____ **4.** Neuron is rare and is the sensory neuron in the eye and nose

D. Functional Classification of Neurons

Identify the neuron type described.

_____ **1.** ⎫
 ⎬ Functional neuron types that are structurally multipolar neurons

_____ **2.** ⎭

_____ **3.** ⎫
 ⎬ Neuron types whose cell bodies are in the spinal cord (CNS)

_____ **4.** ⎭

_____ **5.** Neuron type that is structurally either a unipolar neuron or bipolar neuron whose cell body is found in the PNS

_____ **6.** Functional neuron type most prevalent in the CNS

E. Myelination of Axons and Gray and White Matter

Match the description to the appropriate term.

a. gray matter
b. white matter
c. myelinated fibers
d. unmyelinated fibers
e. nodes of Ranvier

_____ **1.** Contains myelinated fibers

_____ **2.** Contains neuron cell bodies and unmyelinated fibers

_____ **3.** Nerve fibers that are white in color and conduct nerve impulses faster

_____ **4.** Nerve fibers that are gray in color

_____ **5.** Gaps in the myelin sheath

Using Your Knowledge

A. Conduction of a Nervous Impulse

Reorder the following parts of a multipolar neuron in the correct order (1–8) of receiving and sending the nerve impulse. Start with the dendrites as *number 1*.

_____ **1.** axon

_____ **2.** axon hillock

_____ **3.** axon terminal

_____ **4.** cell body

_____ **5.** dendrites

_____ **6.** second neuron or effector

_____ **7.** synapse

_____ **8.** trigger zone

B. Nervous Tissue and Diseases

Using your textbook or other references, identify the nervous tissue cell(s) that is (are) involved in the following diseases:

9. Multiple sclerosis

10. Epilepsy

C. Overview of Communication within the Nervous System

Match the numbers in Figure 16.10 with the description of what is happening at that site. Start with number 1.

__1__ Graded potential starts in a sensory receptor in the skin.

_____ A synapse in the thalamus creates a graded potential followed by an action potential in a secondary interneuron, which reaches the cerebral cortex.

_____ The graded potential triggers an action potential in a sensory neuron.

_____ The lower motor neuron forms a neuromuscular synapse with the hand muscles, which causes the muscles to contract as he writes the letter.

_____ In another synapse, the neurotransmitter creates a graded potential, which triggers an action potential in the lower motor neuron.

_____ A primary interneuron forms an action potential and crosses to the opposite side of the brain.

_____ A stimulus from the brain causes a graded potential and then an action potential to form in an upper motor neuron (which crosses back to the original side of the body).

_____ At a synapse, a presynaptic sensory neuron stimulates the postsynaptic interneuron to form a graded potential in its cell body.

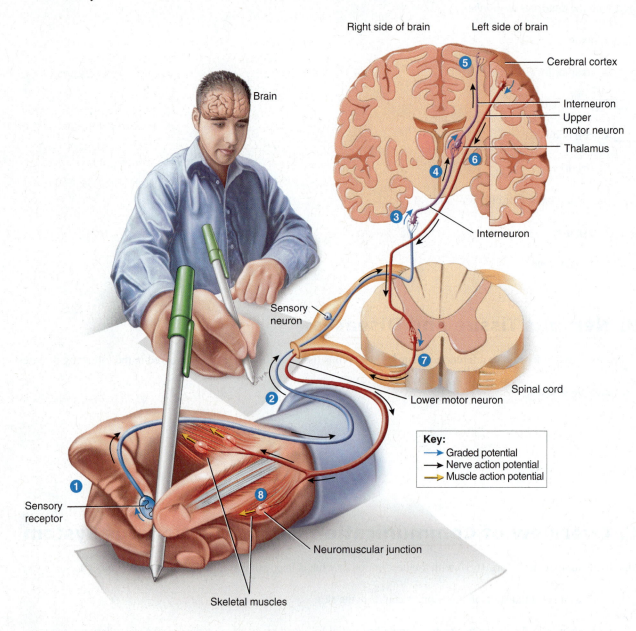

FIGURE 16.10 Overview of nervous system communication.

Spinal Cord Structure and Function

OBJECTIVES

1 Describe the protective structures of the spinal cord
2 Identify and describe the external features of the spinal cord
3 Identify and describe the anatomical features of a spinal cord transverse section

MATERIALS

- models or charts of the complete spinal cord, transverse section of the spinal cord, and vertebral column with spinal cord or use **Real Anatomy** (Nervous)
- compound microscope, lens paper, and prepared microscope slides of spinal cord transverse section
- Dissection: preserved or fresh spinal cord with meninges, dissection equipment, disposable gloves, safety glasses

The spinal cord and the brain make up the central nervous system. Being continuous with the brain, the spinal cord begins at the foramen magnum and terminates between vertebrae L_1 and L_2. It is suspended within the vertebral canal, an area formed by the vertebral foramina of the cervical, thoracic, and lumbar vertebrae. The spinal cord has 2 functions: (1) it carries sensory information to the brain and motor output to nerves, and (2) it mediates spinal reflexes. Spinal reflexes process sensory input from and convey motor output to the spinal nerves.

A. Protective Structures and Spinal Meninges

The **spinal cord** is protected by the bony vertebrae, adipose tissue, spinal meninges, and cerebrospinal fluid. Adipose tissue cushions the spinal cord and is found within the space between the vertebrae and the meninges known as the **epidural space.** The 3 **meninges** (meninx, sing.) or connective tissue membranes cover the spinal cord and are continuous with the cranial meninges that protect the brain. **Dura mater,** the outer meninx, is a tough, single-layered membrane that is deep to the epidural space and superficial to the spider web-like **arachnoid mater.** The inner meninx, the **pia mater,** is delicate and hugs the spinal cord. **Denticulate ligaments** are lateral extensions of pia mater that fuse with arachnoid mater and secure the spinal cord. Between the pia and arachnoid mater is the **subarachnoid space** that contains **cerebrospinal fluid,** which also cushions the spinal cord.

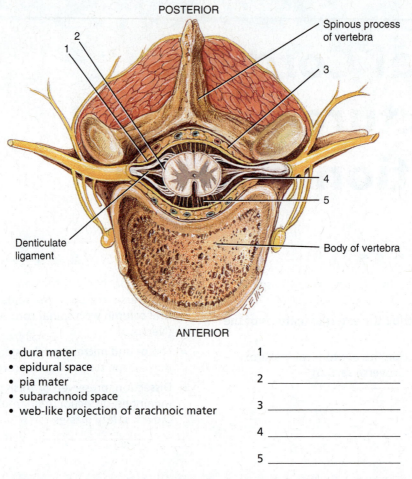

POSTERIOR

2
1
Spinous process
of vertebra

3

4

5

Denticulate
ligament

Body of vertebra

S.E IIIS

ANTERIOR

- dura mater
- epidural space
- pia mater
- subarachnoid space
- web-like projection of arachnoic mater

1 _____

2 _____

3 _____

4 _____

5 _____

FIGURE 17.1 **Transverse section of spinal cord showing meninges.**

Before Going to Lab

1 Label the meningeal structures in Figure 17.1.

LAB ACTIVITY 1 **Spinal Meninges**

1 Identify the meningeal structures in Figure 17.1 on a spinal cord model, chart, or use the search text box in **Real Anatomy** (Nervous) to find these structures. ■

B. External Features of the Spinal Cord

The long, cylindrical spinal cord has 31 pairs of spinal nerves attached to it, with each pair of spinal nerves arising from a different segment of the cord. The cord is wider in the cervical and lumbar regions, forming two enlargements. The **cervical enlargement** is located at levels C_3 or C_4 through T_1. This bulge designates the location of nuclei (collection of neuron cell bodies) for the upper extremities.

The inferior **lumbar enlargement** is located at levels T_9 through T_{12} and contains nuclei for the lower extremities. The spinal cord ends inferiorly as the **conus medullaris** between vertebral levels L_1 and L_2. Nerves arising from the inferior portion of the spinal cord continue inferiorly as a group called the **cauda equina** (*cauda* = tail; *equin-* = horse), or "horse's tail." An extension of the pia mater continues past the conus medullaris as the **filum terminale** (*filum* = filament; *termin-* = terminal) and connects the inferior end of the spinal cord to the coccyx.

Before Going to Lab

1 Label the spinal cord structures in Figure 17.2.

LAB ACTIVITY 2 **External Features of the Spinal Cord**

1 Identify the spinal cord structures in Figure 17.2 on a spinal cord model, chart, or use the search text box in **Real Anatomy** (Nervous) to find these structures. ■

SUPERIOR

Conus medullaris

Dura mater

Posterior (dorsal)
rami of spinal nerves

Cauda equina

Sacrum

Filum terminale

Right coccygeal
nerve

INFERIOR

(b) Posterior view of inferior portion of spinal cord

C1
C2
C3
C4
C5
C6
C7
C8

Cervical
spinal
nerves

1

T1
T2
T3
T4
T5
T6
T7
T8
T9
T10
T11
T12

Thoracic
spinal
nerves

2

3

L1
L2
L3
L4
L5

Lumbar
spinal
nerves

4

S1
S2
S3
S4
S5

Sacral
spinal
nerves

5

(a) Posterior view

- cauda equina (CAU-da ee-QUI-na)
- cervical enlargement
- conus medullaris (CO-nus med-u-LAR-is)
- filum terminale (FI-lum ter-min-AL-ee)
- lumbar enlargement

1 _____

2 _____

3 _____

4 _____

5 _____

FIGURE 17.2 Posterior view of longitudinal spinal cord.

C. Transverse Section of the Spinal Cord

The most obvious parts of the spinal cord in cross-section are the anterior median fissure, the posterior median sulcus, and the gray and white matter. The **anterior median fissure** is a wide, deep groove on the anterior surface of the spinal cord, and the **posterior median sulcus** is a narrow groove on the posterior surface. The **gray matter** looks like a butterfly or a modified "H" and is more centrally located than the white matter. The gray matter is divided into the anterior, lateral, and posterior gray horns and consists of nerve cell bodies and dendrites. Somatic motor neuron cell bodies are located in the **anterior (ventral) gray horns,** whereas the **lateral gray horns** (not present in cervical cord segments) contain cell bodies of autonomic motor neurons. The **posterior (dorsal) gray horns** contain neuron cell bodies that receive impulses from sensory neurons. The **gray commissure** is a narrow bridge of gray matter that connects the right and left sides of gray matter in the middle of the spinal cord. The **central canal** is in the center of the gray commissure and contains cerebrospinal fluid. **White matter** surrounds the gray matter and forms the **anterior, lateral,** and **posterior white columns.** These columns or funiculi are made up of white, myelinated fibers (axons) that are either sensory or motor fibers.

A **spinal nerve** is formed from a **posterior (dorsal) root** and an **anterior (ventral) root.** Roots are collections of axons that are going to and leaving the spinal cord. The posterior (dorsal) root carries sensory fibers, whereas the anterior (ventral) root carries motor fibers. Because the sensory and motor roots merge to form the spinal nerve, these nerves are called **mixed nerves.** Near the spinal cord, there is a bulge in the posterior (dorsal) root called the **posterior (dorsal) root ganglion.** The posterior (dorsal) root ganglion consists of somatic sensory neuron cell bodies that synapse onto interneuron and/or motor neuron cell bodies in the spinal gray matter.

Before Going to Lab

1 Label Figures 17.3 and 17.4.

2 Label the photomicrograph of a transverse section of the spinal cord in Figure 17.5.

LAB ACTIVITY 3 Transverse Section of Spinal Cord

1 Identify the spinal cord structures in Figures 17.3 and 17.4 on a transverse section model or chart of the spinal cord, or use the search text box in **Real Anatomy** (Nervous) to find these structures.

2 Examine a prepared microscope slide of a transverse section of the spinal cord.
 • Using the low-power objective lens, identify the structures listed in Figure 17.5.
 • Using the high-power objective, observe myelinated axons in the white matter and unmyelinated processes, neuron cell bodies, and neuroglia in the gray matter.

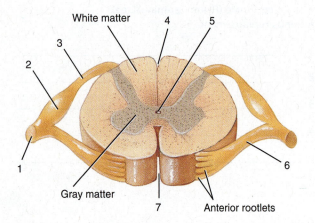

POSTERIOR

White matter

Gray matter

Anterior rootlets

ANTERIOR

FIGURE 17.3 Transverse section of spinal cord.

• anterior median fissure 7
• anterior (ventral) root 6
• central canal 5
• posterior (dorsal) root 3
• posterior (dorsal) root ganglion (GANG-li-on) 2
• posterior median sulcus 4
• spinal nerve 1

1 _____
2 _____
3 _____
4 _____
5 _____
6 _____
7 _____

Transverse section of the thoracic spinal cord

- anterior gray horn
- anterior white column
- gray commissure (COM-mis-sure)
- lateral gray horn
- lateral white column
- posterior gray horn
- posterior white column

1 _____ 5 _____

2 _____ 6 _____

3 _____ 7 _____

4 _____

FIGURE 17.4 **Transverse section of spinal cord with areas of gray and white matter.**

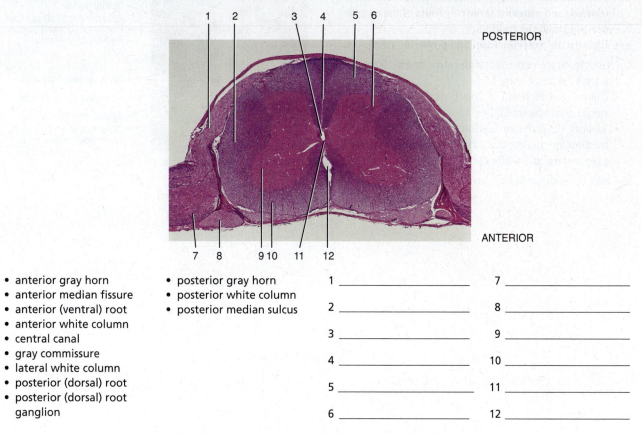

- anterior gray horn
- anterior median fissure
- anterior (ventral) root
- anterior white column
- central canal
- gray commissure
- lateral white column
- posterior (dorsal) root
- posterior (dorsal) root ganglion

- posterior gray horn
- posterior white column
- posterior median sulcus

1 _____ 7 _____

2 _____ 8 _____

3 _____ 9 _____

4 _____ 10 _____

5 _____ 11 _____

6 _____ 12 _____

FIGURE 17.5 **Photomicrograph of transverse section of spinal cord with spinal nerve.**

D. Dissection of the Spinal Cord

Use either a preserved cow or sheep spinal cord or a fresh spinal cord from a butcher. Remember: If the specimen is preserved, it will be firmer and look different than a fresh specimen.

> **SAFETY NOTE:** *Use safety glasses and gloves when handling preserved or fresh tissue. Always wash your hands thoroughly with soap and water when you are done.*

LAB ACTIVITY 4 **Spinal Cord Dissection**

1 Observe the posterior structures of the spinal cord in Figure 17.6(a) and then the anterior structures in Figure 17.6(b).
 • After putting on your gloves, place the spinal cord in the dissection pan.
 • Use a blunt probe or forceps to separate the **spinal meninges.**
 • Identify the **dura** and **arachnoid mater.**
 • Use a pointed dissection probe or pin to detach the **pia mater** from the spinal cord.
 • Identify the **denticulate ligaments.**
 • Peel back the meninges to uncover the **posterior (dorsal)** and **anterior (ventral) roots** of the spinal nerve. Locate the **rootlets.**
 • Identify the **anterior rami** and **posterior rami.**

2 Observe transverse section structures. Refer to Figure 17.4 and Figure 17.5.
 • Cut a 1 × 4 inch to 1 × 2 inch section from your spinal cord specimen.
 • Identify the **anterior median fissure, posterior median sulcus, central canal, gray commissure, gray horns,** and **white columns.**

3 Clean up as directed by your instructor.

SUPERIOR

— Cerebellum of brain (cut)
— Fourth ventricle

— Occipital bone (cut)
— Posterior median sulcus
— Vertebral artery

— Posterior rootlets

— Denticulate ligaments

— Dura mater

INFERIOR
(a) Posterior view

— Denticulate ligament
— Spinal nerve
— Anterior (ventral) ramus
— Posterior (dorsal) ramus
— Pedicle of vertebra (cut)

— Anterior (ventral) root
— Posterior (dorsal) root
— Spinal nerve
— Dura mater and arachnoid mater

(b) Anterior view and oblique section of spinal cord

FIGURE 17.6 **Spinal cord.**

Reviewing Your Knowledge

A. Meninges

Write the name of the structure described.

_____ **1.** Middle meninx; web-like

_____ **2.** Tough, outer meninx

_____ **3.** Space filled with adipose tissue

_____ **4.** Thin meninx intimate with spinal cord

_____ **5.** Contains cerebrospinal fluid

_____ **6.** Extension of pia mater attaching to dura

B. Spinal Cord Structures

Write the terms that match the description.

_____ **1.** Contains neuron cell bodies and unmyelinated processes

_____ **2.** Shallow groove on dorsal side

_____ **3.** Connects right and left halves of gray matter in spinal cord

_____ **4.** Sensory branch of spinal nerve entering spinal cord

_____ **5.** Tapered end of spinal cord

_____ **6.** Motor branch of spinal nerve exiting spinal cord

_____ **7.** Contains sensory neuron cell bodies

_____ **8.** Collection of spinal nerves that arise from inferior end of spinal cord

_____ **9.** Contains myelinated axons

_____ **10.** Contains somatic motor neuron cell bodies

_____ **11.** Space in center of spinal cord that contains cerebrospinal fluid

_____ **12.** Bulge in spinal cord containing cell bodies of motor neurons supplying upper limb

_____ **13.** Wide, deep groove on ventral side

_____ **14.** Extension of pia mater that attaches spinal cord to coccyx

_____ **15.** Bulge in spinal cord at T_9–T_{12}

Using Your Knowledge

A. Spinal Cord Transverse Section

Label Figure 17.7.

POSTERIOR

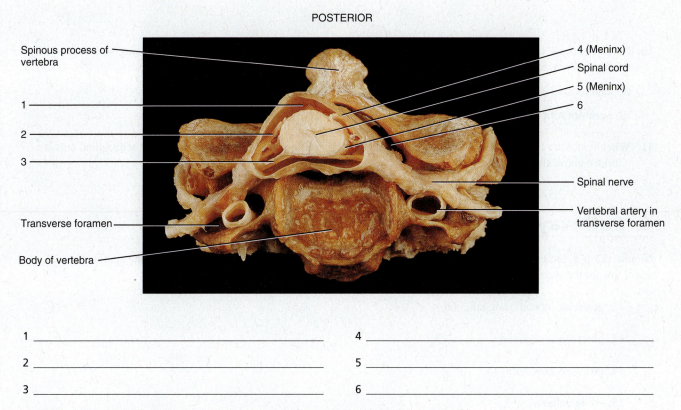

Spinous process of vertebra

1

2

3

Transverse foramen

Body of vertebra

4 (Meninx)

Spinal cord

5 (Meninx)

6

Spinal nerve

Vertebral artery in transverse foramen

1 _____	4 _____
2 _____	5 _____
3 _____	6 _____

FIGURE 17.7 **Photographic cross-section of spinal cord and cervical vertebra.**

B. Spinal Cord Questions

7. The polio virus can cause skeletal muscle paralysis by destroying neuron cell bodies. Identify the area of the spinal cord that is destroyed.

8. Shingles is a condition characterized by pain, discoloration of the skin, and eruption of skin blisters along a sensory nerve. It is caused by the herpes zoster virus, the same virus that causes chicken pox. Using your textbook or another source, identify the spinal cord structure to which herpes virus retreats after chicken pox. The virus remains dormant there until it is activated and causes a shingles outbreak.

9. Identify the structural class(es) of neurons whose cell bodies are present in the spinal cord—unipolar, bipolar, or multipolar.

10. Nerve fibers are classified according to diameter and presence or absence of a myelin sheath. Using your textbook or another reference, name the nerve fiber types (Type A, B, C) present in the:

(a) anterior root

(b) posterior root

11. When removing cerebrospinal fluid during a spinal tap, the needle is inserted below L_2. Explain why spinal taps are not done above this level.

C. Pathway of Sensory and Motor Impulses

Numbering 1–5, indicate the order of structures through which sensory impulses pass as they enter the spinal cord and travel toward the brain.

_____ **12.** posterior (dorsal) root ganglion

_____ **13.** posterior (dorsal) root

_____ **14.** posterior gray horn

_____ **15.** white column

_____ **16.** spinal nerve

Numbering 1–4, indicate the order of structures through which motor impulses pass as they descend from the brain and leave the spinal cord.

_____ **17.** spinal nerve

_____ **18.** white column

_____ **19.** anterior (ventral) gray horn

_____ **20.** anterior (ventral) root

Spinal Nerves

1 Describe the connective tissue coverings of the spinal nerves
2 Identify the rami that carry impulses to and away from the spinal cord
3 Describe the organization and distribution of spinal nerve divisions and the formation of the spinal plexuses
4 Identify the 4 spinal plexuses and the major nerves arising from each plexus

MATERIALS

- compound microscope, lens paper, prepared microscope slide of peripheral nerve cross-section, and prepared microscope slide of peripheral nerve longitudinal section
- model or chart of vertebral column with spinal cord and spinal nerves or use **Real Anatomy** (Nervous)
- model or chart with spinal nerves of the upper and lower extremities or use **Real Anatomy** (Nervous)
- Dissection: preserved cat or fetal pig, dissection equipment, dispoable gloves, safety glasses, and dissection manual
- **Real Anatomy:** Virtual Cadaver Dissection

Spinal nerves send information from peripheral **sensory receptors** to the spinal cord and information from the spinal cord to **effectors** (muscles and glands). The 31 pairs of spinal nerves emerge from each side of the spinal cord through the **intervertebral foramina** and are named for the vertebral region and level from which they emerge. Spinal nerves connect to the spinal cord via a **posterior (dorsal) root** and an **anterior (ventral) root** and are called **mixed nerves** because each nerve contains **sensory** and **motor axons**.

A. Connective Tissue Coverings of Spinal Nerves

Each spinal nerve has 3 protective connective tissue layers: the **epineurium** (*neuri-* = nerve) that surrounds the whole nerve, the **perineurium** that encases each **fascicle** (*fasciculus* = little bundle), and the **endoneurium** that covers myelinated and unmyelinated axons.

1 Label Figures 18.1 and 18.2.

LAB ACTIVITY 1 **Connective Tissue Coverings of Spinal Nerves**

1 Examine a prepared microscope slide of a cross-section of a spinal nerve.
- Using the low-power objective lens, identify the **epineurium, fascicles,** and **perineurium.**
- Using the high-power objective lens, identify the **endoneurium, axons,** and the **myelin sheath** of **myelinated neurons.**

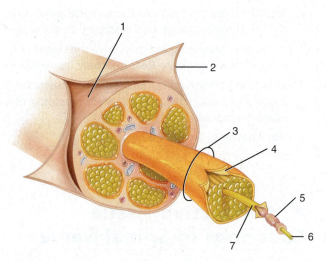

FIGURE 18.1 **Transverse section showing the coverings of a spinal nerve.**

B. Rami of the Spinal Nerves

The spinal nerves branch lateral to the intervertebral foramen. These branches or rami (*rami* (pl.) = branches; *ramus* (sing.) = branch) are the posterior (dorsal) ramus, the anterior (ventral) ramus, the meningeal branch, and the rami communicantes. The **posterior (dorsal) ramus** curves around to the dorsal surface and innervates the skin and deep muscles of the back or trunk. The **anterior (ventral) ramus** supplies the muscles and skin of all four limbs, as well as the anterior and lateral parts of the body. The **meningeal** branch serves the vertebrae, vertebral ligaments, blood vessels of the spinal cord, and the meninges. The 2 **rami communicantes** (*communicans* = communicating) connect to the **sympathetic ganglion** (*sympathein* = to feel with) of the autonomic nervous system.

- axon (AX-on)
- endoneurium (endo-NEUR-i-um)
- epineurium (epi-NEUR-i-um)
- fascicle (FAS-i-cul)
- myelin sheath
- perineurium (peri-NEUR-i-um)
- spinal nerve

1 _____

2 _____

3 _____

4 _____

5 _____

6 _____

7 _____

- axon
- endoneurium
- myelin sheath
- perineurium

1 _____

2 _____

3 _____

4 _____

FIGURE 18.2 **Photomicrograph of a transverse section through a fascicle of a spinal nerve.**

1 Label Figure 18.3.

LAB ACTIVITY 2 **Rami of the Spinal Nerves**

1 Identify the structures from Figure 18.3(a) and (b) on a model or chart. ■

POSTERIOR

Posterior (dorsal) root

Anterior (ventral) root

3 _____

2 _____

1 _____

Meningeal branch

4

Sympathetic ganglion

ANTERIOR

(a) Transverse section

- anterior (ventral) ramus (RAY-mus)
- posterior (dorsal) ramus
- rami communicantes (RAY-my com-mun-i-CAHN-tayce)
- spinal nerve

1 _____

2 _____

3 _____

4 _____

6

7

8

9

5

(b) Lateral view

- anterior (ventral) ramus
- intervertebral foramen
- posterior (dorsal) ramus
- rami communicantes
- sympathetic ganglion

5 _____

6 _____

7 _____

8 _____

9 _____

FIGURE 18.3 Spinal cord with branches of a spinal nerve.

C. Spinal Nerve Divisions and the Four Spinal Plexuses

Of the 31 pairs of spinal nerves, there are **8 cervical, 12 thoracic, 5 lumbar, 5 sacral, and 1 coccygeal.** Many spinal nerves join with other spinal nerves to form a braided network or **plexus** before they innervate body structures. This occurs in 4 regions of the body where the networks form the **cervical, brachial, lumbar,** and **sacral plexuses.** The **thoracic** (intercostal) spinal nerves (T_2–T_{12}) do not form a plexus.

Before Going to Lab

1 Label the cervical, thoracic, lumbar, sacral, and coccygeal nerves in Figure 18.4.

2 Label the 4 spinal plexuses also in Figure 18.4.

LAB ACTIVITY 3 **Spinal Nerve Divisions and Spinal Plexuses**

1 Identify the spinal nerve divisions and plexuses from Figure 18.4 on a model, chart, or use the search text box in **Real Anatomy** (Nervous) to find these structures. ■

- brachial plexus (PLEX-us)
- cervical nerves
- cervical plexus
- coccygeal (cox-sih-GEAL) nerve
- lumbar nerves
- lumbar plexus
- sacral nerves
- sacral plexus
- thoracic nerves

1 _____

2 _____

3 _____

4 _____

5 _____

6 _____

7 _____

8 _____

9 _____

FIGURE 18.4 **Posterior view of the four spinal plexuses.**

D. Major Nerves from the Cervical and Brachial Plexuses

The **cervical plexus** is formed from the anterior (ventral) rami of C_1–C_5 on both the right and left sides of the spinal cord. An important paired nerve from this plexus is the **phrenic nerve** (*phreni-* = relating to the diaphragm), C_3–C_5, which innervates the diaphragm and is important for breathing. Remember the saying, *"Cervical nerves 3, 4, and 5 keep the diaphragm alive."* Other cervical nerves mainly supply the scalp, neck, shoulders, and chest. The **brachial plexus** is formed from the ventral rami of C_5–T_1. This plexus serves the shoulders and upper limbs. The main nerves that arise from the brachial plexus are the **axillary, median, musculocutaneous** (*musculo-* = muscle; *cutan-* = skin), **radial,** and **ulnar.**

> **Before Going to Lab**
>
> **1** Label the spinal nerves in Figure 18.5.

LAB ACTIVITY 4 **Major Nerves from the Cervical and Brachial Plexuses**

1 Identify the nerves from Figure 18.5 on a model, chart, or use the search text box in **Real Anatomy** (Nervous) to find these structures.

2 Identify approximate location of nerves in Table 18.1 on yourself and demonstrate muscle action that occurs when each nerve is stimulated.

3 Test nerve conduction.
- Flex your forearm to 90°.
- Palpate a cord-like nerve between your medial epicondyle and olecranon process.
- Place pressure on the nerve with your fingers until you feel a difference in your forearm, hand, and digits.
- Describe what sensations you felt and what nerve was compressed.
- Describe what digits are involved and what produced your symptoms.

4 Discuss your results with the class.

ANTERIOR VIEW
5 (Posterior to bone)

- axillary nerve
- median nerve
- musculocutaneous (mus-cu-lo-cu-TAYN-e-ous) nerve
- phrenic (FREN-ic) nerve
- radial nerve
- ulnar nerve

1 _____
2 _____
3 _____
4 _____
5 _____
6 _____

FIGURE 18.5 **Major nerves from the cervical and brachial plexuses.**

TABLE 18.1 Motor Function of Major Nerves from Cervical and Brachial Plexuses

NERVE	MUSCLES INNERVATED BY NERVE
Phrenic	Diaphragm
Axillary	Deltoid and teres minor muscles
Musculocutaneous	Anterior muscles of the arm
Ulnar	Flexor carpi ulnaris, medial-half of flexor digitorum profundis, and most hand muscles
Median	Muscles of anterior forearm (excluding flexor carpi ulnaris and other muscles supplied by ulnar nerve) and some of the muscles of the hand
Radial	Muscles of posterior arm and forearm

E. Major Nerves from the Lumbar and Sacral Plexuses

The **lumbar plexus** is made up of anterior (ventral) rami from L_1–L_4. The lumbar plexus supplies the skin and muscles of the abdominal wall, external genitalia, and part of the lower limbs. The major nerves are the **femoral** and **obturator nerves.**

The **sacral plexus** is formed from the anterior (ventral) rami from L_4–S_4. This plexus supplies the buttocks, perineum, and most of the lower limbs. Its major nerves are the **pudendal** and **sciatic.** The sciatic is composed of the **tibial** and **common fibular (peroneal)** nerves.

Before Going to Lab

1 Label the nerves listed in Figure 18.6.

LAB ACTIVITY 5 Major Nerves from the Lumbar and Sacral Plexuses

1 Identify the nerves from Figure 18.6 on models, charts, or use the search text box in **Real Anatomy** (Nervous) to find these structures.

2 Identify approximate location of nerves in Table 18.2 on yourself and demonstrate muscle action that occurs when each nerve is stimulated. ▪

SAFETY NOTE: *Wear safety glasses and gloves when using fresh or preserved tissue. Wash hands thoroughly with soap and water when you are done.*

F. Dissection of Nerve Plexuses and Major Nerves

If you are dissecting an animal to observe nerves, refer to the dissection manual. Major nerves of animal plexuses are similar to human nerves and will reinforce your knowledge of human nerves. This dissection will allow you to observe the fascia that surrounds and protects the nerves as well as showing you how nerves dive between skeletal muscles to surface in other locations. **Real Anatomy,** a virtual cadaver dissection, can be used to complement or substitute for animal dissection of nerves.

ANTERIOR VIEW POSTERIOR VIEW

(b) Distribution of nerves from the lumbar and sacral plexuses

- femoral (FEM-o-rul) nerve 1 _____
- obturator (OB-tur-a-tor) nerve 2 _____
- common fibular (peroneal) 3 _____
- pudendal (pyoo-DEN-dal) 4 _____
- sciatic (sci-A-tic) 5 _____
- tibial 6 _____

FIGURE 18.6 Major nerves of the lumbar and sacral plexuses.

TABLE 18.2 Motor Function of Nerves from the Lumbar and Sacral Plexuses

NERVE	MUSCLES INNERVATED BY NERVES
Femoral	Iliacus, quadriceps femoris, sartorius, pectineus
Obturator	Adductor longus, adductor brevis, and part of adductor magnus, gracilis
Common fibular	Fibularis longus muscle, tibialis anterior, extensor digitorum longus
Pudendal	Muscles of perineum
Sciatic	Biceps femoris, semimembranosus, semitendinosus (hamstrings)
Tibial	Gastrocnemius, soleus, tibialis posterior, flexor digitorum longus, plantaris, flexor hallucis longus

Reviewing Your Knowledge

A. Connective Tissue Coverings

Identify the connective tissue covering.

_____ **1.** Covers unmyelinated or myelinated axons

_____ **2.** Covers fascicles

_____ **3.** Covers nerves

B. Rami of Spinal Nerves

Identify the spinal nerve ramus.

_____ **1.** Branch that serves the deep muscles and skin of the back

_____ **2.** Branches that belong to the sympathetic nervous system

_____ **3.** Branch that forms nerves serving the limbs

C. Major Nerves of the Nerve Plexuses

Identify the nerve plexus from which each nerve originates.

_____ **1.** femoral n.

_____ **2.** sciatic n.

_____ **3.** phrenic n.

_____ **4.** ulnar n.

_____ **5.** axillary n.

_____ **6.** tibial n.

_____ **7.** obturator n.

_____ **8.** radial n.

_____ **9.** common fibular n.

_____**10.** pudendal n.

D. Spinal Nerves

Complete the sentences with the correct term about the spinal nerves.

1. The sciatic nerve is composed of two nerves, the _____.

2. and the _____.

3. There are _____ pairs of spinal nerves.

4. The nerve that supplies the posterior thigh _____.

5. There is (are) _____ pair(s) of coccygeal nerves.

6. If the anterior (ventral) ramus of a spinal nerve is severed, there is a _____ (motor and/or sensory) loss.

7. There is (are) _____ pair(s) of thoracic nerves.

8. The nerve that supplies the anterior thigh is the _____.

9. There is (are) _____ pair(s) of sacral nerves.

10. There is (are) _____ pair(s) of lumbar nerves.

11. If the posterior (dorsal) ramus of a spinal nerve is severed, the functional loss is _____ (motor and/or sensory).

12. There is (are) _____ pair(s) of cervical nerves.

13. The _____ nerve supplies the deltoid and teres minor muscles.

14. The _____ nerve supplies the triceps brachii muscles and the extensor digitorum longus.

15. The _____ nerve supplies the biceps brachii muscles.

16. The _____ nerve supplies the diaphragm.

17. The _____ nerve supplies most hand muscles.

18. The _____ nerve supplies the flexor carpi radialis.

Using Your Knowledge

Identify the correct nerve for each question.

_____ **1.** A broken forearm resulted in an inability to pronate the forearm and loss of finger movement in digits 1–3. Name the nerve that was injured.

_____ **2.** An injection into the shoulder results in an inability to extend the wrist and fingers. Name the nerve that was injured.

_____ **3.** Health care professionals are taught how to properly administer gluteal injections to avoid pain and injury caused by inadvertently striking a major nerve. Name the nerve to avoid.

_____ **4.** Hitting the medial epicondyle results in a tingling sensation in part of the hand. Name the nerve hit and the part of the hand that tingles.

_____ **5.** John Jones injured his spinal cord. He has use of his serratus anterior muscle, biceps brachii, and deltoid. However, he does not have movement in most muscles of his hand and digits 4 and 5. Where is his spinal cord injury?

_____ **6.** Following the birth of her daughter, Mary had trouble adducting her lower limbs. Which nerve was injured during childbirth?

_____ **7.** Charles broke his leg playing softball. The fracture was a compound fracture of the fibula. After the cast was removed, he experienced difficulty dorsiflexing his foot. Which nerve was affected?

Identify the nerve that would carry pain impulses from each of the following injured areas.

_____ **8.** Greenstick fracture of the tibia

_____ **9.** Sunburn on the skin over deltoid muscles

_____ **10.** Splinter in digit 5 of the hand

Somatic Reflexes

OBJECTIVES

1 Identify and describe the five components of a somatic reflex arc

2 Describe how monosynaptic and polysynaptic reflex arcs differ

3 Test and describe somatic reflexes

4 Describe the effect of distractions on reflexes

MATERIALS

- reflex hammer (with rubber head)
- cross-section spinal cord model
- **BIOPAC** Systems, Inc. WWW.BIOPAC.COM **Biopac Laboratory Guide Experiments:**
 - Effect of Physical and Mental Distractions on Patellar Reflex (Knee Jerk) Response

Reflexes are rapid, involuntary motor responses to an environmental stimulus detected by sensory receptors. A nerve impulse travels from the receptor through a neural **reflex arc** pathway to an effector. If the motor response is contraction of skeletal muscle, the reflex is a **somatic reflex.** If the motor response involves cardiac muscle, smooth muscle, or glands, the reflex is an **autonomic (visceral) reflex.** Reflexes mediated by spinal nerves are called **spinal reflexes,** whereas reflexes mediated by cranial nerves are called **cranial reflexes.** Most reflexes help our bodies maintain homeostasis and therefore have a protective function.

A. Reflex Arcs

There are 5 components of a reflex arc:

1. **Sensory receptor.** If the stimulus to the sensory receptor is strong enough, an action potential is generated in the sensory neuron.

2. **Sensory neuron.** The sensory neuron propagates the action potential and synapses with neurons in the spinal cord or brain stem.

3. **Integrating center.** The integrating center is located within the gray matter of the central nervous system (CNS) and transfers information from the sensory neuron to the motor neuron. The integrating center of a **monosynaptic reflex arc** is a single synapse between a sensory and motor neuron. In a **polysynaptic reflex arc**, the integrating center consists of multiple synapses involving one or more interneurons (association neurons) between a sensory and a motor neuron.

4. **Motor neuron.** The motor neuron carries the action potential initiated by the integrating center to the effector.

5. **Effector.** An effector can be skeletal muscle (somatic reflex), cardiac muscle, smooth muscle, or glands (autonomic reflexes). A **reflex** is the response of the effector to stimulation by the motor neuron of the reflex arc.

Reflex arcs that involve sensory receptors, sensory nerve fibers, motor neurons, and effectors all on the same side of the body are **ipsilateral. Contralateral** reflex arcs involve sensory receptors and neurons on one side of the body and motor neurons and effectors on the opposite side. **Bilateral** (consensual) reflexes involve both sides of the body simultaneously.

Before Going to Lab

1 Label Figure 19.1.
2 Identify whether the reflexes in Table 19.1 are somatic reflexes or autonomic reflexes, and whether a spinal nerve or a cranial nerve mediates it.

LAB ACTIVITY 1 Reflexes

1 Using the cross-section spinal cord model or chart, trace the pathway of the reflex arc illustrated in Figure 19.1.
2 Answer the Discussion Questions with your lab group.

DISCUSSION QUESTIONS
Reflex Arc

1 Is the reflex arc in Figure 19.1 monosynaptic or polysynaptic?

2 Is the reflex arc in Figure 19.1 ipsilateral or contralateral?

TABLE 19.1 Types of Reflexes

REFLEX ACTION	SOMATIC OR AUTONOMIC REFLEX	SPINAL NERVE REFLEXES OR CRANIAL NERVE REFLEXES
Plantar flexion of foot when Achilles tendon is stretched		
Salivation in response to lemon juice on tongue		
Blinking in response to touching the cornea		
Flexing of arm in response to touching a hot object		

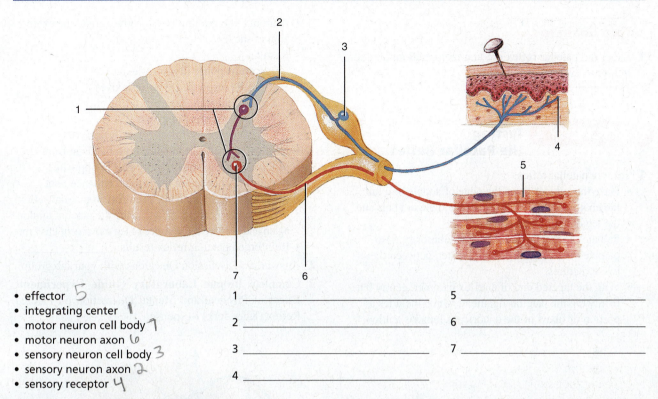

- effector 5
- integrating center 1
- motor neuron cell body 7
- motor neuron axon 6
- sensory neuron cell body 3
- sensory neuron axon 2
- sensory receptor 4

1 _____ 5 _____

2 _____ 6 _____

3 _____ 7 _____

4 _____

FIGURE 19.1 Reflex arc components.

B. Reflex Testing

A series of reflex tests are used clinically to evaluate the nervous system and to diagnose an abnormality or dysfunction that may cause an inhibition, exaggeration, or absence of reflexes.

1. Patellar Reflex (Knee Jerk)

The **patellar reflex** is the extension of the knee that occurs when the quadriceps femoris tendon is stretched. A stretched tendon will stretch the muscle and cause it to contract. This reflex helps prevent injury by preventing muscles from overstretching. The stretch reflex helps maintain posture and equilibrium.

The components of the **patellar reflex arc** are:

- Sensory receptors: **Muscle spindles** located in the quadriceps femoris muscle group. Tapping the quadriceps tendon stretches this muscle group and stimulates muscle spindles (stretch receptors), initiating nerve impulses in axons of sensory neurons.

- Sensory neuron: Sensory axons carry nerve impulses to the integrating center (gray matter) in the spinal cord.
- Integrating center: The **monosynaptic integrating center** (two neurons, one synapse) is located in the anterior gray horn of the spinal cord. The sensory axon synapses with and initiates nerve impulses in the motor neuron that innervates the quadriceps femoris muscle group.
- Motor neuron: Axons of the motor neuron travel in the femoral nerve to the quadriceps femoris muscle group.
- Effector: Quadriceps femoris muscle group (agonists) contracts and extends the leg when stimulated. The sensory neuron associated with the monosynaptic stretch reflex arc is also a component of a **polysynaptic reflex arc** that has 3 neurons and 2 synapses. This reflex arc inhibits the motor neuron to the antagonist muscle group and results in its relaxation. This is referred to as **reciprocal innervation**—stimulation of contraction in agonist muscles with simultaneous inhibition of antagonistic muscles.

Before Going to Lab

1 Label the patellar reflex arc and reciprocal innervation in Figure 19.2.

LAB ACTIVITY 2 Testing the Patellar Reflex

1 Test the patellar reflex.
- Have the subject sit on the edge of a table or a tall lab chair with his/her knee off the table or chair and the leg dangling and relaxed.
- Palpate the patella and the tibial tuberosity; also palpate the patellar ligament located between these two structures.
- With the tapered end of a reflex hammer, gently but firmly tap the patellar ligament, which includes a portion of fibers of the quadriceps femoris tendon.

- Although the grading is subjective, grade the extension of the leg:

 Results: _____

 0 = no response (hypo-reflexive)
 +1 = little response (hypo-reflexive)
 +2 = normal response
 +3 = above normal response
 +4 = exaggerated reflex (hyper-reflexive)
- Now ask the subject to clasp both hands in front of the chest and isometrically pull in opposite directions. This action leads to an enhancement of spinal reflexes causing a reinforced patellar reflex.
- While the subject is pulling, strike the patellar tendon again and observe the distance of leg movement this time.
- Reinforced patellar reflex results: _____

2 Answer the Discussion Questions with your lab group.

3 Complete **Biopac Laboratory Guide Experiment:** Effect of Physical and Mental Distraction on Patellar Reflex (Knee Jerk) Response.

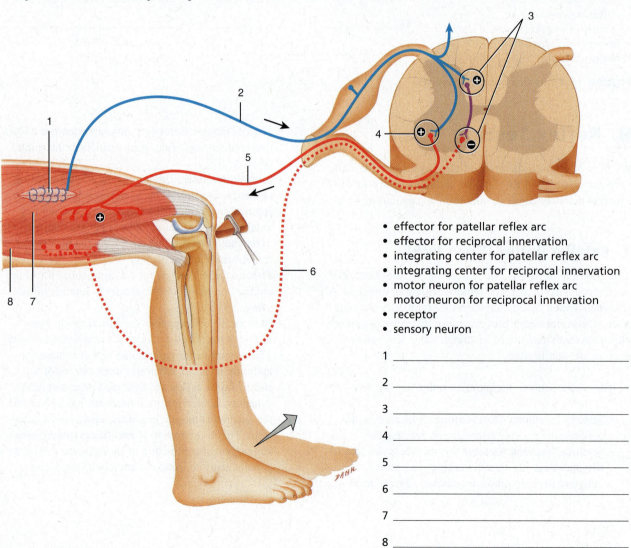

- effector for patellar reflex arc
- effector for reciprocal innervation
- integrating center for patellar reflex arc
- integrating center for reciprocal innervation
- motor neuron for patellar reflex arc
- motor neuron for reciprocal innervation
- receptor
- sensory neuron

1 _____

2 _____

3 _____

4 _____

5 _____

6 _____

7 _____

8 _____

FIGURE 19.2 Patellar reflex arc.

DISCUSSION QUESTIONS
Patellar Reflex

1 Is the patellar reflex ipsilateral or contralateral?

2 During the patellar reflex, are the motor neurons supplying the hamstrings stimulated or inhibited?

3 Describe the effect that clasping and pulling the hands has on the patellar reflex compared with the first patellar reflex.

2. Biceps Reflex (Biceps Jerk)

The **biceps reflex** is the contraction of the biceps brachii muscle that occurs when the biceps tendon is stretched. Although the biceps brachii flexes the forearm, in this test you may only see the contraction or tension in the biceps brachii muscle because the forearm is resting on the table.

LAB ACTIVITY 3 **Testing the Biceps Reflex (Biceps Jerk)**

1 Test the biceps reflex.
- Have the subject stand with an arm completely relaxed and hanging down at the side.
- Ask the subject to isometrically contract the biceps brachii so you can palpate the tendon in the antecubital fossa.
- After locating the tendon, have the subject relax and place your thumb over the tendon.
- Gently but firmly tap your thumb with the tapered end of the reflex hammer.
- Biceps reflex results: _____
 - 0 = no response
 - +1 = little response
 - +2 = normal response
 - +3 = above normal response
 - +4 = exaggerated reflex
- If the reflex is absent, have the subject clench his/her teeth and try the reflex test again.
- Reinforced biceps reflex results: _____

2 Answer the Discussion Questions with your lab group.

DISCUSSION QUESTIONS
Biceps Reflex

1 Name the nerve that carries the sensory and motor axons for this reflex arc.

2 Name the antagonistic muscles that are inhibited by reciprocal innervation.

3. Triceps Reflex (Triceps Jerk)

The triceps reflex is the contraction of the triceps brachii muscle that occurs when the triceps brachii tendon is stretched.

LAB ACTIVITY 4 **Triceps Reflex (Triceps Jerk)**

1 Test the triceps reflex.
- Have the subject stand with an arm completely relaxed and hanging down at the side.
- Palpate the triceps tendon (proximal to olecranon process) as the subject isometrically contracts the triceps muscles.
- Keep your finger on the tendon and have the subject bend the arm across the front of the body.
- Have the subject support the arm by holding it in his/her other hand.
- Use the tapered end of the reflex hammer to tap the triceps tendon.
- Triceps tendon reflex results: _____
 - 0 = no response
 - +1 = little response
 - +2 = normal response
 - +3 = above normal response
 - +4 = exaggerated reflex
- If the reflex is absent, have the subject clench his/her teeth and try the reflex again.
- Reinforced triceps tendon results: _____

2 Answer the Discussion Questions with your lab group.

DISCUSSION QUESTIONS
Triceps Reflex

1 Name the nerve that carries the sensory and motor axons for this reflex arc.

2 Name the antagonistic muscles that are inhibited by reciprocal innervation.

4. Achilles Reflex (Ankle Jerk)

The response of this reflex is plantar flexion when the Achilles (calcaneal) tendon is tapped with the reflex hammer.

LAB ACTIVITY 5 **Achilles Reflex (Ankle Jerk)**

1 Test the Achilles reflex.
- Have the subject keep one foot on the floor and place the knee of the other leg on a chair with the foot dangling over the edge.
- Ask the subject to slightly dorsiflex the foot to stretch the tendon a little, but be relaxed.
- Tap the calcaneal tendon with the tapered end of a reflex hammer and observe the response of the foot.
- Calcaneal reflex results: _____
 - 0 = no response
 - +1 = little response
 - +2 = normal response
 - +3 = above normal response
 - +4 = exaggerated reflex

2 Answer the Discussion Questions with your lab group.

DISCUSSION QUESTIONS
Achilles Reflex

1 Name the nerve that carries the sensory and motor neurons for this reflex arc.

2 Name the antagonistic muscles that are inhibited by reciprocal innervation.

5. Plantar Flexion

Plantar flexion is a superficial cord reflex that is an important neurological test. This test stimulates the cutaneous receptors and involves the brain in addition to the spinal cord. In adults, **plantar flexion** and flexed (curled) toes occurs when the plantar surface of the foot is stroked. A reaction of dorsiflexion with extended flared toes (Babinski's sign) is seen in infants because all the nerve fibers are not myelinated. The Babinski's sign is abnormal in adults.

LAB ACTIVITY 6 **Plantar Flexion**

1 Test the plantar flexion reflex.
- Have the subject seated with a foot propped up and relaxed.
- Using the metal end of the reflex hammer, stroke the plantar surface of the foot starting at the heel, extending up the lateral side of the foot, and crossing over to the great toe area.
- Plantar flexion results: _____
 - 0 = no response
 - +1 = little response
 - +2 = normal response
 - +3 = above normal response
 - +4 = exaggerated reflex

2 Answer Discussion Questions with your lab group.

DISCUSSION QUESTIONS
Plantar Flexion

1 Name the nerve that carries the sensory and motor neurons for this reflex arc.

2 Name the antagonistic muscles that are inhibited by reciprocal innervation.

Reviewing Your Knowledge

A. Reflex Arc

1. Name the 5 components of a reflex arc in order.

a. _____

b. _____

c. _____

d. _____

e. _____

2. How many neurons are in a monosynaptic reflex arc? _____
How many synapses are in the integrating center? _____

3. How many neurons are in a polysynaptic reflex arc? _____
How many synapses are in a polysynaptic reflex arc containing two interneurons in the integrating center? _____

4. Which type of neuron does the sensory neuron synapse with in a monosynaptic reflex arc? _____

5. Which type of neuron does the sensory neuron synapse with in a polysynaptic reflex arc? _____

B. Reflex Tests

Name the nerve that is tested in each of the following reflexes.

1. Achilles reflex _____

2. Biceps reflex _____

3. Patellar reflex _____

4. Plantar flexion reflex _____

5. Triceps reflex _____

6. Is a Babinski's sign normal in adults? _____

C. Reflexes

1. Define reflex.

2. Describe the difference between a somatic and visceral reflex.

3. Describe the difference between a cranial and spinal reflex.

4. Describe the difference between an ipsilateral and contralateral reflex arc.

5. Define reciprocal innervation.

Using Your Knowledge

A. Flexor and Crossed Extensor Reflex

You are walking along the beach barefoot and step on a sharp object with your right foot. You immediately flex the right leg and balance yourself by extending the left leg. Label the following neurons in Figure 19.3.

1. Sensory neuron

2. Interneuron sending impulses up and down the spinal cord

3. Interneuron synapsing with motor neurons

4. Motor neuron causing flexion

5. Motor neuron causing extension

Right leg Left leg

FIGURE 19.3 **Flexor and crossed extensor reflex arc.**

For the reflex arc in Figure 19.3, name the:

_____ **6.** Nerve carrying motor information causing right leg flexion

_____ **7.** Nerve carrying motor information causing left leg extension

_____ **8.** Agonistic muscles for right leg flexion

_____ **9.** Agonistic muscles for left leg extension

B. Modification of Reflex Activity

10. Reflex activity can be modified by the cerebral cortex. For example, you are baking a roast in the oven. As you are moving the roast from the oven to the counter, hot juices splash you. Reflex activity would cause you to drop the pan, but you hold on to it. Explain how the cerebral cortex overrides the reflex.

Brain Structure and Function

OBJECTIVES

1 Identify the major external and internal structures of the brain

2 Describe the basic functions of the principal structures of the brain

3 Identify brain waves and describe the effect of stimulation on brain wave patterns

4 Identify the 3 main cranial fossae

5 Name the 3 meninges and describe their similarities and differences

6 Identify the 4 ventricles of the brain and describe their function

7 Trace the cerebrospinal fluid circulation

8 Compare the anatomy of the sheep brain with that of the human brain

MATERIALS

- human brain models, charts, or use **Real Anatomy** (Nervous)
- preserved human brain, skull
- ventricular system model
- Dissection: preserved sheep brains, dissection equipment, disposable gloves, safety glasses
- **BIOPAC** Systems, Inc. WWW.BIOPAC.COM **Biopac Laboratory Guide Experiment:**
 - The Effect of Mental and Sensory Stimulations on Brain Wave Patterns

The human brain simultaneously conducts and coordinates a variety of incredible processes with which no computer is presently able to compete. The expansive development of the human brain distinguishes it from that of all other creatures. One of the largest organs in the human body, our brain is responsible for our memory, intellect, ideas, and behavior. The brain is the center for cataloging sensory information, integrating this information with previously recorded information, and producing actions based on the results of the information synthesis.

The brain is well protected, located within the cranium of the skull. During development, the brain and skull are growing simultaneously, so the shape of the cranial cavity mirrors the shape of the brain.

A. Major Brain Regions

The brain is composed of 4 main regions: the brain stem, cerebellum (*cerebel-* = little brain), diencephalons (*di-* = through; *encephal-* = brain), and cerebrum (*cerebr-* = brain). The **brain stem** is connected to the superior part of the spinal cord, the **cerebellum** is posterior to the brain stem, and the **diencephalon** is superior to the brain stem in the center of the brain. The large **cerebrum** is the dominant brain structure, with many folds and crevices enveloping the diencephalon.

1 Label the major brain regions in Figure 20.1.

Identification of Major Brain Regions

1 Identify the major brain regions on human brain models, charts, or use the search text box in **Real Anatomy** (Nervous) to find these structures. ■

Sagittal plane

CEREBRUM

Corpus callosum

Fornix

DIENCEPHALON:
Thalamus
Hypothalamus
Epithalamus
Pineal gland

BRAIN STEM:
Midbrain
Pons
Medulla oblongata

CEREBELLUM

Infundibulum
Pituitary gland

Spinal cord

POSTERIOR

ANTERIOR

Sagittal section of brain, medial view

SUPERIOR

POSTERIOR

ANTERIOR

1

3

4

• brain stem
• cerebellum (cer-e-BELL-um)
• cerebrum (ce-REE-brum)
• diencephalon

1 _____

2 _____

3 _____

4 _____

2

Spinal cord

INFERIOR

FIGURE 20.1 Four major brain regions.

B. The Brain Stem

The brain stem is composed of 3 structures: the medulla (*medull-* = marrow) oblongata, the pons, and the midbrain.

1. The Medulla Oblongata

The first brain structure, the **medulla oblongata,** is immediately superior to the spinal cord and is the most vital part of the brain because it houses the respiratory and cardiovascular control centers. The respiratory center controls the rate and depth of breathing, and the cardiovascular center is responsible for the rate and force of the heartbeat and blood pressure reflexes. Other reflex centers in the medulla are for coughing, vomiting, and sneezing.

2. The Pons

The **pons** (*pons* = bridge) is an expanded structure located superior to the medulla oblongata and anterior to the cerebellum, and has respiratory centers that assist the medulla oblongata in controlling breathing. The pons also relays information to the diencephalon and the cerebellum.

3. The Midbrain

The **midbrain** is a smaller area superior to the pons and inferior to the diencephalon, consisting of cerebral peduncles (*ped-* = foot) and the corpora quadrigemina (*corpora* = body; *quad-* = four; *gemin-* = twin; i.e., 4 twin bodies). The **cerebral peduncles** are white fibers that connect the upper and lower brain areas, and the **corpora quadrigemina** are composed of 2 superior colliculi (*colliculus* = small mound) and 2 inferior colliculi. The **superior colliculi** have reflex centers involved in eye, head, and neck movements with visual stimulation, whereas the **inferior colliculi** have reflex centers involved in auditory stimuli that result in head and trunk movements. The midbrain is observed best from a sagittal section, medial view. It spans from the dorsal to the ventral side of the brain stem.

Before Going to Lab

1 Label the brain stem structures on Figure 20.2(a) and (b).

LAB ACTIVITY 2 The Brain Stem

1 Identify the brain stem structures on a human brain model, chart, or use the search text box in **Real Anatomy** (Nervous) to find these structures. ■

SUPERIOR

POSTERIOR

ANTERIOR

Superior colliculus ⎤ 1
Inferior colliculus ⎦
5
2
3
4

Spinal cord

INFERIOR

(a) Sagittal section, medial view

- cerebral peduncle
- corpora quadrigemina (cor-POR-a quad-ri-GEM-i-na)
- medulla oblongata
- midbrain
- pons

1 _____

2 _____

3 _____

4 _____

5 _____

FIGURE 20.2 Brain stem.

ANTERIOR

Midbrain

Pons

Middle cerebellar peduncle

Medulla oblongata

Brain stem

Spinal cord

POSTERIOR

- medulla oblongata (meh-DEW-la ob-lon-GAH-tah)
- midbrain
- pons
- spinal cord

6 _____

7 _____

8 _____

9 _____

6 _____

Middle cerebellar peduncle

7 _____

8 _____

9 _____

(b) Inferior aspect of brain

FIGURE 20.2 **Brain stem,** *continued.*

C. The Cerebellum

The **cerebellum** (*cerebellum* = little brain) is second in size to the cerebrum and is located inferior to it and posterior to the medulla and pons. There are 2 **cerebellar hemispheres,** with a central area, the **vermis,** connecting them. When cut in sagittal section, gray matter can be observed on the exterior, with deeper white matter called the **arbor vitae** (*arbor* = tree; *vitae* = living) appearing as branches of a tree. The outer layer of gray matter is called the **cerebellar cortex.** Like the cerebral cortex, the cerebellar cortex has folds that increase the surface area allowing for more neuron cell bodies. The cerebellar folds are slender, pleated gyri or **folia** (*folia* = leaves) that look similar to pages in a book. The cerebellum regulates posture and balance and, in addition, smoothes and coordinates skilled skeletal muscle movements. The cerebellum is connected to the brain stem by the superior, middle, and inferior cerebellar peduncles. Do not confuse them with the cerebral peduncles.

Before Going to Lab

1 Label the structures of the cerebellum in Figure 20.3(a) and (b).

LAB ACTIVITY 3 Cerebellum

1 Identify the structures of the cerebellum on a human brain model, chart, or use the search text box in **Real Anatomy** (Nervous) to find these structures.

(a) Posterior view
- cerebellar (cer-e-BELL-ar) hemispheres
- folia (FO-lia)
- vermis (VER-mis)

1 _____

2 _____

3 _____

(b) Sagittal view
- arbor vitae (white matter)
- cerebellar cortex (gray matter)
- inferior colliculus
- medulla oblongata
- pons
- superior colliculus

4 _____

5 _____

6 _____

7 _____

8 _____

9 _____

(a) Superior view

(b) Midsagittal section

FIGURE 20.3 The cerebellum.

D. The Diencephalon

The **diencephalon** (*di-* = two; *-cephalon* = brain) is located in the brain's central area and has 3 main regions: the thalamus, the hypothalamus, and the epithalamus.

1. The Thalamus

The **thalamus** (*thala-* = inner chamber) is composed of paired, egg-shaped bodies centrally located in the diencephalon and makes up approximately 80% of this structure. Each cerebral hemisphere contains half of the thalamus, which is connected by a small bridge called the **intermediate mass.** The thalamus is the brain's "Grand Central relay station" because it is the principal relay station for sensory fibers and some somatic motor fibers. Sensory fibers that synapse in the thalamus are relayed to a particular area of the cerebral cortex to be interpreted, and other fibers relay messages to the somatic motor cortex. The thalamus also filters out unnecessary sensory information and is involved in consciousness, emotions, learning, and memory.

2. The Hypothalamus

The **hypothalamus** (*hypo-* = below) is located below the thalamus and is a quadrangular-shaped structure. The hypothalamus has important nuclei (a group of nerve cell bodies) that control many body functions and homeostasis. Some of the major functions include integrating and controlling the pituitary gland and hormonal functions, autonomic functions, emotions and behavior, body temperature, eating, and drinking.

The hypothalamus includes the infundibulum and mammillary bodies. The **infundibulum** (*infundibulum* = funnel) is a stalk that connects the pituitary gland to the hypothalamus. The **mammillary bodies** (*mammilla* = nipple)—two small, round masses located just posterior to the infundibulum—are relay stations for smell and taste reflexes.

Other structures that are observed in this area are the optic chiasm and the pituitary gland. The **optic chiasm** (*chi* = χ, Greek letter chi, a crossing over), the area where the optic nerves cross, is anterior to the infundibulum. The **pituitary gland** looks like a large pea and is attached to the end of the infundibulum. The hypothalamus controls the pituitary gland.

3. The Epithalamus

The **epithalamus** (*epi-* = above) is superior and posterior to the thalamus and includes the **pineal** (*pineas* = pinecone) **gland** or **body,** a small endocrine gland that secretes the hormone melatonin.

Before Going to Lab

1 Label the diencephalon structures in Figure 20.4(a), (b), and (c). Refer to Figure 20.1 if necessary.

LAB ACTIVITY 4 **The Diencephalon**

1 Identify the diencephalon structures listed on the human brain model, chart, or use the search text box in **Real Anatomy** (Nervous) to find these structures. ■

CEREBRUM

CEREBELLUM

Spinal cord

1
2 — 4
3

BRAIN STEM:
— Midbrain

— Pons

— Medulla oblongata

(a)
- diencephalon
- hypothalamus (hypo-THAL-a-mus)
- pineal (pi-NEE-al) gland (part of epithalamus)
- thalamus

1 _____

2 _____

3 _____

4 _____

(a) Sagittal section, medial view

FIGURE 20.4 The diencephalon.

(b)

- hypothalamus
- infundibulum (in-fun-DIB-u-lum)
- intermediate mass of thalamus
- mammillary (MAM-mil-lary) body
- pineal gland
- pituitary gland
- thalamus (THAL-a-mus)

5 _____

6 _____

7 _____

8 _____

9 _____

10 _____

11 _____

Optic chiasm

(b) Sagittal section, magnified

(c)

- hypothalamus
- lateral ventricles
- thalamus
- third ventricles

12 _____

13 _____

14 _____

15 _____

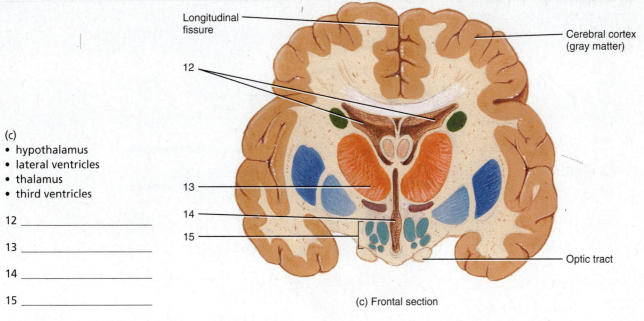

Longitudinal fissure

Cerebral cortex (gray matter)

Optic tract

(c) Frontal section

FIGURE 20.4 **The diencephalon,** *continued.*

E. The Cerebrum

The **cerebrum** is made up of right and left **cerebral hemispheres** and is the largest and most complex division of the brain. The cerebrum is superior to and surrounds the diencephalon and part of the brain stem. The cerebrum is the center of higher mental processes such as intelligence, communication, learning and memory, reasoning, and emotions. In addition, it interprets sensory input and initiates skeletal muscle contraction.

1. Organization of Cerebral Gray and White Matter

The 3 main regions of the cerebrum are the **cerebral cortex** (*cortex* = bark), **white matter,** and the deep **basal nuclei.** The **cerebral cortex** (also known as the cortical area) is the superficial cerebral gray matter on the exterior of the cerebrum composed of nerve cell bodies and dendrites. The cerebral cortex integrates sensory information, initiates motor output, and is also involved in emotions and intellectual processes. **Basal nuclei** (*ganglia* = knot) are areas of cerebral gray matter composed of paired nuclei (clusters of neuron cell bodies in the CNS) that are found deep within each cerebral hemisphere. The basal nuclei control automatic skeletal muscle movement and are involved with the limbic system or emotional brain.

 Cerebral white matter lies deep to the outer cortex and is composed mostly of myelinated axons that give it the white appearance. These axons are organized into three

kinds of fiber tracts in the cerebrum that are named according to the direction of the fibers [Figure 20.5(a)]. **Association fibers** transmit nerve impulses within the same hemisphere, whereas **commissural** (*commisura* = connection) **fibers** transmit nerve impulses between the two hemispheres. **Projection fibers** are ascending and descending tracts that project nervous impulses from inferior to superior brain areas or vice versa. The **corpus callosum** (*corpus* = body; *callosus* = hard), a prominent commissural fiber tract that is readily observable in midsagittal sections of the brain, connects the two cerebral hemispheres. The **fornix** looks like a group of commissural fibers but is actually a tract of arched association fibers. The **internal capsule,** a large group of projection fibers, contains sensory and motor tracts that connect the cerebral cortex to the brain stem and spinal cord.

Before Going to Lab

1 Label the gray and white matter in Figure 20.5(b).

LAB ACTIVITY 5 Organization of Gray and White Matter in Cerebrum

1 Identify the cerebral gray and white matter on a human coronal section of brain, brain model, chart, or use the search text box in **Real Anatomy** (Nervous) to find these structures.

Commissural fibers
(Corpus callosum)

Association fibers

Projection fibers

Lateral ventricles

Third ventricle

Cerebellum

(a) Frontal section

FIGURE 20.5 **Gray and white matter in the cerebrum.**

Frontal section

INFERIOR

(b) Frontal section

(b)
- basal nuclei (gray matter)
- cerebral cortex (gray matter)
- cerebral white matter
- corpus callosum (commissural fibers)
- fornix (association fibers)
- internal capsule (projection fibers)

1	_____	4	_____
2	_____	5	_____
3	_____	6	_____

FIGURE 20.5 **Gray and white matter in the cerebrum,** *continued.*

2. Surface Features of the Cerebrum

There are 4 lobes that compose the exterior of the cerebrum, which are mainly named for the overlying cranial bones. These lobes are the **frontal, parietal, occipital, and temporal lobes.** An inner lobe, the **insula,** lies deep to the lateral cerebral fissure and is not visible from the exterior. Other obvious external anatomical features of the cerebrum include the following:

- **gyrus** (*gyros* = circle; **gyri,** pl.)—elevation or fold in the cerebral cortex; increases the surface area for neuron cell bodies
- **sulcus** (*sulcus* = furrow; **sulci,** pl.)—shallow groove between elevations
- **central sulcus**—shallow groove separating frontal lobe from parietal lobe
- **precentral gyrus**—elevation located just anterior to the central sulcus
- **postcentral gyrus**—elevation located just posterior to the central sulcus

- **lateral cerebral sulcus**—shallow groove separating frontal and temporal lobes
- **parieto-occipital sulcus**—shallow groove separating parietal and occipital lobes
- **longitudinal fissure**—deep groove separating the 2 cerebral hemispheres at the midline
- **transverse fissure**—deep groove separating the cerebrum from the cerebellum in the posterior/inferior part of the brain

Before Going to Lab

1 Label the structures listed for Figure 20.6 and locate the longitudinal fissure on Figures 20.2(b) and 20.5(a).

LAB ACTIVITY 6 External Features of the Cerebrum

1 Identify the external features of the cerebrum on a brain model, chart, or use the search text box in **Real Anatomy** (Nervous) to find these structures. ◼

Postcentral gyrus
Parietal lobe (blue area)
Parieto-occipital sulcus
Occipital lobe
Transverse fissure
Cerebellum

Central sulcus
Precentral gyrus
Frontal lobe
Insula (gold area)
Lateral cerebral sulcus
Temporal lobe

Right lateral view

POSTERIOR ANTERIOR
1 (space)
2
3 (lobe)
4 (lobe)
5 (space)
Cerebellum

6
7 (lobe)
8 (lobe)
9 (lobe)

Right lateral view with temporal lobe cut away

- central sulcus
- frontal lobe
- insula
- occipital lobe
- parietal lobe
- postcentral gyrus
- precentral gyrus
- temporal lobe (cut)
- transverse fissure

1 _____
2 _____
3 _____
4 _____
5 _____
6 _____
7 _____
8 _____
9 _____

FIGURE 20.6 External features of the cerebrum.

3. Functional Areas of the Cerebral Cortex

The cerebral cortex is composed of 3 types of functional areas: motor, sensory, and association areas. **Sensory areas** receive and interpret impulses from sensory receptors, while **motor areas** initiate impulses to skeletal muscles. **Association areas,** which perform complex integrative functions, receive and send information to multiple areas of the cortex via association fibers. The majority of the cortex is composed of association areas.

Motor Areas
- **primary motor area**—located in the precentral gyrus of each frontal lobe; initiates impulses to skeletal muscles
- **Broca's speech area**—located superior to the lateral sulcus and anterior to the primary motor cortex, usually in the left hemisphere; initiates impulses that result in speech

Sensory Areas
- **primary somatosensory area**—located in the postcentral gyrus of each parietal lobe; receives nerve impulses for touch, proprioception, pain, and temperature
- **primary auditory area**—located in each temporal lobe across the lateral sulcus from the gustatory area; receives impulses when the auditory receptors of the ear are stimulated

- **primary gustatory area**—in each postcentral gyrus, just superior to the lateral sulcus; receives impulses when the taste buds are stimulated
- **primary olfactory area**—located on the medial side of each temporal lobe; cannot be seen from the lateral view; receives impulses when the olfactory receptors of the nose are stimulated
- **primary visual area**—in the posterior occipital lobe; receives impulses from the thalamus when the retina is stimulated

Selected Association Areas
- **Wernicke's area**—located in left temporal and parietal lobes; recognizes spoken words, translates words into thoughts, and possibly helps us sound out strange or new words
- **somatosensory, visual, and auditory association areas**—larger areas adjacent to the corresponding sensory cortex; integrate sensory information from the sensory cortex with past experiences allowing us, for example, to identify objects by touch or to identify sound as music or speech

Before Going to Lab

1 Label the functional areas of the cortex in Figure 20.7.

LAB ACTIVITY 7 Functional Areas of Cerebral Cortex

1 Identify functional areas of the cerebral cortex on a brain model or chart. ■

- Broca's speech area (dotted area)
- central sulcus
- primary auditory area
- primary gustatory (GUS-tah-tory) area
- primary motor area
- primary somatosensory (so-ma-to-SEN-sory) area
- primary visual area
- Wernicke's area (dotted area)

1 _____

2 _____

3 _____

4 _____

5 _____

6 _____

7 _____

8 _____ ANTERIOR

Left lateral view POSTERIOR

FIGURE 20.7 Functional areas of the cerebral cortex.

4. Encephalography

Millions of brain neurons close to the surface of the cerebral cortex collectively create action potentials (nerve impulses or electrical currents) that can be detected by electrodes positioned on the scalp. The electrodes are connected with wires to a computer, and the electrical currents are displayed on a screen as **brain waves**. The record of brain waves is called an **electroencephalogram (EEG)**. The four main types of brain waves in an EEG, **Alpha, Beta, Theta,** and **Delta,** can be recognized by their characteristic patterns (Figure 21.8). These brain waves differ in frequency (cycles per second) in the various areas of the brain. Neurologists use EEGs diagnostically for many situations, including epilepsy, coma, brain disease, dementia, brain death, and to study sleep disorders and monitor brain activity during general anesthesia.

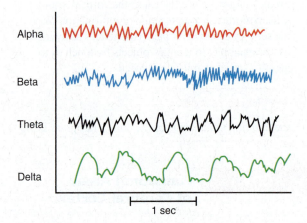

FIGURE 20.8 **Types of brain waves recorded in an electroencephalogram (EEG).**

LAB ACTIVITY 8 **Encephalography**

1 Complete **Biopac Laboratory Guide Experiment:** The Effect of Mental and Sensory Stimulation on Brain Wave Patterns. ■

F. Protection of the Brain

The brain is protected both physically and chemically by cranial bones, the blood-brain barrier, the cranial meninges, and cerebrospinal fluid.

1. Cranial Bones

The cranial bones form a vault called the cranium, which surrounds and protects the brain. The floor of the cranium contains depressions—the anterior, middle, and posterior cranial fossae—which support parts of the brain. The

anterior cranial fossae support the frontal lobes of the cerebrum; the **middle cranial fossae** support portions of the temporal and parietal lobes of the cerebrum and the diencephalon; the **posterior cranial fossae** support portions of the temporal, parietal, and occipital lobes of the cerebrum, the cerebellum, and the brain stem.

2. Cranial Meninges

There are 3 cranial meninges (connective tissue membranes): the dura mater, arachnoid mater, and pia mater. The **dura mater** (*dura* = hard; *mater* = mother) is the first meninx (sing.) located deep to the cranial bones. It is composed of 2 layers: the **periosteal layer,** which is a tough membrane attached to the cranial bones, and the **meningeal layer,** which is exterior to the arachnoid mater. The 2 dural layers split to form the **dural sinuses** that eventually drain cranial blood into the jugular veins. The **superior sagittal sinus,** located superior to the longitudinal fissure, is one of the main dural sinuses. The double-layered dura mater extends deep into the longitudinal fissure forming the **falx** (*falx* = sickle-shaped) **cerebri,** into the transverse fissure forming the **tentorium** (tent) **cerebelli,** and between the cerebellar hemispheres forming the **falx cerebelli.**

The **arachnoid** (*arachnoid* = spider-like) **mater** is the 2nd meninx located deep to the dura mater. Projections of the arachnoid mater into the dural sinuses are called **arachnoid villi** (*villi* = tiny projections). The **pia** (*pia* = delicate) **mater** is the thin, inner meninx. It hugs and overlays the cerebral cortex, following each gyrus and sulcus. Between the arachnoid and pia is the **subarachnoid** space.

Before Going to Lab

1 Observe the cranial meninges in Figure 20.9(a).

2 Label the structures in Figure 20.9(b).

LAB ACTIVITY 9 **Cranial Bones and Cranial Meninges**

1 Identify the cranial fossae in the cranial floor of the skull.

2 Using a preserved human brain or the search text box in **Real Anatomy** (Nervous), identify the meninges.

3 Using a brain model, preserved human brain, or the search text box in **Real Anatomy** (Nervous), show where the falx cerebri and tentorium cerebelli would be found. ■

Frontal plane

Subarachnoid space

Falx cerebri

Superior sagittal sinus
Skin
Parietal bone of cranium
CRANIAL MENINGES:
Dura mater
Arachnoid mater

Pia mater

Cerebral cortex

(a) Frontal section

- arachnoid mater (a-RAK-noid MAH-ter)
- arachnoid villus (VIL-us)
- cerebral (ce-REE-bral) cortex
- dura mater
- falx cerebri (falks ce-REE-bree)

- parietal bone
- pia (PEE-ah) mater
- subarachnoid (sub-a-RAK-noid) space
- superior sagittal (SA-jih-tahl) sinus
- white matter

1 _____ 6 _____

2 _____ 7 _____

3 _____ 8 _____

4 _____ 9 _____

5 _____ 10 _____

Frontal plane

5

Skin

6

7

8

9

10

1

2

3

4

(b) Frontal section through skull.

FIGURE 20.9 Cranial meninges.

3. Cerebrospinal Fluid and Ventricles of the Brain

Cerebrospinal fluid (CSF) constantly bathes the brain and spinal cord with oxygen, nutrients, and vital chemicals. Although different in content, CSF is made from blood plasma that leaks out of specialized, tiny blood vessels (capillaries) called the **choroid plexus** (*choroid* = membrane-like; *plexus* = pleated) and passes through ependymal cells into 4 small brain cavities or ventricles. The ependymal cells have cilia that move the CSF in one direction. There are choroid plexuses in the roof of all four ventricles. A **lateral ventricle** is located in each **cerebral hemisphere** with a thin membrane, the **septum pellucidum** (*septum* = partition; *pellucid* = transparent), separating the 2 ventricles anteriorly. Each arched lateral ventricle has an **interventricular foramen** that opens medially into the third ventricle. The **third ventricle** is medially located between the paired masses of the thalamus and is narrower and smaller than the other ventricles. Connecting the third ventricle to the fourth ventricle is a thin tube, the **cerebral aqueduct** (*aqua* = water; *duct* = way; i.e., waterway). The **fourth ventricle** is located between the pons and the **cerebellum. Lateral** and **median apertures** or openings allow the CSF to flow from the fourth ventricle into the **subarachnoid space** surrounding the brain and the spinal cord. CSF also flows through the central canal of the spinal cord and back out into the subarachnoid space. Just as CSF enters the brain ventricles from the bloodstream, it is returned to the blood by reabsorption through the **arachnoid villi** (tiny projections) located in the dural sinuses, especially the **superior sagittal sinus.**

Before Going to Lab

1 Label the structures in Figure 20.10(a), (b), and (c).

LAB ACTIVITY 10 CSF Circulation

1 Identify these structures on a human brain model, chart, or use the search text box in **Real Anatomy** (Nervous) to find these structures.

2 With your laboratory partner, trace a drop of CSF from its origin in a lateral ventricle through the other ventricles and subarachnoid space, and follow it until it is reabsorbed into the bloodstream (Figure 20.10c). ■

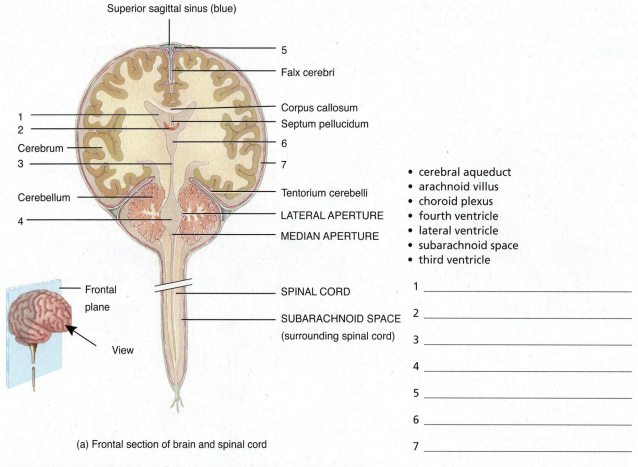

Superior sagittal sinus (blue)

5
Falx cerebri

Corpus callosum
Septum pellucidum

1
2
Cerebrum
3

6
7

Cerebellum
4

Tentorium cerebelli
LATERAL APERTURE
MEDIAN APERTURE

Frontal plane

View

SPINAL CORD
SUBARACHNOID SPACE
(surrounding spinal cord)

(a) Frontal section of brain and spinal cord

- cerebral aqueduct
- arachnoid villus
- choroid plexus
- fourth ventricle
- lateral ventricle
- subarachnoid space
- third ventricle

1 _____

2 _____

3 _____

4 _____

5 _____

6 _____

7 _____

FIGURE 20.10 **Ventricles of the brain and cerebral spinal fluid.**

- central canal of the spinal cord
- cerebral aqueduct (AH-que-duct)
- fourth ventricle
- interventricular (in-ter-ven-TRIK-u-lar) foramen
- lateral ventricles
- third ventricle

8 _____

9 _____

10 _____

11 _____

12 _____

13 _____

(b) Right lateral view (ventricles superimposed)

- arachnoid villus (a-RACH-noid VIL-us)
- central canal
- cerebral aqueduct
- choroid plexus (CHOR-oid PLEX-us)
- fourth ventricle
- lateral ventricle
- subarachnoid (sub-ah-RAK-noid) space
- superior sagittal sinus
- third ventricle

14 _____

15 _____

16 _____

17 _____

18 _____

19 _____

20 _____

21 _____

22 _____

(c) Sagittal section of brain and spinal cord

FIGURE 20.10 Ventricles of the brain and cerebrospinal fluid, *continued.*

G. Sheep Brain Dissection

> **SAFETY NOTE:** *Use disposable gloves, safety glasses, and a lab coat when handling preserved material.*

Remember that preserved material does not look like a fresh specimen. Usually more detail may be observed in a preserved brain because the tissue is firmer.

LAB ACTIVITY 11 Dissection of Sheep Brain

1 Rinse the sheep brain to remove preservative.

2 Observe meninges and main brain regions.
- Examine the brain to see if the tough, outer **dura mater** is present. If present, note the **falx cerebri** and **tentorium cerebelli.** Carefully remove the dura mater.
- Now look for the stringy, web-like **arachnoid mater** beneath and adhering to the dura mater. Deep to this membrane is the very thin **pia mater** that follows the contours of the gyri and sulci.
- Compare the sheep brain with the main external regions of the human brain. Identify the **cerebrum, brain stem (medulla** and **pons), and cerebellum.**

3 Identify dorsal structures (Figure 20.11).
- With the dorsal side up, identify the **cerebral hemispheres, gyri, sulci, longitudinal fissure,** and **transverse fissure.**
- Identify the 4 main **lobes** of the brain—**frontal, parietal, occipital,** and **temporal.**
- At the longitudinal fissure, gently separate the 2 parts and look down between them for the thick band of white fibers, the **corpus callosum.**

4 Identify ventral structures (Figure 20.12).
- Place the sheep brain ventral side up.
- Identify the **olfactory bulb, olfactory tract, optic nerve, optic chiasm,** and **optic tract.**
- Posterior to these structures, locate the *one* large **mammillary body.**
- If present, identify the **infundibulum** and **pituitary gland.**
- Look at the three parts of the **brain stem**—the **midbrain, pons,** and **medulla oblongata.**
- The large **trigeminal nerves** (V) are located laterally at the junction of the pons and medulla.

- The small **abducens nerves** (VI) are found medial and slightly posterior to the trigeminal nerves, arising from the pons.

5 Identify midbrain structures (Figure 20.13).
- Carefully pull the cerebellum away from the cerebrum. Identify the **pineal body, superior colliculi,** and **inferior colliculi.**

6 Identify midsagittal section structures (Figure 20.14).
- Using a sharp knife or scalpel, carefully make a midsagittal section.
- Locate the **brain stem** components: the **medulla oblongata,** the **pons,** and the **midbrain.** Compare this with your human brain model.
- Identify the **arbor vitae** in the **cerebellum.**
- Note the **cerebral aqueduct** and the **fourth ventricle.**
- Identify the **thalamus, corpus collosum, septum pellucidum, fornix,** and **lateral ventricles.**

7 Observe a coronal section.
- Observe the coronal section your instructor may have as a demonstration. Note the **gray matter, white matter, longitudinal fissure, corpus callosum, thalamus, lateral ventricles,** and **third ventricle.**
- Observe the transverse or horizontal section your instructor may have as a demonstration. Identify the **gray matter, white matter,** and **ventricles.**

8 Clean up as directed by your instructor.

9 Answer Discussion Questions with your lab group.

DISCUSSION QUESTIONS
Sheep Brain Dissections

1 Why are the olfactory tracts not called olfactory nerves?

2 How does the human mammillary body look different from the sheep?

3 How do the three parts of the sheep brain stem compare to the human brain stem?

4 Does the arbor vitae in the cerebellum of the sheep brain look similar or different from the arbor vitae of the human brain?

FIGURE 20.11 Dorsal view of the sheep brain.

FIGURE 20.12 Ventral view of the sheep brain.

FIGURE 20.13 **Posterior view of the midbrain structures of the sheep brain.**

Cerebrum

Occipital lobe

Pineal body

Superior
colliculi

Inferior
colliculi

Cerebellum

FIGURE 20.14 **Midsagittal section of the sheep brain.**

Pineal body

Thalamus

Lateral
ventricle

Fornix

Cerebrum

Transverse fissure

Corpus
collosum

Corpora quadrigemina

Cerebellum

Optic
chiasm

Fourth ventricle

Pituitary gland

Pons

Mammillary body

Medulla oblongata

Reviewing Your Knowledge

A. Brain Structure

Label the structures in Figures 20.15(a) and (b).

(a) Sagittal section, medial view

(b) Inferior aspect of brain

FIGURE 20.15 Brain structure.

(a)

1 _____

2 _____

3 _____

4 _____

5 _____

6 _____

7 _____

8 _____

9 _____

10 _____

11 _____

12 _____

(b)

13 _____

14 _____

15 _____

16 _____

17 _____

18 _____

19 _____

20 _____

21 _____

22 _____

B. Functions of Brain Regions

Give the brain region for the functions described.

_____ **1.** Contains vital centers that regulate heartbeat, breathing, blood pressure, vomiting, coughing

_____ **2.** Smoothes and coordinates skilled skeletal muscle movement; also posture and balance or equilibrium

_____ **3.** Secretes melatonin that controls the sleep-wake cycle

_____ **4.** Controls and integrates the autonomic nervous system; regulates hormones, emotional behavior, temperature, eating, and drinking behavior

_____ **5.** Interprets sensory input, controls skilled skeletal muscle movements, and is involved in emotional and intellectual processes

_____ **6.** Helps control breathing; conducts impulses to and from the cerebellum, midbrain, and medulla

_____ **7.** Relays all sensory input to the cerebral cortex; involved in skeletal muscle actions and memory processing

_____ **8.** Coordinates visual and auditory reflexes

_____ **9.** Coordinates gross, automatic muscle movements; also involved with the limbic system

_____ **10.** White fiber tracts communicating between hemispheres

C. Flow of Cerebrospinal Fluid

Fill in the numbered blanks with the name of the structure that corresponds with the number in the following paragraph.

Fluid for the CSF is derived from the bloodstream. The sites of CSF formation are the (1), special tiny capillaries located in the walls of (2), (3), and (4). Cells that line the ventricles have cilia that move the CSF and are called (5). The two lateral ventricles are separated by a thin membrane called the (6). CSF flows by cilia movement from the two lateral ventricles through the interventricular foramen to the (7). From here, the CSF flows through the (8) into the fourth ventricle. The CSF leaves the fourth ventricle through three openings: the median aperture and two lateral apertures. CSF is now located in the (9) space around the brain, and circulates all around the cerebrum and cerebellum. CSF continues to flow into the inner part of the spinal cord by flowing through the tiny (10) of the spinal cord, as well as around the exterior of the spinal cord in the (11) space. Because CSF is continually being made at the rate of about 20 mL/hr, it has to exit back into the bloodstream by being reabsorbed through the (12) that protrude into the dural venous sinuses. The venous sinus that overlies the brain superiorly is called the (13).

1. _____ 6. _____ 11. _____

2. _____ 7. _____ 12. _____

3. _____ 8. _____ 13. _____

4. _____ 9. _____

5. _____ 10. _____

Using Your Knowledge

1
2
3
4
5

FIGURE 20.16 Midsagittal section of the human brain MRI.

A. MRI Scan of the Human Brain

Identify the structures in Figure 20.16, midsagittal view.

1. _____

2. _____

3. _____

4. _____

5. _____

FIGURE 20.17 **Human brain, transverse section.**

B. Transverse Section of a Human Brain

Identify the structures in Figure 20.17, a transverse section.

6. _____

7. _____

8. _____

9. _____

10. _____

FIGURE 20.18 **Human brain, sagittal section.**

C. Sagittal Section of Human Brain

Identify the structures in Figure 20.18, sagittal section.

11. _____

12. _____

13. _____

14. _____

15. _____

SUPERIOR

19
20

16

17

18

Spinal cord

Body of cervical
vertebra

POSTERIOR

ANTERIOR

INFERIOR

Right lateral view

FIGURE 20.19 Human brain, lateral view.

D. Brain Injury

Match the brain injury to the appropriate change in function. Use Figure 20.19 to locate the area injured.

a. cessation of breathing
b. loss of equilibrium
c. loss of use of left arm
d. loss of vision
e. loss of pain localization in the shoulder

_____ **16.** The effect of a blow to the back of the head that damages this area.

_____ **17.** The effect of alcohol on this area.

_____ **18.** The effect of a head injury (i.e., from diving into a pool) that forces the dens into this area.

_____ **19.** The effect of a stroke that damages this area.

_____ **20.** The effect of a stroke that damages this area.

Cranial Nerves

OBJECTIVES

1 Identify the 12 pairs of cranial nerves by name and Roman numeral on brain models and/or preserved human brains
2 State the function of the 12 pairs of cranial nerves
3 Test cranial nerve function
4 Identify specific cranial nerves on the sheep brain

MATERIALS

- human brain models, charts, or use **Real Anatomy** (Nervous)
- skull with or without cranial nerves
- preserved human brain (if available)
- **Testing Cranial Nerve Function:** spice or aromatic oil, penlight, cotton, ice water, sugar, quinine or bitter food (vinegar, turmeric, unsweetened cocoa powder, coffee), tuning fork, tongue depressors

The 12 pairs of cranial nerves are part of the peripheral nervous system. They are numbered from anterior to posterior with Roman numerals I–XII. Cranial nerves originate from various ventral areas of the brain and exit the cranial cavity through foramina to reach their destinations, which are primarily in the head and neck.

A. Name and Location of Cranial Nerves

The cranial nerve's name sometimes indicates the structure it innervates or its function. The names of the cranial nerves I–XII are in order as follows: **olfactory** (*olfactus* = sense of smell), **optic, oculomotor** (*oculo-* = eye; *motor* = mover), **trochlear** (*trochlea* = pulley), **trigeminal** (*tri-* = three; *-gemini* = twins), **abducens** (*ab-* = away; *-ducens* = to lead), **facial, vestibulocochlear** (*vestibulo-* = vestibule of ear/equilibrium; *-cochlear* = cochlea of ear/hearing), **glossopharyngeal** (*glosso-* = tongue; *-pharyngeal* = throat), **vagus** (*vagus* = wandering), **accessory,** and **hypoglossal** (*hypo-* = below; *-glossal* = tongue). The vagus nerve has the distinction of being different from the other 11 cranial nerves in its distribution. As its word derivative states, it is the "wandering nerve" that branches extensively and innervates the viscera of the thoracic and abdominopelvic cavities.

You may want to use the following mnemonic device to help remember the names of the cranial nerves in order: "**O**n **O**ld **O**lympus' **T**owering **T**op **A** **F**riendly **V**iking **G**rew **V**ines **A**nd **H**ops." The bold first letter of each word matches with the first letter in the name of the cranial nerve.

The olfactory nerves are very small and will not be seen in Figure 21.1. The olfactory bulbs and tracts can be observed and are marked instead.

When locating the foramen for the cranial nerves, note that cranial nerves III, IV, VI, and the ophthalmic division of V, all exit the same foramen, the superior orbital fissure. Cranial nerves IX, X, and XI all exit the jugular foramen. Each of the other 7 foramina has one cranial nerve exiting.

Before Going to Lab

1 Learn the mnemonic for recalling the names of the cranial nerves.

2 Label the cranial nerves on Figure 21.1.

3 In Figure 21.2, identify the cranial nerve(s) that exit each foramen by writing the name(s) in the blank.

4 Learn the Roman numeral, name, and function of each cranial nerve.

LAB ACTIVITY 1 Name and Location of Cranial Nerves

1 Identify each cranial nerve on a preserved human brain, human brain model, chart, or use the search text box in **Real Anatomy** (Nervous) to locate these structures.

2 Locate the foramen for each cranial nerve on a skull with or without cranial nerves. Refer to Table 21.1 and Figure 21.2.

3 Quiz your lab partner on cranial nerve names and functions.

TABLE 21.1 Foramen of Cranial Nerves

NUMBER	NAME	FORAMEN
I	Olfactory	Olfactory foramina of cribriform plate
II	Optic	Optic foramen
III	Oculomotor	Superior orbital fissure
IV	Trochlear	Superior orbital fissure
V	Trigeminal	Ophthalmic—Superior orbital fissure
		Maxillary—Foramen rotundum
		Mandibular—Foramen ovale
VI	Abducens	Superior orbital fissure
VII	Facial	Stylomastoid foramen
VIII	Vestibulocochlear	Internal auditory meatus
IX	Glossopharyngeal	Jugular foramen
X	Vagus	Jugular foramen
XI	Accessory	Jugular foramen
XII	Hypoglossal	Hypoglossal canal

View

ANTERIOR

Olfactory bulb

Olfactory tract

Optic nerve

Trochlear nerve

Abducens nerve

Vestibulo-
cochlear nerve

Vagus nerve

Accessory nerve

Oculomotor nerve

Trigeminal nerve

Facial nerve

Glossopharyngeal
nerve

Hypoglossal nerve

POSTERIOR

- abducens (ab-DUE-senz)
- accessory
- facial
- glossopharyngeal
 (gloss-oh-fah-RIN-jeal)
- hypoglossal (hypo-GLOSS-al)
- oculomotor (ok-u-low-MO-tor)
- olfactory (OHL-fac-tory) tract
- optic
- trigeminal (tri-GEM-i-nal)
- trochlear (TROH-klee-ur)
- vagus (VAY-gus)
- vestibulocochlear
 (ves-tib-u-lo-COKE-lee-ur)

1 _____

2 _____

3 _____

4 _____

5 _____

6 _____

7 _____

8 _____

9 _____

10 _____

11 _____

12 _____

CRANIAL NERVES:

1

2

3

4

5

6

7

8

9

10

11

12

Inferior aspect of brain

FIGURE 21.1 **Cranial nerves of the human brain.**

Posterolateral view of dissected brain
with cranial nerves

- abducens
- accessory
- facial
- glossopharyngeal
- hypoglossal
- oculomotor
- olfactory
- optic
- trigeminal
- trochlear
- vagus
- vestibulocochlear

1. foramen ovale _____

2. foramen rotundum _____

3. hypoglossal canal _____

4. internal auditory meatus _____

5. jugular foramen _____

6. axons pass through olfactory foramina _____

7. optic foramen _____

8. stylomastoid foramen _____

9. superior orbital fissure _____

FIGURE 21.2 Cranial nerves and foramina.

B. Testing Cranial Nerve Function

Not all cranial nerves function in the same way. Some cranial nerves are primarily special sensory in function, others are primarily motor in function, and still others have both sensory and motor fibers. A mnemonic device for recalling the function of cranial nerves is: "**S**ome **S**ay **M**arry **M**oney **B**ut **M**y **B**rother **S**ays **B**ad **B**usiness **M**arry **M**oney." The 3 types are: **S** = sensory; **M** = motor; **B** = both sensory and motor (mixed nerve). Motor nerves also have sensory proprioceptors located in the muscles they innervate, but their main function is motor. Cranial nerve VIII has been shown to have some motor activity, but its main function is sensory. Two of the cranial nerves have branches. The trigeminal nerve (V) has 3 branches: ophthalmic, maxillary, and mandibular. The accessory nerve (XI) has a cranial portion and spinal portion. In addition to somatic motor fibers, cranial nerves III, VII, IX, and X also have parasympathetic motor fibers.

An easy way to remember the names of the cranial nerves that innervate the eyeball is by using the following formula: LR_6SO_4 = the lateral rectus muscle (LR) is innervated by cranial nerve VI; the superior oblique muscle (SO) is innervated by cranial nerve IV; the other 4 external muscles of the eyeball are innervated by cranial nerve III.

TABLE 21.2 Cranial Nerve Function, Distribution, and Action

NUMBER AND NAME	DISTRIBUTION	ACTION
I Olfactory	Nasal mucosa	Smell
II Optic	Eye	Vision
III Oculomotor	Levator palpebrae superioris; four extrinsic eye muscles (inferior oblique, superior rectus, medial rectus, inferior rectus); ciliary muscle (intrinsic eye muscle); iris muscles of eye (intrinsic eye muscles)	Movement of eyelid; movement of eyeball; accommodation of lens; pupillary constriction
IV Trochlear	Superior oblique muscle of eyeball	Movement of eyeball
V Trigeminal	Ophthalmic branch (eye and forehead) Maxillary branch Mandibular branch	Cutaneous sensations from ophthalmic, maxillary, and mandibular areas; chewing
VI Abducens	Lateral rectus muscle of eyeball	Movement of eyeball
VII Facial	Anterior 2/3 of tongue; facial, scalp, and neck muscles; lacrimal glands; salivary glands	Taste; facial expression; secretion of tears; salivation
VIII Vestibulocochlear	Semicircular canals and cochlea of ear	Equilibrium and hearing
IX Glossopharyngeal	Posterior 1/3 of tongue; pharyngeal muscles; parotid gland	Taste; swallowing and speech; secretion of saliva
X Vagus	Pharyngeal muscles and epiglottis; smooth muscles of thorax and GI tract; cardiac muscle; glands of GI tract	Taste and somatic sensation from pharynx and epiglottis; swallowing, coughing, and voice production; smooth muscle contraction of GI tract; slows heart rate; secretion by digestive glands
XI Accessory	Cranial portion—muscles of pharynx, larynx, and soft palate	Swallowing
	Spinal portion—sternocleidomastoid and trapezius muscles	Movement of head and shoulders
XII Hypoglossal	Tongue muscles	Speech and swallowing

LAB ACTIVITY 2 **Testing Cranial Nerve Function**

1 Perform the cranial nerve tests to determine if function is normal, consulting Table 21.2 for each nerve action. Record results in Table 21.3 by circling either normal (N) or abnormal (A) for each nerve.

2 Clean up as directed by your instructor.

SAFETY NOTE: *INSTRUCTIONS FOR TESTING CRANIAL NERVE FUNCTION*
Anything that was used in the mouth (tongue depressor or cotton-tipped swab or applicator) needs to be immediately placed in an autoclavable bag after use. Do not place these used items on your lab bench.

TABLE 21.3 Testing Cranial Nerve Function

NUMBER	CRANIAL NERVE QUICK TEST	RESULTS	
I	• Sniff aromatic oil, spice, or coffee grounds (ability to smell is normal)	N	A
II	• Read a Snellen eye chart at 20 ft. (ability to read is normal)	N	A
III, IV, VI	• Observe eyelids (eyelid not drooping is normal)	N	A
	• Look up, down, medially, laterally, upper lateral, lower lateral (ability to move eyeball is normal)	N	A
	• Shine penlight on pupils (constriction is normal)	N	A
V	• Lightly touch the cornea of the subject with a wisp of cotton (blinking is normal)	N	A
	• Subject bites down on tongue depressor; try to pull it out (good resistance to pulling is normal)	N	A
VII	• Have subject smile, raise eyebrows, whistle, and close the eyes (ability to do these actions is normal)	N	A
	• Check tip of tongue for taste with sugar (ability to taste is normal)	N	A
VIII	• Use a tuning fork to check hearing in each ear (ability to hear is normal)	N	A
	• Have the subject stand straight with eyes closed (absence of swaying is normal)	N	A
IX, X	• Have subject say, "Ah"; check uvula position with mouth open (midline is normal)	N	A
	• Using a cotton-tipped applicator, *gently* touch the subject's uvula to elicit a gag reflex (gag response is normal)	N	A
	• The posterior 1/3 of tongue can be tested with cotton applicator dipped in quinine (taste is normal)	N	A
XI	• Have subject shrug shoulders against resistance as you are holding them down (both sides having equal strength is normal)	N	A
XII	• Ask your lab partner to stick out and retract his/her tongue (tongue not deviating to one side is normal)	N	A

Reviewing Your Knowledge

A. Cranial Nerve Numbers

Give the Roman numeral for the 12 pairs of cranial nerves.

_____ **1.** Abducens

_____ **2.** Accessory

_____ **3.** Facial

_____ **4.** Glossopharyngeal

_____ **5.** Hypoglossal

_____ **6.** Oculomotor

_____ **7.** Olfactory

_____ **8.** Optic

_____ **9.** Trigeminal

_____ **10.** Trochlear

_____ **11.** Vagus

_____ **12.** Vestibulocochlear

Identify the cranial nerves by writing the name in the blank.

ANTERIOR

CRANIAL NERVES:

1
2
3
4
5
6
7
8
9
10
11
12

1 _____

2 _____

3 _____

4 _____

5 _____

6 _____

7 _____

8 _____

9 _____

10 _____

11 _____

12 _____

POSTERIOR

FIGURE 21.3 Cranial nerves.

B. Cranial Nerve Function

Identify if each cranial nerve is mainly sensory, motor, or both.
S = sensory M = motor B = both sensory and motor

_____ **1.** Olfactory

_____ **2.** Optic

_____ **3.** Oculomotor

_____ **4.** Trochlear

_____ **5.** Trigeminal

_____ **6.** Abducens

_____ **7.** Facial

_____ **8.** Vestibulocochlear

_____ **9.** Glossopharyngeal

_____ **10.** Vagus

_____ **11.** Accessory

_____ **12.** Hypoglossal

C. Cranial Nerve Action—Matching

Choose the action of the cranial nerve and write the letter next to the cranial number.

_____ I.

_____ II.

_____ III.

_____ IV.

_____ V.

_____ VI.

_____ VII.

_____ VIII.

_____ IX.

_____ X.

_____ XI.

_____ XII.

A. Movement of lateral rectus muscle

B. Speech and swallowing—tongue muscles

C. Taste; facial expression; tears; salivation

D. Equilibrium and hearing

E. Smell

F. Movement of superior oblique muscle

G. Movement of head and shoulders—sternocleidomastoid and trapezius muscles

H. Vision

I. Posterior 1/3 of tongue; taste, swallowing and speech; secretion of saliva

J. Cutaneous sensations from ophthalmic, maxillary, and mandibular areas; chewing

K. Taste; pharynx and epiglottis sensations; swallowing, coughing; voice production; smooth muscle of GI tract; secretion of digestive glands; slows heart rate

L. Movement of 4 extrinsic eye muscles; accommodation of lens; pupillary constriction

Using Your Knowledge

A. Normal Cranial Nerve Function

Give the Roman numeral of the cranial nerve(s) involved with the following functions or activities.

_____ **1.** Hearing the crack of a bat hitting a baseball

_____ **2.** Smelling dinner cooking

_____ **3.** Being dizzy after going on a Tilt-a-Whirl ride at the fair

_____ **4.** Lifting your shoulders while doing warm-up exercises

_____ **5.** Chewing a tender steak

_____ **6.** Swallowing the tender steak

_____ **7.** Salivation and crying

_____ **8.** Reading a book

_____ **9.** Seeing a sunset

_____ **10.** Looking cross-eyed

_____ **11.** Smiling

_____ **12.** Moving your head from side to side

_____ **13.** Feeling a hot curling iron touching the forehead

_____ **14.** Decreased heart rate

_____ **15.** Rolling your eyes

_____ **16.** All of the cranial nerves that are involved with the digestive system in some way, starting with the mouth

_____ **17.** Speaking

B. Cranial Nerve Injury

Write the name of the cranial nerve(s) involved.

18. Injury to this cranial nerve causes Bell's palsy—a loss of taste, decreased salivation, and loss of the ability to close the eyes, even during sleep. Name this nerve.

19. Injury to this cranial nerve causes a loss of taste sensation, decreased salivation, and difficulty in swallowing. Name this nerve.

20. Injury to this cranial nerve causes paralysis of the vocal cords, interferes with swallowing, interrupts sensations from many organs, and causes the heart rate to increase. Name this nerve.

Autonomic Nervous System Structure and Function

OBJECTIVES

1. Compare and contrast the anatomical components of the sympathetic and parasympathetic divisions of the autonomic nervous system (ANS)
2. Name and locate the ganglia of the sympathetic and parasympathetic nervous systems
3. Compare somatic and autonomic reflexes
4. Describe the pupillary light reflex
5. Describe ANS reponses as observed in a lie detector test

MATERIALS

- spinal cord model or chart with autonomic ganglia
- brain model or chart showing the cranial nerves
- transverse section model or chart of the spinal cord with white and gray rami communicantes
- **Pupillary Light Reflex:** penlight
- **Lie Detector Test:** millimeter rulers, temperature strips (forehead) or temple temperature thermometer, stopwatch
- **Stress and Relaxation:** blood pressure cuff, stopwatch
- **Biopac Laboratory Guide Experiments:**
 - Observing ANS Responses to Lying
 - Observing ANS Responses Following Meditation

The **autonomic nervous system (ANS)** is the division of the peripheral nervous system that operates without conscious control in most individuals. The ANS has autonomic sensory neurons that carry nerve impulses from sensory receptors in visceral organs and blood vessels to the central nervous system (CNS), and autonomic motor neurons that carry nerve impulses from the CNS to smooth muscle, cardiac muscle, and glands (effectors). There are 2 autonomic motor neurons between the spinal cord and the effectors. The first motor neuron (**preganglionic**) synapses with the second motor neuron (**postganglionic**) in ANS ganglia.

The ANS has 2 different motor divisions, the sympathetic division and the parasympathetic division. Most ANS effectors are innervated by both the sympathetic and parasympathetic divisions, which have opposing actions.

Activation of the **sympathetic division** occurs during exercise, emergencies, excitement, and embarrassment and results in a series of physiological responses that together are called the fight-or-flight response. The fight-or-flight response includes activities that promote mental alertness, vision, and skeletal muscle activity.

Activation of the **parasympathetic division** promotes rest, reading, restoration, and repair.

A. Anatomy of the Sympathetic Division

1. Preganglionic and Postganglionic Motor Neurons

Cell bodies of **preganglionic neurons** are located in the gray matter of the lateral horn of spinal cord segments T_1–L_2. Because the nerve impulses from the sympathetic preganglionic neuron originate in the thoracic and lumbar segments of the spinal cord, the sympathetic division is also called the **thoracolumbar division.**

Axons of the first motor neurons of this division, **preganglionic neurons,** exit the spinal cord in the anterior (ventral) root of the spinal nerves and travel with somatic motor neurons in the spinal nerves a short distance. These myelinated preganglionic axons quickly leave the spinal nerve as **white rami communicantes** that connect with sympathetic trunk ganglia, paired chains of ganglia that lie parallel to the spinal cord. In these ganglia, the axons either synapse with postganglionic cell bodies, or they *pass through the sympathetic trunk ganglia without synapsing,* emerging to form **splanchnic nerves** (*splanchnic* = viscera) in the abdominopelvic region.

Nerve cell bodies of the second motor neuron, the **postganglionic neuron,** are located in the sympathetic ganglia. Preganglionic axons synapse with postganglionic neuron cell bodies in the ganglia. Some of the axons of postganglionic neurons return to the spinal nerve at the same level via the **gray rami communicantes.** Postganglionic axons are longer than preganglionic axons because they travel a greater distance to innervate the effectors.

2. Sympathetic Ganglia

The sympathetic division has 2 groups of ganglia: the sympathetic trunk (chain) ganglia and the prevertebral (collateral) ganglia.

- **sympathetic trunk** *(chain)* **ganglia**—These paired ganglia chains lie parallel to the spinal cord. Postganglionic neurons from these ganglia innervate organs superior to the diaphragm.
- **prevertebral** *(collateral)* **ganglia: celiac, superior mesenteric,** and **inferior mesenteric**—These 3 single ganglia lie anterior to the abdominal aorta, and their postganglionic neurons innervate organs inferior to the diaphragm. Preganglionic axons traveling in splanchnic nerves synapse with postsynaptic neuron cell bodies in the prevertebral ganglia or in the adrenal medullae.

Before Going to Lab

1 Label the structures in Figure 22.1.

2 Identify the following structures on a chart showing the sympathetic nervous system: **right** and **left sympathetic trunk ganglia, prevertebral ganglia, splanchnic nerves,** and **preganglionic** and **postganglionic motor neurons.**

LAB ACTIVITY 1 Anatomy of the Sympathetic Division

1 Identify the **sympathetic trunk ganglia** on a longitudinal spinal cord model.

2 Identify the **rami communicantes** and the **sympathetic trunk ganglia** on a transverse section spinal cord model or chart. ■

Trachea

Arch of aorta

Left primary bronchus

Esophagus

Thoracic aorta

Diaphragm

Left kidney

Abdominal aorta

Anterior view

- left sympathetic trunk ganglia
- prevertebral ganglia
- right sympathetic trunk ganglia
- splanchnic nerves

1 _____

2 _____

3 _____

4 _____

FIGURE 22.1 Location of sympathetic motor ganglia and nerves.

B. Anatomy of the Parasympathetic Division

Cell bodies of the preganglionic motor neurons of this division are located in the nuclei of 4 cranial nerves in the brain stem and in the lateral horns of spinal cord segments S_2–S_4. Because the outflow is from these areas, this division is also known as the **craniosacral division.**

Axons of the *cranial* **preganglionic neurons** exit the brain in cranial nerves III, VII, IX, and X and synapse with postganglionic neurons in **terminal** (intramural) **ganglia** located near their target organs in the head region. The major terminal ganglia in the head region are: **ciliary, pterygopalatine, submandibular,** and **otic ganglia.**

Axons of the *sacral* preganglionic neurons leave the spinal cord in the anterior (ventral) roots, form nerves called **pelvic splanchnic nerves,** and synapse with postganglionic neurons in **terminal ganglia** that are within the wall of the target organ.

Nerve cell bodies of **postganglionic neurons** (2nd motor neurons) are in the terminal ganglia. Axons of postganglionic neurons are short because they travel a short distance to their effectors.

Before Going to Lab

1 Label the terminal ganglia in Figure 22.2. Use the name of the ganglia to find the location.

LAB ACTIVITY 2 Anatomy of the Parasympathetic Division

1 Identify the following structures on a model or chart showing the parasympathetic nervous system: **terminal ganglia, pelvic splanchnic nerves,** and **preganglionic** and **postganglionic motor neurons.** ◼

- ciliary ganglion
- otic ganglion
- pterygopalatine ganglion
- submandibular ganglion

1 _____

2 _____

3 _____

4 _____

FIGURE 22.2 Parasympathetic nerves and ganglia.

C. Autonomic Reflexes

Autonomic reflexes control many body functions such as blood pressure, respiration, and digestion. The **autonomic reflex arc** contains the same components as the somatic reflex arc: receptor, sensory neuron, integrating center, motor neurons, and effector. Although most autonomic sensory receptors are located in visceral organs, some autonomic reflexes are responses to changes in the external environment. Autonomic integrating centers are **polysynaptic** and most are located in the hypothalamus and brain stem. However, the autonomic reflex arcs that control urination and defecation have integrating centers in the spinal cord. ANS reflex arcs have *2 motor neurons*—preganglionic and postganglionic neurons—which synapse in the autonomic ganglia. ANS effectors are cardiac muscle, smooth muscle, and glands.

Most ANS effectors receive motor neurons from both the sympathetic and the parasympathetic branches of the ANS. Since sympathetic and parasympathetic responses usually oppose each other, the hypothalamic integrating center determines which response is appropriate.

Before Going to Lab

1 Label the parts of the autonomic reflex arc in Figure 22.3.

1. Pupillary Light Reflex

When photoreceptors in the retina of the eye are stimulated by light, nerve impulses are sent via the optic nerve to integrating centers in the brain. Interneurons from the integrating centers send impulses to motor neurons in the midbrain. Axons from these preganglionic motor neurons travel along cranial nerve III (oculomotor nerve) to the ciliary ganglion where they synapse with the postganglionic motor neuron. The postganglionic axons travel a short distance to smooth muscles in the iris of the eye. When these muscles contract, the pupil constricts.

LAB ACTIVITY 3 Pupillary Light Reflex

1 Test the pupillary light reflex.
 - This experiment works best if conducted in a dimly lit room using a subject with light-colored eyes.
 - Have the subject cover his or her right eye by holding a hand over the eye.
 - Using a penlight, shine a light into the subject's left eye. Observe the pupil and describe the change in the pupil size. _____
 - While shining a light into the subject's left eye, uncover the right eye and immediately observe the size of the right pupil. Is the right pupil the same size as the left pupil? _____
 - Shut off the penlight and observe the change in size of the subject's pupils. _____

2 Answer the Discussion Questions with your lab group.

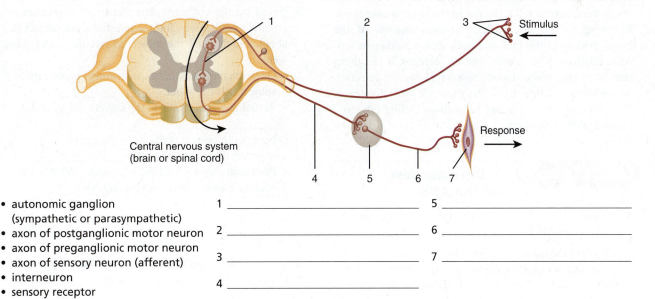

- autonomic ganglion (sympathetic or parasympathetic)
- axon of postganglionic motor neuron
- axon of preganglionic motor neuron
- axon of sensory neuron (afferent)
- interneuron
- sensory receptor
- visceral effector

1 _____ 5 _____

2 _____ 6 _____

3 _____ 7 _____

4 _____

FIGURE 22.3 **Autonomic reflex arc.**

DISCUSSION QUESTIONS
Pupillary Light Reflex

1 Is this reflex a somatic or autonomic reflex?

2 Identify the nerve that is carrying the sensory information from the eye to the brain.

3 Identify the cranial nerve carrying the preganglionic parasympathetic axons to the ciliary ganglia.

4 Is this a monosynaptic or polysynaptic reflex?

5 Is the pupillary light reflex ipsilateral, contralateral, or both?

2. The Effect of Higher Brain Centers on ANS Functions

The cerebral cortex, limbic system (emotional brain), and thalamus can also alter the ANS responses. For example, when some people tell a lie, the limbic system causes the following sympathetic ANS responses: dilation of the blood vessels in the face and neck area resulting in redness, dilation of the pupils, and an increase in breathing rate, heart rate, and/or blood pressure. However, detection of autonomic changes by observation is a subjective exercise. Also, some people can lie without feeling emotion, avoiding ANS responses.

LAB ACTIVITY 4 Lie Detector Test and the ANS

1 Decide who will be the subject, the interrogator, the observer, and the pulse-taker.

2 Use 10 of the questions below or other questions that the group devises and has the instructor's approval. Assign each question a number (1–10) indicating the order in which the questions are to be asked.

How old are you? _____
You were born in what city and state? _____
How many brothers and sisters do you have? _____
When driving a car, do you purposely exceed the speed limit over twice a week? _____
Have you ever flunked a college or university class? _____
Are you single or married? _____
Do you like this anatomy and physiology class? _____
Do you study for this class as much as you should? _____
Are you getting the grade in this class that you want? _____
What is your major? _____
Have you ever cheated on your income taxes? _____
What is your weight? _____
What is your height? _____
Have you ever cheated on an exam? _____

3 Have the subject privately write down which three questions are going to be lies. In a real polygraph test, the subject sees all questions before the test commences.

4 Perform the lie detector test.
- Have the observer obtain baseline information on: the subject's redness of face and neck, pupil size, pulse rate in 15 seconds, nervousness of hands or fingers, or any other nervous changes that might indicate lying. You may take the temperature on the forehead with temperature strips, if available. Record the results in Table 22.1.
- Have the interrogator ask the 10 chosen questions.
- Observe ANS changes after each question and record in Table 22.1 before continuing to the next question.

5 Based on the collected data, have the experimenters privately predict which three questions the subject lied about, and then ask the subject to give the answers that were lies.
- Question numbers that experimenters predict the subject is lying about. _____
- Actual question numbers the subject did lie about. _____
- Number of correct predictions chosen by experimenters in your group. _____
- Pool your data with all the lab groups. What is the average number of correct predictions? _____

6 Answer Discussion Questions with your lab group.

7 Discuss your results with the class.

8 Complete the **Biopac Laboratory Guide Experiment:** Observing ANS Responses to Lying.

TABLE 22.1 ANS Responses and Lie Detector Test

PRE-QUESTION OBSERVATIONS	ANS RESPONSES	ANS RESPONSES FOR QUESTION									
		1	2	3	4	5	6	7	8	9	10
Color _____	Color No Change (N) or Redness (R)										
Pupil Size (mm) _____	Pupil Size No Change (N), Increased (I), or Decreased (D)										
Pulse Rate (beats/15 sec) _____	Pulse Rate (beats/15 sec) N, I, or D										
Temperature (°F or °C) _____	Temperature (°F or °C) N, I, or D										
	Nervousness Yes (Y) or No (N)										
	Prediction: Was the answer the Truth (T) or a Lie (L)?										

DISCUSSION QUESTIONS
Lie Detector Test

1 Based on the pooled class data, give reasons why using this type of lie detector test to solve a crime may or may not be advantageous.

2 A professional **polygraph test** is administered by someone who is trained to perform the test and lasts 2–3 hours, with additional time to analyze the results. Four reactions are recorded by four sensors attached to a person's body that record chest movement during respiration, abdominal movement during respiration, skin conductance (sweating), and heart rate/blood pressure (Figure 22.4). Note the response (truthful answer) on the left side of the figure, compared with the response (lie) on the right side.

3 Give reasons why the use of a polygraph test given by a professional may or may not reveal if the person is telling the truth.

Chest movement during respiration

Abdominal movement during respiration

Skin conductance

Heart rate and blood pressure

FIGURE 22.4 Polygraph test reactions.

3. The Effect of Stress and Relaxation on ANS Functions

Breathing rate is an ANS response that can be controlled. Decreasing breathing rate and increasing depth of breathing can cause a person to relax. This state of relaxation is characterized by diminished sympathetic stimulation and increased parasympathetic stimulation. Alternately, increasing breathing rate can lead to arousal caused by increased sympathetic stimulation.

LAB ACTIVITY 5 Effect of Relaxation on ANS Functions

1 Decide who will be the subject and who will measure and record pulse rate and blood pressure.

2 Have the subject sit down and measure resting pulse rate and blood pressure after 2 minutes. Record measurements in Table 22.2.

3 Give the subject a series of multiplication questions that the subject must answer in a short period of time.

4 Measure pulse rate and blood pressure immediately after doing the math problems and record in Table 22.2.

5 Have subject take deep slow breaths (8 breaths/min) for 10 minutes.

6 Measure pulse rate and blood pressure and record in Table 22.2.

7 Answer Discussion Questions with your lab group.

DISCUSSION QUESTIONS

Effect of Stress and Relaxation on Pulse and Blood Pressure

1 State whether stress increased or decreased blood pressure and pulse rate. Which division of the ANS caused this change?

2 State whether relaxation increased or decreased blood pressure and pulse rate. Which division of the ANS caused this change?

3 The ANS regulation of blood pressure is involuntary, yet biofeedback techniques enable individuals to reduce their pulse rate and blood pressure. Explain how this occurs.

TABLE 22.2 Effect of Stress and Relaxation on Pulse and Blood Pressure

ACTIVITY	PULSE RATE	BLOOD PRESSURE
Resting		
Stress		
Deep, slow breathing		

Reviewing Your Knowledge

A. Comparison of Somatic and Autonomic Nervous Systems

Answer each of the following questions:

1. Compare the number of motor neurons in a somatic and autonomic reflex arc.

2. Compare the effectors of the somatic and autonomic nervous systems.

B. Comparison of Sympathetic and Parasympathetic Divisions

Match the terms with the appropriate ANS division: P = parasympathetic division; S = sympathetic division. Use one or both choices to answer the question.

_____ **1.** two motor neurons

_____ **2.** short preganglionic axons

_____ **3.** long preganglionic axons

_____ **4.** innervates adrenal medullae

_____ **5.** terminal ganglia

_____ **6.** trunk or chain ganglia

_____ **7.** prevertebral ganglia

_____ **8.** craniosacral division

_____ **9.** thoracolumbar division

_____ **10.** activated during exercise, fighting, or fleeing

_____ **11.** rami communicantes

_____ **12.** promotes digestion

_____ **13.** short postganglionic axons

_____ **14.** long postganglionic axons

C. Autonomic Reflex Arc

Answer each of the following questions:

1. List the components of the autonomic reflex arc. Compare with somatic reflex arc.

2. Where are autonomic integrating centers located?

3. Are the autonomic integrating centers monosynaptic or polysynaptic?

4. Can the autonomic integrating centers be influenced by higher brain area?

Using Your Knowledge

A. Sympathetic Division

Refer to a sympathetic nervous system chart (or textbook) to answer questions 1–6.

Brittany was visibly upset that a class she needed to take was cancelled because of low enrollment. Jim observed that her eyes were dilated. Trace the sympathetic pathway from the lateral gray horn of the spinal cord to the iris of her eye.

1. Identify the spinal cord area (cervical or thoracic) that the preganglionic axons exit in the anterior (ventral) root.

2. Name the sympathetic ganglia in which the preganglionic and postganglionic neurons synapse.

Paul ate a big lunch and then decided to jog a mile, which was not a very good idea. His digestive system was put "on hold" while his sympathetic nervous system was stimulated to run. Trace the sympathetic pathway from the lateral gray horn of the spinal cord to the stomach.

3. Identify the nerve that the preganglionic axons form.

4. Name the effector (the tissue within the organ that is innervated by the postganglionic neuron).

Burke is in the final 100 yards of a race with an opponent at his heels. His sympathetic nervous system is stimulating his adrenal medullae. Trace the sympathetic pathway from the lateral gray horn of the spinal cord to the adrenal medulla.

5. Name the sympathetic ganglia in which the preganglionic axons synapse.

6. What is different about these postganglionic cells?

B. Parasympathetic Division

Refer to parasympathetic nervous system chart (or textbook) to answer questions 7–10.

Tom had anesthesia for surgery and was unable to void for a short time after awakening. Trace the normal parasympathetic pathway from the lateral gray horn of the spinal cord to the urinary bladder.

7. Name the spinal cord segments that the preganglionic axons exit in the anterior (ventral) root.

8. Name the effector (the tissue within the organ that is innervated by the postganglionic neuron).

Michael just ate lunch and his salivary glands are responding with secretions. Trace the parasympathetic pathway from the medulla oblongata to the salivary glands.

9. Name the two nerves that carry the preganglionic axons to the parasympathetic ganglia.

10. Name the parasympathetic ganglia in which the preganglionic and postganglionic neurons synapse.

C. Functions of the Sympathetic and Parasympathetic Nervous Systems

Choose the correct division of the ANS that generates the following results: S = sympathetic; P = parasympathetic.

_____ **11.** sweaty palms

_____ **12.** blushing

_____ **13.** urination

_____ **14.** stomach churning

_____ **15.** salivary gland secretion

_____ **16.** constriction of pupils

_____ **17.** increased blood pressure

_____ **18.** increased respiration

_____ **19.** decreased heart rate

_____ **20.** digestive enzyme secretions

General Senses

OBJECTIVES

1 Differentiate between free and capsulated nerve endings and receptor cells
2 Describe the pathway from the sensory receptor to the cerebral cortex
3 Identify areas of the body that have the greatest tactile discrimination
4 Observe adaptation of sensory receptors
5 Observe the phenomenon of referred pain

MATERIALS

- skin model
- compound microscope, lens paper, prepared microscopic slides of corpuscles of touch (Meissner's corpuscles) and lamellated (Pacinian) corpuscles
- cross-section model or chart of spinal cord
- **Tactile Sensitivity:** toothpicks or calipers or aesthesiometer, millimeter rulers, sandpaper, velvet or synthetic fur, smooth surface such as glass
- **Adaptation of Temperature Receptors:** 3 fingerbowls or 1,000-mL beakers, water (10°C, 25°C, 45°C), thermometer
- **Adaptation to Light Pressure:** 4 pennies per group
- **Referred Pain:** bowl of ice water

Sensory receptors provide information about the environment outside and within our body. General sensory information originates either from somatic sensory receptors (skin and skeletal muscle) or visceral sensory receptors (visceral organs). Special sensory information originates from special sense organs (eye, inner ear, nasal mucosae, or taste buds).

A. Sensory Receptors

Structural classes of general sensory receptors include: free nerve endings, encapsulated nerve endings, or specialized receptor cells. **Free nerve endings** are dendrites of sensory nerves that convey to the brain the general sensations of pain, temperature, tickle, itch, and some touch sensations. **Encapsulated nerve endings** are dendrites of sensory neurons enclosed by a connective tissue capsule that convey the general sensations of touch and pressure to the brain. Sensory receptors of special sense organs are **receptor cells** that form synapses with sensory neurons.

Before Going to Lab

1 Label the somatic sensory receptors in Figure 23.1(a) and (b). Figure 23.1(a) shows somatic sensory receptors found in hairless skin and skin with hair follicles. Refer to Table 23.1, which identifies location, structure (free or encapsulated nerve ending), and stimuli of somatic sensory receptors.

LAB ACTIVITY 1 General Sensory Receptors

1 Point to the sensory receptors identified in Figure 23.1(a) on a skin model or anatomical chart.

2 Examine prepared microscope slides of cutaneous sensory receptors. ■

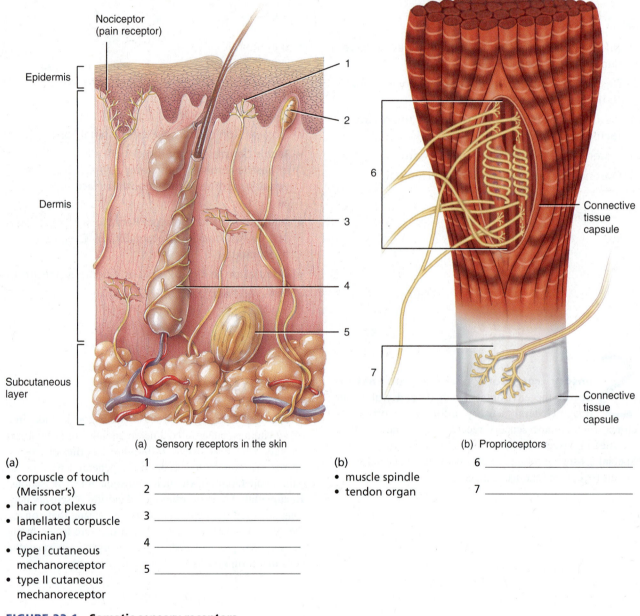

(a) Sensory receptors in the skin

(b) Proprioceptors

(a)
- corpuscle of touch (Meissner's)
- hair root plexus
- lamellated corpuscle (Pacinian)
- type I cutaneous mechanoreceptor
- type II cutaneous mechanoreceptor

1 _____

2 _____

3 _____

4 _____

5 _____

(b)
- muscle spindle
- tendon organ

6 _____

7 _____

FIGURE 23.1 Somatic sensory receptors.

TABLE 23.1 **Somatic Sensory Receptors**

RECEPTOR	LOCATION	STRUCTURE	STIMULI
Corpuscles of touch (Meissner's corpuscles)	Dermal papillae of hairless skin	Encapsulated nerve endings	Touch and pressure
Hair root plexuses	Surrounds hair follicles	Free nerve endings	Touching hair
Lamellated corpuscles (Pacinian corpuscles)	Subcutaneous and submucosal tissue, joints, tendons, and muscles	Encapsulated nerve endings	Touch and pressure
Type I cutaneous mechanoreceptors (Merkel discs)	Associated with Merkel cells in stratum basale layer of epidermis	Free nerve endings	Touch and pressure
Type II cutaneous mechanoreceptors (Ruffini's corpuscles)	Dermis, ligaments, and tendons	Encapsulated nerve endings	Stretching of digits and limbs
Muscle spindles	Found within most skeletal muscles	Encapsulated nerve endings	Respond to changes in muscle length
Tendon organs	Found at junction of tendons and muscles	Encapsulated nerve endings	Respond to changes in joint position and movement
Warm receptors	Dermis	Free nerve endings	Thermoreceptor; responds to temperatures between 32° and 48°C (90°–118°F)
Cold receptors	Stratum basale of epidermis	Free nerve endings	Thermoreceptor; responds to temperatures between 10° and 40°C (50°–105°F)
Nociceptors	Every body tissue	Free nerve endings	Pain receptors; respond to harmful stimuli

B. Sensory Pathways

Sensations that are consciously perceived include touch, pressure, pain, temperature, and some proprioception. These sensations have sensory pathways consisting of a chain of 3 neurons (first-, second-, and third-order neurons) that terminate in the cerebral cortex. **Sensory neurons (first-order neurons)** located in the peripheral nervous system synapse with an **association neuron** (second-order neuron) in the spinal cord or brain. **Second-order neurons** synapse with third-order neurons in the thalamus. **Third-order neurons** transmit information to the primary somatosensory area of the cerebral cortex. These sensory pathways cross to the opposite side of the CNS somewhere during their ascent.

Sensations that do not reach the cerebral cortex are not consciously perceived and include some proprioception and information from visceral sensory receptors. These pathways do not contain third-order neurons.

Before Going to Lab

1 Label the sensory pathways in Figure 23.2(a) and (b).

LAB ACTIVITY 2 Sensory Pathways

1 Point out the pathway of the first- and second-order neurons in Figure 23.2(a) and (b) on a cross-section spinal cord model or chart.

2 Answer Discussion Questions with your lab group.

DISCUSSION QUESTIONS
Sensory Pathways

Refer to Exercise 16: Nervous Tissue, if necessary.

1 Are sensory neurons from general sensory receptors unipolar, bipolar, or multipolar neurons?

2 a. Where are the cell bodies for the sensory neurons (first-order neurons) located?

b. Where are their axon terminals located?

3 a. Where are the cell bodies for the second-order neurons located?

b. Where are their axon terminals located?

4 a. Where are the cell bodies for the third-order neurons located?

b. Where are their axon terminals located?

5 Where does the posterior column-medial lemniscus pathway cross the opposite side of the CNS?

6 Where does the anteriolateral (spinothalamic) pathway cross to the opposite side of the CNS?

(a) Anterolateral (spinothalmic) pathway

(b) Posterior column-medial lemniscus pathway

(a)
• first-order neuron
• second-order neuron
• third-order neuron

1 _____

2 _____

3 _____

(b)
• first-order neuron
• second-order neuron
• third-order neuron

4 _____

5 _____

6 _____

FIGURE 23.2 **Selected sensory pathways.**

C. Tactile Sensitivity of Different Body Areas

Some areas of the skin have greater **tactile sensitivity** than others. The greater the number of cutaneous receptors in an area (touch receptor density), the greater the tactile sensitivity of that area. The size of the somatosensory cortex area receiving sensory information from a specific body area is directly proportional to the cutaneous receptor density.

The **two-point discrimination test** is an indirect measure of cutaneous touch receptor density. A subject is touched by 2 closely spaced points and asked if he or she can feel both points. The objects are moved farther apart until 2 points can be felt. An area of skin with a greater density of touch receptors is more sensitive to touch and can discriminate between 2 points closer together than an area with a lower density of touch receptors.

LAB ACTIVITY 3 **Experiment: Tactile Sensitivity: Two-Point Discrimination**

1 **Prediction:** List the cutaneous receptor density of the following body areas (cheek, fingertip, palm, forearm, back of leg) from the greatest to the least.
- _____ (greatest)
- _____
- _____
- _____
- _____ (least)

2 **Materials:** Obtain materials for the tactile sensitivity experiment.

3 **Data Collection:** Perform the two-point discrimination test on the different body areas and record your results in Table 23.2.
- Decide who will be the activity coordinator, subject, data collector, and data recorder.
- Have the subject's eyes closed during the experiment.
- Put the 2 caliper points (or toothpicks) together.
- Place caliper points on skin area to be tested. Touch both caliper points to the skin at the same time.
- Ask the subject if 1 or 2 points of the caliper can be felt.
- Increase the distance between caliper points. For the fingertip and palm, increase the distance 1 mm. For the cheek, forearm, and back of leg, increase the distance by 2 mm.
- Continue to increase the distance between the caliper points until the subject can feel 2 points. This distance is **the two-point discrimination distance** and is to be measured with a millimeter ruler. Record this value in Table 23.2.

4 With eyes closed, have the subject feel objects of different textures (sandpaper, velvet or synthetic fur, smooth surface such as glass) with fingertips, posterior surface of forearm, and side of leg.

5 Clean up as directed by your instructor.

TABLE 23.2 Tactile Sensitivity: Two-Point Discrimination

AREA TESTED	TWO-POINT DISCRIMINATION DISTANCE (mm)	RECIPROCAL (mm)	CLASS AVERAGE TWO-POINT DISCRIMINATION DISTANCE (mm)	CLASS AVERAGE RECIPROCAL (mm)
Cheek				
Fingertip				
Palm				
Forearm				
Back of leg				

6 Data Analysis:

- Calculate the reciprocal $\left(\dfrac{1}{\text{two-point distance}}\right)$ of the two-point discrimination distance for each area and record the value in Table 23.2. The reciprocal represents the portion of the somatosensory cortex that receives information from sensory receptors for a given body area. Areas with high sensory receptor density are represented by a corresponding greater area of cerebral cortex.

- Collect the individual two-point discrimination distance and reciprocal values for each body area from each lab group.

- Calculate the average two-point discrimination distance and reciprocal for each body area and record in Table 23.2.

- Create a bar graph with the body areas on the X axis and the class averages of the reciprocals on the Y axis. Use the graph in Figure 23.3.

7 Complete the Experimental Report with your lab group.

EXPERIMENTAL REPORT
Tactile Sensitivity

Results:

- State which body area tested has the highest average of cutaneous receptor density and which has the lowest.

- State whether variation in tactile sensitivity per body area was observed among subjects.

- State which body area tested had the least and most sensitivity to different textures.

Discussion:

- Which body area tested was represented by the largest area of cerebral cortex? Discuss the reason for your conclusion.

- Which body area tested was represented by the smallest area of cerebral cortex? Discuss the reason for your conclusion.

- Discuss why cutaneous receptor density varied or did not vary among individuals.

Conclusion:

- Write a sentence that postulates a correlation between the average two-point discrimination and the number of cutaneous receptors (receptor density) in a body area.

- Write a sentence that states which body area has the greatest and the least density of cutaneous receptors.

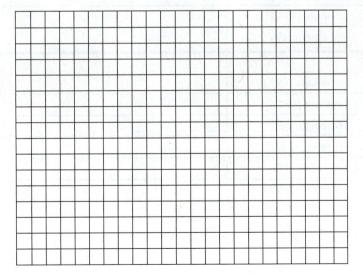

FIGURE 23.3 Tactile sensitivity of different body areas.

D. Adaptation of Sensory Receptors

Adaptation (or fatigue) occurs when the continued stimulus of sensory receptors results in a decrease in nerve impulse generation in the sensory neuron(s) associated with the receptors. This phenomenon may occur with the length of time a stimulus is given and with the stimulus intensity. Receptors may adapt rapidly or slowly to continued stimulation. During this adaptation, conscious awareness dissipates. In this exercise, we will examine adaptation of temperature receptors that are initially rapidly adapting but maintain some response to stimuli.

LAB ACTIVITY 4 Adaptation of Temperature Receptors

1 Part 1: Observe adaptation for warm temperature receptors.
- Decide who will be the subject, the timer, and the data collector and recorder.
- Obtain a bowl or 1,000-mL beaker of 45°C warm water.

- Have the subject immerse the *left hand* into the water and record the immediate sensations in Table 23.3.
- Keep the left hand immersed for 1 minute; then also immerse the *right hand* into the same water. Now record the temperature sensation of both hands in Table 23.3.
- In which hand did adaptation occur? _____
- Have the subject wash hands in room temperature water and then wait for 5 minutes before beginning the next procedure.

2 Part 2: Observe adaptation for warm and cold temperature receptors.
- Obtain 3 bowls or 1,000-mL beakers of water: one room temperature (25°C), one warm (45°C), and one ice water (10°C).
- Have the subject place the *left hand* in 45°C warm water and the *right hand* in 10°C ice water simultaneously. Record the immediate temperature sensation of each hand (hot, cold, or no change) in Table 23.3.
- Record the sensations of each hand after 1 minute.
- Have the subject simultaneously put both hands into the room-temperature (25°C) bowl. Record the immediate temperature sensation of each hand in Table 23.3.

3 Clean up as directed by your instructor.

4 Answer the Discussion Questions with your lab group.

TABLE 23.3 Adaptation of Temperature Receptors

BODY AREA	TIME	TEMPERATURE SENSATION
Part 1 Left hand	Immediate (45°C)	
	1 minute (45°C)	
Right hand	Immediate (45°C)	
Part 2 Left hand	Immediate (45°C)	
	1 minute (45°C)	
	Immediate (25°C)	
Right hand	Immediate (10°C)	
	1 minute (10°C)	
	Immediate (25°C)	

DISCUSSION QUESTIONS
Adaptation of Temperature Receptors

1 How long did it take for the cold and warm temperature receptors to adapt for the subject in your group?

2 When sensory receptors adapt, does the cerebral cortex receive an increased or decreased number of sensory nerve impulses?

3 Indicate whether each of the following general sensory receptors adapt quickly, slowly, or not at all. Use your own experience.

 a. Cutaneous touch receptors (stimulated when skin is touched by an object, for example, when clothes touch the skin).

 b. Cutaneous pressure receptors (stimulated when pressure is applied to skin).

 c. Proprioceptors associated with skeletal muscles and tendons that are important for balance (proprioceptors maintain muscle contraction needed for standing erect and keeping head erect).

 d. Pain receptors.

LAB ACTIVITY 5 Adaptation to Light Pressure

1 Observe adaptation for light pressure receptors.
- Rest your forearm on the desk and close your eyes. Your lab partner will place one penny on the anterior surface of your forearm. Determine the length of time that the pressure sensation persists and record.
- Duration of pressure sensation = _____ sec
- Your lab partner will now stack 3 more pennies on top of the first penny.
- Does the pressure sensation return? _____ Determine the length of time that this new pressure sensation persists and record.
- Duration of pressure sensation = _____ sec

2 Repeat the activity for each member of your lab group.

3 Answer the Discussion Questions with your lab group.

DISCUSSION QUESTIONS
Adaptation of Light Pressure

1 How long did it take for the light pressure receptors to adapt to one penny versus 4 pennies on the forearm of your subject?

2 How did the results differ among members of your lab group? Was there a very large difference in adaptation time?

3 Discuss what the difference is between the stimulation of one penny versus the stimulation of 4 pennies. What do you conclude from the results?

E. Referred Pain

Pain that is felt in areas that are not injured or stimulated is called **referred pain.** Sometimes this pain originates in visceral organs and is perceived in the skin that is innervated by the same spinal segments. The pain of a heart attack is usually felt in the skin over the heart and down the left arm.

LAB ACTIVITY 6 Referred Pain

1 Observe referred pain in the elbow.
- Choose a subject, a timer, and a data collector and recorder. This experiment works best on a subject who has thin arms. The subject will be asked to immerse his/her elbow in ice water and describe the sensation (pain, discomfort, or tingling) and where the sensation is located.
- Obtain a bowl of ice water (10°C).
- Have the subject immerse an elbow in ice water and immediately describe the sensation and where it is located. Record results in Table 23.4. Have the subject keep the elbow in the water until directions indicate to remove it.

- After 1 minute, ask the subject to describe the sensation and where it is located. Record results in Table 23.4.
- After 2 minutes, ask the subject to describe the sensation and where it is located. Have the subject remove elbow from water. Record results in Table 23.4.

2 Clean up as directed by your instructor.

3 Answer the Discussion Questions with your lab group.

DISCUSSION QUESTIONS
Referred Pain

1 The location of the sensation changed over time. Name the areas in order where the sensation was felt.

2 Which nerve supplies the areas where the sensation occurred?

TABLE 23.4 Referred Pain

TIME (min)	SENSATION	LOCATION OF SENSATION
Immersion of elbow (0 min)		
1 minute after immersion		
2 minutes after immersion		

Reviewing Your Knowledge

A. Sensory Receptors

1. Describe the function of sensory receptors.

2. Name the general sensory receptors that belong to each structural class.

 (a) Free nerve endings

 (b) Encapsulated nerve endings

B. Sensory Pathways

Match each term to the correct numbered phrase.

cerebral cortex
first-order neuron
posterior (dorsal) root ganglion
second-order neuron
thalamus
third-order neuron

_____ **1.** Receives input from third-order neurons; sensory information is consciously perceived.

_____ **2.** Cell body located in spinal cord or brain stem.

_____ **3.** Gateway to cerebral cortex; location of synapse between second- and third-order neurons.

_____ **4.** Neuron associated with sensory receptor.

_____ **5.** Relays sensory nerve impulses from thalamus to cerebral cortex.

_____ **6.** Cell body for first-order neuron located here.

C. Tactile Sensitivity

1. Describe correlation between tactile density of a body area and the size of the cerebral cortex receiving information from those receptors.

D. Adaptation

1. Define adaptation.

2. Do warm receptors adapt quickly?

3. Do cold receptors adapt quickly?

E. Referred Pain

1. Define referred pain.

2. After the subject's elbow had been immersed in water for 2 minutes, where was the referred pain located? Explain.

Using Your Knowledge

Questions 1–3.

Figure 23.4 shows a map of the specific areas of the primary somatosensory cortex that represent areas of the body. The cerebral cortex localizes the precise area of the body that sends sensory nerve impulses. Based on the size of the area represented on the cortex and the size of the body area, circle which body area has the greatest density of receptors:

1. The *lips* or the *hip*

2. The *tongue* or the *toes*

3. The *foot* or the *hand*

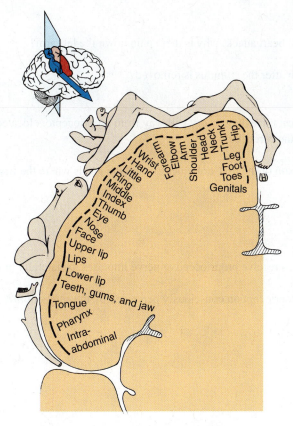

FIGURE 23.4 Somatosensory cortex.

Answer the following questions.

4. Using your knowledge, compare the size of the somatosensory cortex representing the fingertips for a visually impaired person who reads Braille (increased tactile sensitivity) to that of someone with normal vision who has normal tactile sensitivity.

5. Leprosy often results in loss of pain to infected body areas. Describe the hazards of this.

6. When a hair on your head is moved, the hair follicle touch receptors are stimulated. If a person put his or her hair in a ponytail, what would be the result if there was no adaptation to these touch receptors?

7. When you hit your funny bone (olecranon), where do you experience referred pain? *Hint:* What nerve would you hit?

8. (a) When a person experiences a heart attack, why is there pain down the left arm?

 (b) Why would pain persist even after the stimulus is removed?

Using your textbook or other reference, trace the sensory pathway from proprioceptors located in the quadriceps muscle to the spinocerebellar tract traveling to the cerebellum.

9. Identify the nerve carrying the axon of the first-order neuron and the levels where the first-order neuron enters the spinal cord.

10. (a) Do the cerebellar hemispheres receive proprioceptive nerve impulses ipsilaterally or contralaterally?

 (b) Using your knowledge, is proprioception consciously perceived by the cerebellum?

Special Senses

OBJECTIVES

The Eye and Vision

1 Identify the accessory structures of the eye and the structures of the eyeball

2 Identify the 3 principal layers of the retina

3 Perform visual tests that are used to determine visual acuity, near point of vision, astigmatism, location of blind spot, and presence of red-green color blindness

The Ear, Hearing, and Equilibrium

1 Identify the structures of the external, middle, and internal ear

2 Identify the major structures of the cochlea, vestibule, and semicircular canals

3 Perform tests that are used in determining the cause of hearing loss

4 Perform tests that exhibit the function of the receptors for dynamic and static equilibrium

The Nose and Olfaction

1 Identify the location of receptors for olfaction

2 Identify olfactory receptor structures

3 Determine time of adaptation for olfactory receptors

The Taste Buds and Gustation

1 Identify the type of taste buds and their location

2 Identify gustatory structures

3 Observe the influence of smell and texture on taste

MATERIALS

The Eye and Vision

- eye models or charts
- Dissection: cow eyes, dissection equipment, disposable gloves, safety glasses
- compound microscope, lens paper, prepared slides of an eyeball
- ophthalmoscope
- metric ruler, Snellen eye chart

The Ear, Hearing, and Equilibrium

- ear models or charts, skull
- compound microscope, lens paper, prepared slides of the cochlea, macula, and crista
- tuning forks, cotton balls
- swivel chair or stool

The Nose and Olfaction

- skull
- stopwatch; cotton balls; clove oil, peppermint oil, cinnamon oil, or any three aromatic substances

The Taste Buds and Gustation

- compound microscope, lens paper, prepared slides of taste buds
- mirror
- 1/2-inch cubes of apple, banana, cheese, carrot, and raw potato

A. Structure of the Eye and Vision

1. Accessory Eye Structures

Accessory structures of the eye include the eyebrows, eyelids or **palpabrae,** eyelashes, conjunctiva, lacrimal apparatus, and the extrinsic eye muscles. The **conjunctiva** is the thin, protective mucous membrane that covers the anterior eye and folds to cover the inner eyelid. The conjunctival fold forms a pocket that keeps contacts from moving toward the posterior part of the eyeball. The **palpebral conjunctiva** covers the interior of the eyelid, and the **bulbar conjunctiva** covers the anterior part of the white of the eye, but not the cornea.

The **lacrimal apparatus** (*lacrim-* = tears) is a group of structures involved in producing and draining tears. The **lacrimal gland** produces and secretes tears onto the eye surface, and the **lacrimal canals** drain the tears from the eyes into the enlarged **lacrimal sac.** The enlarged **nasolacrimal duct** receives tears from the lacrimal sac and drains the tears into the nasal cavity.

The **extrinsic eye muscles** are six skeletal muscles that insert on the exterior of the eyeball to move the eyeball in all directions. The **superior, inferior, medial,** and **lateral** rectus muscles are parallel to the long axis of the eyeball. The **superior** and **inferior oblique** muscles attach to the eyeball at an angle.

> **Before Going to Lab**
>
> **1** Label the parts of the conjunctiva in Figure 24.1(a) and structures of the lacrimal apparatus in (b).
>
> **2** Label the extrinsic eye muscles in Figure 24.2(a) and (b).

(a) Lateral view of eyeball

(b) Lacrimal apparatus

(a)
- bulbar conjunctiva (con-junk-TIE-va)
- conjunctival fold
- palpebral (PAL-pub-bral) conjunctiva

1 _____

2 _____

3 _____

(b)
- lacrimal (LAK-rih-mal) canals
- lacrimal gland
- lacrimal sac
- nasolacrimal duct

4 _____

5 _____

6 _____

7 _____

FIGURE 24.1 Accessory structures of the eye.

LAB ACTIVITY 1 Accessory Structures of the Eye

1 Identify all the accessory structures of the eye, including all parts of the lacrimal apparatus, on an eye model or chart.

2 Identify the extrinsic eye muscles on an eye model.

3 Determine how each muscle moves the eye by observing where it inserts on the eye and how the eye would move if the insertion is moved toward the origin.

4 Write the function of each extrinsic eye muscle in Table 24.1. Choose one of the following eye movements for each muscle: *moves eye superiorly, inferiorly, medially, laterally, superiorly and laterally, or inferiorly and laterally.*

(a) Lateral view of right eyeball

(b) Superior view of left eyeball

- inferior oblique
- inferior rectus
- lateral rectus
- medial rectus
- superior oblique
- superior rectus

1 _____ 4 _____

2 _____ 5 _____

3 _____ 6 _____

FIGURE 24.2 Extrinsic eye muscles.

TABLE 24.1 Function of the Extrinsic Eye Muscles

EXTRINSIC EYE MUSCLE	FUNCTION
Inferior oblique	
Inferior rectus	
Lateral rectus	
Medial rectus	
Superior oblique	
Superior rectus	

2. Structure of the Eyeball

The wall of the eyeball has 3 layers: the outer fibrous tunic, the middle vascular tunic, and the inner retina. The **fibrous tunic** is composed of the cornea and sclera. The **cornea** is the transparent anterior portion that covers the iris and pupil, and the **sclera** (*scler-* = hard) is the tough, white part of the eye that forms the majority of the eyeball. The **scleral venous sinus** (canal of Schlemm) is an opening found at the junction of the cornea and the sclera.

The middle **vascular tunic** is composed of the iris, ciliary body, and choroid. The **iris** is the most anterior portion of the vascular tunic and contains pigmented cells. It is made of circular and radial smooth muscle and controls the pupil size. The **pupil** is the opening in the middle of the iris that allows light to enter the eyeball and changes size in response to the intensity of light. The **ciliary body** begins posterior to the iris at the junction of the cornea and sclera and consists of the ciliary muscle and ciliary processes. The **ciliary muscle** is a circular smooth muscle that contracts to control the shape of the lens. **Ciliary processes** are folds that protrude from the ciliary body toward the lens. They contain capillaries that secrete aqueous humor, the fluid in the anterior chamber of the eyeball. **Suspensory ligaments (zonular fibers)** are thin fibers that attach the lens to these processes. The **choroid** is the most posterior part of the vascular tunic that lines most of the interior of the sclera. It contains many blood vessels that nourish the retina.

The **retina** (sensory tunic) is the inner coat that begins at the **ora serrata,** the serrated boundary between the ciliary muscle and the retina. The retina continues posteriorly, lining the interior of the choroid. The **pigmented layer** of the retina is the outer portion, and the **neural layer** is the inner portion that contains photoreceptors and associated neurons.

- choroid
- ciliary (SIL-ee-air-ee) body
- ciliary muscle
- ciliary process
- cornea (KOR-nee-ah)
- iris
- ora serrata (ser-RAH-tah)
- pupil
- retina
- sclera (SKLER-ah)
- scleral venous sinus
- suspensory ligament (zonular fibers)

1 _____ 7 _____

2 _____ 8 _____

3 _____ 9 _____

4 _____ 10 _____

5 _____ 11 _____

6 _____ 12 _____

FIGURE 24.3 Structure of the eyeball.

1 Label the eyeball structures in Figure 24.3.

LAB ACTIVITY 2 **Structure of the Eyeball**

1 Identify these structures on an eye model or chart. ■

3. Interior of the Eyeball

The interior of the eyeball contains the lens, anterior cavity, and vitreous chamber. The **lens** divides the interior of the eyeball into an anterior cavity and a vitreous chamber (posterior cavity). The **anterior cavity** is a space between the cornea and the lens that is filled with watery **aqueous humor** (*aqua* = water; *humor* = moist). This cavity is subdivided into an **anterior chamber** (between the cornea and the iris) and a **posterior chamber** (between the iris and the lens). The **scleral venous sinus** (canal of Schlemm) is an opening found at the junction of the cornea and sclera that drains aqueous humor back into the bloodstream. The **vitreous chamber** is the larger, posterior cavity located between the lens and the retina. This cavity is filled with a gel-like substance called the **vitreous body** (humor) that holds the retina flat against the choroid.

CLINICAL NOTE: Glaucoma *is caused by an increase in pressure within the eye called intraocular pressure. Blockage of the scleral venous sinus prevents drainage of aqueous humor, increasing the amount in the anterior cavity that causes increased intraocular pressure.*

Before Going to Lab

1 Label the interior spaces and structures of the eyeball in Figure 24.4.

LAB ACTIVITY 3 **Anterior and Posterior Cavities of the Eye**

1 Identify these spaces and structures on a model or chart. ■

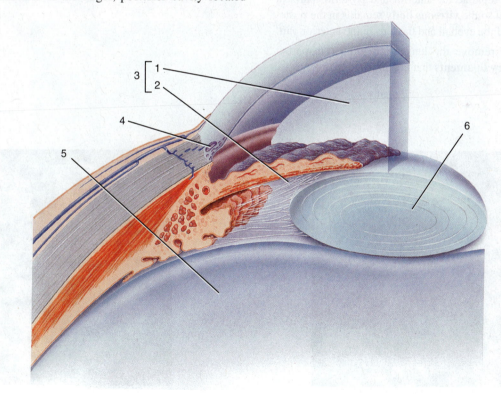

- anterior cavity
- anterior chamber
- lens
- posterior chamber
- scleral venous sinus
- vitreous chamber

1 _____ 4 _____
2 _____ 5 _____
3 _____ 6 _____

FIGURE 24.4 Anterior and posterior cavities of the eyeball.

LAB ACTIVITY 4 Dissection of Cow Eye

> **SAFETY NOTE:** *Wear safety glasses and gloves when using preserved or fresh tissue. Wash hands thoroughly with soap and water when you are done.*

1 Using gloves, obtain a cow eye and rinse it to remove excess preservative.

2 The posterior portion of the eye may be encased in adipose tissue. Carefully remove the adipose tissue protecting the optic nerve.

3 Identify the external eye structures: **cornea, sclera, optic nerve,** and **extrinsic eye muscles.** Refer to Figure 24.5(a). The preservative causes the cornea to change from being transparent and smooth to opaque and wrinkled.

4 Using the point of a scalpel, punch an opening ¼-inch posterior to the cornea through the very tough sclera. Be careful not to squeeze the eyeball too tightly or liquid may squirt out. Use the scissors to cut an incision all the way around the eyeball, separating it into two parts.

5 Carefully separate the anterior and posterior parts of the eyeball so the **vitreous body** remains in the posterior part of the eyeball and the **lens** in the anterior part.

6 Carefully remove the lens, noting the transparent **suspensory ligaments** that are attached to the lens.

7 Using Figure 24.5(b), identify the following structures in the anterior portion of the eyeball: **pupil, iris,** and **ciliary muscle.**

8 Carefully remove the vitreous body from the posterior portion of the eyeball. Refer to Figure 24.5(b) to identify the structures in the posterior portion of the eyeball. The **retina,** a thin beige layer, will probably separate from the choroid. The only point of attachment of the retina to the wall of the eyeball is at the **optic disc.**

9 Pull the retina away from the **choroid,** the dark middle layer of the eyeball that contains the pigment melanin. The choroid of the cow's eye has an iridescent reflecting surface called the **tapetum lucidum** that is not found in humans. The tapetum lucidum reflects light within the eye and enables animals to see in low-light conditions.

10 Separate the choroid from the **sclera.** Observe how the three coats or tunics—the **retina, choroid,** and **sclera**—form the wall of the eyeball.

11 Clean up as directed by your instructor. ■

(a) External structures

(b) Anterior and posterior sections

FIGURE 24.5 Cow eye dissection.

4. Anatomy of the Retina

The neural portion of the retina is an outgrowth of the brain and contains three layers of neurons: the photoreceptor layer (deepest cell layer), the bipolar cell layer (middle layer), and the ganglion cell layer (the superficial cell layer). The **photoreceptor cell layer** contains the rods and cones. **Rods** are used in night vision and respond to low levels of light, allowing us to perceive shades of gray, black, and white. Visual acuity with rods is low. **Cones** require brighter light for stimulation, but allow us to see color and provide high visual acuity. The rods and cones synapse with the bipolar neurons in the **bipolar cell layer,** which synapses on the ganglion cells in the **ganglion cell layer.** Axons from the ganglion cells extend through the optic disc and leave the eyeball as the optic nerve. The **optic disc** does not contain photoreceptors and forms the **blind spot** of the retina. It is also the site where the central retinal artery and vein enter and leave the retina, and the only place where the retina is secured to the other layers of the eyeball.

The **macula lutea** (*macula* = flat spot; *lutea* = yellow), the site of macular degeneration, is in the center of the neural portion of the retina. In the middle of the macula lutea is the **central fovea** (fovea centralis). This area of the retina has the highest density of cones of any area of the retina and is not covered by ganglion and bipolar cell layers. Therefore, this area has the highest visual acuity (sharpness of vision) of any area of the retina. When we look at an object, the light rays reflected from the object are focused onto the central fovea.

The retina can be viewed with an **ophthalmoscope.** The ophthalmoscope illuminates the interior of the eye, and the retina appears red from the many blood vessels. Blood vessels can be seen branching from the optic disc while the circular macula lutea appears dark because of the absence of blood vessels. The central fovea is in the middle of the macula lutea.

Before Going to Lab

1 Label the structures of the retina as observed with an ophthalmoscope in Figure 24.6.

2 Label the structures of the eyeball in Figure 24.7(a), and (b).

3 Label the structures of the retina in Figure 24.8(a) and (b).

LAB ACTIVITY 5 Anatomy of the Retina

1 Examine a prepared microscope slide of a cross-section through an eyeball.
- Using the low-power objective lens, find the posterior surface of the eyeball and identify the **sclera, choroid,** and **retina.** Place the retina in the center of the field of view and switch to the high-power objective lens.
- Using the high-power objective lens, identify the **pigmented epithelium of the retina** and the **neural layer of the retina.** Within the neural layer of the retina, identify the **ganglion layer,** the **bipolar layer,** and the **photoreceptor layer.**

2 View the retina with an ophthalmoscope. Instructions will be provided by your instructor. ■

- blood vessel
- central fovea
- macula lutea
- optic disc

1 _____

2 _____

3 _____

4 _____

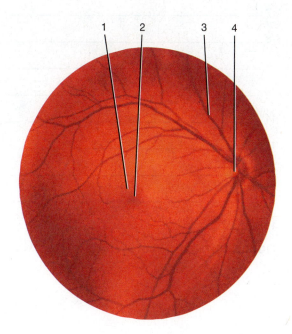

FIGURE 24.6 Normal retina viewed with an ophthalmoscope.

(a)
- central fovea
- central retinal artery
- central retinal vein
- optic disc
- optic nerve

1 _____

2 _____

3 _____

4 _____

5 _____

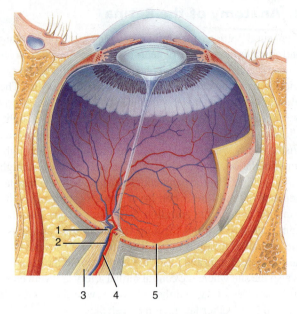

(a) Transverse section of eyeball

(b)
- choroid
- ciliary body
- cornea
- iris
- lens
- optic disc
- optic nerve
- pupil
- retina
- sclera

6 _____

7 _____

8 _____

9 _____

10 _____

11 _____

12 _____

13 _____

14 _____

15 _____

(b) Photomicrograph of eyeball, 5×

FIGURE 24.7 Section of eyeball.

(a)
- bipolar cell layer
- ganglion cell layer
- neural portion of retina
- optic nerve fibers (axons)
- photoreceptor layer
- pigmented epithelium of retina

1 _____

2 _____

3 _____

4 _____

5 _____

6 _____

(a) Microscopic structure of retina

(b) Photomicrograph of retina

450×

(b)
- bipolar cell layer
- ganglion cell layer
- neural portion of retina
- photoreceptor layer
- pigmented epithelium of retina

7 _____

8 _____

9 _____

10 _____

11 _____

FIGURE 24.8 **Microscopic anatomy of the retina.**

LAB ACTIVITY 6 Locating Blind Spot

1 Looking at Figure 24.9, hold the lab manual in both hands and extend both arms in front of you. Close your left eye. Keep your right eye open and focus on the dot. *Slowly* move the figure toward you. The X will disappear when it crosses your blind spot, but if you move the figure too fast, you will miss it.

2 Have a lab partner measure in centimeters the distance the book is from your face when this happens. Then move the figure closer to your face, and the X will reappear.

Blind spot distance for right eye: _____

3 Close your right eye. Keep your left eye open and focus on the X. *Slowly* move the figure toward you until the dot disappears. Measure the distance in centimeters. Then move the figure closer to your face, and the dot will reappear.

Blind spot distance for left eye: _____

4 Explain why the X and the dot disappear and reappear. ◼

FIGURE 24.9 Blind spot test.

5. Visual Acuity Tests

Visual acuity tests measure the ability of the lens to focus light reflected from an object on the central fovea of the retina. The lens can **accommodate** or change shape to bend light rays to focus them on the central fovea. At 20 feet, light rays from an object are nearly parallel and do not have to bend as much to focus on the central fovea. At this distance, the lens is flattened and the refractive power (ability to bend light rays) of the lens is lowest. To observe objects closer than 20 feet, the lens must change shape or accommodate to focus the light rays on the central fovea. The lens bulges to increase the refractive power. Individuals who have normal distance vision and near vision are **emmetropic,** individuals who have normal distance vision but blurry near vision are **hyperopic** (farsighted), and individuals who have blurry distance vision but normal near vision are **myopic** (nearsighted).

As we age, the ability of the lens to accommodate diminishes and the ability to focus on very close objects decreases, a condition called **presbyopia** (*presby-* = old). The **near point of vision** is the closest distance that a person can focus on an object. The average near point of vision is 10 cm for young adults, 20 cm for adults in their 40s, and 80 cm for people in their 60s.

Astigmatism is caused by irregularities in the curvature of the cornea or lens. This causes parts of an image to be blurry.

LAB ACTIVITY 7 Visual Acuity Tests

1 **Distance visual acuity** is measured using a **Snellen eye chart** (provided by your instructor). *If you wear eyeglasses or contact lenses, remove them to determine visual acuity without correction or wear them to determine visual acuity with correction.*

- Have the subject stand 20 feet from the Snellen eye chart that is placed in a well-lighted area and cover the left eye with a hand.
- Have the subject read the smallest line of letters they can see clearly without squinting. If the subject can correctly read half of the letters or more, then ask the subject to read the letters on the next, smaller line.
- Record the number of the line with the smallest size letters read with half or greater accuracy in Table 24.2.
- Have the subject cover the right eye and repeat the procedure.
- A value of 20/20 indicates that the subject has normal vision. A value of 20/40 indicates that the subject sees at 20 feet what a person who has normal vision sees at 40 feet. This is not as good as normal vision. A value of 20/15 indicates that the subject sees at 20 feet what a person with normal vision sees at 15 feet. This is better than normal vision.

2 **Near visual acuity** is measured using a **Snellen acuity card** (Fig. 24.10).

- Have the subject hold the Snellen acuity card 14 inches from his or her face. The card should be illuminated with the light source either behind or above the subject. Instruct the subject to cover the left eye with his or her hand.
- Repeat the steps used to measure distance visual acuity. Start with the second bulleted step.

> **CLINICAL NOTE:** *To obtain an accurate acuity test, professionals also count the number of letters a patient missed on the line recorded as the best vision. For example, if the best vision was 20/40, and the patient missed two letters on that line, it would be recorded as 20/40 $^{(-2)}$. If a patient read two letters on the 20/15 line, it would be recorded as 20/15 $^{(+2)}$.*

3 Measure the **near point of vision** and compare it with the average for the age of your subject.

- Have the subject hold the Snellen visual acuity card (Figure 24.10) 14 inches from his or her face.
- Have the subject cover the left eye and read letters from a line above his or her near visual acuity.

- Instruct the subject to slowly move the chart closer to his or her face until the letters are blurry.
- Measure the distance from the card to the subject's eye in centimeters.
- Record the value in Table 24.2.

TABLE 24.2 Visual Test Results

	RESULTS	
TEST	**LEFT EYE**	**RIGHT EYE**
Distance visual acuity		
Near visual acuity		
Near point of vision		
Astigmatism	Yes or No	Yes or No

4 Determine whether you have **astigmatism.**
- Remove corrective lenses.
- Cover the right eye and look at the center of the astigmatism chart in Figure 24.11.
- If all the radiating lines are equally sharp and dark, you do not have astigmatism. However, if some lines are lighter or less distinct than others, you have astigmatism.
- Cover the left eye and repeat the procedure.

5 Answer the Discussion Questions with your lab group.

FIGURE 24.11 Astigmatism chart.

DISCUSSION QUESTIONS
Visual Tests

1 Why must the subject stand 20 feet from the Snellen eye chart to test distance vision?

2 Why must the subject hold the Snellen acuity card close to his or her eyes to test near vision?

FIGURE 24.10 Snellen acuity card.

6. Red-Green Color Blindness

Red-green color blindness in an inherited disorder and is the most common form of color blindness. This form of color blindness is due to the absence of either red or green cones. There are 3 types of cones in the human retina, and these cones differ in the type of photopigment present. The approximate distribution of cones are 64% red, 34% green, and 2% blue. Perception of different colors occurs when different wave lengths of color selectively activate different photopigments. In order to differentiate between red and green, both red and green cones must be present.

LAB ACTIVITY 8 **Red-Green Color Blindness Tests**

1 Look at Figure 24.12(a) and (b) to see if you may have red-green color blindness.

2 Record the number of students in your class that have:
- normal color vision _____
- red color blindness _____
- green color blindness _____
- total color blindness _____

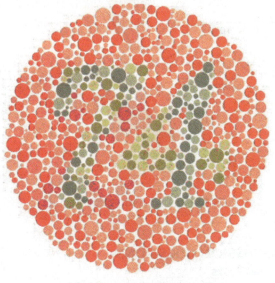

(a) Ishihara Color Blindness Test
Plate 7

What did you see?

Those with normal color vision see a **74**.

Those with red-green color blindness see a **21**.

Those with total color blindness see **spots**.

(b) Ishihara Color Blindness Test
Plate 16

What did you see?

Those with normal color vision should see a **26**.

Red color blind people will see a **6**.

Green color blind people will see a **2**.

FIGURE 24.12 **Red-Green Color Blindness Test.**

B. The Ear, Hearing, and Equilibrium

1. Anatomy of the Ear

The ear is divided into 3 regions: the external (outer) ear, the middle ear, and the internal (inner) ear. The **external ear,** consisting of the auricle, external auditory canal, and tympanic membrane, extends from the auricle to the tympanic membrane. The **auricle,** the flexible external structure that is commonly called the ear, collects sound waves and directs them toward the external auditory canal. The rim of the auricle is called the **helix** and the fleshy, inferior portion is the **lobule.** The **external auditory canal** conducts sound waves from the auricle to the tympanic membrane. The **tympanic membrane** (*tympan-* = drum) or eardrum converts sound waves to vibrations that are transferred to middle ear structures.

The **middle ear** is an air-filled cavity within the temporal bone that extends from the tympanic membrane to the oval window. Middle ear structures include auditory ossicles, oval window, round window, and auditory tube. **Auditory ossicles** are small bones within the cavity that are connected by synovial joints. These bones transfer vibrations from the tympanic membrane to the oval window. The **malleus** is the outermost bone and is attached to the **tympanic** membrane. The **incus** is the middle bone and connects to the stapes. The innermost bone is the **stapes,** which connects to the incus and oval window. The **oval window** is the membrane-covered opening that separates the middle and inner ear and transfers vibrations to the inner ear. The **round window** is a membrane-covered opening between the middle ear and cochlea. The **auditory tube** (pharyngotympanic or Eustachian tube) connects the middle ear to the nasopharynx (part of the throat near the nasal cavity) and equalizes the air pressure of the middle ear with atmospheric air.

Before Going to Lab

1 Label the structures of the external and middle ear in Figure 24.13.

LAB ACTIVITY 9 Anatomy of the External and Middle Ear

1 Identify the external and middle ear structures on a model or chart.

2 Identify the area of the temporal bone that houses the middle and internal ear on the skull. ■

- auditory tube
- auricle
- external auditory canal
- external ear
- helix (HEE-liks)
- incus (INK-us)
- internal ear
- lobule
- malleus (MAL-ee-us)
- middle ear
- stapes (STAY-peez) attached to oval window
- tympanic (tim-PAN-ik) membrane

1 _____ 7 _____

2 _____ 8 _____

3 _____ 9 _____

4 _____ 10 _____

5 _____ 11 _____

6 _____ 12 _____

FIGURE 24.13 Anatomy of the external and middle ear.

The internal ear is housed within the temporal bone. It consists of cavities within the bone called the **bony labyrinth** that encloses a series of connected membranous sacs, the **membranous labyrinth.** The bony labyrinth contains a fluid called **perilymph** that surrounds the membranous labyrinth. **Endolymph** is the fluid within the membranous labyrinth. The bony labyrinth has 3 main regions: the vestibule, the semicircular canals, and the cochlea.

The **vestibule** is the middle area of the bony labyrinth that encircles 2 sections of membranous labyrinth, the utricle and the saccule. The **utricle** (*utricle* = little bag) is the posterior section of the membranous labyrinth within the vestibule, and it houses equilibrium receptors. The **saccule** (*saccule* = little sac) is the anterior section of the membranous labyrinth within the vestibule. The saccule is continuous with the utricle and also houses equilibrium receptors.

Semicircular canals are 3 bony canals posterior to the vestibule that project posteriorly, laterally, and superiorly from the vestibule. Each canal is at right angles to the other two. **Semicircular ducts** are sections of membranous labyrinth within the semicircular canals that contain equilibrium receptors and connect with the utricle. The **ampulla** is the widened end of each semicircular canal and duct.

The **cochlea** is the spiral area of the bony labyrinth anterior to the vestibule. The **cochlear duct** is the section of membranous labyrinth within the cochlea. The cochlear duct contains the hearing receptors and is connected to the saccule.

Hearing and equilibrium receptors initiate nerve impulses that are carried by the **vestibulocochlear nerve** (cranial nerve VIII) to the brain. The vestibulocochlear nerve has two branches: the **vestibular branch** that carries nerve impulses generated by equilibrium receptors and the **cochlear branch** that carries nerve impulses generated by the hearing receptors.

Before Going to Lab

1 Label the structures of the internal ear in Figure 24.14.

LAB ACTIVITY 10 Anatomy of the Internal (Inner) Ear

1 Identify the internal ear structures on a model or chart. ■

• ampulla of semicircular canal and duct
• anterior semicircular canal
• cochlea (COKE-lee-uh)
• cochlear duct
• lateral semicircular canal
• membranous semicircular duct
• oval window
• posterior semicircular duct
• round window
• saccule (SAK-yool)
• utricle (YOU-trih-cul)
• vestibule

1 _____ 7 _____

2 _____ 8 _____

3 _____ 9 _____

4 _____ 10 _____

5 _____ 11 _____

6 _____ 12 _____

FIGURE 24.14 Anatomy of the internal ear.

2. Microscopic Anatomy of the Cochlea and the Spiral Organ of Corti

The cochlea is a spiral cavity that resembles the space within a snail shell. The cochlea makes 3 turns around a bony core. A section through the cochlea shows 3 channels: the scala vestibuli, the cochlear duct (scala media), and the scala tympani. The **scala vestibuli** is part of the cochlea and is superior to the cochlear duct. It is separated from the cochlear duct by the **vestibular membrane.** The **scala tympani** is also part of the cochlea and is posterior to the cochlear duct. It is separated from the cochlear duct by the **basilar membrane.** The scala vestibuli and scala tympani are continuous with one another and are filled with perilymph. The **cochlear duct,** part of the membranous labyrinth, houses the spiral organ of Corti and is filled with endolymph.

The **spiral organ of Corti** sits on the basilar membrane. It contains hair cells (receptors for hearing) and supporting cells. The hair cells have a **hair bundle** composed of stereocilia at their apical end. Superior to and in contact with the stereocilia is the **tectorial membrane** (*tector* = covering). The basal end of the hair cells synapse with sensory and motor neurons from the cochlear branch of the vestibulocochlear nerve.

Movement of the basilar membrane, caused by peri-lymph movement, forces the hair bundle into the tectorial membrane and bends the stereocilia. This results in a generation of nerve impulses in the sensory neurons.

Before Going to Lab

1 Label the structures of the cochlea and spiral organ of Corti in Figure 24.15(a) and (b).

LAB ACTIVITY 11 Microscopic Anatomy of the Cochlea and Spiral Organ of Corti

1 Identify the cochlear structures on a model or chart.

2 Examine a prepared microscope slide showing a cross-section through the cochlea.

- Using the low-power objective, identify the **basilar membrane, cochlear duct, spiral organ of Corti, scala tympani, scala vestibuli,** and **vestibular membrane.**

- Using the high-power objective, identify the **basilar membrane, hair bundle, hair cells,** and **tectorial membrane.**

20×

(a) Cross-section through cochlea

(a)
- basilar (BAY-sih-lur) membrane
- cochlear duct
- scala (SCAY-lah) tympani
- scala vestibuli
- spiral organ of Corti
- vestibular membrane

1 _____

2 _____

3 _____

4 _____

5 _____

6 _____

110×

(b) Cross-section through spiral organ of Corti

(b)
- basilar membrane
- cochlear duct
- hair cells
- tectorial membrane
- vestibular membrane

7 _____

8 _____

9 _____

10 _____

11 _____

FIGURE 24.15 **Photomicrographs of the cochlea and spiral organ of Corti.**

3. Microscopic Anatomy of the Equilibrium Receptors of the Internal Ear

There are 2 types of equilibrium receptors: the maculae (macula, sing.) located in the utricle and saccule, and the cristae (crista, sing.) located in the membranous semicircular ducts within the ampullae. The **maculae** provide information on head position (static equilibrium), as well as linear acceleration and deceleration, a type of dynamic equilibrium. The maculae consist of **hair cells** with **hair bundles** and **supporting cells.** The hair bundles are in contact with a gelatinous membrane, the **otolithic membrane** (*oto* = stone), which contains calcium carbonate crystals called **otoliths.** Movement of the head causes movement of the otoliths and otolithic membrane, which bends the hair bundles. The direction

of movement will determine if the hair cells release more or less neurotransmitter to the associated sensory neurons.

The **crista** (ampullaris) detects rotational acceleration and deceleration, a type of dynamic equilibrium. Each crista consists of **hair cells** and **supporting cells.** The hair bundles of the hair cells are covered by a gelatinous structure called the **cupula.** When the head moves, movement of endolymph pushes the cupula causing the hair cells to bend. Bending of the hair bundles results in generation of nerve impulses in the vestibular branch of the vestibulocochlear nerve.

> **Before Going to Lab**
>
> **1** Label the structures of the macula and crista in Figure 24.16(a) and (b).

> **LAB ACTIVITY 12** **Microscopic Anatomy of the Macula and Crista**

1 Identify macula and crista structures on a model or chart.

2 Examine a prepared microscope slide showing a cross-section through the crista with the high-power objective lens. Identify the **cupula** and **hair cells.** ■

(a) Macula

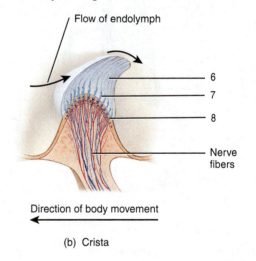

(b) Crista

(a)
- hair cell
- otoliths (OH-toe-liths)
- otolithic membrane
- stereocilium in hair bundle
- vestibular branches of vestibulocochlear nerve

1 _____

2 _____

3 _____

4 _____

5 _____

(b)
- cupula (KYou-pyul-uh)
- hair bundle
- hair cell

6 _____

7 _____

8 _____

FIGURE 24.16 **Microscopic anatomy of the macula and crista.**

4. Auditory and Equilibrium Tests

Hearing loss can be described as either conduction deafness or sensorineural deafness. **Conduction deafness** occurs when there is a decreased ability to conduct the energy of sound waves through the external and middle ear to hearing receptors in the inner ear. Ear wax buildup, damage to the tympanic membrane, or fusion of auditory ossicles may cause conduction deafness. **Sensorineural deafness** is caused by damage to hearing receptors, damage to the cochlear branch of the vestibulocochlear nerve, or damage of the neural pathways to the auditory cortex.

Equilibrium receptors provide information that enables the body to maintain balance. There are two types of equilibrium receptors, static and dynamic. **Static equilibrium receptors** provide information about body position relative to the force of gravity (standing upright vs. being upside down). **Dynamic equilibrium receptors** provide information about body position in response to sudden movement such as rotation, acceleration, and deceleration (spinning, going faster, stopping). Inflammation of or injury to equilibrium receptors results in an inability to maintain body position, vertigo, and/or dizziness. **Vertigo** is the sensation of circular motion either of oneself or external objects, while dizziness is often used to describe faintness, unsteadiness, or lightheadedness. Severe vertigo may be accompanied by **nystagmus**—rapid, involuntary movement of eyeballs.

NOTE: *In the Weber test, sound is louder in the ear with conduction deafness. This is not a typo error. Research this and you will find several explanations from simple to complex. This simplest explanation is that bone conducts sound better than air. The ear with conduction deafness hears only sound conducted through bone, whereas the normal ear hears sound conducted through air in the external ear canal and sound conducted through bone. In the normal ear, the sound conducted through air in the external ear canal acts like background noise.*

LAB ACTIVITY 13 Auditory and Equilibrium Tests

1 Test for unilateral (one side only) deafness using the **Weber test.**
- Choose a subject from your group. Have the subject sit with the head erect and facing forward.
- Strike a tuning fork (middle C preferably) and place it medially on the subject's forehead (bone conduction).
- Ask the subject if the sound is equally loud in both ears or louder in one ear. Circle the result in Table 24.3.
- If the sound is equally loud in both ears, the subject either has normal hearing or bilateral deafness (equal hearing loss in both ears). If the subject hears the sound louder in one ear, then the subject may have unilateral deafness (either conduction deafness in that ear or sensorineural deafness in the opposite ear).
- If the subject has unilateral deafness, conduct the Rinne test to determine if hearing loss is conduction deafness.
- Mimic unilateral conduction deafness by placing a cotton ball (or finger) in one external auditory canal and repeat the Weber test.

TABLE 24.3 Results of Weber, Rinne, and Barany Tests

TEST	RESULTS (CIRCLE YOUR RESULTS)		
Weber test	Equal loudness in both ears = normal hearing or equal hearing loss in both ears *or* Sound is louder in right ear = conduction deafness in right ear or sensorineural deafness in left ear *or* Sound is louder in left ear = conduction deafness in left ear or sensorineural deafness in right ear		
Rinne test	Conduction deafness *or* no conduction deafness		
Balance test	Static equilibrium receptors: functioning *or* not functioning		
Barany test		Direction of eye movement	Name semicircular canal stimulated
	Head slightly forward:	lateral *or* vertical *or* rotational	lateral *or* anterior *or* posterior
	Head toward shoulder:	lateral *or* vertical *or* rotational	lateral *or* anterior *or* posterior
	Head on chin:	lateral *or* vertical *or* rotational	lateral *or* anterior *or* posterior

2 Test for conduction deafness using the **Rinne test.**

- This test will be conducted on the ear that may have conduction deafness as indicated by the Weber test.
- Have the subject sit with head erect and facing forward.
- Strike a tuning fork and place it on the subject's mastoid process to test hearing by **bone conduction.**
- Ask the subject to tell you when the sound can no longer be heard. Immediately place the still-vibrating tuning fork close to the subject's ear to test hearing by air conduction. If the subject can hear the tuning fork again when it is placed next to his or her ear, the subject does not have conduction deafness in that ear. If the subject cannot hear the tuning fork again, then the subject *may* have conduction deafness in that ear.
- Circle the results in Table 24.3. If the subject does not have conduction deafness, the test is complete.
- To verify conduction deafness, test the same ear again. This time you are testing hearing by **air conduction** first.
- Strike the tuning fork again and place the tuning fork close to the subject's ear.
- Ask the subject to tell you when the sound can no longer be heard, then place the tuning fork on the subject's mastoid process (bone conduction). If the subject hears the sound again, there is conduction deafness in that ear. Record whether the subject has conduction deafness or no conduction deafness in that ear by circling the results in Table 24.3.

3 Conduct a balance test to evaluate **static equilibrium** receptors.

- Have the subject stand in front of a whiteboard or chalkboard with arms at the sides. The subject may not lean against the wall or support him- or herself in any manner.
- Tell the subject to stand perfectly still. Mark the outline of the shoulders to help determine when he or she sways.
- Tell the subject to close his or her eyes. Observe movement in the subject's shoulders. Notice that, although the subject may sway slightly, posture is always corrected. Signals from the static equilibrium receptors are helping the subject to maintain posture. If the static equilibrium receptors are not functioning,

the subject will not be able to maintain posture and will exhibit large swaying movements or will fall. Be prepared to support the subject if necessary.

- Record the results in Table 24.3.

4 Conduct the **Barany test** to evaluate function of semicircular canals and **dynamic equilibrium** receptors.

- Choose a subject from your lab group who does not readily experience dizziness or become nauseated when rotated. If the subject experiences nausea during the demonstration, immediately stop rotation.
- Provide the subject with a chair or stool that can be rotated. Have the subject sit on the chair and hold on to the arms or seat for safety. Decide how the subject will position his or her legs during rotation to ensure safety and prevent interference. Position 3 to 4 students around the chair with their feet firmly against the legs of the chair to prevent the chair from tipping over or the subject from falling off.
- Tell the subject to slightly tilt his or her head forward, focus on a distant object, and keep both eyes open during the rotation.
- Carefully turn the chair or stool clockwise (to the right). Complete 10 turns, one turn per 2 seconds, and stop suddenly. **Be prepared to support the subject until vertigo and/or dizziness has passed.** The subject will still experience rotation, indicating that the semicircular canals are functioning. Endolymph continues to move within the membranous semicircular ducts for a short time after rotation has stopped.
- Observe which way the subject's eyeballs are moving immediately after stopping rotation. Record direction in Table 24.3.
- *Lateral movement of the eyes indicates stimulation of the dynamic equilibrium receptors in the lateral semicircular canals.*
- *Vertical movement of the eyes indicates stimulation of the dynamic equilibrium receptors in the anterior semicircular canals.*
- *Rotational movement of the eyes indicates stimulation of the dynamic equilibrium receptors in the posterior semicircular canals.*
- Repeat the demonstration with the subject's head tilted toward one shoulder, and then again with the subject's chin resting on his or her chest. ∎

C. The Nose and Olfaction

The nose contains the receptors for the sense of smell or olfaction (*olfact-* = smell). Olfactory receptors are found within the olfactory epithelium, a specialized area of the epithelium lining the nasal cavity. The **olfactory epithelium** covers the inferior surface of the cribriform plate, the superior nasal concha, and the upper part of the middle nasal concha.

The olfactory epithelium contains olfactory receptor cells, supporting cells, basal stem cells, and ducts of olfactory glands. The **olfactory receptor cells** are bipolar neurons whose dendritic end is embedded in the mucus layer covering the surface of the olfactory epithelium and whose axons form the **olfactory nerves.** The olfactory receptors are located on **olfactory hairs** that project from the dendrites of the olfactory receptor cells.

Olfactory nerves pass through olfactory foramina in the **cribriform plate** and synapse on neurons in the **olfactory bulb.** Nerve impulses then travel along the **olfactory tract** to the lateral olfactory area of the cerebral cortex. Olfactory receptors adapt to odors very quickly. This explains why when we are trying to determine the source of an odor we often lose the smell before we find it. If we leave the area and return, we can smell the odor again.

> ### Before Going to Lab
>
> **1** Label the structures of the olfactory epithelium in Figure 24.17(a) and (b).

LAB ACTIVITY 14 **Structure of the Olfactory Epithelium**

1 Identify the following areas on a skull: location of **olfactory epithelium** and the **olfactory foramina** in the cribriform plate. ■

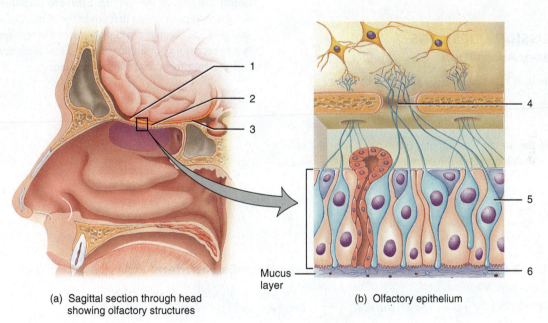

(a) Sagittal section through head showing olfactory structures

(b) Olfactory epithelium

Mucus layer

(a)
- cribriform (CRIB-ri-form) plate
- olfactory (OHL-fak-tore-ee) bulb
- olfactory tract

1 _____

2 _____

3 _____

(b)
- olfactory hair
- olfactory nerve
- olfactory receptor cell

4 _____

5 _____

6 _____

FIGURE 24.17 Olfactory structures.

LAB ACTIVITY 15 **Olfactory Adaptation**

1 Choose a subject, a timer, an experimenter (holds vial under subject's nose), and a recorder.

2 Have the subject plug one nostril with cotton and close both eyes.

3 Noting the time, hold a container of cloves (or another aromatic substance) just under the open nostril and ask the subject to inhale through the open nostril and exhale through the mouth.

4 Ask the subject to tell you when the odor has disappeared, note the time, and record in Table 24.4. Instruct the subject to immediately pull out the cotton in the other nostril and inhale (vial still under nose).

5 Ask the subject if he or she can smell the odor. Record the result in Table 24.4.

6 Repeat the experiment with two distinct smells, such as peppermint or cinnamon. Avoid irritating odors.

7 Clean up as directed by your instructor.

8 Answer the Discussion Questions with your lab group.

DISCUSSION QUESTIONS
Olfactory Adaptation

1 Is the time of adaptation the same for all odors?

2 After the nostril was unplugged, explain why the subject was able to smell the odor again.

D. Taste Buds and Gustation

Taste buds, which are found on the tongue, soft palate, pharynx (throat), and larynx, are microscopic, onion-shaped structures that contain gustatory cells, gustatory hairs, and supporting cells. Each **gustatory** (*gust-* = taste) **cell** has one **gustatory hair** that projects through an opening, the **taste pore,** on the apical end of the taste bud. **Gustatory receptors** are located on the gustatory hairs. The basal end of gustatory cells synapse onto the dendritic end of sensory neurons. Axons from the sensory neurons contribute fibers to the facial nerve (cranial nerve VII), glossopharyngeal nerve (IX), or vagus nerve (X), depending on the location of the taste bud.

Taste buds on the tongue are located in **papillae,** elevated structures that give the tongue its rough appearance. There are 4 types of papillae: vallate (circumvallate), fungiform, foliate, and filiform. **Vallate** (circumvallate) **papillae** are the largest papillae and form an inverted V at the posterior of the tongue. **Fungiform papillae** are mushroom-shaped and are scattered over the surface of the tongue. **Foliate papillae** are present mostly in children and are located in lateral margins of the tongue. **Filiform papillae** are slender, pointed projections that cover the surface of the tongue and give the tongue a rough texture. These papillae have tactile receptors but no taste buds. Taste buds are found in vallate, fungiform, and foliate papillae.

TABLE 24.4 **Results of Olfactory Adaptation Experiment**

ODOR	TIME TO ADAPTATION (sec)	CAN SUBJECT SMELL ODOR AFTER NOSTRIL UNPLUGGED?

There are 4 primary taste sensations: sweet, bitter, salty, and sour, and a possible fifth, MSG (monosodium glutamate). Gustatory receptors most sensitive to sweet and salty sensations are found on the tip of the tongue, while bitter sensations are in the back and sour sensations are on the sides of the tongue. Other taste sensations are a mixture of these four. Smell, temperature, and texture (tactile sensation) contribute to our sense of taste. A person with a cold often has a loss of taste due to a loss of smell. Cold French fries are not as tasty as hot ones, and mushy apples are not as good as crisp ones.

Before Going to Lab

1 Label the gustatory structures in Figure 24.18(a), (b), and (c).

LAB ACTIVITY 16 **Gustatory Structures and Sensations**

1 Examine a prepared microscope slide of a vallate papilla.
- Using the low-power objective lens, identify the taste bud in the wall of the papilla.
- Using the high-power objective lens, identify the **taste pore, gustatory hairs,** and **gustatory receptor cells** forming the wall of the taste bud.

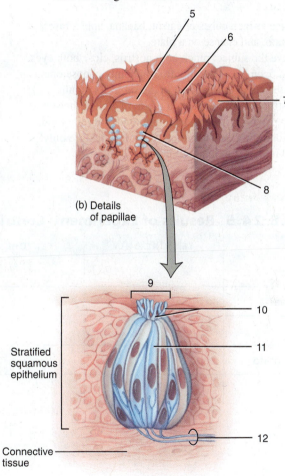

(a) Dorsum of tongue showing location of papillae

(b) Details of papillae

(c) Structure of a taste bud

(a)	(b)	(c)
• filiform papilla	• filiform papilla	• gustatory hairs
• foliate papilla	• fungiform papilla	• gustatory receptor cell
• fungiform papilla	• taste bud	• sensory axons
• vallate (circumvallate) papilla	• vallate papilla	• taste pore

1 _____	5 _____	9 _____
2 _____	6 _____	10 _____
3 _____	7 _____	11 _____
4 _____	8 _____	12 _____

FIGURE 24.18 **Gustatory structures.**

2 Use a mirror to identify fungiform and filiform papillae on your tongue. Try to see vallate papillae on the back of the tongue.

3 Examine the contribution of texture and smell to the sense of taste. Since many labs do not allow eating or drinking, this activity may need to be done in another area or at home.

- Choose a subject, an experimenter (who will give the food cubes to the subject), and a recorder. The subject will first try to identify food by texture only (rolling food on surface of tongue), then by taste (chewing food increases the amount of chemicals dissolved in saliva and capable of interacting with taste receptors), and finally with the addition of smell.
- Place ½-inch cubes of carrot, banana, apple, raw potato, and cheese on a plate.
- Have the subject pinch both nostrils, close both eyes, and open his or her mouth. Randomly choose one of the cubes and place it in the subject's mouth.
- Instruct the subject to roll the food around the surface of the tongue and attempt to identify the food. If identification is correct, check the "texture only" column next to the food in Table 24.5.
- Instruct the subject to chew the food and attempt to identify the food. If identification is correct, check "texture and taste" column in Table 24.5.
- Instruct the subject to open both nostrils and attempt to identify the food. If identification is correct, check the "texture, taste, and smell" column in Table 24.5.
- Repeat procedure with the other food cubes.

4 Clean up as directed by your instructor.

5 Answer the Discussion Questions with your lab group.

DISCUSSION QUESTION
Gustatory Structures and Sensations

1 Compare the number of foods identified by texture, identified by taste, and identified by the sense of smell.

TABLE 24.5 Results of Experiment: Contribution of Texture and Smell to Taste

FOOD	TEXTURE ONLY	TEXTURE AND TASTE	TEXTURE, TASTE, AND SMELL
Carrot			
Banana			
Apple			
Raw potato			
Cheese			

Reviewing Your Knowledge

A. Accessory Eye Structures

Name the structure that corresponds to each statement.

_____ **1.** drains tears into nasal cavity

_____ **2.** produces tears

_____ **3.** membrane that covers the inner surface of eyelid

_____ **4.** tears from surface of eye drain into here

_____ **5.** membrane that covers the anterior surface of the sclera

_____ **6.** drains tears into the nasolacrimal duct

B. Extrinsic Eye Muscles

Name the muscle that applies to each statement.

_____ **1.** moves eyeball superiorly

_____ **2.** moves eyeball laterally

_____ **3.** moves eyeball inferiorly

_____ **4.** moves eyeball medially

_____ **5.** moves eyeball laterally and inferiorly

_____ **6.** moves eyeball laterally and superiorly

C. Eye Structures

Choose the structure that applies to each statement. More than one structure may apply to a statement, and a structure may be used more than once.

a. anterior cavity
b. anterior chamber
c. choroid
d. ciliary body
e. ciliary muscle
f. ciliary process
g. cornea
h. iris
i. lens

j. ora serrata
k. posterior chamber
l. pupil
m. retina
n. sclera
o. scleral venous sinus
p. suspensory ligaments

_____ **1.** produces aqueous humor

_____ **2.** structures that are part of vascular tunic

_____ **3.** contains photoreceptors

_____ **4.** controls the size of the pupil

_____ **5.** drains aqueous humor from the anterior chamber

_____ **6.** structures that are part of the fibrous tunic

_____ **7.** most anterior part of the eyeball

_____ **8.** anterior boundary of retina

_____ **9.** attaches lens to ciliary body

_____ **10.** changes shape to focus light on retina

_____ **11.** location of aqueous humor

_____ **12.** white, tough outer layer of eyeball

Answer the following questions:

13. Why can the retina pull away from the back of the eyeball?

14. Name the instrument used to view the retina during a physical exam.

15. Name the two layers of the retina.

D. Anatomy of the Retina

Choose the structure that corresponds to each statement.

a. bipolar cell layer e. macula lutea
b. central fovea f. optic disc
c. cones g. photoreceptor layer
d. ganglion cell layer h. rods

_____ **1.** has the highest density of cones in the retina

_____ **2.** axons form optic nerve

_____ **3.** does not contain photoreceptors; blind spot

_____ **4.** photoreceptor that allows us to see color

_____ **5.** contains rods and cones

_____ **6.** the center of the neural portion of the retina

_____ **7.** photoreceptor used in night vision

_____ **8.** rods and cones synapse on these cells

E. Visual Acuity Tests

Answer the following questions:

1. Define accommodation.

2. A 10-year-old patient's distance visual acuity was tested and determined to be 20/80.
 (a) Explain what that means.

 (b) Is 20/80 better or worse than 20/20? Explain.

3. A 60-year-old man has to hold his newspaper at arm's length to read it. What condition does he have?

4. A person with _____ has blurry near vision but normal distance vision.

5. A person with _____ does not see everything within a visual field clearly; some parts are blurry.

6. A person with _____ has normal near vision but blurry distance vision.

F. External and Middle Ear Structures

Choose the structure that applies to each statement. More than one structure may apply to a statement, and a structure may be used more than once.

a. auditory tube
b. auricle
c. external auditory canal
d. external ear
e. helix
f. incus
g. lobule

h. malleus
i. middle ear
j. oval window
k. stapes
l. tympanic membrane

_____ **1.** auditory ossicle attached to tympanic membrane

_____ **2.** equalizes air pressure in middle ear with external air pressure

_____ **3.** external ear structures

_____ **4.** ear drum

_____ **5.** external feature of ear that contains the helix and lobule

_____ **6.** stapes is attached to this membrane-covered opening

_____ **7.** middle auditory ossicle

_____ **8.** small bones of middle ear that are connected by synovial joints

G. Internal Ear Structures

Choose the structure that applies to each statement. More than one structure may apply to a statement, and a structure may be used more than once.

a. ampulla
b. cochlea
c. cochlear duct
d. perilymph
e. saccule

f. semicircular canals
g. semicircular ducts
h. utricle
i. vestibule

_____ **1.** interconnected components of membranous labyrinth

_____ **2.** fluid found within all bony labyrinth structures

_____ **3.** sections of the membranous labyrinth found within vestibule

_____ **4.** section of the membranous labyrinth that contains hearing receptors

_____ **5.** sections of the membranous labyrinth that contain equilibrium receptors

_____ **6.** interconnected components of bony labyrinth

H. Microscopic Anatomy of the Cochlea, Spiral Organ of Corti, Macula, and Crista

Choose the structure that applies to each statement. More than one structure may apply to a statement, and a structure may be used more than once.

a. basilar membrane	f. hair cells	k. spiral organ of Corti
b. cochlear duct	g. macula	l. supporting cells
c. crista	h. otolithic membrane	m. tectorial membrane
d. cupula	i. scala tympani	n. vestibular membrane
e. hair bundle	j. scala vestibuli	

_____ **1.** receptor for hearing

_____ **2.** receptor(s) that contain(s) hair bundles, hair cells, and supporting cells

_____ **3.** components of macula

_____ **4.** membrane separating the superior chamber of cochlea from cochlear duct

_____ **5.** structure(s) that bend(s) stereocilia of hair cells

_____ **6.** spiral organ of Corti sits on this membrane

_____ **7.** equilibrium receptors

_____ **8.** contains endolymph

_____ **9.** contains perilymph

_____ **10.** equilibrium receptor found within ampullae of semicircular canals

I. Auditory and Equilibrium Tests

Answer the following questions:

1. What causes nystagmus?

2. What causes conduction deafness?

3. What causes sensorineural deafness?

4. If a vibrating tuning fork is placed on the mastoid process, who would "hear" the sound—someone with normal hearing, someone with conduction deafness, or someone with sensorineural deafness? Circle all that apply.

5. Inability to maintain posture while standing still would indicate a problem with which equilibrium receptor?

J. Olfaction

Choose the structure that applies to each statement.

a. cribriform plate
b. superior portion of nasal cavity
c. olfactory hair
d. olfactory nerve
e. olfactory receptor cells

_____ **1.** bipolar neurons

_____ **2.** part of olfactory receptor cell that contains the olfactory receptors

_____ **3.** formed of bipolar neuron axons

_____ **4.** olfactory nerves pass through this structure before synapsing onto olfactory bulb

_____ **5.** location of the olfactory epithelium

K. Taste

Choose the structure that applies to each statement. More than one structure may apply to a statement, and a structure may be used more than once.

a. filiform papillae f. gustatory hair
b. fungiform papillae g. supporting cell
c. pharynx h. taste buds
d. soft palate i. taste pore
e. gustatory cell j. vallate papillae

_____ **1.** contain taste buds

_____ **2.** part of gustatory cell that contains gustatory receptor

_____ **3.** opening in taste bud through which gustatory hair projects

_____ **4.** components of taste bud

_____ **5.** mushroom-shaped projections on tongue

_____ **6.** projections that form inverted V on back of tongue

Answer the following questions:

7. Identify where on the tongue the receptors most sensitive to each of the four taste sensations usually are located.

8. What other sensations contribute to the sensation of taste?

Using Your Knowledge

A. The Eye and Vision

1. Strabismus, a misalignment of the eyeball caused by an imbalance in the extrinsic eye muscles, is also called lazy eye or wandering eye. Identify the extrinsic eye muscle that would cause lateral strabismus of the right eye.

2. Why do photographers use a red light in the darkroom when developing black and white film?

3–4. Name all the accessory eyeball structures, fluids, and retinal layers that light passes through before it hits the photoreceptors.

5. Explain why the retina can detach from the eyeball with just a blow to the back of the head.

6. Is the retinal area in Figure 24.19 the macula lutea, fovea centralis, or optic disc? Explain your choice.

Retina
Choroid
Sclera

FIGURE 24.19 Retinal area.

(a) Normal (b) Abnormal

FIGURE 24.20 Visual fields.

7. Explain why a pituitary tumor may affect vision.

8. A 10-year-old boy has 20/20 distance vision and 20/60 near vision. What refraction abnormality does he have?

9. Using your textbook or another source, research how and why glaucoma and macular degeneration affect vision. Compare Figure 24.20(a) and (b) and identify whether the visual field in (b) is representative of a person with macular degeneration or glaucoma.

10. Explain why the condition represented in Figure 24.20(b) causes loss of vision in the center of the field of vision and blurriness in the periphery.

11. What structure keeps a contact lens on the cornea and prevents it from becoming lodged on the posterior surface of the eyeball?

B. The Ear, Hearing, and Balance

12–13. Sound waves enter the external auditory canal and cause vibrations in structures and fluids of the ear. Identify in order the structures and fluids that vibrate in the pathway from the external auditory canal to the spiral organ of Corti.

14. Diving to the bottom of a deep pool will sometimes cause discomfort in the middle ear. Explain the cause of this discomfort.

15. What do the receptor cells for hearing, static equilibrium, and dynamic equilibrium have in common?

16. Would the receptors for equilibrium work in space at zero gravity?

17. Describe a series of movements that would stimulate all semicircular duct receptors and if repeated may cause vertigo. (*Hint:* Think of a carnival ride that causes vertigo.)

C. Olfaction and Gustation

18. Why do we sniff when we wish to smell something better?

19. When you lick ice cream off an ice cream cone, which papillae are involved in scraping the ice cream off the cone?

20. Licking cold ice cream off an ice cream cone stimulates a variety of sensory receptors associated with the tongue. Name all the sensory receptors involved.

Endocrine Structure and Function

OBJECTIVES

1 Identify major endocrine glands
2 Name the main hormones secreted by each major endocrine gland
3 Identify microscopic structures of endocrine glands
4 Describe the function of each major hormone

MATERIALS

- human torso, brain model, endocrine chart, or use **Real Anatomy** (Endocrine)
- compound microscope, lens paper, prepared slides of selected endocrine glands: hypothalamus, pituitary gland, thyroid gland, parathyroid gland, adrenal gland, pancreas, or use **Real Anatomy** (Histology)
- **Measurement of Blood Glucose:** glucometer, alcohol wipes, lancet, test strips, other material specified by lab group's experimental protocol
- Dissection: preserved cats or fetal pigs, dissecting equipment, disposable gloves, safety glasses, dissection manual
- **Real Anatomy:** Virtual Cadaver Dissection
- **PowerPhys Experiments:**
 - Blood Glucose Regulation
 - Thyroid Function

The endocrine system (*endo-* = within; *krinein* = to secrete) has many glands that secrete hormones into the bloodstream. These chemicals are transported throughout the body in blood and bind to target cells that have cell membrane receptors for a specific hormone. Hormones cause changes in activity in these target cells and direct a variety of cellular activities to keep the body in homeostasis.

pituitary gland, thyroid gland, parathyroid glands, thymus, adrenal glands, pancreas, ovaries, and testes.

Before Going To Lab

1 Use your textbook to label the major endocrine glands in Figure 25.1.

A. Identification of Major Endocrine Glands

The selected, major endocrine glands that we study in this exercise are shown in Figure 25.1. From the head inferiorly, they are the **pineal gland (body), hypothalamus,**

LAB ACTIVITY 1 Identification of the Major Endocrine Glands

1 Identify the major endocrine glands on a human torso, brain model, endocrine chart, or use the search text box in **Real Anatomy** (Endocrine) to locate these structures.

- adrenal (a-DREE-nul) glands
- hypothalamus (hypo-THAL-a-mus)
- ovaries
- pancreas
- parathyroid (para-THY-roid) glands
- pineal (pie-NEE-ul) gland
- pituitary gland or hypophysis (hy-POF-ih-sis)
- testes (TES-teez)
- thymus (THY-mus)
- thyroid gland

1 _____

2 _____

3 _____

4 _____

5 _____

6 _____

7 _____

8 _____

9 _____

10 _____

Anterior view

FIGURE 25.1 **Major endocrine glands.**

B. Hypothalamus and Pituitary Gland

Located inferior to the thalamus in the brain, the **hypothalamus** is an important component of both the nervous and endocrine systems and couples these two regulatory systems together. Although in the past the pituitary gland was called the "master gland," it is now known that the hormones produced by the hypothalamus regulate the pituitary gland. The hypothalamic hormones and several of the anterior pituitary hormones are called **tropic hormones** because they target another endocrine gland.

The **pituitary gland** is also called the **hypophysis** (*hypo-* = under; *phyein* = to grow; translated as "to grow under the hypothalamus") and is composed of the **anterior pituitary** or **adenohypophysis** (*adeno-* = gland) and the **posterior pituitary** or **neurohypophysis** (*neuro-* = nerves). An **intermediate lobe** is present in the fetus but

atrophies during fetal development. The **infundibulum** is the stalk that connects the hypothalamus to the pituitary gland and contains a direct blood supply to the pituitary.

The **anterior pituitary** is composed of glandular epithelial tissue that makes and secretes seven hormones: **human growth hormone** (hGH) or somatotropin, **thyroid-stimulating hormone** (TSH), **follicle-stimulating hormone** (FSH), **luteinizing hormone** (LH), **prolactin** (PRL), **adrenocorticotropic hormone** (ACTH), and **melanocyte-stimulating hormone** (MSH). Secretory or inhibitory hormones secreted by the hypothalamus either stimulate or inhibit secretion of anterior pituitary hormones.

The **posterior pituitary** is composed of nervous tissue that stores and releases into the blood two hormones that are not synthesized in the posterior pituitary: **antidiuretic hormone** (ADH) and **oxytocin** (OT). These two hormones are synthesized within cell bodies of hypothalamic neurons, are packaged into vesicles, and travel down axons that pass through the infundibulum to the posterior pituitary gland.

1 Label the drawing and photomicrographs of the pituitary gland in Figure 25.2(a), (b), and (c).

LAB ACTIVITY 2 Pituitary Gland

1 Examine a prepared microscope slide of the anterior and posterior pituitary with a compound microscope and identify the structures listed with Figure 25.2(c). ■

- anterior pituitary or adenohypophysis (a-den-oh-hy-POF-ih-sis)
- hypothalamus (hypo-THAL-uh-mus)
- infundibulum (in-fun-DIB-u-lum)
- posterior pituitary or neurohypophysis (neur-oh-hy-POF-ih-sis)

1 _____

2 _____

3 _____

4 _____

Mammillary body

Optic chiasm

Sphenoid bone (sella turcica)

(a) Hypothalamus and pituitary gland

10×

(b) Survey photomicrograph of infundibulum and pituitary

- anterior pituitary
- hypothalamus
- infundibulum
- posterior pituitary

5 _____

6 _____

7 _____

8 _____

90×

(c) Photomicrograph of the anterior and posterior pituitary

- anterior pituitary
- posterior pituitary

9 _____

10 _____

FIGURE 25.2 The pituitary gland.

C. Thyroid and Parathyroid Glands

The thyroid gland has two lobes with a connecting isthmus and lies on both sides of the trachea near the thyroid cartilage (Adam's apple) of the larynx. The thyroid gland has a striking microscopic structure with large **follicles** that are filled with a protein **colloid** (large insoluble molecules). Simple cuboidal or simple columnar cells called **follicular cells** form the follicular walls. Follicle cells synthesize **thyroglobulin,** a thyroid hormone precursor stored in colloid. Two **thyroid hormones (TH), thyroxine (T$_4$)** or **tetraiodothyronine** (*tetra-* = four; *iodo-* = iodine) and **triiodothyronine (T$_3$),** are produced from thyroglobulin. Located between the follicles are **parafollicular cells** or **C cells,** which synthesize and secrete **calcitonin.**

Embedded in the posterior surface of the thyroid gland are typically four small, round **parathyroid glands.** Two types of epithelial cells are found in the parathyroid glands: the principal cells and the oxyphil cells. The **principal cells,** which produce **parathyroid hormone (PTH),** are more numerous and smaller than the **oxyphil cells,** whose function is unknown.

Before Going to Lab

1 Label the structures of the thyroid and parathyroid glands in Figure 25.3(a) and (b).

2 Label the photomicrographs of the sections of thyroid and parathyroid glands in Figure 25.4(a) and (b).

LAB ACTIVITY 3 Thyroid and Parathyroid Glands

1 Place your fingers on each side of the trachea, three fingers' breadth superior to the suprasternal (jugular) notch, and feel the thyroid gland move upward as you swallow.

2 Examine a prepared microscope slide of the thyroid and parathyroid glands, and identify the structures listed in Figure 25.4(a) and (b), or identify the structures using **Real Anatomy** (Histology).

3 Answer the Discussion Question with your lab group.

DISCUSSION QUESTION
Thyroid and Parathyroid Glands

1 Examine Figure 25.4. Thyroglobulin is stored in the colloid, and calcitonin and PTH are stored in the parafollicular cells and principal cells, respectively. Compare the storage space for thyroglobulin (TH precursor) to the storage space for calcitonin and PTH.

Trachea —
Thyroid gland

(a)
- isthmus of thyroid gland
- left lobe of thyroid gland
- right lobe of thyroid gland
- thyroid cartilage of larynx
- trachea

1 _____

2 _____

3 _____

4 _____

5 _____

(a) Anterior view of thyroid gland

FIGURE 25.3 **The thyroid and parathyroid glands.**

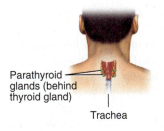

Parathyroid glands (behind thyroid gland)

Trachea

(b) Posterior view

(b)
- isthmus of thyroid gland
- left lobe of thyroid gland
- left parathyroid glands
- right lobe of thyroid gland
- right parathyroid glands
- trachea

6 _____

7 _____

8 _____

9 _____

10 _____

11 _____

FIGURE 25.3 **The thyroid and parathyroid glands,** *continued.*

(a) Survey photomicrograph of thyroid and parathyroid glands

(b) Photomicrograph of thyroid gland

(a)
- parathyroid gland
- thyroid gland

1 _____

2 _____

(b)
- colloid (COL-oid)-filled follicle
- follicular (fohl-LIK-u-lar) cell
- parafollicular (para-fohl-LIK-u-lar) cell or C cell

3 _____

4 _____

5 _____

FIGURE 25.4 **Sectional views of the thyroid and parathyroid glands.**

D. Adrenal Glands

The **adrenal glands** (*ad-* = addition to; *renal* = kidneys) are also known as the suprarenal glands because of their location on the superior part of the kidneys. Each adrenal gland is surrounded by a **capsule** and is composed of an outer cortex and a central medulla. The cortex and medulla are formed of different tissue types, and therefore produce and secrete different types of hormones.

The **adrenal cortex** is glandular and can be divided histologically into three layers, each layer secreting a different hormone. The **zona glomerulosa** (*zona* = belt; *glomerulosa* = little ball) is the outermost layer whose cells are arranged in little columns. The middle layer, the **zona fasciculata** (*fasciculata* = little bundle), has long cords and is the largest layer. The **zona reticularis** (*reticul-* = network) is the innermost layer and has branching cords.

Hormones secreted by the adrenal cortex are mineralcorticoids, glucocorticoids, and androgens (gonadocorticoids). The zona glomerulosa secretes **mineralcorticoids,** with **aldosterone** being the main hormone secreted. The zona fasciculata secretes **glucocorticoids,** with **cortisol** being the main hormone secreted. In males and females, the zona reticularis mainly secretes a small amount of the male androgen **DHEA** (dehydroepiandrosterone), which has weak hormonal effects. Androgens (*andros* = man) are

hormones that increase male characteristics. Some of the peripheral target tissues of DHEA convert this hormone to a more powerful hormone, testosterone, while other peripheral target tissues convert DHEA into estrogen.

The **adrenal medulla** is made up of nervous tissue that is stimulated by the sympathetic nervous system to secrete two hormones: **epinephrine** and **norepinephrine (NE).** Like the posterior pituitary gland, the chemicals secreted by this nervous tissue are called hormones because they are secreted into the blood and travel to another part of the body to target cells.

Before Going to Lab

1 Observe the location of the adrenal glands in the cadaver photos in Figure 25.5.

2 Label the terms in Figure 25.6(a), (b), and (c).

LAB ACTIVITY 4 Adrenal Glands

1 Examine a prepared microscope slide of the adrenal cortex and medulla and identify the structures listed with Figure 25.6(c), or identify the structures using **Real Anatomy** (Histology). ■

Diaphragm

Adrenal glands

Inferior vena cava

Descending aorta

FIGURE 25.5 The adrenal glands.

Adrenal glands

Kidney

(a)
- kidney
- left adrenal gland
- right adrenal gland

1 _____

2 _____

3 _____

(a) Anterior view

(b) Section through left adrenal gland

(b)
- adrenal cortex
- adrenal medulla
- capsule

(c)
- adrenal cortex
- adrenal medulla
- capsule
- zona fasciculata (ZOH-na fah-sick-you-LAH-ta)
- zona glomerulosa (glow-mare-you-LOH-sa)
- zona reticularis (reh-tik-you-LAIR-is)

4 _____

5 _____

6 _____

7 _____

8 _____

9 _____

10 _____

11 _____

12 _____

65×

(c) Photomicrograph of the adrenal gland

FIGURE 25.6 The adrenal glands.

E. Pancreas

The **pancreas** is composed of a head, body, and tail. Its head is cradled in the curvature of the duodenum, and the body and tail are inferior and posterior to the stomach near the spleen. The pancreas has both endocrine and exocrine (*exo-* = outside; literally "to secrete outside") functions. The exocrine or acini (*acinus* = grape-like) cells vastly outnumber the endocrine cells. Exocrine enzymes are secreted into the duodenum of the small intestine through a duct called the **pancreatic duct** and function in digestion. Endocrine cells, called **pancreatic islets (islets of Langerhans),** are located as little islands among the clusters of acini cells. The **alpha** or **A cells** in the pancreatic islets secrete **glucagon,** and the **beta** or **B cells** secrete **insulin.**

Before Going to Lab

1 Observe the location of the pancreas in the cadaver photo in Figure 25.7.

2 Label the structures on the drawing in Figure 25.8(a).

3 Using the labeled line drawing in Figure 25.8(b), label the structures in (c).

LAB ACTIVITY 5 **Pancreas**

1 Examine a prepared microscope slide of the pancreas and identify the structures listed in Figure 25.8(c), or identify the structures using **Real Anatomy** (Histology). ■

Anterior view

FIGURE 25.7 The pancreas.

(a)
- body of pancreas
- head of pancreas
- tail of pancreas

1 _____

2 _____

3 _____

(c)
- exocrine or acini (AS-ih-nur) cells
- pancreatic islet (EYE-let)

4 _____

5 _____

Duodenum

1

(a) Anterior view

3

2

Spleen (elevated)

4

5

450×

Blood capillary

Exocrine acinus

Alpha cell (secretes glucagon)

Beta cell (secretes insulin)

(b) Pancreatic islet and surrounding acini

(c) Photomicrograph of pancreatic islet and acini cells

FIGURE 25.8 The pancreas.

F. Ovaries, Testes, Pineal Gland, and Thymus

Ovaries are female gonads that not only produce and house the ova as they mature but are also endocrine organs that produce two major hormones: **estrogen** and **progesterone.** These two hormones and the microscopic anatomy of the ovary will be studied further in the Exercise 39. The **testes** are male gonads that not only produce sperm but also produce and secrete androgens, primarily **testosterone.** These hormones and the microscopic anatomy of the testes will be studied in Exercise 38.

The **pineal gland** (*pinea-* = pinecone) or **pineal body** is a small, cone-shaped gland that is located in the brain posterior to the thalamus and superior to the cerebellum. This gland secretes **melatonin** in darkness and, to a lesser degree in daytime, helps set the body's biological clock.

The **thymus** plays a role in immune function and is located anterior and superior to the heart (Figure 25.9). This gland is much larger in babies and children and regresses in size as a person ages. The main hormone produced by this gland is **thymosin.**

G. Hormone Functions

Hormones act by changing target cell activity. Changes in target cell activity include:

- Synthesis of molecules within the cell (example: estrogen)
- Changing plasma membrane permeability (example: aldosterone)
- Altering cellular metabolism (example: thyroxin)
- Secretion of cell products (example: thyroid-stimulating hormone)
- Contraction of smooth muscle (example: oxytocin)
- Contraction of cardiac muscle (example: epinephrine)

LAB ACTIVITY 6 **Hormone Functions**

1 In Table 25.1, identify the hormone(s) produced by each gland and review the target cells and function of each hormone.

2 Using a human torso, brain model, endocrine chart, or the search text box in **Real Anatomy** (Endocrine), locate each endocrine gland and name the hormone(s) it secretes.

3 For each hormone, point to and name its target cells on a torso model, and describe how the hormone stimulates target cell activity.

4 Complete **PowerPhys Experiments:**
 - Blood Glucose Regulation
 - Thyroid Function

TABLE 25.1 Endocrine Glands, Hormones, Target Cells, and Hormone Function

GLAND	HORMONE	LOCATION OF TARGET CELLS	HORMONE FUNCTION
Anterior pituitary	1. _____	Cartilage, bone, skeletal muscle, liver, and other body tissues	Stimulates secretion of hormones that stimulate body growth and metabolism.
	2. _____	Thyroid gland	Stimulates growth of thyroid gland and secretion of its hormones.
	3. _____	Testes	Stimulates sperm production.
		Ovaries	Stimulates oocyte production and estrogen secretion.
	4. _____	Testes	Stimulates secretion of testosterone.
		Ovaries	Triggers ovulation and stimulates secretion of estrogen and progesterone.
	5. _____	Mammary gland	Stimulates production and secretion of milk.
	6. _____	Adrenal cortex	Stimulates secretion of hormones by adrenal cortex.
	7. _____	Skin	Darkens skin pigmentation.
Posterior pituitary	1. _____	Kidneys	Decreases water lost in urine by returning water to the blood.
	2. _____	Uterus and mammary glands	Stimulates uterine contractions and milk ejection during suckling.
Thyroid gland	1. _____	Most body cells	Increases metabolism and basal metabolic rate (BMR).
	2. _____		
	3. _____	Osteoclast cells in bones	Decreases blood calcium levels by inhibiting osteoclasts.
Parathyroid glands	_____	Osteoclast cells in bones	Increases blood calcium levels by stimulating osteoclasts to break down bone matrix.
Adrenal cortex	1. _____	Kidneys	Decreases sodium and water loss in urine by returning sodium and water to the blood.
	2. _____	Liver, muscle, and cells involved in body defenses	Increases resistance to stress, increases blood glucose levels, and decreases inflammation.
	3. _____	Uterus, mammary glands, and other body cells involved in secondary sex characteristics	Insignificant in males; increases sex drive in females.

TABLE 25.1 Endocrine Glands, Hormones, Target Cells, and Hormone Function *(Continued)*

GLAND	HORMONE	LOCATION OF TARGET CELLS	HORMONE FUNCTION
Adrenal medulla	1. _____ 2. _____	Body cells involved in fight-or-flight response	Promotes fight-or-flight response.
Pancreas	1. _____	Most body cells	Decreases blood glucose levels by transporting glucose into body cells.
	2. _____	Liver	Increases blood glucose levels by stimulating liver to break down glycogen into glucose.
Ovaries	1. _____ 2. _____	Uterus, mammary glands, and other body cells involved in female sexual characteristics	Stimulates development of female sex characteristics; helps regulate menstrual cycle.
Testes	_____	Testes, muscle, and other body cells involved in male sexual characteristics	Stimulates development of male sex characteristics; stimulates male sex drive; regulates sperm production.
Pineal gland	_____	Brain	Helps to set biological clock.
Thymus	_____	T cells (type of white blood cell involved in immune response)	Promotes the maturation of T cells for the immune response.

LAB ACTIVITY 7 Measurement of Blood Glucose

1 Devise an experiment with your group or with the whole class that will show a change in blood glucose levels.

2 Things to consider:
- Amount of class time available to complete the experiment
- Time since last meal or snack
- Relative amount of carbohydrate, fat, and protein in last meal or snack
- Variable that can be changed in lab that will either increase or decrease blood glucose
- Time required for variable to change blood glucose
- When and how often blood glucose should be measured

3 Decide who will be the subjects and who will record the data.

4 Read the instructions for your glucometer.

5 Take baseline blood glucose measurement.

6 Take a glucometer, a test strip, a lancet, and an alcohol wipe to your lab area.
- Wash your hands thoroughly to prevent infection.
- Decide which finger you will use to obtain blood.
- Turn on the glucometer. When the glucometer is ready, insert the test strip.
- Watch the indicator for placing the blood to the strip.
- Wipe the fingertip you will pierce with an alcohol wipe.
- Using the lancet, pierce your fingertip.
- Apply blood to the test strip according to the instructions for your glucometer.
- Record the blood glucose reading when it appears.

7 Clean up as directed by your instructor.

8 Discuss your experiment and results with the class. ■

SAFETY NOTE: *Use safety glasses and gloves when handling preserved or fresh tissue. Always wash your hands thoroughly with soap and water when you are done.*

H. Dissection of Endocrine Glands

If you are dissecting a cat or fetal pig to observe endocrine organs, refer to the dissection manual. The major endocrine organs of the cat and fetal pig have a similar location and structure compared with humans.

Real Anatomy, a virtual cadaver dissection, can be used to complement or substitute for animal dissection of endocrine glands.

Reviewing Your Knowledge

A. Hormone Abbreviations

Write the name of the hormone next to the abbreviation.

1. ACTH _____

2. ADH _____

3. DHEA _____

4. FSH _____

5. hGH _____

6. LH _____

7. NE _____

8. OT _____

9. PRL _____

10. PTH _____

11. T_3 _____

12. T_4 _____

13. TSH _____

14. MSH _____

15. TH _____

B. Main Endocrine Glands and Their Hormones

Write the name of the endocrine gland that secretes the following hormones.

Hormone **Endocrine Gland**

1. ACTH _____

2. ADH _____

3. aldosterone _____

4. androgens (DHEA) _____

5. calcitonin _____

6. cortisol; cortisone _____

7. epinephrine/NE _____

8. estrogen; progesterone _____

9. FSH _____

10. glucagon _____

11. hGH _____

12. insulin _____

13. LH _____

14. melatonin _____

15. OT _____

16. PRL _____

17. PTH _____

18. T_3 and T_4 _____

19. testosterone _____

20. TSH _____

21. thymosin _____

22. MSH _____

C. Hormone Function

Write the name of the hormone that matches its function.

ACTH	glucagons	progesterone
ADH	hGH	PTH
aldosterone	insulin	T_3
androgens	LH	T_4
calcitonin	melatonin	testosterone
cortisol	MSH	thymosin
epinephrine	NE	TSH
estrogen	OT	
FSH	PRL	

Hormone Function

_____ **1.** Stimulates uterine contractions and milk ejection during suckling.

_____ **2.** Stimulates secretion of hormones by the adrenal cortex.

_____ **3.** Decreases water loss by increasing reabsorption of water into blood and decreasing urine production.

_____ **4.** Increases sex drive in females.

_____ **5.**
_____ **6.** } Increases metabolism and BMR.

_____ **7.** Triggers ovulation and stimulates secretion of estrogen and progesterone.

_____ **8.** Increases blood calcium levels by stimulating osteoclast activity.

_____ **9.**
_____ **10.** } Promotes fight-or-flight response.

_____ **11.** Stimulates production and secretion of milk.

_____ **12.** Increases resistance to stress, increases blood glucose levels, and decreases inflammation.

_____ **13.** Helps to set the biological clock.

_____ **14.** Darkens skin pigmentation.

_____ **15.** Stimulates secretion of hormones that stimulate body growth and metabolism.

_____ **16.** Decreases blood calcium levels by inhibiting osteoclasts.

_____ **17.** Promotes the maturation of T cells for the immune response.

_____ **18.** Stimulates oocyte production and estrogen secretion.

_____ **19.** Decreases blood glucose levels by transporting glucose into body cells.

_____ **20.** Increases blood glucose levels by stimulating the liver to break down glycogen into glucose.

_____ **21.**

Stimulates development of female sex characteristics and helps regulate

_____ **22.** menstrual cycle.

_____ **23.** Regulates sperm development, stimulates development of male sex characteristics, and stimulates male sex drive.

_____ **24.** Stimulates secretion of testosterone.

_____ **25.** Stimulates secretion of thyroid hormones.

_____ **26.** Stimulates sperm production.

_____ **27.** Increases reabsorption of sodium and water into blood and decreases urine output.

Using Your Knowledge

A. Endocrine Glands of the Thorax

Label the structures in Figure 25.9. For question 3, name the hormone secreted by the endocrine gland named in question 2.

SUPERIOR

1

2

INFERIOR

FIGURE 25.9 Neck and thorax region, anterior view.

1. _____

2. _____

3. _____

B. Hormone Imbalances

Using your textbook or another reference book, identify the gland and hormone(s) affected by each of the following operations or conditions.

4. Oophorectomy

5. Orchiectomy

6. Cushing's disease

C. Hormones

Answer the following questions:

7. Name two hormones that are also neurotransmitters.

8. Explain why hormones are able to affect only certain cells and not other cells.

Sometimes more than one tropic hormone is involved in a cascade of events involved in negative feedback regulation. In Figures 25.10 and 25.11, a stimulus disrupts homeostasis and initiates release of hormones in a specific order to return to homeostasis. Answer the questions associated with each figure.

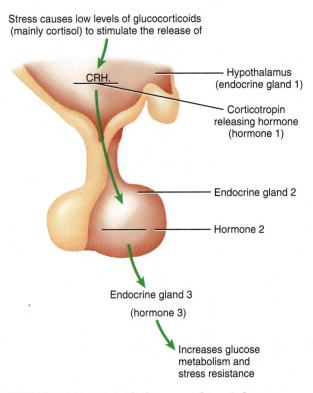

Stress causes low levels of glucocorticoids (mainly cortisol) to stimulate the release of

CRH.

Hypothalamus (endocrine gland 1)

Corticotropin releasing hormone (hormone 1)

Endocrine gland 2

Hormone 2

Endocrine gland 3

(hormone 3)

Increases glucose metabolism and stress resistance

9. _____ Identify endocrine gland 2.

10. _____ Identify hormone 2.

11. _____ Identify endocrine gland 3.

12. _____ Identify hormone 3.

13. _____ ⎫
14. _____ ⎬ Name the 2 tropic hormones

FIGURE 25.10 **Hypothalamus and tropic hormones.**

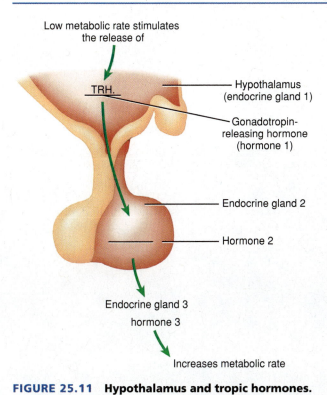

Low metabolic rate stimulates
the release of

TRH.

Hypothalamus
(endocrine gland 1)

Gonadotropin-
releasing hormone
(hormone 1)

Endocrine gland 2

Hormone 2

Endocrine gland 3

hormone 3

Increases metabolic rate

FIGURE 25.11 Hypothalamus and tropic hormones.

15. _____ Identify endocrine gland 2.

16. _____ Identify hormone 2.

17. _____ Identify endocrine gland 3.

18. _____ Identify hormone 3.

19. _____ } Name the 2
20. _____ tropic hormones

Blood Components and Blood Tests

OBJECTIVES

1 Describe the functions of blood
2 Name the components of blood
3 Describe the structure, characteristics, and function of red blood cells (RBCs), each type of white blood cell (WBC), and platelets
4 Identify RBCs, each type of WBC, and platelets
5 Describe the importance of and perform the following blood tests: differential WBC count, hematocrit, hemoglobin, and coagulation time
6 Describe how blood is typed and perform ABO and Rh blood typing

MATERIALS

- **WBC Identification and Differential WBC Count:** compound microscopes, lens paper, hand counter, prepared slides of normal human blood (with Wright's stain), prepared slides of abnormal blood (allergies or infections), or **Virtual Differential WBC Count Activity** (Instructors: download from Wiley Instructors' Companion Site)
- materials needed if using human or animal blood: biohazardous waste container; sharps container; safety glasses; disposable gloves; 10% bleach solution in spray bottle; sterile, disposable lancets or Autolets; alcohol swabs; cotton balls
- **Hematocrit:** human or animal blood, heparinized capillary tubes, sealing clay, microcentrifuge, mm rulers (or hematocrit reader)
- **Hemoglobin:** human or animal blood, Tallquist paper or hemoglobinometers
- **Coagulation Time:** human or animal blood, nonheparinized capillary tubes, small metal file
- **Blood Typing:** blood typing kit or antisera (anti-A; anti-B; anti-D), new test cards or glass slides (two per student), toothpicks, wax marking pencil, Rh warming tray
- **PowerPhys Experiment:** Hematocrit and Hemoglobin Concentration and Blood Typing

The adult cardiovascular system contains approximately 5.5 liters (1.5 gallons) of blood. Blood transports oxygen, nutrients, and hormones to body tissues and transports carbon dioxide, heat, and metabolic wastes away from body tissues. It regulates pH, body temperature, and cell water content and also provides protection from blood loss through clotting and against disease through phagocytic white blood cells and antibodies.

A. Components of Blood

When centrifuged, **blood** in a heparinized tube separates visually into two main components: plasma and formed elements. The clear, straw-colored liquid is called **plasma** and the dark-red and buff-colored portions are the **formed elements** (Figure 26.1). The formed elements include **red blood cells** (RBCs), **white blood cells** (WBCs), and **platelets** (cell fragments). RBCs are also called **erythrocytes** (*erythro-* = red; *-cytes* = cells), WBCs are known as **leukocytes** (*leuko-* = white), and platelets are known as **thrombocytes** (*thrombo-* = clot). The formed elements constitute approximately 45% of whole blood volume, and plasma composes about 55%. Plasma is about 91.5% water and 8.5% solutes. The solutes are mostly plasma proteins but also include nutrients (glucose, amino acids, and lipids), blood gases (oxygen and carbon dioxide), electrolytes, hormones, enzymes, and waste materials.

Serum is plasma minus clotting proteins. It is the watery fluid created when blood is allowed to sit and clot in an unheparinized tube.

Uncentrifuged Centrifuged

(a) Plasma and formed elements

(b) Serum and formed elements

FIGURE 26.1 Components of blood in a normal adult.

B. The Structure and Function of Red Blood Cells, White Blood Cells, and Platelets

1. Red Blood Cells (Erythrocytes or RBCs)

Erythrocytes are small, anucleate (without a nucleus) biconcave cells that contain **hemoglobin,** a large molecule used to transport oxygen and carbon dioxide in the blood. About 33% of the total weight of RBCs is composed of hemoglobin. Hemoglobin contains a red pigment called heme that gives blood its red color. Blood is bright red when oxygen-rich and darker red when oxygen-poor. Characteristics of RBCs are given in Table 26.1.

There are approximately 4.5 to 5 million RBCs per microliter of blood; females have lower amounts and males have higher amounts. An abnormally high number of RBCs is called **polycythemia** (*poly-* = many; *cyto-* = cells; *-emia* = blood), and an unusually low number of RBCs is one type of **anemia.**

2. White Blood Cells (Leukocytes or WBCs)

WBCs, nucleated cells that are typically larger than RBCs, attack pathogens and other foreign substances in the body. There are five kinds of leukocytes that are divided into two categories: granular and agranular. **Granular leukocytes** have discernible vesicles (granules) in the cytoplasm that can be seen after staining, and **agranular leukocytes** also have granules, but they cannot be observed with the light microscope. These WBCs were named agranular because the microscopes that were used at that time were not powerful enough to distinguish the granules. Granular leukocytes include **neutrophils, eosinophils,** and **basophils.** Agranular leukocytes include **lymphocytes** and **monocytes.** Characteristics of WBCs are given in Table 26.1.

There are normally 5,000 to 10,000 WBCs per microliter of blood. An abnormally high number of WBCs is called **leukocytosis** (*-osis* = an increase in a pathological condition), and a decrease in the number of WBCs is called **leukopenia** (*-penia* = deficiency).

3. Platelets (Thrombocytes)

Platelets are formed in red bone marrow from large, multinuclear cells called **megakaryocytes** (*mega-* = large; *karyo-* = nucleus). Large megakaryocytes break into tiny cytoplasmic fragments called platelets that do not have nuclei and are not considered to be cells. Characteristics of platelets are given in Table 26.1. These special cell fragments protect the body by forming a platelet plug to stop bleeding when blood vessels rupture and by secreting chemicals that aid in blood clotting. **Thrombocytopenia** (thrombo- = clot; -penia = deficiency) is a deficiency in the number of circulating platelets.

TABLE 26.1 **Characteristics of RBCs, WBCs, and Platelets (Wright's Stain)**

FORMED ELEMENT	SIZE	NUCLEUS	GRANULES	COLOR OF CYTOPLASM
RBCs	7–8 micrometer (μm) diameter	No nucleus	No granules	Light pinkish-red
Neutrophils (*neutr* = neutral; *-phil* = loving)	10–12 μm diameter	Multilobed nucleus with 2–5 or more lobes connected by threads.	Small, nondistinct pale lilac to neutral-staining granules	Usually pink but, depending on the stain used, sometimes has a reddish tinge that makes students mistake these cells for eosinophils
Eosinophils (*eosin-* = red, acidic dye)	10–12 μm in diameter	Bi-lobed nucleus (occasionally 3 lobes)	Many medium, red-orange granules (some stains make them look very dark brownish-red-orange)	Pink
Basophils (*baso-* = dark blue, basic dye)	8–10 μm in diameter	Nucleus is large, varied in shape; generally obscured by large granules	Large, dark blue-purple granules	Purple
Lymphocytes (lymph cell)	Small lymphocyte: 6–9 μm in diameter; large lymphocyte: 10–14 μm in diameter	Large, round, or slightly indented nucleus that stains very dark purple	Granules not obvious with light microscope	Light sky-blue cytoplasm; small cells have only a rim of cytoplasm; more cytoplasm in larger lymphocytes
Monocytes (*mono-* = one; pertains to having only one nucleus)	Very large cell; 12–20 μm in diameter	Large kidney bean or horseshoe-shaped lacy nucleus; sometimes oval and indented	Granules not obvious with light microscope	Light blue-gray
Platelets	2–4 μm in diameter; small cell fragments	No nucleus	Dark purple granules	Difficult to see because of dark purple granules

C. Microscopic Examination of the Formed Elements

Blood cells are difficult to see at low magnifications because the cells are so small. The erythrocytes are the most numerous (700:35:1 ratio of RBCs:platelets:WBCs) and are small, pinkish cells. The WBCs are conspicuous because their nuclei stain dark blue to dark purple with Wright's stain. The majority of WBCs are larger than the RBCs. RBCs are used as a standard to compare the size of WBCs. When looking at WBCs in normal blood cells, the majority of WBCs that you will see are neutrophils (60 to 70% of WBCs). Lymphocytes make up 20 to 25% of WBCs, monocytes 3 to 8%, eosinophils 2 to 4%, and basophils 0.5 to 1%. Often it is difficult to find basophils in normal blood smears. The platelets, which are much smaller than RBCs, stain dark purple and may just look like an extra stain on the slide.

Keep in mind that blood cells and their nuclei have a three-dimensional spherical shape. You will not always find the nucleus appearing as described or as seen in photomicrographs because of the various views that are possible. Figure 26.2 depicts the three-dimensional structure of RBCs, WBCs, and platelets.

LAB ACTIVITY 1 Identification of RBCs, WBCs, and Platelets

1 Examine a prepared blood slide using the low-power lens to locate RBCs and WBCs. The tiny RBCs are difficult to see at first and will look like tiny dots. You may see the blue/purple colored nuclei of the WBCs. See Figure 26.3.

2 Using the high-power or oil immersion lens, locate RBCs, WBCs, and platelets.

- When you begin your identification, the obvious thing you will notice is there are many, many RBCs found everywhere. Then you will notice the most numerous of the WBCs, the neutrophils. The next prevalent WBC type is the lymphocyte. In fact, you may tire of seeing so many of these two types of WBCs; but the other three types of WBCs are a little more elusive and can be found with diligent searching. The numbers below indicate the percentages each specific WBC represents out of the total WBCs present. Use the size of the RBCs (7–8 μm) to estimate the size of the WBCs.

Step 1 Identify the <u>most numerous WBC</u>—the **neutrophil**—60–70% of WBCs (multilobed nucleus; neutral to pink granules; 10–12 μm)

Step 2 Identify the <u>next most numerous WBC</u>—the **lymphocyte**—20–25% of WBCs (smallest WBC; nucleus is 90% of cell; small lymphocyte is 6–9 μm)

Step 3 Identify the <u>largest WBC</u>—the **monocyte**—3–8% of WBCs (horseshoe-shaped nucleus; light blue cytoplasm; 12–20 μm)

Step 4 Identify the <u>red granular WBC</u>—the **eosinophil**—2–4% of WBCs (small red granules; bilobed nucleus; 10–12 μm)

Step 5 Identify the <u>blue granular WBC</u>—the **basophil**—0.5–1% of WBCs (large blue-purple droplets; nucleus is obscured by granules; 8-10 μm)

3 Clean up as directed by your instructor.

CLINICAL NOTE: *Neutrophils have a multilobed nucleus that gives them the nickname **polymorphonuclear leukocytes**, or **PMNs**. Neutrophils are also called "**segs**" because of the segmented look of the nucleus.*

(a) Neutrophils (b) Eosinophils (c) Basophils

(d) Lymphocytes (e) Monocytes **LM** all 1000×

FIGURE 26.2 Photomicrographs of variations of white blood cells.

D. Differential White Blood Cell Count

A **differential WBC count** is performed to determine the percentage of each of the five types of WBCs in a blood sample. When attempting to learn which WBCs are normally present from the greatest to least percentages, it is helpful to remember the following mnemonic: "Never Let Monkeys Eat Bananas" (N = neutrophils; L = lymphocytes; M = monocytes; E = eosinophils; and B = basophils). Because the normal percentage of each WBC type is known, any significant abnormality in these percentages, elevated or depressed, can be indicative of particular disorders. Numerical values for the normal range of WBCs may vary depending on the reference source.

> **SAFETY NOTE:** *When using human or animal blood samples, you must protect yourself from any blood-borne infectious disease (hepatitis, HIV, etc.). Wear gloves and safety goggles while performing blood tests. If a blood spill occurs, cover the spill with paper towels soaked in 10% bleach solution and immediately inform your instructor.*

LAB ACTIVITY 2 Differential WBC Count

1 Complete a differential WBC count using a prepared microscope slide of normal blood, or use the Virtual Differential WBC Count Activity supplied by your instructor. INSTRUCTORS: Download Activity from Wiley Instructor Companion Site.

- Complete this activity individually or with a lab partner as your instructor directs. If done together, one person will identify each WBC with the microscope and call out the type to the lab partner. The other person will mark the type of cells encountered by using tally marks (卌 II) and also keep track of the total number of cells.
- Using a prepared, normal blood slide, focus using the low-power objective lens and then switch to a higher-power objective lens as directed by your instructor.
- To avoid recounting any WBCs, begin at one end of the slide that has a good separation of cells and slowly scan systematically, moving the slide down and over, then up and over, and repeat as seen in Figure 26.3.
- If you encounter a cell that you cannot identify or are not sure of, do not count it and continue on.
- Continue this pattern until you have found and recorded 100 WBCs in Table 26.2.
- Because you are counting 100 cells, the total number of each type of WBC counted is its percentage in the blood sample (i.e., 26 lymphocytes means 26% of the WBCs are lymphocytes).
- Write your results on the board to pool class data.

2 Complete a differential WBC count using a prepared microscope slide of abnormal blood, if available, or use the Virtual Differential WBC Count Activity.

- Follow the instructions for a differential WBC using normal blood (listed above).
- Record your tally counts and totals in Table 26.2.
- Determine the disorder by consulting Table 26.3 and record at the bottom of Table 26.2.
- Write your results on the board to pool class data for comparison.

3 Clean up as directed by your instructor. ■

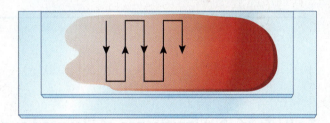

FIGURE 26.3 Procedure for scanning WBC differential count.

TABLE 26.2 Results of Differential WBC Count

A. Normal Differential WBC Count

TYPE OF WBC	OBSERVATION TALLY	% OF TOTAL WBCs	CLASS AVERAGE (% OF TOTAL WBCs)	NORMAL PERCENTAGE OF WBCs
Neutrophil				60–70%
Lymphocyte				20–25%
Monocyte				3–8%
Eosinophil				2–4%
Basophil				0.5–1%
Total WBCs				

B. Pathological Differential WBC Count

Neutrophil				60–70%
Lymphocyte				20–25%
Monocyte				3–8%
Eosinophil				2–4%
Basophil				0.5–1%
Total WBCs				

Type of Disorder

TABLE 26.3 Significance of Elevated and Depressed WBC Counts

WBC TYPE	HIGH COUNT MAY INDICATE	LOW COUNT MAY INDICATE
Neutrophils	Bacterial infection, burns, stress, inflammation	Radiation exposure, drug toxicity, vitamin B_{12} deficiency, systemic lupus erythematosus
Lymphocytes	Viral infections, some leukemias	Prolonged illness, immunosuppression, treatment with cortisol
Monocytes	Viral or fungal infections, tuberculosis, some leukemias, other chronic diseases	Bone marrow depression, treatment with cortisol
Eosinophils	Allergic reactions, parasitic infections, autoimmune diseases	Drug toxicity, stress
Basophils	Allergic reactions, leukemias, cancers, hypothyroidism	Pregnancy, ovulation, stress, hyperthyroidism

E. Hematocrit (Packed Red Cell Volume)

A **hematocrit** determines the volume of RBCs described as the percentage of RBCs in a whole blood sample. A capillary tube of blood is centrifuged to pack the red cells at the outer end of the tube, separating them from the WBCs, platelets, and plasma. After measuring the length of the RBC column and the total length of the blood column, the percentage of RBCs can be calculated. The normal hematocrit range for females is approximately 38 to 46% and for males it is 40 to 54%. An abnormally high hematocrit (generally 65% or above) is indicative of **polycythemia,** and a hematocrit with RBCs below the normal level indicates a type of **anemia.**

> **SAFETY NOTE:** *When using human or animal blood samples, you must protect yourself from any blood-borne infectious disease (hepatitis, HIV, etc.). Wear gloves and safety goggles while performing blood tests. If a blood spill occurs, cover the spill with paper towels soaked in 10% bleach solution and immediately inform your instructor.*

LAB ACTIVITY 3 Hematocrit

1 Calculate the hematocrit for blood in the capillary tubes in Figure 26.4.
- Use a millimeter ruler to measure (in mm) the length of the whole column (the length of packed RBCs, WBCs, and plasma) and record it in Table 26.4.
- Measure the length (in mm) of the RBCs and record in Table 26.4.
- Use the following formula to calculate the percentage of RBCs (the hematocrit) in whole blood:

$$\left(\frac{\text{Length of RBCs in mm}}{\text{Length of the whole column in mm}}\right) \times 100$$

- Record the hematocrit in Table 26.4.

2 Calculate the hematocrit on a blood sample if equipment is available.
- If using your own blood, clean your finger with an alcohol wipe, lance it with a new, sterile lancet, and discard the lancet in the sharps container for biohazardous material only.
- Touch the red-lined end of a heparinized (prevents blood from coagulating) capillary tube to the blood and hold the tube at a downward angle. Taking care that no air is allowed to enter while the tube fills by capillary action, fill the capillary tube about three-fourths full if possible.
- Hold a finger over the end of the tube so blood cannot drain out. Place the blood end of the tube into sealing clay to plug the end.
- Place the sealed tube in a microhematocrit centrifuge with the sealed end on the rubber gasket pointing outward and away from the center.
- Prepare a second tube in the same manner.
- Place the second sealed tube in the centrifuge opposite the first tube for balance. *Always keep the centrifuge balanced.*
- If others are using the centrifuge, make a note of the groove numbers of your tubes.
- Secure the inside and outside covers of the centrifuge and set the timer for 4 to 5 minutes.
- After the centrifuge has stopped, remove your tubes. Note that the RBCs are packed at the bottom near the sealing clay, the WBCs (and platelets) are buff colored next to them, and the clear plasma is on the top.
- Use a microhematocrit reader or a millimeter ruler to measure the length of the whole column, the length of the packed RBCs, WBCs, and plasma in millimeters (Figure 26.4).
- Record the whole column length (in mm) and the RBC length (in mm) in Table 26.4.
- Use the following formula to calculate the percentage of RBCs (the hematocrit) in whole blood:

$$\left(\frac{\text{Length of RBCs in mm}}{\text{Length of the whole column in mm}}\right) \times 100$$

- Record the hematocrit in Table 26.4.

3 Clean up as directed by your instructor.

4 Answer Discussion Questions with your lab group.

FIGURE 26.4 Capillary tubes for calculating a hematocrit.

TABLE 26.4 **Hematocrit Results**

HEMATOCRIT CAPILLARY TUBES	RBC COLUMN (mm)	WHOLE COLUMN (mm)	HEMATOCRIT (PERCENTAGE OF RBCs/TOTAL BLOOD)
Hematocrit a			
Hematocrit b			
Hematocrit (your blood)			
MALE CLASS AVERAGE	**FEMALE CLASS AVERAGE**	**NATIONAL AVERAGE FOR MALES**	**NATIONAL AVERAGE FOR FEMALES**
		40–54%	38–46%

DISCUSSION QUESTIONS
Hematocrit

1 For the capillary tubes in Figure 26.4, is the hematocrit normal, low, or high? If abnormal, do values indicate anemia or polycythemia?

2 Is the class average hematocrit for males higher than females?

3 How does your class compare with the national averages?

F. Determining Hemoglobin Content of Blood

Hemoglobin (Hb) is a protein that carries oxygen in the RBCs. Therefore, hemoglobin concentration in blood determines the oxygen-carrying capacity of the blood. It is possible for you to have a normal hematocrit or RBC volume and still be anemic due to low hemoglobin concentration. Normal values are 12 to 15 g hemoglobin/100 mL blood in females and 13 to 16 g/100 mL in males. In severe anemia, the hemoglobin content can be less than 7 g/100 mL. Clinically, the relationship of the hematocrit (%) is compared with the hemoglobin reading (g/100 mL) and is generally a ratio of 3:1.

Two of the several techniques for determining the hemoglobin content of blood are discussed here. The Tallquist measurement is the simpler but older method that does not always produce accurate results. Hemoglobinometers are accurately calibrated instruments but are more expensive.

LAB ACTIVITY 4 **Determining Hemoglobin Content of Blood**

1 Follow the Tallquist procedure for determining hemoglobin content, or if using a hemoglobinometer, follow the procedure given in its instruction booklet.
- After reviewing the Safety Note before Lab Activity 3 on using blood, obtain a blood sample provided by your instructor or your own blood using methods described previously.
- Place one drop of blood on the absorbent Tallquist paper. Be sure the blood spot is large enough to be seen in all the chart holes simultaneously.
- Allow blood to dry until it is no longer shiny but not so dry as to turn brownish. Match its color with the color scale provided.
- Record the results as grams of hemoglobin/100 mL of blood in Table 26.5.

2 Clean up as directed by your instructor.

3 Answer Discussion Questions with your lab group.

TABLE 26.5 **Hemoglobin Content**

HEMOGLOBIN CONTENT
Tallquist method: _____ g/100 mL
Hemoglobinometer: _____ g/100 mL

DISCUSSION QUESTIONS
Hemoglobin Content of Blood

1 Compare your hematocrit results with your grams of hemoglobin/100 mL. Calculate the ratio. _____ Is this close to the normal ratio? (yes/no)

2 To measure the hemoglobin content, RBCs are hemolyzed. How would the results be different if you did not completely hemolyze the RBCs? Would the hemoglobin concentration be higher or lower?

3 Within your class, are there any students who have the same hematocrit but different hemoglobin concentrations? Explain why this can occur.

G. Coagulation Time

The process of blood clotting or **coagulation** prevents excessive blood loss. There are many blood clotting factors present in the plasma. Other factors are released by injured tissues and platelets to initiate the chemical chain reaction that forms a blood clot. Fibrin is a long, insoluble thread-like protein strand that forms a mesh to trap platelets. Fibrin is formed from soluble fibrinogen molecules during clotting.

The coagulation time is the time it takes for clotting to take place when blood is removed from the body. Normal clotting time is within 2 to 6 minutes.

LAB ACTIVITY 5 Coagulation Time

1 Observe the demonstration of clotting blood.
- Your instructor will place animal or sterile human blood into two test tubes, one with and one without an anticoagulant.
- Observe the clot form and settle to the bottom of the tube without anticoagulant. Observe the straw-colored serum above the clot.
- Compare the tube with the blood clot to the blood in the test tube with an anticoagulant.

2 Determine coagulation time.
- If using human blood, review the Safety Note before Lab Activity 3.

- If using your own blood, use an alcohol swab on a clean finger and prick the finger with a clean, sterile lancet. Have a free flow of blood.
- Hold the nonheparinized capillary tube at an angle to collect the blood and put the upper end in the drop of blood.
- Fill the tube a minimum of two-thirds full and record the time started. Time: _____
- Lay the tube on a paper towel.
- After 30 seconds, scratch the glass with the metal file close to an end of the tube. Holding the tube close to each side of the scratch, gently break the tube away from you.
- Gently pull the tube apart and observe the blood to see if a fibrin thread is present.
- Repeat this process every 30 seconds until fibrin occurs. Record the amount of time required for coagulation in Table 26.6.
- Write results on the board to pool class data. Calculate average coagulation time for your class and record in Table 26.6.

3 Clean up as directed by your instructor.

4 Answer Discussion Questions with your lab group.

DISCUSSION QUESTIONS
Coagulation Time

1 Would a person with hemophilia have higher or lower than normal clotting times? Explain.

2 How would coagulation time be affected if a heparinized capillary tube was used? Explain.

TABLE 26.6 Coagulation Time

YOUR COAGULATION TIME	CLASS AVERAGE OF COAGULATION TIMES

H. ABO and Rh Blood Typing

Blood typing is critical for blood transfusions, transplantations, and maternal-fetal compatibility, but is also used in genetic studies, forensic studies, legal medicine, and anthropology. Although there are many different systems for classifying human blood, we will be studying the **ABO** and **Rh systems** because they are most commonly used. Blood typing is based on the **antigenic** (agglutinogens) molecules that are on the surface of the RBC membranes. An **antigen** is a substance that is able to produce an immune response and will react with a specific antibody. **Antibodies** are plasma proteins that combine with a specific antigen to inhibit or destroy it. In the **ABO system,** there are two types of antigens (A and B) that can be present as surface membrane molecules on RBCs. If the plasma membranes of your RBCs have only the A antigen present, you have **type A** blood; correspondingly, if you have only B antigens present, you have **type B** blood. If you have both antigens A and B present you have **type AB** blood, and if you do not have either A or B antigen present, you have **type O** blood (Figure 26.5).

A or B antibodies appear in babies' blood a few months after birth. If you have type A blood, you do not have its corresponding anti-A antibody. If the two are mixed, they will form a detrimental antigen-antibody complex and cause clumping. People with type A blood have anti-B antibodies that will become cross-linked and **agglutinate** (clump) if type B blood is given to them (Figure 26.6). Agglutination is followed by the activation of another plasma protein that attaches to the recipient's RBCs and hemolyzes or bursts them, releasing hemoglobin that can cause kidney damage. A and B antibodies do not cross the placenta because of their large size.

The **Rh blood system** is different from the ABO system but has some similarities. If you have the Rh antigen as a surface membrane molecule on your RBCs, you are **Rh$^+$**. If you do not have the Rh antigen, you are **Rh$^-$**. An Rh$^-$ person does not spontaneously develop the anti-Rh antibody (as in the ABO system) and does not produce this antibody until the person is exposed to the Rh antigen from Rh$^+$ blood. This can happen through a blood transfusion, by sharing hypodermic needles, or by an Rh$^-$ mother carrying an Rh$^+$ child. During delivery, the baby's blood can leak from the placenta into the mother's bloodstream, causing the mother's body to make Rh antibodies. The first baby would not be affected, but subsequent pregnancies with Rh$^+$ fetuses can result in the small Rh antibodies crossing the placenta and causing hemolysis in the fetuses' blood. This condition is called **hemolytic disease of the newborn.** Rh$^-$ mothers are typically given RhoGAM so they will not make Rh antibodies.

The following activity will use **antisera** (antiserum, sing.), which can be artificial serum or serum from an animal or human containing antibodies against A, B, or D (Rh) antigens. **Serum** is blood plasma without clotting proteins. If anti-A serum clumps a particular blood sample, the blood is type A because of an antibody-antigen complex formed with the A antigens on the RBCs.

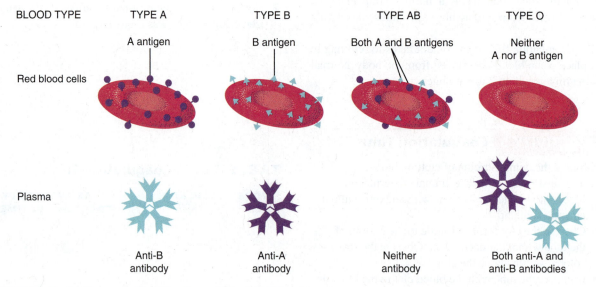

BLOOD TYPE	TYPE A	TYPE B	TYPE AB	TYPE O
	A antigen	B antigen	Both A and B antigens	Neither A nor B antigen
Red blood cells				
Plasma	Anti-B antibody	Anti-A antibody	Neither antibody	Both anti-A and anti-B antibodies

FIGURE 26.5 **Antigens and antibodies of the ABO blood types.**

Before Going to Lab

1 Complete Table 26.7 by writing in the antigen present on the RBCs and the antibody present in plasma for each blood type. Also, determine each compatible and incompatible donor for each blood type.

> **SAFETY NOTE:** *When using human or animal blood samples, you must protect yourself from any blood-borne infectious disease (hepatitis, HIV, etc.). Wear gloves and safety goggles while performing blood tests. If a blood spill occurs, cover the spill with paper towels soaked in 10% bleach solution and immediately inform your instructor.*

LAB ACTIVITY 6 ABO and Rh Blood Typing

1 Your instructor will tell you whether you will be using sterile blood, your own blood, or simulated blood.

2 Procedure for obtaining blood sample.
- Thoroughly wash hands with soap and dry with a clean paper towel.
- Thoroughly clean the tip of index finger with an alcohol swab.
- Open a new, sterile blood lancet exposing the sharp tip only (or use an Autolet). Lance just to the side of the finger pad with the new lancet. *Never* reuse a lancet, even your own.
- Deposit the used lancet in the sharps container for biohazardous materials only.
- Wipe away the first drop of blood with cotton ball and dispose in biohazard waste container. Gently squeeze one drop of blood on each side of a clean prepared slide.

3 Perform ABO blood typing according to the procedure below, using Figure 26.6 as a guide.
- Obtain a clean glass slide (test cards if using a typing kit), two new toothpicks, a wax marking pencil, and anti-A and anti-B sera.
- Divide the glass slide in half with a wax pencil. Mark A on the left side and B on the right side. Place one drop of anti-A serum on the left side and one drop of anti-B serum on the right side.
- Place one drop of blood on sides A and B of slide.
- Using *separate* new toothpicks, mix each sample of blood with the corresponding antiserum.
- Results may take up to 2 minutes. Observe each sample for agglutination or the appearance of granulation (Figure 26.6); interpret the results.
- Record your results in Table 26.8 by writing "Yes" if clumping occurs and "No" if clumping does not occur.
- Write your results on the board to pool class data for comparison.
- Record percentage of ABO blood types for your class in Table 26.8.

TABLE 26.7 Summary of ABO Blood Group Interactions

BLOOD TYPE	A	B	AB	O
Antigen on RBCs				
Antibody in plasma				
Compatible donor blood types (no hemolysis)				
Incompatible donor blood types (hemolysis)				

FIGURE 26.6 Serum agglutination results.

4 Perform Rh blood typing according to the procedure below.
- Obtain a clean glass slide and antiserum D (Rh).
- Place one drop of anti-D on the glass slide and add a drop of blood.
- Mix the two liquids with a new toothpick.
- Place slide on the warm Rh typing box and rock gently to mix. Rh typing requires a higher temperature than ABO typing does.
- Results may take up to 2 minutes. Observe your sample for clumping or the appearance of granulation (Figure 26.6); interpret the results.
- Sometimes, it is more difficult to obtain positive results with the anti-Rh serum, depending on the supply and delivery situations.
- Record your results in Table 26.8 for each sample you tested.
- Write your results on the board to pool class data for comparison.
- Record percentage of Rh blood types for your class in Table 26.8.

5 Clean up as directed by your instructor. ■

LAB ACTIVITY 7 **Hematocrit and Hemoglobin Concentration and Blood Typing**

1 Complete **PowerPhys Experiments:** Hematocrit and Hemoglobin Concentration and Blood Typing.
- Effect of Altitude on Hematocrit and Hemoglobin.
- ABO and Rh Blood Typing. ■

TABLE 26.8 Blood Typing Results

SERUM AGGLUTINATION RESULTS	STUDENT BLOOD SAMPLE	UNKNOWN 1	UNKNOWN 2	UNKNOWN 3	UNKNOWN 4
Clumping with anti-A					
Clumping with anti-B					
Clumping with anti-D (Rh)					
ABO blood type					
Rh type					

BLOOD TYPES	CLASS BLOOD TYPES (PERCENTAGE)
A	
B	
AB	
O	
Rh$^+$	
Rh$^-$	

I. Complete Blood Count (CBC)

A **complete blood count (CBC),** a vital diagnostic tool, screens for abnormalities in the number or structure of formed elements. The CBC includes: a total RBC count, total WBC count, platelet count, differential WBC count, hematocrit, and hemoglobin concentration. The CBC is used along with a battery of blood chemistry tests to put together a comprehensive profile of a person's general level of health based on blood values. Examine Exhibit 26.1 to observe the complexity of a CBC.

EXHIBIT 26.1 Complete Blood Count Results

PATIENT: ACCOUNT NO: MEDICAL RECORD NO.:			DOCTOR: REPORT FOR LOCATION:	
HEMATOLOGY				
DAY: −1 DATE: AUG 20 13 TIME: 1000			REFERENCE	UNITS
WBC	8.0		4.5–11.0	$\times 10^3/\mu L$
RBC	4.12	L	4.70–6.10	$\times 10^6/\mu L$
HGB	14.4		14.0–18.0	g/dL
HCT	41.6	L	42.0–52.0	%
MCV	101	H	83–99	fL
MCH	34.8	H	28–34	pg
MCHC	34.5		31.0–37.0	%
RDW	13.1		11.0–15.5	%
PLT	265		150–450	$\times 10^3/\mu L$
MPV	8.2		7.4–10.4	fL
NEUT%	47.7	L	49.0–90.0	%
LYMPH%	41.0		22.0–51.1	%
MONO%	6.5		0.0–13.0	%
EO%	4.3		0.0–7.0	%
BASO%	0.5		0.0–3.0	%
NEUT#	3.90		1.50–7.06	$\times 10^3/\mu L$
LYMPH#	3.30	H	0.70–3.00	$\times 10^3/\mu L$
MONO#	0.5		0.0–1.4	$\times 10^3/\mu L$
EO#	0.3		0.0–0.8	$\times 10^3/\mu L$
BASO#	0.0		0.0–0.3	$\times 10^3/\mu L$
COAGULATION				
DAY: −1 DATE: AUG 20 13 TIME: 1000			REFERENCE	UNITS
PROTIME:	11.4		10.9–13.3	sec
PTT:	28.3		25.9–35.3	sec

WBC = white blood cell; RBC = red blood cell; HGB = hemoglobin; HCT = hematocrit; MCV = mean corpuscular volume; MCH = mean corpuscular hemoglobin; MCHC = mean corpuscular hemoglobin concentration; RDW = red blood cell distribution index; PLT = platelet; MPV = mean platelet volume; NEUT% = neutrophil percent; LYMPH% = lymphocyte percent; MONO% = monocyte percent; EO% = eosinophil percent; BASO% = basophil percent; NEUT# = neutrophil number; LYMPH# = lymphocyte number; MONO# = monocyte number; EO# = eosinophil number; BASO# = basophil number; L = low; H = high; PTT = partial thromboplastin time.

Reviewing Your Knowledge

A. Characteristics of the Formed Elements and Blood Abnormalities

Fill in the blank with the term that fits the description.

_____ **1.** The oxygen and carbon dioxide carrying cell

_____ **2.** Help the body fight infections and foreign substances

_____ **3.** Form a clot to help the body stop bleeding

_____ **4.** Another name for red blood cells

_____ **5.** Another name for platelets

_____ **6.** Another name for white blood cells

_____ **7.** Large cells that develop into platelets

_____ **8.** A deficiency in number of RBCs or decreased hemoglobin content of blood

_____ **9.** An abnormal increase in RBCs

_____ **10.** An abnormal increase in WBCs

_____ **11.** A deficiency in WBCs

_____ **12.** A deficiency in platelets

B. White Blood Cell Structure and Characteristics

Fill in the blank with the term that fits the description. Terms may be used more than once.

_____ **1.** 60–70% of all WBCs

_____ **2.** 2–4% of all WBCs

_____ **3.** 0.5–1% of all WBCs

_____ **4.** 20–25% of all WBCs

_____ **5.** 3–8% of all WBCs

_____ **6.** 10–12 μm; nucleus with 2–5 connected lobes; pale lilac granules

_____ **7.** 10–12 μm; nucleus with 2 or 3 lobes; red-orange granules

_____ **8.** 8–10 μm; nucleus difficult to see; large deep blue-purple granules

_____ **9.** 6–9 μm; round nucleus that is dark purple; sky blue cytoplasm, no visible granules

_____ **10.** 12–20 μm; kidney-shaped nucleus; blue-gray cytoplasm, no visible granules

_____ **11.** Abbreviation for polymorphonuclear leukocytes

_____ **12.** General name for all of the WBCs

_____ **13.**
$\Bigg\}$ Nicknames for neutrophils
_____ **14.**

C. White Blood Cells

Write the name of the white blood cell that matches the description. Terms may be used more than once.

_____ **1.**

_____ **2.** $\Big\}$ granulocytes

_____ **3.**

_____ **4.**

$\Big\}$ agranulocytes

_____ **5.**

_____ **6.** most numerous leukocyte

_____ **7.** least numerous leukocyte

D. ABO and Rh Blood Typing

Name the antigens present on the RBCs and the antibodies present in plasma of each blood type.

	Antigens on RBCs	**Antibodies in Plasma**
1. O^+	_____	_____
2. A^-	_____	_____
3. B^-	_____	_____
4. AB^+	_____	_____

5. Based on what you know about antigens and antibodies, what blood type is the universal donor? _____ Explain.

6. What blood type is the universal recipient? _____ Explain.

E. Hematocrit

1. Define hematocrit.

2. What is the anticoagulant used in this type of blood test? _____

3. Ron has a hematocrit of 47%. Is this within the normal range? (yes/no)

4. Janey has a hematocrit of 58%. Is this within the normal range? (yes/no)

F. Hemoglobin Content and Coagulation Time

1. Do the hematocrit and hemoglobin content of blood measure the same thing? _____ Explain.

2. What is the importance of coagulation time?

3. Would a hemophiliac have an above or below normal coagulation time?

Using Your Knowledge

A. Formed Elements of Blood

Identify the formed elements in Figure 26.7.

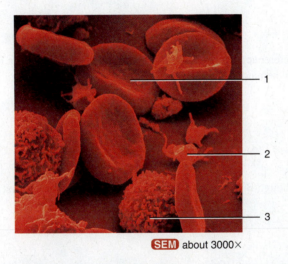

SEM about 3000×

FIGURE 26.7 **Scanning electron micrograph of the formed elements of blood.**

Platelet
Red blood cell
White blood cell

1. _____

2. _____

3. _____

B. White Blood Cell Functions

4. Ken has a bacterial infection with inflammation. What type of WBC increases in number first to help him fight the infection? Which WBC arrives second to help?

C. Differential White Blood Cell Count

5. Lindsey had a differential WBC count that showed 60% neutrophils, 10% eosinophils, 1% basophils, 22% lymphocytes, and 7% monocytes. What could have caused this?

D. Hematocrit

6. Michaela is an athlete who has been training for several years only in Florida. She had an annual physical, and the doctor read her hematocrit as 59%. What could this reading potentially tell the doctor?

E. Hemoglobin

7. Using your textbook or another reference, determine what type of anemia could cause a reduced hemoglobin level while the person still has a normal hematocrit.

F. Blood Typing

8. Blake's doctor found out that, although Blake is Rh^-, he received Rh^+ blood for the first time. Will he have a transfusion reaction? Explain.

9. Jim has type AB^+ blood. What type(s) of blood can he receive in a transfusion?

10. If Brittany is blood type O^- and her husband is blood type AB^-, what effect can the Rh factor have on their second child?

11. Michael is type O^-. He is donating blood today and wants to know what blood types can accept his blood. What would you tell him?

12. Staci has antibody A in her blood with no Rh antibody. What blood type does she have?

13. Leslie has AB^+ blood and had a blood transfusion with AB^- blood. Could this be a life-threatening situation? Why or why not?

Heart Structure and Function

OBJECTIVES

1 Identify and describe the functions of the major heart structures

2 Explain the differences in structure and function of the two types of heart valves

3 Describe the changes that take place in the heart after birth

4 Trace the flow of a drop of blood through the pulmonary and systemic circulations, listing the vessels, chambers, and valves

5 Identify the major vessels involved in coronary circulation

6 Describe and identify the microscopic structures of cardiac muscle

7 Describe and identify the two layers of the pericardium and the three layers of the heart wall

8 Dissect a sheep heart and identify selected heart structures

MATERIALS

- human heart model or use **Real Anatomy** (Cardiovascular)

- **Location of Heart:** skeleton, small removable labels

- human torso or chart showing the heart/pulmonary/systemic circulations, colored pencils (red, blue, purple), index cards

- model or chart of the coronary circulation

- compound microscope, lens paper, prepared microscope slides of cardiac muscle, or **Real Anatomy** (Histology)

- **Simulated Pericardium:** gallon plastic zip bags (one per group)

- Dissection: preserved sheep heart (or other mammalian heart), dissecting instruments, 5-inch blade knife, disposable gloves, safety glasses

The heart is a small double pump that simultaneously pumps blood to and from body cells through the systemic circulation and to and from the lungs through the pulmonary circulation. The blood vessels that carry blood from the heart are called **arteries** and the blood vessels that carry blood to the heart are called **veins.**

A. Location and Surface Features of the Heart

Like all mammalian hearts, the human heart has four chambers and is divided into right and left sides. Each side has an upper chamber called an **atrium** and a lower chamber called a **ventricle.** The atria have pouch-like extensions called **auricles.** From the exterior, the auricles look like flaps with wrinkled edges.

The **heart** is about the size of a fist and is shaped like a cone. In *situ*, it lies on its side in the thoracic cavity within the **mediastinum.** The mediastinum is an area bounded by the lungs laterally, the sternum and ribs anteriorly, the thoracic vertebrae posteriorly, and the diaphragm inferiorly. Approximately two-thirds of the heart lies to the left of the thoracic midline. The right ventricle forms most of the anterior surface of the heart. The left ventricle is also observed on the anterior surface of the heart. The **apex** of the heart is the inferior pointed end of the left ventricle and is located in the 5th intercostal space. The inferior surface of the heart lies primary on the diaphragm between the apex and right lung and is attached to the diaphragm by dense fibrous connective tissue. Both atria and the left ventricle are observed on the posterior surface of the heart. The right and left atria form the **base** of the heart, and the base faces the right shoulder while the apex points to the left hip (see Figure 27.1).

Coronary blood vessels and adipose tissue are found in the **sulci** or grooves that externally mark the boundaries between the four heart chambers. Although a considerable amount of external adipose tissue is present on the heart surface for protection and padding, most heart models do not show this. The **coronary sulcus** is a deep sulcus that externally shows the separation of the atria and the ventricles. The **anterior interventricular sulcus** and the **posterior interventricular sulcus** are shallow grooves that depict the surface boundaries between the two ventricles.

Before Going to Lab

1 Label the heart structures in Figure 27.2(a) and (b).

LAB ACTIVITY 1 Location and Surface Features of the Heart

1 Identify each structure in Figure 27.2(a) and (b) on a model, chart, or use the search text box in **Real Anatomy** (Cardiovascular) to locate these structures.

2 Locate the 4 points that form the outline of the heart by placing small removable labels on the following locations on a skeleton. Refer to Figure 27.1.

- **Superior right point:** about 3 cm to the right of the midline on the superior border of the 3rd costal cartilage
- **Superior left point:** about 3 cm to the left of the midline in the 2nd intercostal space
- **Inferior left point:** about 9 cm to the left of the midline in the 5th intercostal space
- **Inferior right point:** about 3 cm to the right of the midline at the 6th right costal cartilage

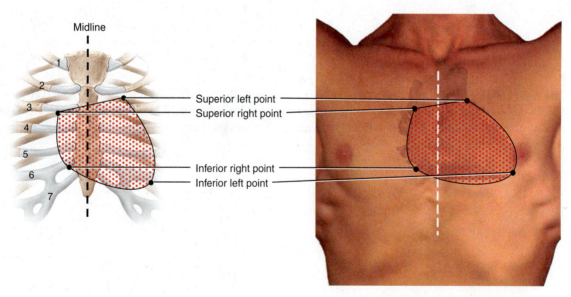

FIGURE 27.1 Location of heart.

- anterior interventricular sulcus
- apex of heart
- coronary sulcus
- left auricle
- left ventricle
- right auricle
- right ventricle

1 _____

2 _____

3 _____

4 _____

5 _____

6 _____

7 _____

1

2 (groove)

3

4

5 (groove)

6

7

(a) Anterior view

- adipose tissue
- coronary sulcus
- left auricle
- left ventricle
- posterior interventricular sulcus
- right auricle
- right ventricle

8 _____

9 _____

10 _____

11 _____

12 _____

13 _____

14 _____

8

9 (groove)

10

11

12 (groove)

13

14

(b) Posterior view

FIGURE 27.2 Surface features of the heart.

B. Great Vessels of the Heart

The great vessels of the heart either return blood to the atria or carry blood away from the ventricles. The superior vena cava, inferior vena cava, and coronary sinus return oxygen-poor blood to the right atrium. The **superior vena cava** returns blood from the head, neck, and arms; the **inferior vena cava** returns blood from the body inferior to the heart; and the **coronary sinus** is a smaller vein that returns blood from the coronary circulation. Blood leaves the right atrium to enter the right ventricle. From here, blood enters the **pulmonary trunk,** the only vessel that removes blood from the right ventricle. This large artery divides into the **right** and **left pulmonary arteries,** which carry blood to the lungs, where it is oxygenated. Oxygen-rich blood returns to the left atrium through two **right** and two **left pulmonary veins.** The blood then passes into the left ventricle, which pumps blood into the large **aorta.** The aorta distributes blood to the systemic circulation. The aorta begins as a short **ascending aorta,** curves to the left to form the **aortic arch,** descends posteriorly, and continues as the **descending aorta.**

The fetal heart contains a short, temporary vascular channel, the **ductus arteriosus** (*ductus* = duct; *arteriosus* = artery), which connects the pulmonary trunk and the aorta. This right heart to left heart shunt re-routes some of the blood destined for the lungs to the systemic circulation via the aorta. In fetal life, oxygen is obtained through the placenta from the mother and not from the lungs. Therefore, it is not detrimental to the baby's health for blood to bypass the lungs. The ductus arteriosus changes into a ligament after birth and remains as the **ligamentum arteriosum.**

Before Going to Lab

1 Label the great vessels of the heart in Figure 27.3(b) and (c).

LAB ACTIVITY 2 Great Vessels of the Heart

1 Identify each great vessel on a model, chart, or use the search text box in **Real Anatomy** (Cardiovascular) to locate these structures. ■

Brachiocephalic trunk

Superior vena cava

Ascending aorta

Right pulmonary veins

RIGHT AURICLE OF RIGHT ATRIUM

RIGHT ATRIUM

CORONARY SULCUS

RIGHT VENTRICLE

Left subclavian artery

Left common carotid artery

Arch of aorta

Ligamentum arteriosum

Left pulmonary artery

Left pulmonary veins

Pulmonary trunk

LEFT AURICLE OF LEFT ATRIUM

LEFT VENTRICLE

ANTERIOR INTERVENTRICULAR SULCUS

(a) Anterior external view

FIGURE 27.3 Great vessels of the heart.

(b) Anterior view

- aortic arch
- ascending aorta
- descending aorta
- inferior vena cava
- left pulmonary artery
- left pulmonary veins
- ligamentum arteriosum
- pulmonary trunk
- right pulmonary artery
- right pulmonary veins
- superior vena cava

1 _____

2 _____

3 _____

4 _____

5 _____

6 _____

7 _____

8 _____

9 _____

10 _____

11 _____

(c) Posterior view

- aortic arch
- ascending aorta
- coronary sinus
- inferior vena cava
- left pulmonary artery
- left pulmonary veins
- ligamentum arteriosum
- right pulmonary artery
- right pulmonary veins
- superior vena cava

12 _____

13 _____

14 _____

15 _____

16 _____

17 _____

18 _____

19 _____

20 _____

21 _____

FIGURE 27.3 **Great vessels of the heart,** *continued.*

C. Internal Features of the Heart

The myocardium of the anterior wall of the right atrium has a honeycombed appearance, and these myocardial ridges, called **pectinate muscles** (*pecten* = comb-like), continue into the auricles. The walls of the right and left atria are separated by the thin **interatrial septum.** In the fetus, there is a hole in the interatrial septum called the foramen ovale. The **foramen ovale** allows blood to bypass the lungs and go from the right atrium to the left atrium, forming another right-heart to left-heart shunt. The **fossa ovalis,** a connective tissue membrane, forms over and closes the fetal foramen ovale after birth.

The ventricles have ridges of muscles called **trabeculae carneae** (*trabecula* = little beam; *carnea* = flesh). The larger of these muscles, the **papillary muscles,** have string-like cords attached to them called the **chordae tendineae** (tendinous strands). The opposite ends of these cords are attached to the valves between the atria and ventricles (atrio-ventricular valves). The **interventricular septum** is a thick wall that separates the right and left ventricles.

The heart has four valves that control the one-way flow of blood through the heart: two **atrioventricular (AV) valves** and two **semilunar valves** (*semi-* = half; *lunar* = moon). Blood passing between the right atrium and the right ventricle goes through the right AV valve, the **tricuspid valve** (*tri-* = three; *cusp* = flap). The left AV valve, the **bicuspid valve,** is between the left atrium and the left ventricle. This valve clinically is called the **mitral valve** (*miter* = tall, liturgical headdress) because the open valve resembles a bishop's headdress. The two AV valves are structurally similar, except the tricuspid valve has three cusps or flaps and the bicuspid valve has two cusps or flaps that prevent blood from flowing back into the atria when the valves are closed.

Blood in the right ventricle goes through the **pulmonary (semilunar) valve** to enter the pulmonary trunk (artery). The **aortic (semilunar) valve** is located between the left ventricle and the aorta. These two semilunar valves are identical, with each having three pockets that fill with blood, preventing blood from flowing back into the ventricles when the valves are closed.

The thinner-walled atria receive the blood returning to the heart from veins. The pressure of blood in the atria opens the AV valves, and most of the blood flows into the ventricles. Both atria then contract simultaneously to pump the remaining blood into the ventricles. The larger, thick ventricular walls are double pumps that also contract simultaneously. When the ventricles contract, the papillary muscles contract and tighten the chordae tendineae. The chordae tendineae prevent the AV valve flaps from opening into the atria. During ventricular contraction, the force of blood in the ventricles pushes the semilunar valves open. Once the semilunar valves open, blood from the right ventricle flows into the pulmonary circulation as blood from the left ventricle simultaneously flows into the systemic circulation. As the ventricles relax, some blood flows back toward the ventricle, causing the semilunar valves to close. The wall of the left ventricle is thicker than the wall of the right ventricle because the left side requires more force to pump blood throughout the systemic circulation.

Before Going to Lab

1 Label the internal heart structures in Figure 27.4(b).

LAB ACTIVITY 3 Internal Features of the Heart

1 Identify each structure on a model, chart, or use the search text box in **Real Anatomy** (Cardiovascular) to locate these structures. ■

- aortic (semilunar) valve 11
- bicuspid valve (mitral) 12
- chordae tendineae 13 (CHOR-dee ten-DIN-ee)
- coronary sinus opening 4
- inferior vena cava opening 5
- interventricular (inter-ven-TRIC-u-lar) septum 14
- left atrium 10
- left ventricle
- left pulmonary vein 9 openings
- papillary (PAP-ih-lary) muscle 15
- pulmonary (semilunar) 2 valve
- right atrium 3
- right ventricle
- superior vena cava 1 opening
- trabeculae carneae 8 (tra-BEC-u-lee CAR-nee)
- tricuspid valve 6

1 _____		9 _____	
2 _____		10 _____	
3 _____		11 _____	
4 _____		12 _____	
5 _____		13 _____	
6 _____		14 _____	
7 _____		15 _____	
8 _____		16 _____	

FIGURE 27.4 Internal features of the heart.

Brachiocephalic trunk

Superior vena cava

Right pulmonary vein

Ascending aorta

RIGHT AURICLE
(cut open)

Pectinate muscles

RIGHT ATRIUM

Cusp of tricuspid valve

Chordae tendineae

Papillary muscle

RIGHT VENTRICLE

Left subclavian artery

Left common carotid artery

Arch of aorta

Ligamentum arteriosum

Pulmonary trunk

Left pulmonary vein

LEFT AURICLE

LEFT VENTRICLE

INTERVENTRICULAR
SEPTUM

TRABECULAE CARNEAE

(a) Anterior view of partially sectioned heart

Frontal plane

1
2

Fossa
ovalis
3

4

5

6

7

8

9

10

11

12

13

14

15

16

(b) Frontal section

FIGURE 27.4 **Internal features of the heart,** *continued.*

D. Systemic and Pulmonary Circulations

As you trace a drop of blood through the heart to the lungs and then to the rest of the body, you will be examining the pulmonary and systemic circulations. The **pulmonary circulation** takes blood from the right ventricle to the lungs and back to the left atrium. The **systemic circulation** takes blood from the left ventricle to the body tissues and back to the right atrium. Note that each circulation begins and ends at the heart, and each circulation is composed of arteries, capillaries, and veins.

Before Going to Lab

1 In Figure 27.5, color the vessels that are carrying oxygen-poor blood blue and the vessels carrying oxygen-rich blood red, being careful to note the color switch in the pulmonary vessels. Color the four capillary beds purple.

2 Trace the pathway of blood in Figure 27.5 through the pulmonary circulation with arrows of one color and the systemic circulation with arrows of another color, starting and ending with the right atrium.

3 Write the names of the blood vessels in the blanks under the appropriate headings.
- aorta
- pulmonary arteries
- pulmonary trunk
- pulmonary veins
- venae cavae

Blood vessels with oxygen-poor blood

a. _____

b. _____

c. _____

Blood vessels with oxygen-rich blood

d. _____

e. _____

4 Trace a drop of blood through the heart and lung by listing *in order* in blanks 2–18 all vessels, heart chambers, and valves through which the blood passes, starting and ending with the right atrium:
- aorta
- aortic (semilunar) valve
- bicuspid (mitral) valve
- left atrium
- left ventricle
- pulmonary arteries

- pulmonary capillaries
- pulmonary (semilunar) valve
- pulmonary trunk
- pulmonary veins
- right atrium
- right ventricle
- systemic arteries
- systemic capillaries
- systemic veins
- tricuspid valve
- venae cavae and coronary sinus

1. _right atrium_____

2. _____

3. _____

4. _____

5. _____

6. _____

7. _____

8. _____

9. _____

10. _____

11. _____

12. _____

13. _____

14. _____

15. _____

16. _____

17. _____

18. _____

LAB ACTIVITY 4 **Systemic and Pulmonary Circulations**

1 Make flash cards from the list of structures given in Before Going to Lab 4.

2 Mix up all the cards and assemble them in order, tracing a drop of blood through the heart and lungs and then out the systemic circulation. Start and end with the right atrium.

3 Check your answers with the class.

FIGURE 27.5 **Systemic and pulmonary circulations.**

E. Coronary Circulation

The walls of the heart have their own blood supply and circulation, the **coronary** (*corona* = crown) **circulation.** These vessels encompass the heart similar to a crown. The endothelium lining the heart chambers is too thick for blood in the chambers to supply nutrients to cardiac muscle tissue. Coronary blood vessels supply blood to cardiac muscle tissue.

On the anterior surface of the heart, the **right** and **left coronary arteries** branch off the base of the ascending aorta just superior to the aortic semilunar valve. These small arteries are supplied with blood when the ventricles are resting. When the ventricles contract, the cusps of the aortic valve open to cover the openings to the coronary arteries. If the coronary arteries were not covered, they would not be able to withstand the blood pressure and would burst like an overinflated balloon.

The **left coronary artery** is shorter than the right coronary artery. As the left coronary artery passes the base of the left auricle, it branches into the anterior interventricular branch (left anterior descending branch, or LAD) and the circumflex branch.

The **anterior interventricular branch** (LAD) supplies both ventricles with oxygen-rich blood and lies within the anterior interventricular sulcus, a shallow depression between the two ventricles. The anterior interventricular branch is commonly occluded, which can result in a myocardial infarct and at times death. The **circumflex branch** continues around the left side of the heart, lying within the coronary sulcus (atrioventricular sulcus), and supplies blood to the left ventricle and left atrium. The circumflex branch forms anastomoses (connections) with the posterior interventricular branch near the posterior interventricular sulcus.

The **right coronary artery,** located in the coronary sulcus, continues inferiorly to the right and branches into the marginal branch and posterior interventricular branch.

The **marginal branch** supplies the anterior right side of the right ventricle. The **posterior interventricular branch** lies in the posterior interventricular sulcus on the posterior surface of the heart, supplying oxygen-rich blood to both ventricles.

Arteries branch into smaller vessels, called arterioles, which penetrate the heart muscle and divide into narrower vessels, called capillaries, which deliver oxygen to the cardiac muscle. Capillaries drain into venules, which exit the heart muscle and connect to veins that receive oxygen-poor blood, returning it to the right atrium. The **great cardiac vein** is the principal vein of the coronary circulation, draining the left anterior portion of the heart. It lies near the anterior interventricular branch in the interventricular sulcus. The **small cardiac vein** drains the right anterior portion of the heart. The **middle cardiac vein,** which lies next to the posterior interventricular branch of the right coronary artery in the posterior interventricular sulcus, drains the posterior portion of the heart. Both of these veins empty into a large, thin-walled venous sinus called the **coronary sinus,** which is located on the posterior surface of the heart. The coronary sinus empties its oxygen-poor blood into the right atrium.

Before Going to Lab

1 Label the coronary vessels in Figure 27.6(c) and (d).

LAB ACTIVITY 5 Coronary Circulation

1 Identify each vessel on a model, chart, or use the search text box in **Real Anatomy** (Cardiovascular) to locate these structures. ■

Ascending aorta

Right auricle

Right coronary artery

Anterior cardiac vein

Right ventricle

Marginal branch

Anterior interventricular branch

Arch of aorta

Left pulmonary artery

Pulmonary trunk

Left auricle

Great cardiac vein

Left coronary artery

Circumflex branch

Left marginal branch

Left ventricle

Tributary to great cardiac vein

(a) Anterior view

FIGURE 27.6 Coronary circulation.

Superior vena cava

Left pulmonary artery

Right pulmonary artery

Right pulmonary veins

Left atrium

Left pulmonary veins

Right atrium

Oblique vein

Circumflex branch of left coronary artery

Inferior vena cava

Middle cardiac vein

Coronary sinus

Right ventricle

Left ventricle

Posterior interventricular branch

Posterior veins of the left ventricle

(b) Posterior view

(c) Coronary blood vessels, anterior view

(d) Coronary blood vessels, posterior view

- anterior interventricular branch (LAD)
- circumflex branch
- coronary sinus
- great cardiac vein
- left coronary artery
- marginal branch
- middle cardiac vein
- posterior interventricular branch
- right coronary artery
- small cardiac vein

1 _____ 6 _____

2 _____ 7 _____

3 _____ 8 _____

4 _____ 9 _____

5 _____ 10 _____

FIGURE 27.6 **Coronary circulation,** *continued.*

F. Pericardium and Layers of the Heart Wall

In the mediastinum, the heart is surrounded and protected by the pericardium (*peri-* = around). The pericardium consists of an outer, tough, fibrous pericardium and an inner, delicate, serous pericardium. The **fibrous pericardium** attaches to the diaphragm and also to the great vessels of the heart, securing the heart in the mediastinum. Like all serous membranes, the **serous pericardium** is a double membrane composed of an outer parietal layer and an inner visceral layer. Between these two layers is the **pericardial cavity** filled with **serous fluid.** The serous fluid reduces friction as the heart moves. The outer **parietal** (*paries* = wall) **pericardium** is attached to the fibrous pericardium, and the **inner visceral** (*viscera* = internal organs) **pericardium** covers the cardiac muscle.

The wall of the heart has three layers: the outer epicardium (*epi-* = on, upon; *cardia* = heart), the middle myocardium (*myo-* = muscle), and the inner endocardium (*endo-* = within, inward). The **epicardium** is the visceral layer of the pericardium. The majority of the heart is the **myocardium** or cardiac muscle tissue. The **endocardium** is a thin layer of endothelium deep to the myocardium that lines the inside chambers of the heart and the valves.

Before Going to Lab

1 Label the structures in Figures 27.7(a) and (b).

LAB ACTIVITY 6 Pericardium and Layers of the Heart Wall

1 Simulate the formation of the pericardium using a 1-gallon zippered plastic bag.
- Figure 27.8 shows the heart and its relationship to the pericardium. In this activity, your fist represents the heart, and the plastic bag represents the pericardium.
- Add a small amount of water to the bag and push out the extra air before sealing the bag.
- Have a lab partner place a fist (simulating the heart) at the bottom of the bag and push the fist into the bag so the bag surrounds the fist.
- Discuss with your lab group what the outer layer of the bag and the water represent. ■

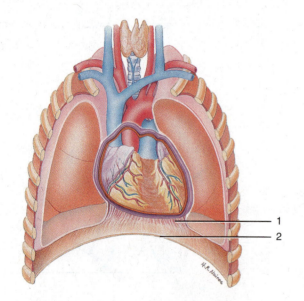

(a) Heart in the mediastinum, anterior view

- diaphragm
- fibrous pericardium (peri-CAR-dee-um)

1 _____

2 _____

(b) Pericardium and layers of heart wall

- endocardium
- fibrous pericardium
- myocardium
- parietal layer of serous pericardium
- pericardial cavity
- visceral layer of serous pericardium (epicardium)

3 _____

4 _____

5 _____

6 _____

7 _____

8 _____

FIGURE 27.7 Pericardium and layers of the heart wall.

FIGURE 27.8 **Simplified relationship of serous pericardium to heart.**

G. Histology of Cardiac Muscle

Cardiac muscle fibers in the myocardium are striated, branching cells with one or two nuclei. The ends of adjacent cardiac muscle fibers are connected by intercalated discs. The **intercalated discs** contain desmosomes that hold cardiac fibers together and gap junctions. **Gap junctions** enable action potentials to spread quickly from cell to cell, allowing cardiac muscle to contract as a unit.

Before Going to Lab

1 Label the photomicrograph of cardiac muscle fibers in Figure 27.9.

LAB ACTIVITY 7 **Histology of Cardiac Muscle**

1 Examine a prepared microscope slide of cardiac tissue or use **Real Anatomy** (Histology).

2 Draw the tissue in the space provided, and label your drawing.

LM 700×

Student drawing

- branching cardiac fibers
- cardiac muscle fiber
- intercalated (in-TER-cal-ated) discs
- nucleus

1 _____

2 _____

3 _____

4 _____

FIGURE 27.9 **Photomicrograph of cardiac muscle fibers.**

H. Dissection of a Sheep Heart

The sheep heart is similar to the human heart in both structure and size. It provides students the opportunity to observe the flexibility of the valves and tissues.

> **SAFETY NOTE:** *Wear safety glasses and gloves when using preserved tissue. Wash hands thoroughly with soap and water when you are done.*

LAB ACTIVITY 8 **Dissection of a Sheep Heart**

1 Observe the pericardial sac surrounding the sheep heart. The outer surface of the pericardium is the fibrous pericardium. Carefully cut through the pericardial sac and remove the outer fibrous pericardium. The parietal layer of the serous pericardium lines the fibrous pericardium. The visceral layer of the serous pericardium (epicardium) will remain on the surface of the heart. Identify the fibrous pericardium, parietal layer of serous pericardium, and the epicardium.

2 Carefully remove adipose tissue so you can identify surface features of the heart.

3 Identify the anterior and posterior surfaces of the sheep heart.

4 Examine the **anterior surface** of the heart. Great vessels are often cut close to the base of the heart and may be difficult to find. Refer to Figure 27.10(a) and (c) and a heart model to identify the following structures:
- epicardium
- base
- apex
- right auricle
- left auricle
- right ventricle
- left ventricle
- pulmonary trunk

5 Examine the **posterior surface** of the heart using Figure 27.10(b) and (d) and identify the following structures:
- coronary sulcus
- left auricle
- left ventricle
- right auricle
- right ventricle

6 Insert a blunt probe into the collapsed superior vena cava and into the right atrium. Maneuver the probe to find the interior opening of the inferior vena cava in the right atrium and push the probe out into this vessel.

(a) Anterior view

(b) Posterior view

FIGURE 27.10 Sheep heart.

Right auricle

Adipose tissue

Right ventricle

Pulmonary trunk

Left auricle

Anterior interventricular sulcus

Left ventricle

Apex of heart

(c) Anterior view

Aortic arch

Right auricle

Left atrium

Coronary sulcus

Adipose tissue

Left ventricle

Right ventricle

(d) Posterior view

FIGURE 27.10 **Sheep heart,** *continued.*

7 Examine the **interior of the heart** by making a coronal section of the heart. Using a knife with about a 5-inch blade, make a coronal cut of the sheep heart starting at the apex and cutting toward the base [Figure 27.10(e)]. Cut through both auricles (to ensure cutting through both atria) but not all the way through the base and the great vessels, so the two halves do not get separated. This cut allows you to observe both atria and both ventricles simultaneously (similar to a heart model) and easily compare the size of the walls of the right and left ventricles.

8 Using Figure 27.10(e), identify the following interior structures on the right side of the heart:
- myocardium
- endocardium
- right atrium
- right auricle
- pectinate muscle
- opening of superior and inferior vena cava
- opening of the coronary sinus
- tricuspid valve
- right ventricle
- chordae tendinae
- papillary muscles
- moderator band (cord between the two walls of the right ventricle)
- interventricular septum
- pulmonary trunk
- pulmonary semilunar valve

9 In the right atrium, insert a blunt probe in the small opening of the coronary sinus that is inferior to the opening of the inferior vena cava. Observe the movement of the probe in the coronary sinus from the posterior view of the heart.

10 In the right ventricle, insert a blunt probe into the opening of the **pulmonary trunk** and push it through to the superior end of the vessel. Remove the probe and take a scalpel to cut the wall of the pulmonary trunk longitudinally to expose the **pulmonary semilunar valve.** Count the three cusps. How does this valve differ from the tricuspid valve?

11 Continue identifying structures on the left side of the heart:
- left atrium
- left auricle
- bicuspid valve
- left ventricle
- aortic semilunar valve
- aorta

12 How many cusps does the bicuspid valve have? _____ Does this valve look similar otherwise to the other AV valve? _____ Are there **chordae tendineae** and **papillary muscles?** _____ Does the left ventricle have a greater or smaller number of papillary muscles compared with the right side? _____ Compare the thickness of the right and left ventricles. Which one is thicker?_____ Why? _____

13 Look just above the cusps of the aortic valve for the openings to the **right** and **left coronary arteries.** Use the blunt probe to push into these small vessels.

14 Clean up as directed by your instructor. ◼

Right auricle

Right atrium

Wall of right ventricle

Tricuspid valve

Interventricular septum

Ascending aorta

Left atrium

Aortic valve

Bicuspid valve

Chordae tendineae

Wall of left ventricle

Papillary muscle

Trabeculae carneae

Apex

(e) Coronal section

FIGURE 27.10 Sheep heart, *continued.*

Reviewing Your Knowledge

A. Major Heart Structures and Great Vessels

Fill in the blank with the term that fits the description.

_____ 1. Arteries that supply blood to cardiac muscle

_____ 2. Layer of heart wall containing cardiac muscle

_____ 3. Extensions of the atria

_____ 4. Heart is located here (area between the lungs)

_____ 5. Lines the heart chambers

_____ 6. Pointed inferior part of the heart

_____ 7. Two heart pumps; lower heart chambers

_____ 8. Superior heart chambers

_____ 9. Another name for visceral pericardium

_____ 10. Atria compose this on the posterior surface of heart

_____ 11. Blood pumped by right ventricle (oxygen-rich or oxygen-poor)

_____ 12. Blood pumped by left ventricle (oxygen-rich or oxygen-poor)

_____ 13. Enlarged muscles in ventricles attached to chordae tendinae

_____ 14. Muscle ridges in ventricles

_____ 15. Ridges in anterior wall of right atrium

_____ 16. Strings attached to AV cusps

_____ 17. Blood vessel that returns blood from the body inferior to the heart

_____ 18. The valve located between the left ventricle and the aorta

_____ 19. A membrane between the atria that closes after birth

_____ 20. The valve located between the right atrium and right ventricle

B. Coronary Circulation

After reviewing the coronary circulation in Figure 27.5, fill in the blank with the word that fits the description.

_____ **1.** Anterior branch of the left coronary artery; lies in anterior interventricular sulcus

_____ **2.** Posterior branch of the right coronary artery; lies in posterior interventricular sulcus

_____ **3.** Branch of the left coronary artery that curves around left side and lies in the coronary sulcus

_____ **4.** Branch of right coronary artery that supplies the anterior right ventricle

_____ **5.** Branch of left coronary artery that supplies both ventricles

_____ **6.** Shorter coronary artery that is hidden anteriorly by pulmonary trunk

_____ **7.** Vein that drains coronary circulation into right atrium

_____ **8.** Vein that drains most of anterior ventricles

_____ **9.** Vein that drains the posterior ventricles

_____ **10.** Vein that drains the right anterior side

C. The Heart and Pulmonary Circulation

Place the following structures *in order,* tracing the blood flow from the *superior vena cava* to the heart, to the pulmonary circulation, and out of the heart to the systemic circulation. Start with superior vena cava as number 1.

- aorta
- aortic valve
- bicuspid valve
- left atrium
- left ventricle
- pulmonary capillaries
- right atrium
- right ventricle
- pulmonary arteries
- pulmonary trunk
- pulmonary valve
- pulmonary veins
- superior vena cava
- tricuspid valve

1. _____

2. _____

3. _____

4. _____

5. _____

6. _____

7. _____

8. _____

9. _____

10. _____

11. _____

12. _____

13. _____

14. _____

Using Your Knowledge

A. The Heart

1. What is the function of the pericardial fluid?

2. If a person has pulmonary valve stenosis, which circulation is affected the most: pulmonary, systemic, or cardiac circulation?

3. A blood clot (thrombus) in which artery of the coronary circulation would be more likely to cause sudden death? (*Circle your answer*.) Explain your choice.

 a. right coronary artery **b.** posterior interventricular branches

 c. circumflex branch **d.** anterior interventricular branch

4. The pulmonary trunk and pulmonary arteries are colored blue in all of the figures. Why are they colored blue like most veins?

5. If the foramen ovale does not fully close after birth, what will happen?

6. If a person has endocarditis, myocarditis, or pericarditis, what areas (*in order*) are infected?

 a. _____ **b.** _____ **c.** _____

7. Athletes have thicker heart walls than non-athletes. Which layer of the heart enlarges? Explain.

8. Atrioventricular valves and semilunar valves have different structures. Explain if high pressure from *above* or *below* each type of valve opens the valves.

 Atrioventricular valves

 Semilunar valves

9. Which ventricular wall is thicker? What is the significance of this?

10. The heart has many anastomoses. What benefit are these to the heart?

11. What is the *significance* of the difference in the vessel wall thickness between the aorta and the pulmonary trunk?

12. Using your knowledge, explain why mitral valve prolapse occurs more often than tricuspid prolapse.

13. How would heart function be affected if the AV valves did not completely close during ventricular contraction?

B. Coronary Circulation

Trace a drop of blood from the ascending aorta to the right coronary artery, and trace the blood back into the heart again.

- ascending aorta
- coronary sinus
- middle cardiac vein
- posterior interventricular branch

- right atrium
- right coronary artery
- right ventricular capillaries

14. _____

15. _____

16. _____

17. _____

18. _____

19. _____

20. _____

Cardiac Cycle

OBJECTIVES

1 Identify the parts of the cardiac conduction system

2 Describe the pathway of action potentials through the cardiac conduction system and explain its association with contractions of the atria and ventricles

3 Identify the components of a normal electrocardiogram (ECG)

4 Describe the association of ECG tracings with electrical events occurring in the heart

5 Identify normal sinus rhythm, tachycardia, and bradycardia on an ECG

6 Describe the relationship of auscultated heart sounds, pulse rate, and heart rate

7 Calculate changes in length of cardiac cycle with exercise and discuss their significance

8 Describe the effect of exercise on cardiac output

MATERIALS

- model or chart of the heart
- **Electrocardiography:** ECG recording equipment (if available), electrodes, cream or jelly
- **Heart Sounds:** stethoscope, alcohol swabs, skeleton, stickers
- **Length of Cardiac Cycle:** stethoscope, alcohol swabs, stopwatch or clock with second hand, calculators
- **PowerPhys Experiment:** Effect of Exercise on Cardiac Output
- **Biopac Laboratory Guide Experiments:**
 - Effect of Body Position and Exercise on ECG
 - Heart Sounds and Events of the Cardiac Cycle

Each cardiac cycle, or one complete heart beat, consists of atrial and ventricular systole (contraction) and diastole (relaxation). The generation of an action potential by special cardiac muscle cells initiates atrial systole followed by ventricular systole. During systole, blood pressure within the heart chamber increases and blood is ejected. Blood flows from areas of higher pressure to areas of lower pressure. During diastole, blood pressure within the heart chambers is low and blood fills the chambers. Flow of blood through the heart is controlled by opening and closing of the AV and semilunar valves.

A. Stimulation of Cardiac Muscle Contraction

Your heart beats without any stimulation from nerves, and if removed from your body, the heart would still beat. The internal stimulation that makes the heart beat by itself is called **intrinsic** (*intrinsic* = inside) **stimulation** and is caused by specialized, noncontractile cells called **autorhythmic cells** (causing a rhythm by themselves).

1. Electrical Conduction System of the Heart

Autorhythmic cells belong to the intrinsic **conduction system** (nodal system) that (1) initiates the action potential that results in contraction of cardiac muscle fibers and (2) provides a pathway for conducting the action potential to all cardiac muscle fibers. The autonomic nervous system and hormones, known as **extrinsic** (*extrinsic* = on the outside) **stimulation,** only increase or decrease the intrinsic pace.

The parts of the conduction system are:

1. **Sinoatrial (SA) node**—called the pacemaker because it initiates action potentials first; located in the wall of the right atrium just inferior to the opening of the superior vena cava; stimulates the atria to contract.

2. **Atrioventricular (AV) node**—receives action potentials from the atrial muscle fibers; located in the lower interatrial septum anterior to the opening of the coronary sinus; sends the action potentials to the AV bundle (bundle of His).

3. **AV bundle (bundle of His)**—located in a membranous septum between the atria and ventricles superior to the interventricular septum; this is the electrical connection between the atria and ventricles; sends action potentials to the bundle branches.

4. **Right and left bundle branches**—located in the interventricular septum; sends action potentials to the Purkinje (conduction) fibers.

5. **Purkinje (conduction) fibers**—located in the apex of the myocardium, as well as in the lateral walls of the right and left ventricles; sends action potentials to the ventricular cardiac muscle fibers and papillary muscles and stimulates them to contract.

Before Going to Lab

1 Label the structures of the conduction system in Figure 28.1.

LAB ACTIVITY 1 Cardiac Conduction System

1 On a model, chart, or **Real Anatomy** (Cardiovascular), identify the location of conduction system structures. ■

- atrioventricular (AV) node 4
- AV bundle 3 (bundle of His)
- left bundle branch 6
- Purkinje (conduction) fibers in left ventricle 7
- Purkinje (conduction) fibers in right ventricle 1
- right bundle branch 2
- sinoatrial (SA) node 5

1 _____
2 _____
3 _____
4 _____
5 _____
6 _____
7 _____

FIGURE 28.1 Cardiac conduction system.

2. Electrocardiography

The conduction of action potentials throughout the heart can be detected as electrical currents. An instrument called an **electrocardiograph** is used to record the electrical changes in the heart. The chart recording of the electrical events that occur before each heartbeat is called an **electrocardiogram** (ECG; originally called an EKG). The electrical events that are recorded are from the entire heart muscle (i.e., the atria and the ventricles) and not just the conduction system activity. The ECG records only voltage changes over time and not the force of contraction. Cardiac muscle will contract after the action potential has occurred.

ECGs are usually recorded by *indirect leads,* with electrodes placed on the skin of the subject's arms and legs rather than on the heart itself. The electrical impulses generated by the depolarization and repolarization of the atria and ventricles are detected on the body surface due to the conductivity of the ions in extracellular fluid. Clinically, 12 leads are used to record an ECG for diagnosing abnormalities of the conduction system, myocardial infarctions, and other clinical situations. However, in anatomy and physiology classes, usually a standard three-lead ECG is utilized (Figure 28.2).

The main parts of a standard three-lead ECG are:

- **P wave**—first wave; small, curved upward deflection; represents **atrial depolarization** that spreads from the sinoatrial (SA) node just before the atria contract.
- **QRS complex**—short downward deflection (Q), tall upward deflection (R), medium downward deflection (S); represents **ventricular depolarization** that spreads from the AV node to the AV bundle, to right and left bundle branches, and to the Purkinje (conduction) fibers just before the ventricles contract.

> **NOTE: Atrial Repolarization** *takes place during the QRS complex but gets overshadowed by the more prominent ventricular depolarization.*

- **T wave**—medium, curved upward deflection; represents **ventricular repolarization** and occurs just before the ventricles relax.

In addition to these three waves, an ECG also contains intervals or segments between the waves:

- **P-Q interval**—the interval between the beginning of the P wave until the beginning of the Q (in the QRS wave); some texts call this the P-R interval; it represents the time it takes for the electrical conduction (excitation) to travel through the atria and AV node to the Purkinje (conduction) fibers.
- **S-T segment**—the segment from the end of the S (in QRS wave) to the beginning of the T wave; it represents the time the ventricular fibers are fully depolarized.
- **Q-T interval**—the interval that begins at the Q (in the QRS wave) to the end of the T wave; it represents the time from the beginning of ventricular depolarization until the end of ventricular repolarization.

FIGURE 28.2 Standard three-lead positions.

LAB ACTIVITY 2 **Electrocardiography**

1 Record an ECG at rest and after exercise if equipment is available.

- Do this activity with a lab partner. Choose who will be the subject and who will do the recording.
- Swab the skin where the electrodes will be placed with alcohol, and then apply electrode paste, jelly, or cream to the same area (see Figure 28.2).
- Attach the electrodes to the anterior surface of each wrist and to the inside area of each ankle.
- Connect the patient cables to the electrodes and also to the recording instrument.
- Follow your instructor's directions for recording the ECG with your equipment.
- Label the P wave, QRS complex, T wave, P-Q interval, S-T segment, and Q-T interval on the recorded ECG.

2 Complete **Biopac Laboratory Guide Experiment**: Effect of Body Position and Exercise on ECGs. ■

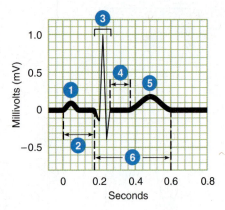

- P wave
- P-Q interval
- QRS complex
- Q-T interval
- S-T segment
- T wave

1 _____

2 _____

3 _____

4 _____

5 _____

6 _____

FIGURE 28.3 **Section of a normal ECG, lead II.**

3. Determining Heart Rate Using an ECG

In an adult, a heart rate of 60 to 100 beats/min is a **normal sinus rhythm** (NSR). A heart rate above 100 beats/min is called **tachycardia,** but in young children, this rate would be considered normal. Heart rates below 60 beats/min are normal for highly conditioned individuals, but in other adults, heart rates below 60 beats/min are called **bradycardia.** Neither condition is considered to be pathological. Prolonged tachycardia can develop into **ventricular fibrillation,** rapid uncoordinated heart contractions that do not pump blood.

LAB ACTIVITY 3 **Heart Rate Calculations**

1 Calculate the heart rate for the ECGs in Figure 28.4(a), (b), and (c).

- Heart rate can be easily calculated from an ECG. Standard ECGs are printed on paper moving at a paper speed of 25 mm/sec. Therefore, the distance of 5 mm (1 large square on standard ECG paper) is equivalent to 0.20 sec and the distance covered by 30 large squares is equivalent to 6 sec.
- Count the number of P waves or R waves in 30 large squares. Multiply this number by 10 to obtain the number of contractions (heartbeats) per minute. This gives you an approximate heart rate.

Figure 28.4(a) = _____ beats/min

Figure 28.4(b) = _____ beats/min

Figure 28.4(c) = _____ beats/min

NOTE: The heart rate calculated using the number of R waves is the ventricular rate, while the heart rate calculated using the number of P waves is the atrial rate. In normal sinus rhythm (NSR), the atrial rate equals the ventricular rate.

2 Determine which ECG in Figure 28.4 illustrates the following heart rates and write the term next to the figure letter.

- normal sinus rhythm Figure 28.4(a) _____
- tachycardia Figure 28.4(b) _____
- bradycardia Figure 28.4(c) _____

3 Calculate the heart rate before and after exercise on the ECGs your lab group recorded in Lab Activity 2.

- Heart rate before exercise = _____ beats/min
- Heart rate after exercise = _____ beats/min

(a)

(b)

(c)

FIGURE 28.4 **Tracings of ECGs, lead II.**

B. Opening and Closing of Heart Valves

Opening of the atrioventricular (AV) valves occurs when the blood pressure in the atria exceeds the blood pressure in the ventricles. This occurs during atrial and ventricular diastole. The semilunar valves open during ventricular systole at the point when the blood pressure in the ventricles exceeds the blood pressure in the pulmonary trunk and aorta.

Closing of the AV valves is due to the flow of blood back toward the atria that occurs at the beginning of ventricular systole. Closure of the semilunar valves is due to the flow of blood back toward the ventricles at the beginning of ventricular diastole.

Closure of the AV and semilunar valves causes the heart sounds heard by auscultation. **Auscultation** (*auscult-* = listening) means listening to body sounds, typically using a stethoscope. Two particular **heart sounds** can be detected during one heartbeat. These sounds occur after the heart valves quietly close and blood strikes against the closed valve, causing turbulence that we can hear with a stethoscope. Although there are four sounds generated during one heartbeat, only the first and second sound can easily be heard without additional amplification. The first sound, **lubb,** is a little longer and louder than **dupp** (or **dubb**), the second sound that occurs shortly after the first. The first sound (S1) occurs with blood turbulence from the closure of the two atrioventricular (AV) valves at ventricular systole.

The second sound (S2) is at ventricular diastole when the two semilunar valves close. The sounds that you will hear are lubb-dupp, pause…lubb-dupp, pause. Remember that these two heart sounds, lubb-dupp, equal one heartbeat.

> **CLINICAL NOTE: Heart Murmurs** *are abnormal heart sounds caused by the restriction of blood flow or backflow of blood that typically indicate an abnormality in valve structure or function. It usually takes special training and a very quiet room to be able to detect heart murmurs.*

Before Going to Lab

1 Find a website that provides heart sounds and listen to them.

2 On yourself, find the location to auscultate the bicuspid (mitral) valve (5th intercostal space just left of the nipple) in the mid-clavicular line and the aortic valve (2nd intercostal space just right of the sternum). Refer to Figure 28.5.

LAB ACTIVITY 4 Heart Sounds

1 On an articulated skeleton, place a sticker on the location where you will auscultate the mitral valve and a sticker on the location where you will ausculate the aortic valve. Red circles on Figure 28.5 are the auscultation locations.

2 Preparation for auscultation of heart sounds.
- Obtain a stethoscope and alcohol swabs.
- Clean the earpieces with an alcohol swab.
- Point earpieces forward to facilitate hearing heart sounds.
- Put the earpieces in place and gently tap the bell to check that you can hear sounds.

3 Auscultate the mitral valve (Figure 28.5).
- Auscultate heart sounds on your chest or your lab partner's while seated.
- To clearly hear the first heart sound (lubb), you will auscultate the mitral valve. Place the large bell of the stethoscope at the apex of the heart. This is located in the 5th intercostal space just left of the nipple.
- If the heart sound is difficult to hear, have the subject bend forward to tip the heart against the anterior thoracic wall and try the auscultation again.

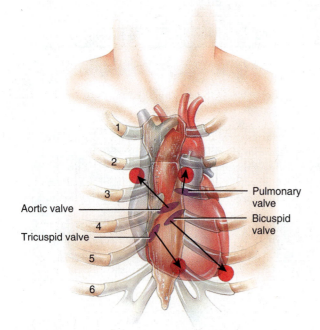

Aortic valve

Tricuspid valve

Pulmonary valve

Bicuspid valve

FIGURE 28.5 Locations for auscultation of the heart.

4 Auscultate the aortic valve (Figure 28.5).
- Palpate the suprasternal (jugular) notch and then the sternal angle (marks the insertion of the 2nd rib).
- Now drop down another 1 inch on the sternum from the sternal angle and then move right 1 inch.
- You should be in the 2nd intercostal space just to the right of the sternum where you can clearly hear dupp, or the second heart sound.

5 Answer Discussion Question with your lab group.

6 Complete **Biopac Laboratory Guide Experiment:** Heart Sounds and Events of the Cardiac Cycle.

DISCUSSION QUESTION
Heart Sounds

1 At rest, which heart sound is louder, lubb or dupp? Explain.

C. Cardiac Cycle Length and Cardiac Output

Heart rate is the number of heartbeats per minute. Each heartbeat represents one **cardiac cycle,** which includes atrial and ventricular **systole** (contraction) and atrial and ventricular **diastole** (relaxation). The heart rate can be measured by counting the number of heartbeats per minute or by the number of pulses per minute. A **pulse** is the blood pressure wave that travels through the arteries when the ventricles contract. Pulses are commonly felt in the radial and carotid arteries. The number of heartbeats per minute (heart rate) will be very close to, but not necessarily equal to, the number of pulses per minute.

The length of one cardiac cycle in seconds can be calculated by dividing 60 seconds by the heart rate (heartbeats/ min). Therefore the length of the cardiac cycle varies with heart rate.

Cardiac output is equal to heart rate × stroke volume. An increase in stroke volume results in more blood being ejected from the ventricles and causes an increase in the blood pressure wave traveling through arteries. Therefore an increase in stroke volume is reflected by an increase in the force of the pulse.

LAB ACTIVITY 5 **Experiment: Effect of Exercise on Cardiac Cycle Length and Cardiac Output**

1 **Prediction:** Exercise causes the length of the cardiac cycle (seconds) to *increase* or *decrease*. (*Circle the correct choice in italics.*)

2 **Materials:** Obtain materials for Length of Cardiac Cycle (see Materials list).

3 Locate the radial and carotid pulses.
- **Radial pulse:** Locate the groove in your lab partner's wrist between the radius and the tendon just medial to the radius. Palpate the radial artery by pressing down with your index and middle fingers. Because the thumb has a noticeable pulse of its own, it is not used.
- **Carotid pulse:** Locate your lab partner's Adam's apple (thyroid cartilage of larynx). Using your index and middle fingers again, palpate the carotid artery on either side of the larynx.

4 **Data Collection:** Measure heart rate and pulse rate at rest and after exercise and record your results in Table 28.1. Compare force of pulse before and after exercise.
- Decide who will be the subject, who will count heartbeats and pulses, and who will be the recorder and the timer. The subject should not have heart problems and should be in good health.
- Locate the radial and carotid pulses on the subject and decide which pulse will be used for this experiment.

Heart beats and pulses at rest
- Listen for the subject's heartbeat and count the number of heartbeats in 15 seconds. Record your data in Table 28.1.
- Count the number of pulses in 15 seconds and record your data in Table 28.1.
- Feel the force of each pulse at rest.

Heart beats and pulses after exercise
- Have the subject run in place for 2 minutes.
- Immediately count the number of heartbeats and pulses in 15 seconds and record your data in Table 28.1.

TABLE 28.1 Heart Rate and Pulse Rate Before and After Exercise

ACTIVITY	BEATS/15 SEC	BEATS/MIN
Heart rate at rest		
Pulse rate at rest		
Heart rate after exercise		
Pulse rate after exercise		

- Repeat the pulse count every minute until the pulse rate returns to the initial resting rate. Record the time for recovery: _____ minutes.
- Did exercise cause the force of pulse to increase or decrease?

5 Data Analysis:

- Calculate the resting and exercising heart rates (beats/min) and record them in Table 28.1. Multiply the number of heartbeats measured in 15 seconds by 4 to convert to beats/min.
- Calculate the resting and exercising pulse rates (pulses/min) and record them in Table 28.1. Multiply the number of pulses measured in 15 seconds by 4 to convert to pulses/min.
- Calculate resting and exercising cardiac cycle lengths and record them in Table 28.2.

$$\frac{60 \text{ sec}}{\text{\# of heartbeats/min}} = \text{length of 1 cardiac cycle in sec}$$

- Calculate the length of diastole and systole at rest and record in Table 28.2.
 At rest, systole = 1/3 of cardiac cycle.
 At rest, diastole = 2/3 of cardiac cycle.
 The change in length of systole and diastole with exercise cannot be calculated with a simple formula. The length of systole and diastole both decrease with exercise.

6 Clean up as directed by your instructor.

7 Complete Experimental Report with your lab group. Use your textbook for reference.

8 Complete **PowerPhys Experiment:** Effect of Exercise on Cardiac Output.

TABLE 28.2 Cardiac Cycle Length

AT REST	LENGTH/SECONDS
Cardiac cycle	
Systole	
Diastole	
AFTER EXERCISE	
Cardiac cycle	

EXPERIMENTAL REPORT
Cardiac Cycle Length and Cardiac Output

Results:

- State whether heart rate and pulse rate *increase* or *decrease* immediately after exercise.
- State whether heart rate and pulse rate at rest are the same and whether exercising heart rate and pulse rates are the same.
- State the time for recovery after exercise.
- State whether the length of the cardiac cycle increases or decreases immediately after exercise.
- State whether the force of pulse increased or decreased with exercise.
- Based on the force of pulse, did stroke volume increase or decrease with exercise?
- Based on your heart rate and force of pulse, did cardiac output increase or decrease with exercise?

Discussion:

- Discuss what causes the change in heart rate and pulse rate with exercise.
- Discuss why heart rate equals pulse rate at rest and after exercise.
- Discuss how the change in cardiac cycle length with exercise affects heart muscle perfusion and time for ventricular filling.
- Discuss factors that affect pulse rate and recovery time.
- Discuss how changes in filling time and venous return with exercise affect end diastolic volume (EDV).
- Discuss how changes in EDV and force of contraction with exercise affect stroke volume.

Conclusion:

- Write a statement that describes the correlation among heart rate, length of the cardiac cycle, and length of diastole.
- Write a statement that correlates changes in pulse rate, force of pulse, and cardiac output.

Reviewing Your Knowledge

A. Heart Sounds

Fill in the blank with the correct term.

1. The first heart sound heard is _____ that is due to blood hitting against the _____ valves.

2. The second heart sound heard is _____ that is due to blood hitting against the _____ valves.

3. What is a heart murmur?

4. Which heart sound is the loudest sound when auscultated, the lubb or dupp? Why?

5. Valerie's radial pulse rate is 70/min. Assuming her heart and arteries are in good health, what is her approximate heart rate (beats/min)?

6. Jaxton's heart rate is 68/min. What is the length of his cardiac cycle in seconds?

B. Electrical Conduction System of the Heart

Circle True or False for the following questions.

1. The heart beats without extrinsic stimulation from the autonomic nervous system. True or False

2. Autorhythmic cells are located only in the interventricular septum. True or False

3. Each heart chamber contracts separately, first the right atrium, then the right ventricle, left atrium, and left ventricle. True or False

4. The ECG records the electrical stimulation of cardiac muscle by the conduction system and not the contraction of the muscle itself. True or False

5. The normal pacemaker of the heart is the AV node. True or False

6. Number the following structures of the cardiac conduction system in the normal order of depolarization (1 to 5).

_____ AV bundle (of His)
_____ AV node
_____ Purkinje (conduction) fibers
_____ Right and left bundle branches
_____ SA node

C. Electrocardiography

Match the following descriptions with the correct terms. A term may be used more than once.

> bradycardia
> fibrillation
> P wave
> QRS complex
> T wave
> tachycardia

_____ **1.** depolarization of the atria

_____ **2.** depolarization of the ventricles

_____ **3.** fast heart rate above 100 beats/min

_____ **4.** repolarization of the atria

_____ **5.** repolarization of the ventricles

_____ **6.** slow heart rate below 60 beats/min

_____ **7.** uncontrolled, rapid heart contractions

Measurements Using the ECG

Using the ECG tracing in Figure 28.4(b), calculate the:

8. QRS complex length: _____ sec

9. P-Q interval: _____ sec

10. Q-T interval: _____ sec

11. Heart rate: _____ beats/min

Using Your Knowledge

A. Auscultations of Heart Sounds

1. Sam counted his lab partner's heartbeats as 12 beats in 15 seconds. What is the lab partner's heart rate? What other information would you like to have about Sam's lab partner to decide if this heart rate is abnormal?

B. Electrical Conduction System

2. Larry had damage to his SA node. What will happen now that the "pacemaker" is not functional?

3. In a normal cardiac cycle, what two chambers of the heart contract first?

4. What two structures directly receive an electrical impulse from the SA node?

5. Valerie's heart rate is 80 beats/min. What is the length of one cardiac cycle?

C. Electrocardiography

6. Fibrillation is an asynchronous contraction of cardiac muscle fibers. Is atrial fibrillation or ventricular fibrillation more serious? Explain.

7. If a myocardial infarction damaged the left bundle branch, how would this affect the spread of the action potential?

8. How would damage to the left bundle branch affect blood flow in the systemic circulation?

Use the abnormal ECG tracing in Figure 28.6 to answer questions 9 and 10.

9. Does the atrial rate equal the ventricular rate in this ECG tracing? Explain.

10. Would you be able to feel a pulse during atrial systole only? Explain.

FIGURE 28.6 Tracing of abnormal ECG, lead II.

Blood Vessel Structure and Function

OBJECTIVES

1 Compare and contrast the structure of arteries, capillaries, and veins
2 Identify layers or tunics of an artery and vein
3 Measure systolic and diastolic blood pressure at rest and after exercise
4 Discuss how exercise and body position affect blood pressure

MATERIALS

- compound microscope, lens paper, prepared microscope slide of transverse section through artery and vein
- **Blood Pressure Measurements:** sphygmomanometer, stethoscope, alcohol wipes, meterstick, exercise mat
- **Effect of Exercise on Blood Pressure:** sphygmomanometer, stethoscope, alcohol wipes, stopwatch or watch with second hand, graph paper
- **PowerPhys Experiment:** Effect of Exercise on Blood Pressure and Vascular Resistance
- **BIOPAC** Systems, Inc. WWW.BIOPAC.COM **Biopac Laboratory Guide Experiment:**
 - Effect of Body Position on Resting Blood Pressure
 - Effect of Exercise on Blood Pressure

Arteries carry blood away from the heart and divide into smaller vessels called **arterioles** that branch into the tiniest vessels called **capillaries.** At the capillary level, an exchange of nutrients, wastes, and gases occurs between blood and interstitial fluid (fluid surrounding tissue cells). Capillaries join to form small **venules** that will merge to form larger **veins** that carry blood back to the heart. The structure of these vessels reflects their functions.

There are two main circulatory routes within the body: the systemic circulation and the pulmonary circulation. In the **systemic circulation,** arteries carry oxygen-rich blood to the body tissues, and veins return oxygen-poor blood to the heart where it will then enter the pulmonary circulation. In the **pulmonary circulation,** arteries carry oxygen-poor blood from the right ventricle to the lungs where gas exchange occurs, and veins return oxygen-rich blood back to the left atrium.

A. Blood Vessel Structure

1. Arteries

Arterial walls have three layers or tunics: the tunica externa, tunica media, and tunica interna. The outer **tunica externa** is composed mainly of elastic and collagen fibers (proteins) that provide support and protection. The **tunica media** is the middle and thickest layer that contains elastic fibers and smooth muscle fibers (cells) encircling the diameter of the vessel. The sympathetic nervous system regulates the diameter of the blood vessel by innervating smooth muscle fibers. Contraction of smooth muscle fibers causes a decrease in lumen diameter or **vasoconstriction,** whereas relaxation causes **vasodilation,** an increase in lumen diameter. Elastic fibers in the tunica media allow the vessel to be stretched and return to its original shape. The **tunica interna (intima)** contains simple squamous

epithelium (**endothelium**), a basement membrane, and elastic tissue (**internal elastic lamina**) that is adjacent to the tunica media. The endothelium lines the lumen of all blood vessels and forms a smooth, slick barrier that promotes blood flow. Damage to the endothelium causes platelets to stick to the blood vessel wall, forming plaques that narrow the lumen and impede blood flow.

The thickness and exact composition of each tunic varies according to vessel type. **Elastic arteries,** such as the aorta, are large-diameter arteries that have more elastic fibers in their tunica media than other arteries. Elastic fibers stretch under greater pressure when blood is pumped into them during systole. During diastole, the recoiling of elastic fibers in elastic arteries continues blood flow. **Muscular arteries,** smaller in diameter than elastic arteries, have more smooth muscle and fewer elastic fibers in their tunica media than elastic arteries. However, they have a layer of elastic tissue, the **external elastic lamina,** between the tunica externa and the tunica media. Muscular arteries distribute blood to different areas of the body (e.g., brachial artery distributes blood to arm) and control blood flow to these areas by vasoconstriction or vasodi-lation. Muscular arteries branch into smaller and smaller arteries that eventually form arterioles, small blood vessels from which capillaries branch. **Arterioles** have a tunica media containing mainly smooth muscle and few elastic fibers. These vessels control blood flow into capillaries and play a large role in controlling blood pressure.

2. Veins

Blood flows from capillaries into **venules** (*venule* = little vein) that drain into veins. Venule walls contain a tunica interna of endothelium only and a tunica media with a few smooth muscle fibers.

The walls of **veins** contain a tunica interna, a tunica media, and a tunica externa. Compared with arterial vessels, the tunica interna and tunica media are thinner, and the tunica media contains fewer smooth muscle fibers and elastic fibers. Veins lack an internal and external elastic lamina. The tunica externa is the thickest of the three layers and like arterial vessels is composed of elastic and collagen fibers. The lumen of veins is larger than arteries and often appears collapsed in tissue sections.

(a) (b)

FIGURE 29.1 **Transverse sections through a frog arteriole.**

The blood pressure gradient in the veins is very small and often is not enough to overcome gravity. Muscular activity squeezes the veins and pushes blood toward the heart while venous **valves,** found in many veins, prevent the backflow of blood, especially in the limbs. Valves are tunica interna folds whose flap-like cusps point toward the heart. Faulty or incompetent valves allow backflow of blood and result in pooling of blood in veins, a condition called varicose veins.

Before Going to Lab

1 In Figure 29.1 compare the lumen diameter, thickness of tunics, and shape of tunica interna in two sections of a frog arteriole. Norepinephrine (NE) was applied to one area of the arteriole to induce vasoconstriction. Identify which cross-section is vasoconstricted, (a) or (b).

2 Label the components of a muscular artery wall in Figure 29.2.

3 Label the components of the wall of a vein in Figure 29.3.

- basement membrane
- endothelium
- external elastic lamina
- internal elastic lamina
- smooth muscle
- tunica externa
- tunica interna
- tunica media

1 _____

2 _____

3 _____

4 _____

5 _____

6 _____

7 _____

8 _____

FIGURE 29.2 Structure of a muscular artery.

- basement membrane
- endothelium
- tunica externa
- tunica interna
- tunica media (smooth muscle)
- valve

1 _____

2 _____

3 _____

4 _____

5 _____

6 _____

FIGURE 29.3 Structure of a vein.

Before Going to Lab

1 Identify and label the tunics of the blood vessels in Figure 29.4.

LAB ACTIVITY 1 Arteries and Veins

1 Examine a prepared microscope slide containing a transverse section of a small artery and vein.
- Using the low-power objective lens, focus and center one blood vessel in the field of view and switch to high power.
- Using the high-power objective lens, focus with fine adjustment knob only and examine the three tunics in the vessel. Note the thickness of the tunica media.
- Repeat procedure with other blood vessel.
- Decide which vessel is the artery and which is the vein. ■

- lumen of artery
- lumen of vein
- tunica externa of artery
- tunica externa of vein
- tunica interna of artery
- tunica interna of vein
- tunica media of artery
- tunica media of vein

1 _____

2 _____

3 _____

4 _____

5 _____

6 _____

7 _____

8 _____

FIGURE 29.4 Transverse section through small artery and vein.

3. Capillaries

Capillaries have the smallest diameter and thinnest walls of any blood vessels. Their lumen is so small that red blood cells can pass through only one at a time. Capillary walls are composed of a single layer of endothelial cells (simple squamous endothelium) supported by a basement membrane. The endothelial cells of capillaries can exhibit:

- **tight junctions**—fusion of plasma membranes of adjacent cells
- **intercellular clefts**—spaces between cells
- **fenestrations** (*fenestra* = window)—holes in plasma membrane covered by a diaphragm (thin membrane)
- **vesicles**

Only substances that are lipid soluble or have membrane carriers can cross capillary walls with tight junctions between cells. However, many other substances can cross capillary walls containing fenestrations, intercellular clefts, and vesicles. The size of the fenestrations and intercellular clefts between endothelial cells determines which molecules can cross. Molecules can also be transported across capillary walls by vesicles, a process called **transcytosis** (*trans-* = across; *cyto* = cell).

Hydrostatic pressure (blood pressure) at the arterial end of the capillaries causes plasma to be filtered across the capillary wall into the interstitial fluid. The filtered plasma containing nutrients, hormones, and other substances is called interstitial fluid once it leaves the capillaries and enters the space around tissue cells. Blood osmotic pressure draws interstitial fluid into the venous end of the capillary (reabsorption) bringing fluid, metabolic wastes, hormones, or other substances into the blood. The combination of these opposing pressures causes some filtered plasma to remain in the interstitial fluid to bathe the tissue cells.

Before Going to Lab

1 Label the blood vessels in Figure 29.5. Identify the arteriole and a capillary.

2 Observe the capillaries in Figure 29.6(a), (b), and (c) and label the capillary structures.

- arteriole 1 _____
- capillary 2 _____

FIGURE 29.5 **Photomicrograph showing capillary network.**

(a)

- fenestration 1 _____
- intercellular cleft
- tight junctions 2 _____
- vesicles involved in trans-
 cytosis. 3 _____

 4 _____

(b)

19,000×

(c)

FIGURE 29.6 **Transmission electron micrographs showing capillary walls.**

B. Blood Pressure

Blood pressure, the pressure exerted by blood against blood vessel walls, is highest in the aorta and large elastic arteries, and decreases as the arteries branch and blood travels farther from the heart. Blood pressure drops significantly in the arterioles and steadily decreases through capillaries, venules, and veins, and drops to zero in the right atrium. The difference in blood pressure between two areas of the circulatory system is the **blood pressure gradient.** The blood pressure gradient between the aorta and right atrium causes blood to flow through the systemic circulation, and the blood pressure gradient between the pulmonary trunk and the left atrium causes blood to flow through the pulmonary circulation.

With each ventricular contraction, blood pressure fluctuates in the large arteries (i.e., the aorta, pulmonary trunk, and other elastic and large muscular arteries). Blood pressure during ventricular systole, **systolic blood pressure,** is higher than blood pressure during ventricular diastole, or **diastolic blood pressure.** During diastole, the ventricles are in relaxation, and blood is not being ejected. These fluctuations diminish within the arterioles, and little or no fluctuations in blood pressure between systole and diastole are observed in the capillaries and venous vessels.

The blood pressure gradient between veins and the right atrium is low, making it difficult for venous blood flow to overcome the force of gravity. Therefore, when standing with arms hanging down, blood may pool in the large veins of the limbs.

1. Blood Pressure Measurements

Arterial blood pressure is reported in millimeters of mercury (mm Hg) and can be measured directly by inserting a pressure transducer into a large artery or indirectly with a sphygmomanometer. A **sphygmomanometer** (SFIG-moe-mah-NOH-meh-ter), or blood pressure cuff, can be used to measure systolic and diastolic blood pressure in any large artery. Clinically, the brachial artery is used most often. The sphygmomanometer contains a pressure gauge attached to an inflatable rubber cuff that is connected by a rubber tube to a hand pump (rubber bulb) or automatic pump. Clinically, it is important to use the right size cuff—pediatric, standard, or large arm—to obtain an accurate reading. The pump is used to inflate the rubber cuff to a pressure greater than the systolic pressure. This puts pressure on the artery, flattens it, and stops blood flow in the artery. A stethoscope is placed over the brachial artery in the antecubital area, and the pressure in the cuff is slowly released by opening a valve. When blood pressure is greater than the pressure in the cuff, the artery opens and blood flow returns. The examiner listens for the sound caused by the turbulent flow of blood that can

be heard until blood flow returns to normal. These sounds are called the **Korotkoff** (koh-ROT-koff) **sounds.** Although there is a range of values considered normal, the average normal systolic pressure (the first sound heard) is 120 mm Hg, and normal diastolic pressure (the last, faint sounds heard) is 80 mm Hg, usually written as 120/80.

Venous blood pressure can also be measured directly with a pressure transducer inserted into a venous vessel or indirectly. Indirect measurement of venous pressure, described in Lab Activity 2, is an estimate at best because blood pressure in veins is very low. The average venous blood pressure is 16 mm Hg.

LAB ACTIVITY 2 **Blood Pressure Measurements**

1 Measure resting systemic arterial blood pressure.
- Work in teams of two, alternating who will be the subject and who will measure blood pressure.
- The subject should sit and rest for 5 minutes before blood pressure is measured.
- Wipe the earpieces of the stethoscope with alcohol and completely deflate the blood pressure cuff.
- Place the cuff of the sphygmomanometer around the arm as shown in Figure 29.7. If the cuff does not fit the arm, the blood pressure measurements will probably be inaccurate. The inflatable portion of the cuff should be over the anterior surface of the arm. If there is an arrow on the cuff, place it over the brachial artery. The brachial artery is located in the anterior arm, immediately superior to the antecubital fossa and lateral to biceps brachii. The bottom of the cuff should be approximately 1 inch above the elbow.
- Close the valve on the rubber bulb.
- Insert earpieces; place the large bell of the stethoscope over the brachial artery inferior to the cuff, and listen for the brachial pulse.
- Inflate the cuff to 160 to 180 mm Hg.
- Immediately use the valve on the hand pump to slowly release air to deflate the cuff, and listen for the first sound.
- The first sound, the systolic pressure, is the return of blood flow through the partially occluded brachial artery. Watch the pressure gauge and continue to listen; the sound will increase, then muffle, and stop. The diastolic pressure is the pressure when the last, faint sound is heard.
- Record blood pressure measurements in Table 29.1.
- Deflate cuff and have subject rest for 2 minutes before repeating the blood pressure measurement. Record second value in Table 29.1.

FIGURE 29.7 **Measuring arterial blood pressure with a sphygmomanometer.**

TABLE 29.1 Resting Blood Pressures

SUBJECT	SYSTOLIC PRESSURE	DIASTOLIC PRESSURE	PULSE PRESSURE	MAP	VENOUS PRESSURE
Subject 1	1.	1.	1.	1.	1.
	2.	2.	2.	2.	2.
	Avg.	Avg.	Avg.	Avg.	Avg.
Subject 2	1.	1.	1.	1.	1.
	2.	2.	2.	2.	2.
	Avg.	Avg.	Avg.	Avg.	Avg.

2 Calculate pulse pressure and mean arterial pressure.
- The **pulse pressure** is the difference between the systolic and diastolic blood pressure. The larger the pulse pressure, the greater the volume of blood ejected from the ventricle. **Mean arterial pressure** (MAP or MABP) is the average blood pressure (BP) over the course of the cardiac cycle.
- Calculate the pulse pressure and record in Table 29.1. Pulse pressure = systolic pressure − diastolic pressure.

- Calculate the mean arterial pressure (MAP) and record in Table 29.1.

$$MAP = \text{diastolic BP} + \frac{(\text{pulse pressure})}{3}$$

3 Estimate systemic venous pressure.
- Decide who will be the subject, the observer, and the recorder.

- Have the subject stand with arms hanging at the side. Observe the blood pooling in the veins on the dorsal surface of the hands.
- Have the subject abduct one arm and observe the veins become smaller in size and flatten (collapse) as the arm is raised above the level of the right atrium. When the force of gravity exceeds venous pressure, the blood in the veins drains away quickly and does not exert enough pressure on the walls of the veins to keep them open.
- Lower the arm to pool blood in the veins of the hand again.
- Have the subject stand in front of a white (or black) board. With the subject's hands at the sides, locate the level of the right atrium (approximately 2 inches above the nipple). Mark on the board where the subject's hand touches the board.
- Have the subject abduct the arm and hand until the veins on the surface of the hand flatten (collapse). Mark where the hand touches the board. Using a meterstick, measure the vertical distance above the right atrium that the hand has traveled in millimeters (1 cm = 10 mm): _____ mm. Each 12.88 mm elevation of the hand above heart level represents an approximately 1 mm rise in Hg.
- Venous pressure (mm Hg) =

$$\frac{\text{mm measured}}{12.88 \text{ mm elevation/mm Hg}}$$

- Repeat measurement and record in Table 29.1.

4 Observe the effect of body position on blood pressure.
- Decide who will be the subject, who will do the measuring, and who will record.
- Have subject lie on a mat (supine position) for 2 minutes. Measure BP and heart rate (HR). Record in Table 29.2.
- Have subject stand and *immediately* measure BP and HR. Record in Table 29.2.
- Take the BP and HR again after the subject stands for 2 minutes. Record in Table 29.2.

5 Answer Discussion Questions with your lab group.

6 Complete **Biopac Laboratory Guide Experiment:** Effect of Body Position on Resting Blood Pressure.

DISCUSSION QUESTIONS
Blood Pressure Measurements

1 Did the subject's BP change after altering body position?

2 Did the subject's HR change after altering body position?

3 Explain what happens in regard to BP when a person changes from a supine to a standing position. Why does this occur?

2. Regulation of Blood Pressure

The body maintains blood pressure (BP) to ensure adequate blood flow to body tissues (tissue perfusion). If blood pressure is too low, tissue perfusion may not be sufficient to provide oxygen and nutrients to cells, especially brain neurons, or to remove metabolic wastes. If blood pressure is too high, it may cause damage to capillaries and the endothelial lining of blood vessels. Eye exams can often detect early hypertension. Damaged retinal vessels can be observed with an ophthalmoscope.

During exercise, cardiac output increases to provide adequate tissue perfusion. The increased stroke volume puts more pressure on the blood vessel walls, causing arterial systolic blood pressure to increase. During aerobic exercise, diastolic blood pressure stays the same or decreases slightly due to vasodilation. Even though there is a greater volume of blood in the arteries due to increased stroke volume, the amount of vasodilation causes diastolic pressure to remain the same as resting or to decrease slightly. During resistance exercise (weight training), diastolic pressure may increase. The skeletal muscle contractions required to lift weight squeezes the arteries in muscles, effectively causing vasoconstriction. This causes peripheral resistance to increase, resulting in an increase in both systolic and diastolic blood pressure during exercise. The changes in blood pressure that occur during exercise are temporary; however, exercising regularly will reduce resting systolic and diastolic pressures.

TABLE 29.2 Effect of Body Position on Blood Pressure

SUBJECT	SYSTOLIC/DIASTOLIC BP (mm Hg)	HR
Supine		
Immediately after standing		
After standing for 2 minutes		

Mean arterial blood pressure increases when cardiac output increases and when resistance to blood flow increases. Blood viscosity, total blood vessel length, and blood vessel diameter affect resistance to blood flow. However, blood vessel diameter is the only variable that can be changed to make immediate changes in blood pressure. Increasing blood vessel diameter (vasodilation) decreases resistance and lowers BP, while decreasing blood vessel diameter (vasoconstriction) increases resistance and raises BP.

LAB ACTIVITY 3 ### Experiment: Effect of Exercise on Blood Pressure

1 **Prediction:** Circle the correct choice (*in italics*) in each statement.
- Exercise will cause systolic and diastolic BP to *increase* or *decrease*.
- Exercise will cause pulse pressure to *increase* or *decrease*.
- Exercise will cause MAP to *increase* or *decrease*.

2 **Materials:** Obtain materials for the Regulation of Blood Pressure experiment from Materials list.

3 **Data Collection:**
- Decide who will be the subject, who will measure BP and HR, and who will be the recorder and timer. The subject should not have heart problems and should be in good health.
- Measure the effect of exercise on BP and record the results in Table 29.3.

BP and HR at rest
- Have the subject stand still at rest for 2 to 3 minutes.
- Measure systolic and diastolic BP, and record the results in Table 29.3.
- Measure the number of heartbeats (or pulses) in 15 seconds and multiply by 4 to obtain HR. Record the results in Table 29.3.

BP and HR immediately after exercise
- Have the subject run in place for 5 minutes.
- Immediately measure systolic and diastolic BP, and record the results in Table 29.3.

- Immediately count the number of heartbeats (or pulses) in 15 seconds and multiply by 4 to obtain HR. Record the results in Table 29.3.

4 **Data Analysis:**
- Calculate resting and post-exercise pulse pressures and MAP. Record in Table 29.3.
- Pool heart rate and MAP class data, calculate averages, and record in Table 29.3.
- If graph paper is available, graph your data.

5 Clean up your lab area as directed by your instructor.

6 Complete the Experimental Report with your lab group.

7 Complete the **PowerPhys Experiment:** Effect of Exercise on Blood Pressure and Vascular Resistance.

8 Complete **Biopac Laboratory Guide Experiment:** Effect of Exercise on Blood Pressure.

EXPERIMENTAL REPORT
Effect of Exercise on Blood Pressure

Results:
- State whether systolic and diastolic BP increased or decreased after exercise.
- State whether pulse pressure and MAP increased or decreased after exercise.
- State whether HR increased or decreased after exercise.

Discussion:
- Why was the resting blood pressure taken standing rather than sitting?
- Discuss how the body changes BP during exercise.
- Discuss why the body changes BP during exercise.

Conclusion:
- Write a statement that postulates how increasing levels of exercise affect BP, pulse pressure, and MAP.

TABLE 29.3 Effect of Exercise on Blood Pressure

	SYSTOLIC BP (mm Hg)	DIASTOLIC BP (mm Hg)	HR	PULSE PRESSURE (mm Hg)	MAP (mm Hg)
Subject At rest					
Post-exercise					
Class Average At rest					
Post-exercise					

Reviewing Your Knowledge

A. Structure of Arteries and Veins

Name the components of each blood vessel tunic. More than one term may apply to each description, and terms may be used more than once.

smooth muscle
endothelium
elastic fibers
collagen fibers

_____ **1.** Tunica media of arteries

_____ **2.** Tunica media of veins

_____ **3.** Tunica interna of arteries

_____ **4.** Tunica interna of veins

_____ **5.** Tunica externa of arteries

_____ **6.** Tunica externa of veins

Circle True or False for the following questions. If false, underline and change word(s) that are incorrect to make the statement true.

7. The tunica media of veins is thicker than the tunica media of arteries. True or False

8. The tunica media of elastic arteries contains more elastic fibers than muscular arteries. True or False

9. Venous valves are folds of the tunica externa. True or False

10. Venous valves prevent backflow of blood. True or False

11. Walls of veins are thicker than the walls of arteries of the same size. True or False

12. The lumen of a vein is larger than the lumen of an artery of the same size. True or False

B. Structure of Capillaries

Match the term to the correct description. More than one term may apply to each description. Terms may be used more than once.

fenestrations
intercellular clefts
tight junctions
transcytosis

_____ **1.** Holes in plasma membrane through which molecules pass across capillary walls

_____ **2.** Fusion of plasma membranes of adjacent endothelial cells; forms very selective barrier

_____ **3.** Vesicles transport substances across capillary wall

_____ **4.** Spaces between cells through which substances pass

Circle True or False for the following questions. If false, underline and change word(s) that are incorrect to make the statement true.

5. Capillary walls are composed of an endothelium and a basement membrane only. True or False

6. Hydrostatic pressure forces plasma across capillary walls at venous end of capillary, and interstitial fluid enters arterial end of capillary by osmotic pressure. True or False

C. Blood Pressure

Fill in the blank with the correct term.

_____ **1.** Term used for arterial pressure during ventricular systole

_____ **2.** Term used for arterial pressure during ventricular diastole

_____ **3.** Device used to measure arterial blood pressure in the brachial artery

_____ **4.** Average normal adult arterial blood pressure

_____ **5.** Average venous blood pressure

_____ **6.** Sounds of turbulent blood flow that occur when blood flow resumes in an artery that has been occluded

Circle True or False for the following questions. If false, underline and change word(s) to make the statement true.

7. The blood pressure gradient from the aorta to the capillaries is greater than the blood pressure gradient from the venules to the right atrium. True or False

8. The blood pressure gradient from the aorta to the capillaries is less than the blood pressure gradient from the arterial end of the capillary to the venous end of the capillary. True or False

9. Kate's systolic blood pressure is 115 and diastolic is 72. Her pulse pressure is 47. True or False

10. Scott's blood pressure is 126/84; therefore, his MAP is 96. True or False

D. Regulation of Blood Pressure and Blood Flow

Circle True or False for the following questions. If false, underline and change word(s) that are incorrect to make the statement true.

1. Increasing heart rate increases blood pressure. True or False

2. Systemic vasoconstriction decreases blood pressure. True or False

3. Increasing arterial blood pressure increases blood flow. True or False

4. Recoil of muscular arteries maintains blood flow during ventricular diastole. True or False

5. Muscular arteries control the blood flow to different body areas. True or False

6. Vasoconstriction of the renal arteries (arteries supplying blood to kidneys) would decrease blood flow to the kidneys. True or False

7. Blood pressure is higher in the supine position than in the standing position. True or False

8. The greater the pulse pressure, the lower the pressure gradient driving blood from the aorta through the systemic circulation. True or False

9. Exercise increases MAP. True or False

Using Your Knowledge

A. Blood Vessel Structure and Function

1. In coronary bypass surgery, a section of vein is used to replace (bypass) occluded coronary arteries. Over time, the vein wall becomes more like an arterial wall. Describe the changes that would occur in the vein wall.

2. The smooth muscle fibers within the wall of an artery do not receive their nutrients from the blood within the lumen of the artery. Why not?

3. Blood flow within the capillaries is slower than within arterial or venous vessels. Explain why this is important.

4. Blood flow to the skeletal muscles increases with exercise. Explain how the autonomic nervous system mediates the increase in blood flow to skeletal muscles.

5. Explain how blood flow to the kidneys is decreased with exercise.

6. Explain why the differences in the amount of elastic fibers and smooth muscle fibers in the tunica media of elastic arteries and muscular arteries is important to their function.

7. Why does the tunica interna of the vasoconstricted section of the arteriole in Figure 29.1 exhibit folds?

8. Capillaries within the brain have tight junctions, whereas capillaries within the anterior pituitary and other endocrine glands contain fenestrations, intercellular clefts, and vesicles for transcytosis. Explain how these structural differences are important to the function of the brain and the anterior pituitary.

9. Compare the strength of arterial and capillary walls.

10. Marfan's syndrome is an inherited genetic disorder that results in malformed elastic fibers that are weak. What effect would this have on the aorta?

B. Blood Pressure Homeostasis

Use the terms in bulleted list to complete the schematic diagram of blood pressure homeostasis in Figure 29.8.

- ADH
- aldosterone
- blood pressure

- blood volume
- cardiac output
- sympathetic nerves

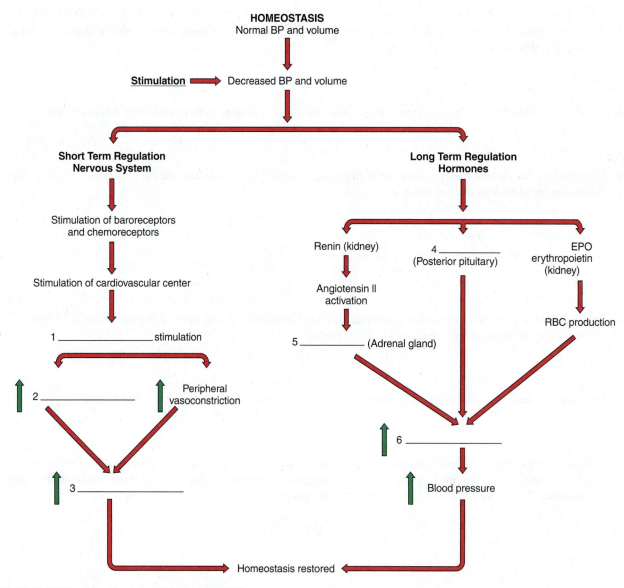

FIGURE 29.8 **Blood pressure homeostasis.**

Blood Vessel Identification

OBJECTIVES

1 Identify the major arteries and veins of the systemic circulation

2 Identify the major vessels of the cerebral arterial circle (circle of Willis), pulmonary circulation, hepatic portal system, and fetal circulation

3 Palpate main arterial pulses

4 Trace blood flow through the cardiovascular system

5 Dissect a cat, fetal pig, or cadaver and identify selected blood vessels

MATERIALS

- models or charts showing human arteries, veins, and circulatory pathways or **Real Anatomy** (Cardiovascular)
- Dissection: preserved cats or fetal pigs, dissecting equipment, disposable gloves, safety glasses, and dissection manual
- **Real Anatomy:** Virtual Cadaver Dissection

Arteries carry blood from the heart to capillaries where gas and nutrient exchange occurs, while veins carry blood back to the heart. There are two circulatory routes within the body: the pulmonary circulation and the systemic circulation. In the **pulmonary circulation,** arteries carry oxygen-poor blood from the right ventricle to the lungs where gas exchange occurs, and veins carry oxygen-rich blood back to the left atrium. In the **systemic circulation,** arteries carry oxygen-rich blood to body tissues, and veins return oxygen-poor blood to the heart where it enters the pulmonary circulation.

Anastomoses provide alternative routes or detours for blood flow in case the main route is blocked or damaged. Arterial anastomoses are blood vessels that connect arteries, while venous anastomoses connect veins.

A. Systemic Arteries

All major arteries of the systemic circulation branch off the aorta. These arteries transport blood to different regions of the body where blood is distributed to smaller arteries and finally to arterioles that connect to capillaries.

1. Major Arteries of the Ascending Aorta and Aortic Arch

Blood is ejected from the left ventricle and into the **ascending aorta.** The **right** and **left coronary arteries** branch off the ascending aorta near its origin and travel to the heart. The ascending aorta curves superiorly toward the left and becomes the **aortic arch.** The superior portion of the aortic arch is posterior to the sternal angle. Three major arteries branch off the aortic arch in the following order from right to left: the **brachiocephalic trunk,** the **left common carotid artery,** and the **left subclavian artery.** The brachiocephalic trunk divides to form the **right common carotid** and **right**

subclavian arteries. The left and right **vertebral arteries** branch off their respective subclavian arteries.

> ### Before Going to Lab
>
> **1** Review the arteries in Figure 30.1(a).
>
> **2** Observe the locations of these arteries in Table 30.1.
>
> **3** Label the arteries in Figure 30.1(b) using the locations described in Table 30.1.

LAB ACTIVITY 1 **Major Arteries of the Ascending Aorta and Aortic Arch**

1 Identify the arteries on a model or chart, or use the search text box in **Real Anatomy** (Cardiovascular) to find these structures.

2 Palpate your sternal angle. The superior portion of the aortic arch is posterior to it. ■

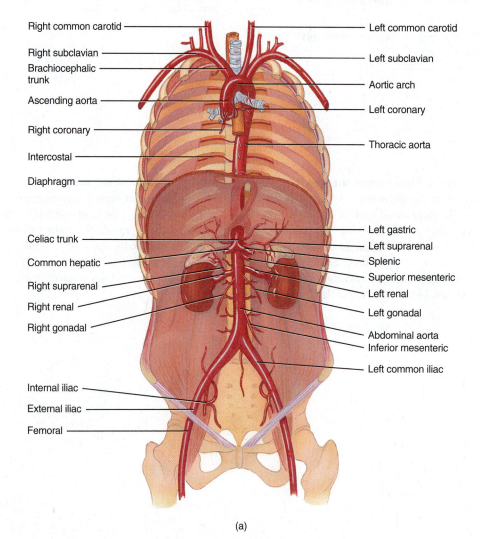

(a)

FIGURE 30.1 **Major arteries of the ascending aorta and aortic arch.**

TABLE 30.1 Major Arteries of the Ascending Aorta and Aortic Arch

ARTERY	LOCATION
Ascending aorta	Begins at left (aortic) semilunar valve and becomes the aortic arch at the level of the sternal angle
Right and left coronary arteries	Branch off ascending aorta just superior to the left aortic valve and end when they divide into their respective arterial branches
Aortic arch	Begins at the level of the sternal angle and becomes the thoracic aorta at the level of the intervertebral disc between the fourth and fifth vertebrae
Brachiocephalic trunk	First branch off aortic arch, divides into right subclavian artery and right common carotid artery at level of sternoclavicular joint
Right common carotid artery	Formed from division of brachiocephalic trunk and divides into the right external and right internal carotid arteries at the superior border of the larynx
Right subclavian artery	Formed from division of brachiocephalic trunk and is renamed axillary artery within axillary region
Left common carotid artery	Second branch off aortic arch, divides into the left external and left internal carotid arteries at the superior border of the larynx
Left subclavian artery	Third branch off aortic arch, renamed axillary artery within the axillary region

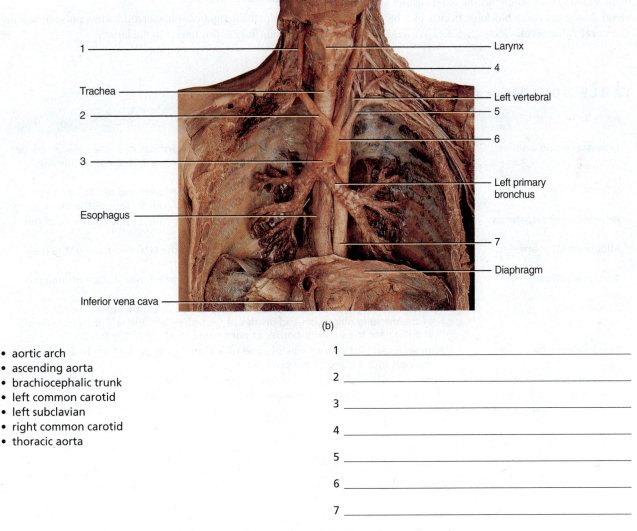

(b)

- aortic arch
- ascending aorta
- brachiocephalic trunk
- left common carotid
- left subclavian
- right common carotid
- thoracic aorta

1 _____

2 _____

3 _____

4 _____

5 _____

6 _____

7 _____

FIGURE 30.1 Major arteries of the ascending aorta and aortic arch, *continued.*

2. Major Arteries Supplying the Head (Including Circle of Willis)

The common carotid arteries branch to form the internal and external carotid arteries. The **internal carotid arteries** travel through the neck and enter the skull through the carotid canal within each temporal bone and supply structures within the skull. Each internal carotid artery branches to form an **anterior cerebral artery** and a **middle cerebral artery** that supply blood to these areas of the brain. The **external carotid arteries** ascend along the lateral surface of the neck and terminate as two arteries near the temporomandibular joint that supply structures external to the skull.

There are two **vertebral arteries,** one branching off each subclavian artery. The vertebral arteries travel superiorly through the transverse foramen of the cervical vertebrae and then through the foramen magnum to reach the base of the brain, where they merge to form the **basilar artery.** The **posterior cerebral artery** is a branch of the basilar artery supplying the posterior part of the brain.

Blood supply to the brain is vital, and the configuration of the blood vessels supplying the brain ensures continuous blood flow even when blockage occurs in a blood vessel. The **cerebral arterial circle** (circle of Willis) is a ring of blood vessels formed by three anastomoses—the anterior communicating artery and the right and left posterior communicating arteries. The **anterior communicating artery** connects the two anterior cerebral arteries, and the **posterior communicating arteries** connect the internal carotid arteries to the posterior cerebral arteries.

Before Going to Lab

1 Label the arteries in Figure 30.2(a) using the locations described in Table 30.1 and Table 30.2.

2 Label the arteries forming the cerebral arterial circle (circle of Willis) in Figure 30.2(b) using the locations described in Table 30.2.

LAB ACTIVITY 2 Major Arteries Supplying the Head

1 Identify the arteries on a model or chart, or use the search text box in **Real Anatomy** (Cardiovascular) to find these structures.

2 Palpate the right common carotid artery pulse by placing your fingers just lateral to the larynx. ■

TABLE 30.2 Major Arteries Supplying the Head

ARTERY	LOCATION
External carotid arteries	Begin as terminal branches off the common carotid arteries near the superior border of the larynx, anterior to the sternocleidomastoid muscle, and end within the parotid gland
Internal carotid arteries	Begin as terminal branches off the common carotid arteries and divide into the anterior and middle cerebral arteries below the inferior surface of the brain
Anterior cerebral arteries	Branch off internal carotid arteries that pass anterior to the optic chiasma and run along the midline of the base of the brain
Middle cerebral arteries	Branch off internal carotid arteries that run toward the lateral surface of the base of the brain
Vertebral arteries	First major branch off subclavian artery that ends near inferior surface of brain, where it forms the basilar artery
Basilar artery	Formed when right and left vertebral arteries merge along the inferior surface of the medulla oblongata and divides to form the right and left posterior cerebral arteries near the superior border of pons along the base of the brain
Posterior cerebral arteries	Formed when basilar artery divides and travels along the base of the brain toward the posterior surface of the cerebrum
Anterior communicating artery	Connects right and left anterior cerebral arteries anterior to the optic chiasma
Right and left posterior communicating arteries	Connect posterior cerebral arteries and internal carotid arteries

Right internal carotid

- brachiocephalic trunk
- common carotid
- external carotid
- internal carotid
- vertebral

1 _____

2 _____

3 _____

4 _____

5 _____

(a) Right lateral view of major arteries of the head and neck

- anterior cerebral
- anterior communicating
- basilar
- internal carotid
- middle cerebral
- posterior cerebral
- posterior communicating
- vertebral

6 _____

7 _____

8 _____

9 _____

10 _____

11 _____

12 _____

13 _____

(b) Inferior view of brain showing cerebral arterial circle

FIGURE 30.2 **Arteries of the head and neck.**

3. Major Arteries of the Upper Extremities

The subclavian artery is renamed the **axillary artery** within the axilla and continues as the **brachial artery** as it enters the arm. As the brachial artery enters the forearm, it divides to form the **radial artery** and the **ulnar artery.** The **superficial palmar arch** and the **deep palmar arch,** both of which supply blood to the fingers and palms, connect the radial and ulnar arteries at their distal ends and are examples of arterial anastomoses.

Before Going to Lab

1 Review the arteries in Figure 30.3(a).

2 Observe the locations of these arteries in Table 30.3.

3 Label the arteries in Figure 30.3(b) using the locations described in Table 30.3.

LAB ACTIVITY 3 **Arteries of the Upper Extremities**

1 Identify the arteries on a model or chart, or use the search text box in **Real Anatomy** (Cardiovascular) to find these structures.

2 Flex your arm and palpate the brachial artery by placing the second and third fingers in the brachial groove (between the biceps and triceps).

3 Palpate the radial artery by placing the second and third fingers medial to the distal end of the radius. ■

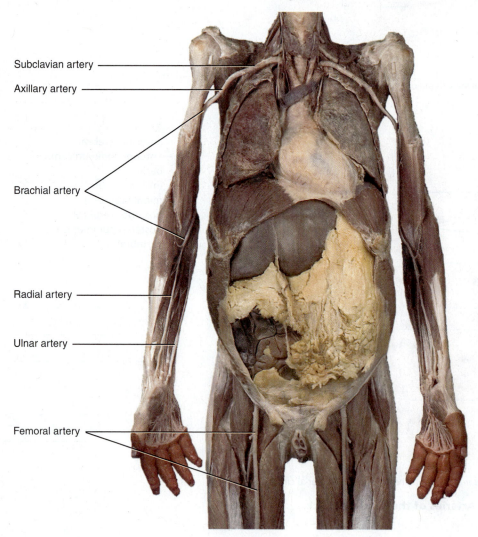

Subclavian artery

Axillary artery

Brachial artery

Radial artery

Ulnar artery

Femoral artery

(a)

FIGURE 30.3 **Arteries supplying the upper extremities.**

TABLE 30.3 Arteries Supplying the Upper Extremities

ARTERY	LOCATION
Axillary arteries	Continuation of subclavian artery that begins at the outer border of first rib and is renamed the brachial artery as it enters the arm
Brachial arteries	Continuation of axillary artery that begins near the tendon of the teres major muscle and ends just distal to the elbow where it divides into the radial and ulnar arteries
Radial arteries	Terminal branch of brachial artery that runs along lateral surface of forearm into the wrist and hand
Ulnar arteries	Terminal branch of brachial artery that runs along medial surface of forearm into the wrist and hand
Deep palmar arch	Anastomoses that connect radial and ulnar arteries; located deep to the tendons of flexor muscles and proximal to superficial palmar arch
Superficial palmar arch	Anastomoses that connect radial and ulnar arteries; located superficial to flexor tendons and distal to deep palmar arch

- axillary
- brachial
- radial
- subclavian
- ulnar

1 _____

2 _____

3 _____

4 _____

5 _____

Deep palmar arch

Superficial palmar arch

(b)

FIGURE 30.3 **Arteries supplying the upper extremities,** *continued.*

4. Major Arterial Branches of the Descending Aorta

The diaphragm divides the descending aorta into thoracic and abdominal sections. The **thoracic aorta** lies to the left of the midline, just anterior to the vertebral column, and has branches that supply thoracic structures. Nine pairs of **posterior intercostal arteries** branch off the thoracic aorta to follow the ribs. The abdominal aorta is also anterior to the vertebral column but closer to the midline. It supplies the abdomen, pelvis, and lower extremities.

The major branches of the abdominal aorta are the **celiac trunk** (which divides into the **left gastric artery, splenic artery,** and **common hepatic artery**), the **superior mesenteric artery,** the paired **suprarenal arteries,** the paired **renal arteries,** the paired **gonadal arteries** (**ovarian arteries** in females and **testicular arteries** in males), the **inferior mesenteric artery,** and four pairs of **lumbar arteries.** Within the pelvis, the abdominal aorta bifurcates into the **common iliac arteries** that are the terminal branches of the abdominal aorta.

Before Going to Lab

1 Label the arteries in Figures 30.4 and 30.5 using the locations described in Table 30.4.

LAB ACTIVITY 4 — Major Arterial Branches of the Descending Aorta

1 Identify the arteries on a model or chart, or use the search text box in **Real Anatomy** (Cardiovascular) to find these structures.

2 Palpate the anterior superior iliac spines. The abdominal aorta bifurcates into the common iliac arteries at this level.

Diaphragm

Inferior vena cava

Right adrenal (suprarenal) gland

Right kidney

1

2

3

4

Hepatic veins

5

6

7

8

9

10

Ureter

- abdominal aorta
- celiac trunk
- common hepatic
- inferior mesenteric
- right common iliac
- right external iliac
- right internal iliac
- left renal
- splenic
- superior mesenteric

1 _____

2 _____

3 _____

4 _____

5 _____

6 _____

7 _____

8 _____

9 _____

10 _____

FIGURE 30.4 **Descending aorta and major arterial branches.**

- celiac trunk
- common hepatic
- left gastric
- splenic

1 _____

2 _____

3 _____

4 _____

FIGURE 30.5 Anterior view of celiac trunk and its arterial branches.

TABLE 30.4 Major Arterial Branches of the Descending Aorta

ARTERY	LOCATION
Thoracic aorta	Continuation of aortic arch that starts at the level of the intervertebral disc between the fourth and fifth thoracic vertebrae and becomes the abdominal aorta at the diaphragm
Abdominal aorta	Continuation of the thoracic aorta that begins at the diaphragm and ends at the level of the fourth lumbar vertebra where it divides into the right and left common iliac arteries
Celiac trunk	Short branch off abdominal aorta just inferior to the diaphragm that immediately divides into three arteries: the left gastric, splenic, and common hepatic arteries
Left gastric artery	Smallest of the three celiac trunk branches, it comes off the left anterior side and courses to the left toward the esophagus and stomach
Splenic artery	Largest of the three celiac trunk branches, it comes off the left anterior side and courses to the left toward the spleen and branches into three arteries: the pancreatic artery, the left gastroepiploic artery, and the short gastric artery
Common hepatic artery	Intermediate of three celiac trunk branches, it comes off the right anterior side and courses toward the right and branches into three arteries
Superior mesenteric artery	Arises off the anterior surface of the abdominal aorta just inferior to the celiac trunk at the level of the first lumbar vertebra; it has many anastomoses and branches into five arteries
Right and left suprarenal arteries	Branch off the lateral surface of the abdominal aorta at the level of the first lumbar vertebra and course toward the adrenal glands
Right and left renal arteries	Branch off the lateral surface of the abdominal aorta at the level of the second lumbar vertebra and course toward the kidneys
Right and left gonadal arteries	Branch off the anterior surface of the abdominal aorta inferior to the renal arteries; ovarian arteries course toward the ovaries and testicular arteries pass through the inguinal canal toward the testes
Inferior mesenteric artery	Branches off the abdominal aorta at the level of the third lumbar vertebra; it has many anastomoses and branches into three arteries
Right and left common iliac arteries	Begin at division of abdominal aorta at level of fourth lumbar vertebra and terminate terminate into internal and external iliac arteries

5. Major Arteries of the Pelvis and Lower Extremities

The **common iliac arteries** divide into the internal iliac and external iliac arteries. The **internal iliac arteries** are medial terminal branches off the common iliac arteries just anterior to the lumbosacral joint. They course posteriorly to the pelvis. The **external iliac artery,** the larger of the two arteries, becomes the femoral artery as it enters the thigh. The **femoral artery** descends along the middle of the anterior two-thirds of the thigh and travels to the posterior thigh, becoming the popliteal artery as it enters the posterior knee area. The **popliteal artery** divides into the **anterior** and **posterior tibial arteries.** The anterior tibial artery travels to the anterior surface of the leg and descends to the ankles where it becomes the **dorsalis pedis artery.** The **posterior tibial artery** descends along the posterior aspect of the leg and terminates at the ankle where it divides into the **medial** and **lateral plantar arteries.** The **fibular (peroneal) artery** branches off the posterior tibial artery and travels down the lateral side of the leg to the foot.

Before Going to Lab

1 Label the arteries in Figure 30.6(a) and (b) using the locations described in Table 30.5.

LAB ACTIVITY 5 — Major Arteries of the Pelvis and Lower Extremities

1 Identify the arteries on a model or chart, or use the search text box in **Real Anatomy** (Cardiovascular) to find these structures.

2 Palpate the femoral artery pulse by placing two fingertips over the inguinal area, specifically in the crease between the thigh and the trunk.

3 Palpate the popliteal artery pulse by placing two fingertips on the posterior surface of the knee. It is best to palpate in the supine position with the knee slightly flexed.

4 Palpate the pulse of the dorsal artery of the foot (dorsalis pedis artery) by placing two fingertips over the dorsal surface of the medial side of the foot near the ankle.

TABLE 30.5 Major Arteries of the Pelvis and the Lower Extremities

ARTERY	LOCATION
Internal iliac arteries	Terminal branches of common iliac arteries that begin at the level of the intervertebral disc between vertebrae L5 and S1 and travel posteriorly until they divide in the pelvis
External iliac arteries	Terminal branches of common iliac arteries that run along the psoas major muscles and become femoral arteries posterior to the inguinal ligaments
Femoral arteries	Begin posterior to inguinal ligaments; descend along the midline of the anterior two-thirds of the thighs; travel to the posterior thigh and become the popliteal arteries in the popliteal region of the lower extremity
Popliteal arteries	Continuation of femoral arteries that begin in the popliteal region and end in the upper leg where they divide into the anterior and posterior tibial arteries
Anterior tibial arteries	Terminal branches of popliteal arteries that begin in the upper posterior leg and travel to the anterior leg; end in the ankles where they become the dorsal arteries of the foot (dorsalis pedis arteries)
Dorsal arteries of the foot (dorsalis pedis arteries)	Continuation of anterior tibial arteries that begin in the ankles and form branches within the foot
Posterior tibial arteries	Terminal branches of popliteal arteries that begin in the upper leg and terminate at the medial malleolus of the tibia where they divide into the medial and lateral plantar arteries
Fibular (peroneal) arteries	Branch off the posterior tibial artery slightly distal to the point where the popliteal artery divides into the anterior and posterior tibial arteries; descends down the lateral surface of the leg, and terminates within the foot

(a) Anterior view

(b) Posterior view

Medial plantar

Lateral plantar

- anterior tibial
- common iliac
- dorsal artery of the foot (dorsalis pedis)
- external iliac
- femoral
- fibular (peroneal)
- internal iliac
- popliteal
- posterior tibial

1 _____

2 _____

3 _____

4 _____

5 _____

6 _____

7 _____

8 _____

9 _____

FIGURE 30.6 **Major arteries of the pelvis and lower extremities.**

B. Systemic Veins

Veins carry blood from the capillaries back toward the heart. Veins can be found under the skin (superficial veins) or deep within the body, typically near an artery. Veins that accompany arteries usually have the same name as the artery. Veins carrying blood from the lungs to the left side of the heart carry oxygen-rich blood and are part of the pulmonary circulation, whereas veins carrying blood from all other body tissues carry oxygen-poor blood and are part of the systemic circulation.

CLINICAL NOTE: *A thrombus in a deep vein can cause a pulmonary embolus if not treated with anticoagulant, whereas a thrombus in a superficial vein does not tend to form an embolus.*

1. Veins Carrying Blood to the Right Atrium of the Heart

Three large veins carry oxygen-poor blood into the right atrium—the **superior vena cava**, the **inferior vena cava,** and the **coronary sinus.**

Before Going to Lab

1 Label the veins in Figure 30.7 using the locations described in Table 30.6.

LAB ACTIVITY 6 Veins Carrying Blood to the Heart

1 Identify the arteries on a model or chart, or use the search text box in **Real Anatomy** (Cardiovascular) to find these structures. ■

- coronary sinus
- inferior vena cava
- superior vena cava

1 _____

2 _____

3 _____

FIGURE 30.7 **Systemic veins carrying blood to the heart.**

TABLE 30.6 Veins Carrying Blood to the Heart

VEIN	LOCATION
Superior vena cava	Formed from the merger of the right and left brachiocephalic veins at the level of the first costal cartilage and ends where it enters the right atrium at the level of the third costal cartilage
Coronary sinus	Formed from the merger of the great, middle, and small cardiac veins, located in the coronary sulcus, and empties into the right atrium
Inferior vena cava	Formed from the merger of the common iliac veins at the level of the fifth lumbar vertebra and ascends behind the peritoneum just right of the midline and empties into the right atrium

2. Major Veins Draining the Head and Neck

Three large paired veins—the internal jugular veins, the external jugular veins, and the vertebral veins—drain blood from the head and neck. The **internal jugular veins** run lateral to the internal carotid and common carotid arteries, the **external jugular veins** are superficial veins descending along the lateral surface of the neck, and the **vertebral veins** descend through the transverse foramina of the vertebral column with the vertebral arteries.

The **external jugular veins** drain the scalp and face and empty into the subclavian veins. The internal jugular and subclavian veins join to form the **brachiocephalic veins.** The right and left brachiocephalic veins merge to form the **superior vena cava.**

Before Going to Lab

1 Label the veins in Figure 30.8 using the locations described in Table 30.7.

LAB ACTIVITY 7 Major Veins Draining the Head and Neck

1 Identify the arteries on a model or chart, or use the search text box in **Real Anatomy** (Cardiovascular) to find these structures.

2 Observe the external jugular vein. This vein is located lateral to the sternocleidomastoid muscle. To observe this vein, look in the mirror while clenching your teeth and place your fingers above the clavicle, pressing firmly to prevent blood from draining the external jugular vein. This should cause the vein to enlarge. ■

- brachiocephalic
- external jugular
- internal jugular
- subclavian
- vertebral

1 _____

2 _____

3 _____

4 _____

5 _____

FIGURE 30.8 **Major veins of the head and neck.**

TABLE 30.7 Major Veins of the Head and Neck

VEIN	LOCATION
Internal jugular veins	Begin at the jugular foramina; run lateral to the internal carotid and common carotid arteries; merge with the subclavian veins to form the brachiocephalic veins
External jugular veins	Begin in the occipital region and the parotid gland; descend through the neck along the sternocleidomastoid muscle and end near the middle of the clavicle where they drain into the subclavian veins
Vertebral veins	Originate inferior to the occipital condyles; travel through the transverse foramina of the first through sixth cervical vertebrae and drain into the area where the subclavian and internal jugular veins merge to form the brachiocephalic veins in the root of the neck

3. Major Veins Draining the Upper Extremities

Superficial veins are visible running beneath the skin of the arm. As we age, these veins become more visible. The major superficial veins of the upper extremities are the basilic and the cephalic veins. Each arm has a **cephalic vein** that travels along the anteriolateral surface of the entire limb and merges with the axillary vein inferior to the clavicle. Each **basilic vein** travels along the medial surface of the forearm and the anterior surface of the arm. Each **median cubital vein,** a common site for obtaining blood samples, is an anastomosis anterior to the elbow that connects the basilic and cephalic veins.

The *major deep veins* of the arm and forearm run along with arteries and are named for the arteries they accompany. The **radial** and **ulnar veins,** found in each forearm, merge to form the **brachial veins.** The brachial veins ascend the arm and join with the basilic vein to form the **axillary vein.** The axillary vein becomes the **subclavian vein** as it leaves the axillary region.

Before Going to Lab

1 Label the veins in Figure 30.9(a) and (b) using the descriptions supplied for superficial veins in Table 30.8 and deep veins in Table 30.9.

LAB ACTIVITY 8 **Major Veins Draining the Upper Extremities**

1 Identify the arteries on a model or chart, or use the search text box in **Real Anatomy** (Cardiovascular) to find these structures.

2 Identify the median cubital vein on yourself.

3 Observe the cephalic and basilic veins on yourself by clenching your fist and extending your arm by your side. ■

(a) Anterior view of superficial veins

(b) Anterior view of deep veins

(a)
- basilic
- cephalic
- median cubital

1 _____
2 _____
3 _____

(b)
- axillary
- brachial veins
- radial veins
- subclavian
- ulnar veins

4 _____
5 _____
6 _____
7 _____
8 _____

FIGURE 30.9 **Major veins of the right upper extremity.**

TABLE 30.8 **Superficial Veins Draining the Upper Extremities**

VEIN	LOCATION
Cephalic veins	Begin at lateral edge of the dorsal venous arches (superficial veins of the hand); travel laterally along the anterior surface of the entire limb and end inferior to the clavicle where they drain into the axillary vein
Basilic veins	Begin at the medial edge of the dorsal venous arches and travel medially along the posterior surface of the forearm and the anterior surface of the arm; in the middle of the arm, this vein travels deep and eventually merges with the brachial vein in the axillary region to form the axillary vein
Median cubital veins	Travel laterally across the anterior surface of the elbow to connect the basilic and cephalic veins

TABLE 30.9 **Deep Veins Draining the Upper Extremities**

VEIN	LOCATION
Radial veins	From venous palmar arches, travel along lateral forearm with the radial arteries and end at brachial veins
Ulnar veins	From superficial palmar venous arches, travel along medial forearm with ulnar arteries and end at brachial veins
Brachial veins	From merger of radial and ulnar veins, run superiorly until merge with basilic veins to form axillary veins
Axillary veins	From merger of basilic and brachial veins, run with axillary arteries until they become subclavian veins at the border of first ribs
Subclavian veins	Continuation of axillary veins that begin at the border of first ribs; end at the sternal end of the clavicle where they join with the internal jugular veins to form the brachiocephalic veins

4. Major Veins Draining the Thorax, Abdomen, and Pelvis (Including Hepatic Portal Circulation)

The **azygos system** of veins includes the **azygos vein,** the **hemiazygos vein,** and the **accessory hemiazygos vein.** These veins drain most thoracic structures and the abdominal wall. They also form anastomoses with the large veins draining the lower limbs and abdomen. Therefore, the azygos system can serve as a bypass if the inferior vena cava becomes obstructed.

The **internal** and **external iliac veins** in the pelvis merge to form the **common iliac** veins that unite to form the **inferior vena cava.** The **lumbar veins, gonadal veins, renal veins, suprarenal veins,** and **hepatic veins** drain directly into the inferior vena cava.

The veins draining the stomach, intestines, spleen, pancreas, and gallbladder do not drain directly into the inferior vena cava but enter the hepatic portal circulation first. The major veins of the hepatic circulation are the inferior mesenteric vein, the splenic vein, the superior mesenteric vein, and the hepatic portal vein. The **inferior mesenteric vein,** which drains blood from the large intestine, joins the

splenic vein, which carries blood from the stomach, pancreas, and spleen.

The **superior mesenteric vein** drains blood from the small intestine and merges with the splenic vein to form the **hepatic portal vein,** which carries nutrient-rich blood to the liver for processing. The liver is drained by the **hepatic veins** that empty into the inferior vena cava.

Before Going to Lab

1 Label the veins in Figure 30.10 using the descriptions supplied in Tables 30.10 and 30.11.

2 Label the veins in Figure 30.11 using the locations described in Table 30.11.

LAB ACTIVITY 9 Major Veins Draining the Thorax, Abdomen, and Pelvis

1 Identify the arteries on a model or chart, or use the search text box in **Real Anatomy** (Cardiovascular) to find these structures.

- accessory hemiazygos
- azygos
- hemiazygos
- hepatic veins
- inferior vena cava
- left brachiocephalic
- left common iliac
- left external iliac
- left gonadal
- left internal iliac
- left renal
- left suprarenal
- right brachiocephalic
- right common iliac
- right external iliac
- right gonadal
- right internal iliac
- right renal
- right suprarenal
- superior vena cava

1 _____

2 _____

3 _____

4 _____

5 _____

6 _____

7 _____

8 _____

9 _____

10 _____

11 _____

12 _____

13 _____

14 _____

15 _____

16 _____

17 _____

18 _____

19 _____

20 _____

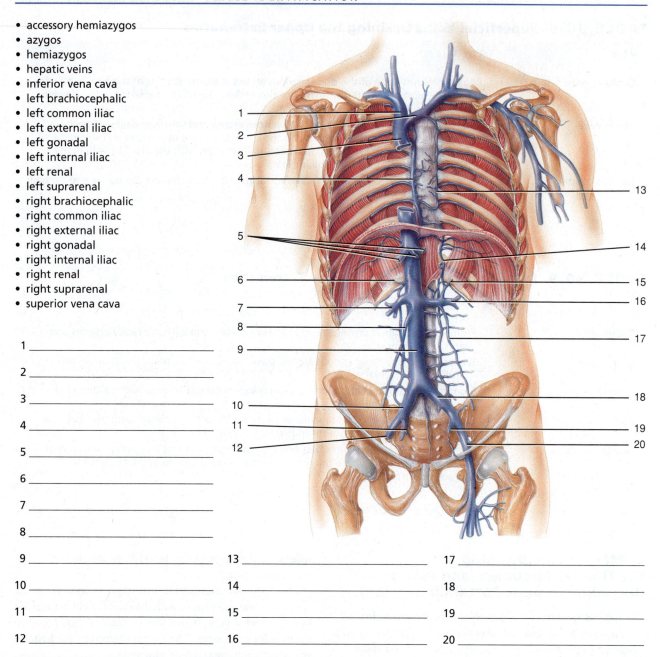

FIGURE 30.10 **Major veins draining the thorax, abdomen, and pelvis.**

TABLE 30.10 Veins Draining the Thorax

VEIN	LOCATION
Right and left brachio-cephalic veins (*brachi-* = arm; *cephalic* = head)	Formed by the merger of the subclavian and internal jugular veins and unite to form the superior vena cava
Azygos vein	Forms near the diaphragm; runs anterior and to the right of the vertebral column and ends in the superior vena cava
Hemiazygos vein	Begins below the diaphragm; runs anterior and to the left of the vertebral column and ends at the level of the ninth thoracic vertebra, where it joins the azygos vein
Accessory hemiazygos vein	Superior to the hemiazygos vein; runs anterior and to the left of the vertebral column from the level of the second thoracic vertebra to the eighth, joining the azygos vein

TABLE 30.11 Veins Draining the Abdomen and Pelvis

VEIN	LOCATION
Common iliac veins	Formed by the merger of the internal and external iliac veins near the sacroiliac joint and end when both common iliac veins form the inferior vena cava
Internal iliac veins	Begin near the greater sciatic notch and drain into the common iliac veins
External iliac veins	Continuations of femoral veins that are renamed within the inguinal region; drain into the common iliac veins
Gonadal veins	Run adjacent to the gonadal arteries in the posterior abdomen; left gonadal veins drain into the left renal vein; right gonadal veins drain into the inferior vena cava
Renal veins	Run anterior to renal arteries and drain into the inferior vena cava
Suprarenal veins	Run adjacent to the suprarenal arteries and drain into the inferior vena cava
HEPATIC PORTAL CIRCULATION	
Inferior mesenteric vein	Formed from the fan-like network of veins from the inferior mesentery; drains into splenic vein and joins the splenic vein near the pyloric region of the stomach to form the hepatic portal vein
Splenic vein	Drains the spleen and joins the superior mesenteric vein near the inferior region of the stomach to form the hepatic portal vein
Superior mesenteric vein	Formed from the fan-like network of veins of the superior mesentery; joins the splenic vein near the inferior region of the stomach to form the hepatic portal vein
Hepatic portal vein	Begins at merger of splenic vein and superior mesenteric vein and ends in the liver
Hepatic veins	Multiple hepatic veins form in the liver and drain into the inferior vena cava slightly inferior to the diaphragm

1
2
3

Gastric veins

4

5

Vein network
of inferior
mesenteric vein

Vein network
of superior
mesenteric vein

- hepatic
- hepatic portal
- inferior mesenteric
- splenic
- superior mesenteric

1 _____

2 _____

3 _____

4 _____

5 _____

FIGURE 30.11 Hepatic portal circulation.

5. Major Veins Draining the Lower Extremities

The great saphenous vein and the small (short) saphenous vein are the major *superficial veins* of the leg. The **great saphenous vein** is the longest vein in the body, running along the medial surface of the leg and thigh, and contains between 10 and 20 valves to prevent backflow of blood. The **small saphenous vein** ascends along the lateral posterior surface of the leg. Varicose veins are common in both of these veins.

Deep veins of the leg ascend in the leg adjacent to the arteries of the same name. The **anterior** and paired **posterior tibial veins** ascend in the anterior and posterior leg, and unite inferior to the popliteal fossa to form the popliteal vein. The paired **fibular (peroneal) veins** travel superiorly along the lateral leg and join the posterior tibial veins. The **popliteal vein** ascends along the posterior surface of the knee and becomes the **femoral vein,** which travels up the posterior thigh and becomes the **external iliac vein** in the pelvis.

Before Going to Lab

1 Label the veins in Figure 30.12 using the descriptions of veins in Table 30.12.

LAB ACTIVITY 10 — Major Veins Draining the Lower Extremities

1 Identify the arteries on a model or chart, or use the search text box in **Real Anatomy** (Cardiovascular) to find these structures. ■

(a)
- anterior tibial
- external iliac
- femoral
- great saphenous
- small saphenous

1 _____

2 _____

3 _____

4 _____

5 _____

(b)
- femoral
- fibular (peroneal)
- great saphenous
- popliteal
- posterior tibial
- small saphenous

6 _____

7 _____

8 _____

9 _____

10 _____

11 _____

(a) Anterior view (b) Posterior view

FIGURE 30.12 **Veins of the right lower limb.**

TABLE 30.12 Veins of the Lower Extremities

VEIN	LOCATION
Great saphenous veins	Superficial veins that form at the medial end of the dorsal venous arches (venous networks on dorsal surface of foot); run anterior to the medial malleolus, ascend along the medial leg and thigh, and drain into the femoral veins in the groin area
Small saphenous veins	Superficial veins that form at the lateral end of the dorsal venous arches; run posterior to the lateral malleolus, ascend along the lateral posterior leg, and empty into the popliteal veins
Posterior tibial veins	Paired deep veins formed from plantar veins draining the foot posterior to the medial malleolus; ascend deep within the posterior leg accompanying the posterior tibial arteries and merge with the anterior tibial veins just inferior to the popliteal fossa to form the popliteal veins
Fibular (peroneal) veins	Paired deep veins that run lateral to the posterior tibial veins; drain into the posterior tibial veins two-thirds of the way up the leg
Anterior tibial veins	Form from the dorsal venous arch (venous network on dorsal surface of foot); travel up the anterior surface of the leg accompanying the anterior tibial arteries; merge with the posterior tibial veins to form the popliteal veins just inferior to the popliteal fossa
Popliteal veins	Form from the merger of the anterior and posterior tibial veins just inferior to the popliteal fossa; become the femoral veins just superior to the knee
Femoral veins	Continuation of popliteal veins that begin just superior to the knee; accompany the femoral arteries up the posterior surface of the thigh; become the external iliac veins as they enter the pelvis

C. Pulmonary Circulation

The pulmonary circulation carries oxygen-poor blood from the right ventricle to the capillaries of the lung, where oxygen is added and carbon dioxide is removed. The **pulmonary trunk** carries oxygen-poor blood from the right ventricle and divides to form the **right** and **left pulmonary arteries,** which carry blood to the lungs. Oxygen-rich blood leaves the lungs via the **pulmonary veins,** which drain into the left atrium.

> **Before Going to Lab**
>
> **1** Label the blood vessels in Figure 30.13.

LAB ACTIVITY 11 **Major Blood Vessels of the Pulmonary Circulation**

1 Identify the arteries on a model or chart, or use the search text box in **Real Anatomy** (Cardiovascular) to find these structures. Note that all oxygen-rich blood vessels are red and oxygen-poor blood vessels are blue.

- left pulmonary artery
- left pulmonary vein
- pulmonary trunk
- right pulmonary artery
- right pulmonary vein

1 _____

2 _____

3 _____

4 _____

5 _____

FIGURE 30.13 **Blood vessels of the pulmonary circulation.**

D. Fetal Circulation

Because the fetus cannot breathe or ingest food, the mother's circulatory system provides the fetus with oxygen and nutrients, and eliminates carbon dioxide and other wastes from fetal blood. This exchange occurs across the **placenta,** which forms within the uterus during early pregnancy. Uterine blood vessels enter the placenta, and substances are exchanged by diffusion between maternal and placental capillaries without direct mixing of maternal and fetal blood. Two **umbilical arteries** carry oxygen-poor fetal blood to the placenta, and one **umbilical vein** carries the oxygen-rich blood back to the fetus. The **ductus venosus** allows blood to enter the inferior vena cava. Several circulatory modifications in the fetus allow blood to bypass the nonfunctioning fetal lungs and gastrointestinal tract (Table 30.13). The **foramen ovale** and the **ductus arteriosus** allow blood to bypass the fetal lungs, and after birth, vascular changes occur to allow blood to enter the lungs.

Before Going to Lab

1 Label the structures in Figure 30.14(a) and (b) using the description in Table 30.13.

LAB ACTIVITY 12 Fetal Circulation

1 Identify the arteries on a model or chart, or use the search text box in **Real Anatomy** (Cardiovascular) to find these structures. ■

TABLE 30.13 Fetal Circulation

STRUCTURE	LOCATION AND FUNCTION	STRUCTURAL MODIFICATION AFTER BIRTH
Umbilical arteries	Branch off internal iliac arteries; carry fetal blood to the placenta	Close and fill with connective tissue and become the medial umbilical ligaments
Umbilical vein	Formed at the placenta and travels to the liver, where it forms two branches, the ductus venosus and another vein that drains into the hepatic portal vein	Vessel collapses and renamed ligamentum teres (round ligament)
Ductus venosus	Major branch of umbilical vein that carries most of fetal blood from the placenta to the inferior vena cava	Vessel collapses and renamed ligamentum venosum
Foramen ovale	Opening in atrial septum that allows blood to pass from right atrium to left atrium	Closes and becomes the fossa ovalis
Ductus arteriosus	Arterial vessel that connects pulmonary trunk with the aorta	Closes and becomes the ligamentum arteriosum

SAFETY NOTE: *Wear safety glasses and gloves when using fresh or preserved tissue. Wash hands thoroughly with soap and water when you are done.*

E. Dissection of Blood Vessels

If you are dissecting a cat or fetal pig to observe blood vessels, refer to the appropriate dissection manual.

The blood vessels of a cat or fetal pig are similar to those of the human. The dissection will illustrate how blood vessels are enfolded in connective tissue and dive deep to disappear between skeletal muscles. Sometimes you will discover that an artery, vein, and nerve will be traveling together and will separate to journey to separate destinations.

Real Anatomy, a virtual cadaver dissection, can be used to complement or substitute for animal dissection of blood vessels.

(a) Fetal circulation

(b) Modifications after birth

(a)

- ductus arteriosus
- ductus venosus
- foramen ovale
- placenta
- umbilical arteries
- umbilical vein

1 _____

2 _____

3 _____

4 _____

5 _____

6 _____

(b)

- fossa ovalis
- ligamentum arteriosum
- ligamentum teres
- ligamentum venosum
- medial umbilical
 ligaments

7 _____

8 _____

9 _____

10 _____

11 _____

FIGURE 30.14 Fetal circulation and modifications after birth.

Reviewing Your Knowledge

A. Arteries and Veins

1. Which arteries carry oxygen-poor blood, and which veins carry oxygen-rich blood?

2. What are the external iliac arteries called when they pass the inguinal ligaments?

3. Varicose veins in the legs are common, but are there also varicose arteries in the legs? Why or why not?

4. What are the names of the paired veins that drain the head and neck?

5. In the arm and forearm, is it the deep veins or superficial veins that accompany arteries with the same names as the veins?

6. Which vein of the upper limb is typically used to draw a blood sample?

7. Where is the azygos vein located, and what area does it drain?

8. Usually arteries carry nutrient-rich blood, but there are veins that do also. What are their names?

B. Review of Arteries

Identify each artery in Figure 30.15 and write its name next to the number.

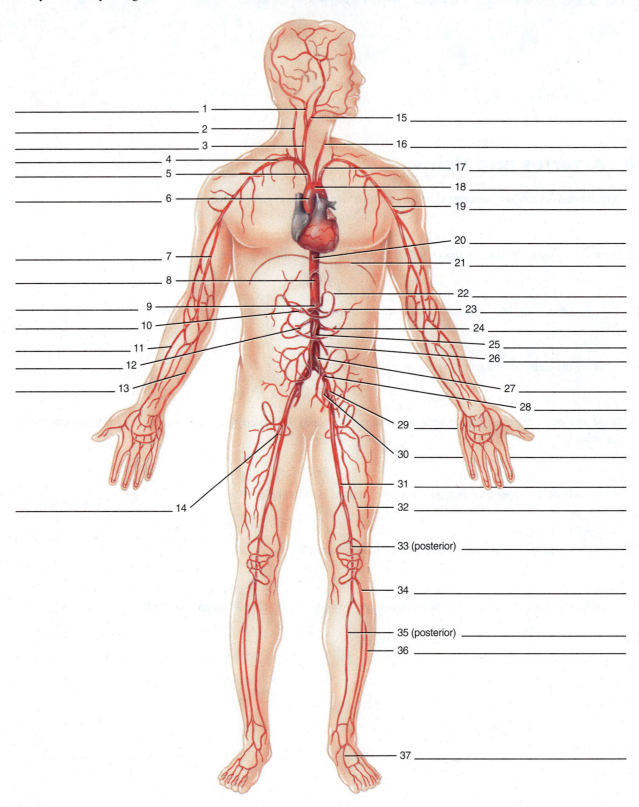

FIGURE 30.15 **Major arteries.**

C. Review of Veins

Identify each vein in Figure 30.16 and write its name next to the number.

FIGURE 30.16 **Principal veins.**

D. Arterial Blood Distribution

Complete the schematic diagrams of arterial blood distribution in Figure 30.17.

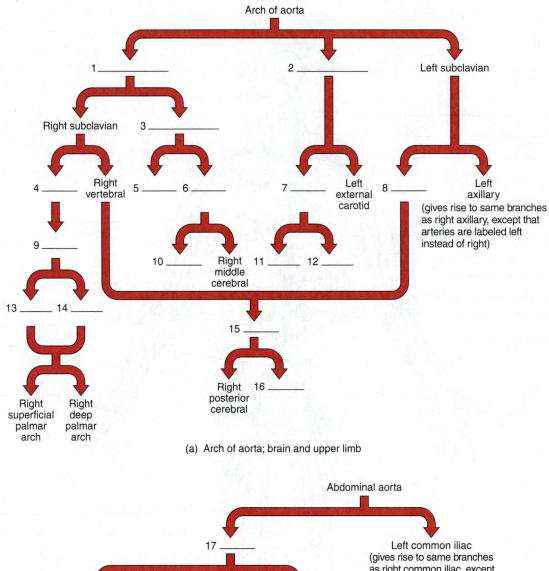

(a) Arch of aorta; brain and upper limb

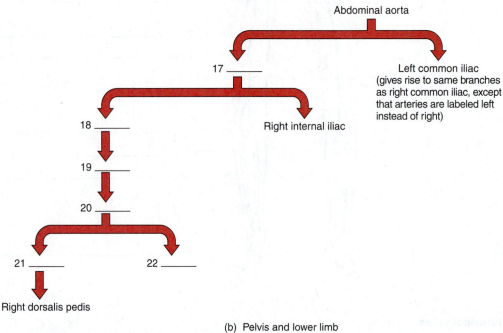

(b) Pelvis and lower limb

FIGURE 30.17 **Schematic diagram illustrating arterial blood distribution.**

Using Your Knowledge

A. Blood Vessel Identification

1. Balloon catheters are inserted into the femoral artery to unblock coronary vessels. Name all the arterial structures, *in order*, that the balloon catheter passes through from the right femoral artery to the right coronary artery.

2. Trace a drop of blood from the aorta to the *small intestine* and back to the right atrium. Include, *in order*, the major arteries and veins.

3. Trace a drop of blood from the aorta to the *skin of the face* and back to the right atrium. Include, *in order*, the major arteries and veins.

4. A blockage in the right internal carotid does not necessarily cause a loss of blood supply to the brain.
 (a) Identify the area of the brain supplied by the right internal carotid, and name the vessels that would supply blood to this area if there is a blockage in the right internal carotid artery.

 (b) Is a thrombus in the external carotid life-threatening? Why or why not?

Identify the blood vessel that supplies blood to each of the structures in questions 5–7:

5. Radius (bone) and radial nerve

6. Skin covering anterior thigh

7. Rectus femoris muscle

Use the angiogram in Figure 30.18 to answer questions 8–10.

FIGURE 30.18 Angiogram.

8. Identify the blood vessel marked in the angiogram. (*Note:* Only one branch of the aorta can be seen. The other two cannot be observed because of the angle of the angiogram.)

9. Identify the blood vessel marked in the angiogram.

10. Identify the blood vessel marked in the angiogram.

Lymphatic System Structure and Immune System Function

OBJECTIVES

1 List the functions of the lymphatic system
2 Describe lymph formation
3 Describe the lymphatic vessels and the flow of lymph through the body
4 Name the organs and tissues of the lymphatic system and list their functions
5 Describe the gross anatomy and histology of the lymph nodes, thymus, and spleen
6 Describe the immune cells involved in body defense
7 Dissect the cat or fetal pig and identify the main lymphatic structures

MATERIALS

- human torso or chart with lymphatic organs or use **Real Anatomy**
- compound microscope, lens paper, prepared slides of the lymph nodes, thymus, and spleen or use **Real Anatomy** (Histology)
- Dissection: preserved cats or fetal pigs, dissection equipment, disposable gloves, safety glasses, and cat or fetal pig dissection manual

The components of the lymphatic system are **lymphatic organs** and **tissues, lymphatic vessels,** and **lymph.** The **lymphatic system** is the structural location where much of the immune response takes place. The lymphatic system also collects excess interstitial fluid and returns this fluid to the bloodstream. In addition, it delivers dietary lipids and lipid-soluble vitamins (A, D, E, and K) absorbed from the small intestine to the bloodstream.

A. Lymph and Lymphatic Vessels

Lymph, the fluid found within lymphatic vessels, is formed from blood plasma. As blood flows through blood capillaries, hydrostatic and osmotic pressures filter more plasma out of the blood capillary and into the interstitial spaces than is drawn back inside. The excess fluid in the interstitial spaces (between blood capillaries and tissues) is now called **interstitial fluid.** When the interstitial fluid enters the **lymphatic capillaries,** which lie near blood capillaries and are closed at one end, it is called **lymph.** Interstitial fluid and lymph are identical in composition but different in location. Blood plasma, however, differs from both of these fluids in that plasma contains many plasma proteins (including albumin) that are too large to cross the blood capillary endothelium into the interstitial fluid.

Lymph flows from **lymphatic capillaries** into many **lymphatic vessels** that take lymph toward the neck region. Lymphatic vessels follow the pathway of veins in the body and also have a similar structure, but they have thinner walls and more valves than veins. Just as a body massage is helpful for moving blood through the veins and toward the heart, a light massage is also helpful for moving lymph through the lymphatic vessels and nodes to cleanse the lymph.

Lymphatic vessels flow into a series or chain of lymph nodes where lymph is filtered. Within a body region, the lymph vessels exiting the last lymph nodes in each chain of nodes merge to form **lymph trunks.** Lymph trunks merge to form two main ducts, the **thoracic duct** (left lymphatic duct) and the **right lymphatic duct.** In the abdominal area, lymph trunks merge to form a sac-like reservoir called the **cysterna chyli.** The long thoracic duct begins at the cysterna chyli and continues superiorly to drain the lymph from the legs, abdomen, left arm, and left side of the thorax, neck, and head into the left subclavian vein. The **right lymphatic duct** is the lesser and very short duct that drains lymph from the right arm via the right jugular trunk and right side of the thorax, neck, and head via the right subclavian trunk. Each duct enters the subclavian vein at its juncture with an internal jugular vein to drain lymph into venous blood.

Before Going to Lab

1 Label the lymphatic structures in Figures 31.1(a) and (b) and 31.2(a–c).

2 Examine the lymphatic vessel with valve in Figure 31.1(c).

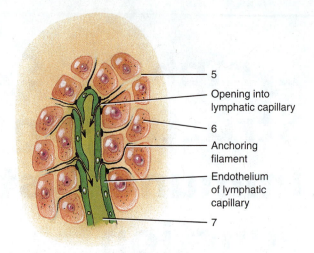

(b) Lymphatic capillary

- interstitial fluid
- lymph in lymphatic capillary
- tissue cell

5 _____

6 _____

7 _____

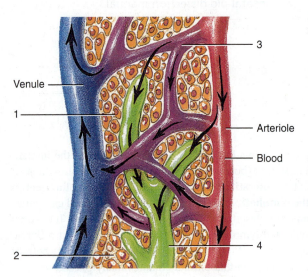

(a) Lymphatic and blood capillaries

- blood capillary
- interstitial (inter-STIH-shul) fluid
- lymphatic capillary
- tissue cell

1 _____

2 _____

3 _____

4 _____

(c) Photomicrograph of a lymphatic vessel and valve

FIGURE 31.1 **Lymphatic capillaries.**

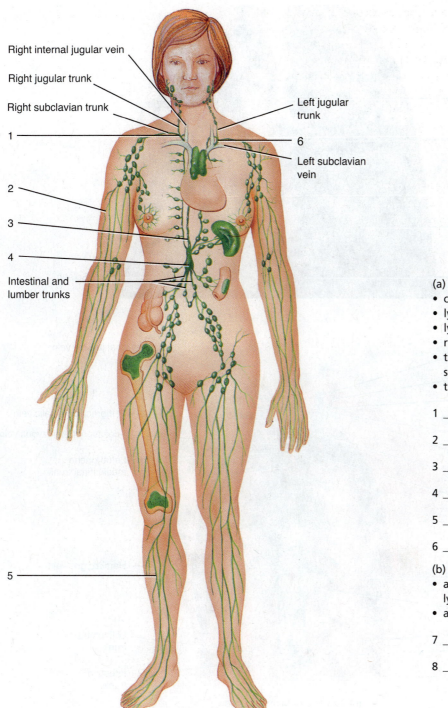

Right internal jugular vein

Right jugular trunk

Right subclavian trunk

1

2

3

4

Intestinal and
lumber trunks

5

Left jugular
trunk

6

Left subclavian
vein

(a) Overview of lymphatic vessels

7

8

(b) Lymph drainage

(a)
- cisterna chyli
- lymphatic vessel in arm
- lymphatic vessel in leg
- right lymphatic duct
- thoracic duct draining into
 subclavian vein
- thoracic duct in thorax

1 _____

2 _____

3 _____

4 _____

5 _____

6 _____

(b)
- area drained by right
 lymphatic duct
- area drained by thoracic duct

7 _____

8 _____

FIGURE 31.2 Lymphatic vessels.

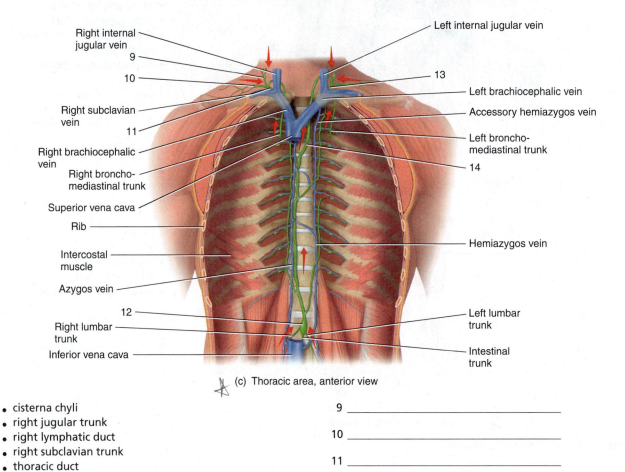

(c) Thoracic area, anterior view

- cisterna chyli
- right jugular trunk
- right lymphatic duct
- right subclavian trunk
- thoracic duct
- thoracic (left lymphatic) duct in thorax

9 _____

10 _____

11 _____

12 _____

13 _____

14 _____

FIGURE 31.2 **Lymphatic vessels,** *continued.*

B. Lymphatic Organs and Tissues

Lymphatic organs and tissues are located throughout the body. **Lymphatic organs** are surrounded by a capsule and include the thymus, lymph nodes, and spleen. **Lymphatic tissues** are not surrounded by a capsule and are called **lymphatic nodules.** Because their functions differ, lymphatic organs and tissues are divided into two categories based on function: primary lymphatic organs, and secondary lymphatic organs and tissues.

Primary lymphatic locations contain stem cells that produce lymphocytes and are the site where these lymphocytes become immunocompetent (recognize and mount immune system response). The two primary lymphatic organs are **red bone marrow** and the **thymus.** The thymus, which is located in the mediastinum just superior to the heart, is much larger in children and decreases considerably in size as they get older.

Secondary lymphatic organs and **tissues** are sites for defense against invading agents and cancer cells and include the **lymph nodes, spleen,** and **lymphatic nodules** (follicles). Lymph nodes are named for their location in the body. Single lymphatic nodules located in the connective tissue of mucous membranes of the gastrointestinal, respiratory, urinary, and reproductive tracts are called **mucosa-associated lymphoid tissue (MALT).** Aggregations of large nodules (follicles) include **Peyer's patches** in the small intestine, lymphoid follicles in the **appendix** (*appendix* = appendage), five tonsils (one **pharyngeal tonsil** and paired **palatine** and **lingual tonsils**), and **bronchial nodules** located in the walls of the respiratory tract. The pharyngeal tonsil is located in the posterior nasopharynx, while the palatine tonsils are in the posterior portion of the oral cavity. The lingual tonsils are located under the tongue. All of these nodules protect the body from foreign substances that are ingested, inhaled, or otherwise enter body openings.

Before Going to Lab

1 Label the lymphatic structures in Figure 31.3(a) and (b).

LAB ACTIVITY 1 Lymphatic Organs and Tissues

1 Identify the lymphatic structures in Figure 31.3(a) and (b) on a human torso or chart.

2 Use the **Real Anatomy** search text box to find lymphatic structures in Figure 31.3(b). ■

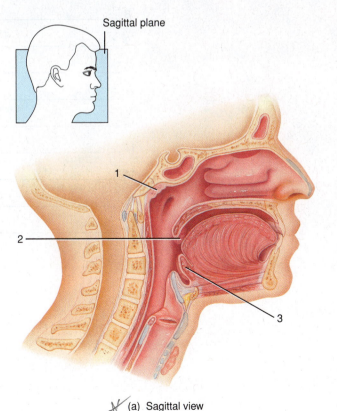

Sagittal plane

- lingual (LING-wal) tonsil
- palatine (PAL-ih-tine) tonsil
- pharyngeal (fuh-RIN-gee-al) tonsil

1 _____

2 _____

3 _____

(a) Sagittal view

FIGURE 31.3 Lymphatic organs and tissues.

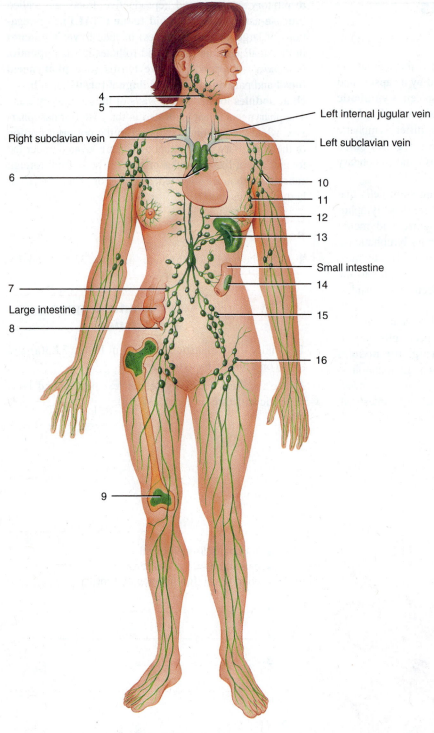

4

5

Left internal jugular vein

Right subclavian vein

Left subclavian vein

6

10

11

12

13

Small intestine

14

7

Large intestine

15

8

16

9

- appendix (vermiform)
- axillary lymph node
- cervical lymph node
- iliac lymph node
- inguinal lymph node
- intestinal lymph node
- mammary lymph node
- Peyer's patches (aggregated lymphatic follicles)
- red bone marrow
- spleen
- submandibular lymph node
- thoracic lymph node
- thymus (THY-mus)

4 _____

5 _____

6 _____

7 _____

8 _____

9 _____

10 _____

11 _____

12 _____

13 _____

14 _____

15 _____

16 _____

(b) Anterior view

FIGURE 31.3 **Lymphatic organs and tissues,** *continued.*

LAB ACTIVITY 2 Flow of Lymph

1 With your lab group, trace the flow of lymph from the thigh to the heart in Figure 31.4. Note that lymph moves superiorly to return to the circulatory system.
- brachiocephalic vein
- cisterna chyli
- iliac lymph nodes
- inguinal lymph nodes
- junction of jugular and subclavian veins
- lumbar trunk
- lymphatic capillaries
- lymphatic vessels (use term twice)
- superior vena cava
- thoracic duct

14. heart _____

↑

13. _____

↑

12. _____

↑

11. _____

↑

10. _____

↑

9. _____

↑

8. _____

↑

7. _____

↑

6. _____

↑

5. _____

↑

4. _____

↑

3. _____

↑

2. interstitial fluid _____

↑

1. blood capillaries _____

FIGURE 31.4 Flow of lymph.

C. Structure of the Lymph Node

Lymph nodes are bean-shaped structures that filter lymph and are usually found in groups. Clusters of these nodes are generally found in different body regions (i.e., the cervical, axillary, and inguinal areas) and are named for their location. Each node has a **capsule** with extensions called **trabeculae** (*trabecula* = little beam) that compartmentalize the node.

The nodes are composed of **reticular** (*reticulum* = process) **tissue,** with reticular fibers forming the net-like support for **reticulocytes** (specialized **fibroblasts**). Two main regions of lymph nodes are the **cortex** (*superficial region*) and **medulla** (*deep region*). There are many lighter-colored oval **lymphatic nodules** in the outer cortex that contain many immune system cells.

Lymph enters each node via many **afferent vessels;** travels through the following spaces, **subcapsular sinus, trabecular sinuses,** and **medullary sinuses;** and exits the node via fewer **efferent vessels.** The **hilum** (*hila* = depression or pit) is the indented area of the node where the efferent vessels leave the node. This design causes the lymph to move slowly through the lymph nodes, allowing time for immune system cells (lymphocytes and macrophages) to attack pathogens, cancer cells, and foreign molecules.

Enlarged lymph nodes could be caused by, but the causes are not limited to, infection, cancer, or scarring from a chronic lesion.

> **Before Going to Lab**
>
> **1** Label the lymph node structures in Figure 31.5(a), (b), and (c).
>
> **2** Observe lymph nodes and lymphatic vessel in Figure 31.6.

LAB ACTIVITY 3 Lymph Node Structure

1 Examine a prepared slide of a lymph node, or use **Real Anatomy** (Histology).

- Using low power, identify the structures listed in Figure 31.5(b).
- Using high power, identify the **reticulocytes** and **reticular fibers** of the lymph node [Figure 31.5(c)]. ■

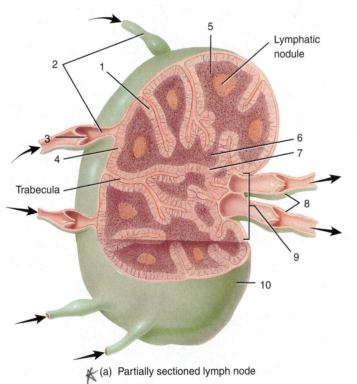

5 — Lymphatic nodule

2

1

3

4

Trabecula

6

7

8

9

10

(a) Partially sectioned lymph node

FIGURE 31.5 **Lymph node.**

- afferent vessels
- capsule
- cortex
- efferent vessels
- hilum
- medulla
- medullary sinus
- subcapsular sinus
- trabecular sinus
- valve

1 _____

2 _____

3 _____

4 _____

5 _____

6 _____

7 _____

8 _____

9 _____

10 _____

14

11

12

15

13

Blood
vessel

Medullary
sinus

LM 17×

(b) Photomicrograph of a
portion of a lymph node

17

16

LM 400×

(c) Photomicrograph of reticular
tissue within a lymph node

(b)
- capsule
- cortex
- lymphatic nodule
- medulla
- trabecula

11 _____

12 _____

13 _____

14 _____

15 _____

(c)
- reticular fiber
- reticulocyte

16 _____

17 _____

FIGURE 31.5 Lymph node, *continued*.

Inguinal ligament

Femoral artery

Sartorius

Lymph nodes

Vastus lateralis

Lymphatic vessels

Rectus femoris

Great saphenous vein

FIGURE 31.6 Lymph nodes and lymphatic vessels of left thigh.

D. Structure of the Thymus

The thymus is bi-lobed organ that is one of the primary lymphatic organs (see Figure 31.7a). Each lobe of the thymus is covered by a connective tissue **capsule. Trabeculae,** extensions of the capsule, divide each lobe into lobules. Each lobule has an outer **cortex** and inner **medulla.** It is located in the mediastinum between the sternum and the superior vena cava. The thymus is larger in infants and children and begins to atrophy during puberty and continues in adulthood. Not much active thymic tissue remains in an elderly person, and most of the thymus is infiltrated with fat.

The main purpose of the thymus gland is to support and stimulate immature T lymphocytes (T cells) that migrate to this gland from their origin in red bone marrow. With the help of thymic hormones, thymosin and others, T cells go through a maturation process that begins when they enter the lobule that "educates" them to recognize and enable an immune response against microbes and foreign molecules that do not belong to your body.

> **Before Going to Lab**
>
> **1** Label the structures of the thymus in Figure 31.7(b).

LAB ACTIVITY 4 **Thymus Structure**

1 Examine a prepared slide of the thymus, or use **Real Anatomy** (Histology).

2 Using low power, identify the structures listed in Figure 31.7(b). ■

(a) Thymus of adolescent

LM 30×

(b) Thymic lobules

- capsule
- cortex of lobule
- medulla of lobule
- trabecula of lobule

1 _____

2 _____

3 _____

4 _____

FIGURE 31.7 **Thymus.**

E. Structure of the Spleen

The largest lymphatic organ, the **spleen,** filters blood of foreign organisms and particles, eliminates aged erythrocytes, and is a blood reservoir for platelets. The spleen is the main organ that filters blood, similar to the lymph nodes being the main organs that filter the lymph. This highly vascular organ lies posterior and lateral to the stomach in the left upper quadrant (LUQ). The **splenic artery** enters the spleen and the **splenic vein** and **efferent lymphatic vessels** exit the spleen at the **hilum.**

The spleen is an encapsulated organ with extensions of the capsule forming **trabeculae.**

Histologically, the spleen has regions of white pulp and red pulp that are named for their appearance in fresh specimens. On prepared slides you will see stained specimens with different colors. **White pulp** resembles nodules with many lymphocytes and macrophages and appears dark purple when stained. Both B and T lymphocytes carry out immune functions here, and macrophages engulf any aged blood cells, foreign material, or debris. **Red pulp** stores

platelets and contains reticular fibers that cannot be seen at this magnification. It stains a lighter color because its many red blood cells are anucleate. Many macrophages can also be found here to phagocytize foreign substances in the blood.

Before Going to Lab

1 Label the splenic structures in Figure 31.8(a) and (b).

LAB ACTIVITY 5 Structure of the Spleen

1 Examine a prepared slide of the spleen, or use **Real Anatomy** (Histology).
- Using low power, identify the capsule, trabecula, red pulp, and white pulp.
- Using high power, identify reticular fibers, reticulocytes, and lymphocytes in white pulp and the reticular fibers, reticulocytes, and red blood cells (RBCs) in the red pulp.

SUPERIOR

1
Gastric impression
2
Colic impression
3
Renal impression

(a) Gross structure

4
Red pulp
5
Central artery
6

LM 25×

(b) Portion of the spleen

(a)
- hilum
- splenic artery
- splenic vein

1 _____
2 _____
3 _____

(b)
- capsule
- trabecula
- white pulp

4 _____
5 _____
6 _____

FIGURE 31.8 The spleen.

F. Immune Cells Involved in Body Defense

The cells involved in body defense are the **T cells** (lymphocytes), **B cells** (lymphocytes), and **macrophages** (*macro-* = big; *phago-* = to eat). T and B cells (immune cells) and macrophages are originally produced in red bone marrow from *pluripotent stem cells,* which are immature stem cells that can develop into various mature blood cells.

B cells complete their maturation and become immunocompetent in the red bone marrow. Immature T cells migrate to the thymus through the bloodstream to complete maturization and become immunocompetent there.

After being programmed for their work, mature T and B cells are now immunocompetent and are released into the bloodstream. They travel to secondary lymphatic organs and tissue to be activated and cloned in the immune response. Secondary lymphatic organs include the lymph nodes and the spleen. Lymphatic tissue (nodules) includes MALT, Peyer's patches, appendix, tonsils, and bronchial nodules.

When stimulated and activated by specific antigens, **B cells** transform into a specialized group of cloned **plasma cells** that make **antibodies** against foreign intruders. This type of immune response is called **antibody-mediated immunity.** B cells also make **B memory cells** that will become active if the same antigen is introduced into your body.

T cells are involved in **cell-mediated immunity** by bringing into play three types of **T cells.** When stimulated by specific antigens, **helper T cells** clone and activate both cytotoxic T cells and B cells. **Cytotoxic T cells** clone and directly attack foreign cells with toxic chemicals that destroy them. After B and T cells either destroy or render foreign cells and molecules inactive, **macrophages** phagocytize debris. Memory T cells are also produced to become activated later on.

Before Going to Lab

1 Observe a macrophage (pink) phagocytizing a bacterium (green) in Figure 31.9(a).

LAB ACTIVITY 6 Cells Involved in Body Defenses

1 With your lab group, identify where T and B cells are produced and programmed, and where active T and B cells are located in Figure 31.9(b). Note that one letter is used more than once. ■

G. Dissection of Lymphatic Organs

You may have already looked at the lymphatic organs in your previous dissections. As you dissected the blood vessels, you may have noticed small, 1/8- to 1/4-inch **lymph nodes** in the axillary or inguinal areas. Because the nodes are small, they are easy to miss if you do not know their structure or location.

Consult the appropriate cat or fetal pig dissection manual for your dissection instructions.

Macrophages

Bacterium

(a) Phagocytosis by a macrophage, human lung

FIGURE 31.9 **Cells of the lymphatic and immune systems.**

(a) origin of T cells
(b) origin of B cells
(c) T cells become immunocompetent
(d) B cells become immunocompetent
(e) active T cells, B cells, and macrophages

1 Palatine tonsil

2 Submandibular node

3 Cervical node

4 Thymus gland

5 Intestinal node

6 Appendix

7 Red bone marrow

Axillary node 8

Spleen 9

Peyer's patches 10

Iliac node 11

Inguinal node 12

(b) Location of immune system cells

FIGURE 31.9 **Cells of the lymphatic and immune systems,** *continued.*

Reviewing Your Knowledge

A. Structure of Lymphatic Organs and Tissues

Write the name of the organ or tissue that matches the description. Terms may be used more than once.

_____ 1. Regresses considerably with age

_____ 2. Largest lymphatic organ; red and white pulp

_____ 3. Bean-shaped; has efferent and afferent vessels

_____ 4.

_____ 5. } Primary lymphatic organs

_____ 6.

_____ 7. } Secondary lymphatic organs

_____ 8. Secondary lymphatic nodules associated with the nasal and oral cavities

_____ 9. Secondary lymphatic nodules located in connective tissue of mucous membranes

_____ 10. Secondary lymphatic nodules located in the small intestine

B. Physiology of Lymphatic Tissues and Organs

Write the name of the organ or tissue that matches the description. Terms may be used more than once.

Peyer's patches
appendix (vermiform)
lymph nodes
mucosa-associated lymphoid tissue (MALT)
red bone marrow
spleen
thymus
tonsils

_____ **1.** Filters lymph

_____ **2.**

_____ **3.**

_____ **4.** } Sites for the immune response

_____ **5.**

_____ **6.**

_____ **7.**

_____ **8.** Filters blood and contains red and white pulp

_____ **9.** Programs T cells for immunocompetence

_____ **10.** Programs B cells for immunocompetence

C. Formation of Lymph

Write the name of the term that matches the phrase.

blood
interstitial fluid
plasma

_____ **1.** Fluid that becomes interstitial fluid

_____ **2.** Fluid that becomes lymph

_____ **3.** Interstitial fluid plus blood cells and large proteins

D. Lymphatic Vessels

Write the name of the vessel that matches the description. Terms may be used more than once.

cisterna chyli
lymphatic vessels
right lymphatic duct
thoracic duct
lymphatic trunks

_____ 1. Sac-like vessel

_____ 2. Left lymphatic duct

_____ 3. Merge to form lymphatic trunks

_____ 4. Merge to form lymphatic ducts

_____ 5. The thoracic duct begins here

_____ 6. Drains the legs, abdominal area, and left side of the body

_____ 7. Drains the right side of thorax, arm, head, and neck

_____ 8. The longer lymphatic duct

E. Immune Cells Involved in Body Defense

Write the name of the immune cells that matches the description. Terms may be used more than once.

_____ 1. Maturation and immunocompetency in red bone marrow

_____ 2. Maturation and immunocompetency in the thymus gland

_____ 3.⎱
 ⎰ Located in secondary organs and tissues
_____ 4.⎰

F. Immune Response

Complete the flow chart in Figure 31.10 with the correct terms from the bulleted list below.

- antibodies
- antibody-mediated immunity
- cell-mediated immunity
- cytotoxic T cells
- mature T cells

- memory B cells
- memory T cells
- plasma cells
- T helper cells
- thymus

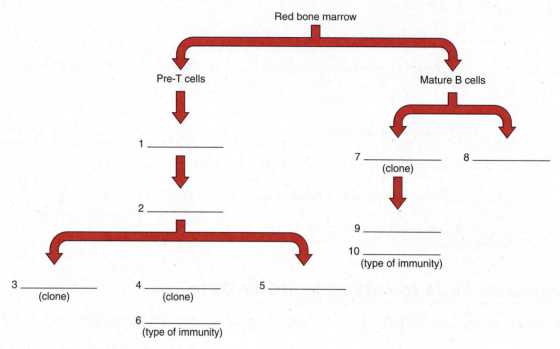

FIGURE 31.10 Cells of the immune response.

Using Your Knowledge

A. Lymphatic and Immune Systems

1. How would the lymph that is leaving the cisterna chyli differ from lymph draining into the right lymphatic duct?

2. If you look into a child's mouth, there are "golf balls" puffing from each side of the oral cavity. What are these structures?

3. How do lymphatic capillaries differ from blood capillaries?

4. If you dissected a fetal pig or a young cat, what difference would you expect in the thymus compared to the adult?

5. Why are the walls of lymphatic vessels thin, like veins?

6. Bacteria and viruses in infected tissues easily enter lymph vessels. Explain why.

7. If a woman had a radical mastectomy (removal of a cancerous breast, surrounding tissues with axillary lymph nodes, and anterior thoracic muscles), would you expect the arm on that side to be edematous (have edema)? Explain.

8. Would you describe the flow of lymph through lymph nodes as being fast or slow compared with blood capillaries? Explain.

9. What would be missing in lymph exiting a lymph node compared with lymph entering the node?

10. A patient has enlarged right inguinal lymph nodes that are very tender to the touch. Examination of the feet reveals a small cut between the right 3rd and 4th toes, which are warm and red. What would your initial impression be, and why are the lymph nodes enlarged?

11. Figure 36.11 depicts a person with elephantitis. How does a person contract this disease?

12. How does this disease affect the lymphatic system?

FIGURE 31.11 **Person with elephantitis.**

Respiratory System Structure and Function

OBJECTIVES

1 Locate and identify the organs of the respiratory system
2 Describe the structure and function of the respiratory system organs
3 Trace the path of air from the nose to the alveoli
4 Dissect a cat, fetal pig, or cadaver, and identify the main respiratory organs

MATERIALS

- models or charts with respiratory organs or use **Real Anatomy** (Respiratory)
- sagittal section model or chart of the human head, neck, and thorax; larynx model or use **Real Anatomy** (Respiratory)
- compound microscope, lens paper, prepared slides of trachea and lung or use **Real Anatomy** (Histology)
- Dissection: preserved cats or fetal pigs, dissection equipment, disposable gloves, safety glasses, and cat or fetal pig dissection manual
- **Real Anatomy:** Virtual Cadaver Dissection

The respiratory system provides an airway for movement of air into and out of the body. It is also the site where atmospheric oxygen diffuses into the bloodstream to be delivered to all body cells and carbon dioxide produced by these cells diffuses out of the bloodstream to be exhaled into the atmosphere. This gas exchange takes a coordinated effort of the respiratory and cardiovascular systems.

A. Gross Anatomy of the Respiratory Organs

The main organs of the respiratory system are the nose, pharynx, larynx, trachea, bronchi, and lungs. Clinically, the nose and pharynx are organs of the upper airways, and the larynx, trachea, bronchi, and lungs constitute the lower airways. Functionally the respiratory system is divided into the **conducting zone** (nose to the terminal bronchioles) and the **respiratory zone** (respiratory bronchioles to the alveoli).

Before Going to Lab

1 Label the structures of the respiratory system in Figure 32.1(b).

LAB ACTIVITY 1 **Structures of the Respiratory System**

1 Identify the respiratory structures on a model or chart, or use the search text box in **Real Anatomy** (Respiratory) to find these structures. ■

1. Nose

The framework of the **nose** is composed of bone and cartilage. Bones make up the base of the nose, and hyaline cartilage the anterior portion. Two **external nares** (*naris,* sing.) are the openings for air to enter a space called the **nasal cavity,** which is lined with a mucous membrane. The **nasal septum** separates this cavity into right and left portions. The nasal septum is highly vascularized and is the site of nosebleeds. A deviated septum may result from trauma, such as a blow to the nose, and can cause difficulty in breathing. On either side of the nose, three curved bony structures extend into the nasal cavity: the

Larynx

Trachea

Lungs

Right primary (main) bronchus

Right phrenic nerve

(a)

- larynx (LAIR-inks) 6
- lungs 3
- nasal cavity 5
- nose 4
- pharynx (FAIR-inks) 1
- right primary or main bronchus (BRON-kus) 2
- trachea (TRAY-key-a) 7

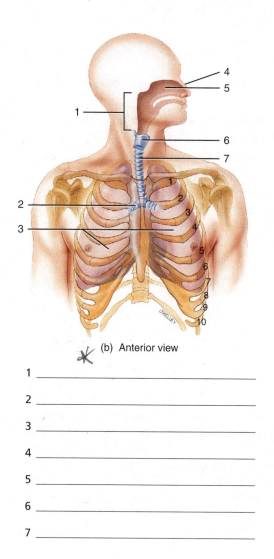

(b) Anterior view

1 _____

2 _____

3 _____

4 _____

5 _____

6 _____

7 _____

 FIGURE 32.1 **Organs of the respiratory system.**

superior, middle, and **inferior nasal conchae** (*concha* = shell-shaped), or **turbinates** (*turbinum* = scroll-shaped). Beneath each nasal concha is a concavity called a **meatus** (*meatus* = passage) that increases surface area in the nose. When incoming air passes over the vascularized mucous membranes of the meatuses, the air is warmed and humidified. This is the reason that the pharynx gets dry if a person's nose is stopped up and breathing is done through the mouth. The bony **hard palate** forms the floor of the nasal cavity. At the junction of the hard and soft palates are the **internal nares,** two openings that lead into the tubular nasal pharynx posteriorly. When the skull was discussed in the axial skeleton exercise, it was mentioned that the nasal cavity is surrounded by four paranasal sinuses.

The frontal, maxillary, ethmoidal, and sphenoidal bones have mucous membrane-lined sinuses or cavities that also warm and moisten the air. These paranasal cavities have ducts that drain into the nasal cavity.

Before Going to Lab

1 Label the structures of the nose in Figure 32.2(a) and (b).

LAB ACTIVITY 2 Nasal Structures

1 Identify the nasal structures on a sagittal section model or chart, or use the search text box in **Real Anatomy** (Respiratory) to find these structures.

(a) Sagittal section

(b) Frontal section

(a)
- external naris (NEH-ris) 8
- hard palate 9
- inferior nasal concha or turbinate (CON-cha; 4 TUR-bin-it)
- inferior nasal meatus 7 (mee-AY-tus)
- internal naris 1
- middle nasal concha or turbinate 3
- middle nasal meatus 6
- superior nasal concha or turbinate 2
- superior nasal meatus 5

1 _____
2 _____
3 _____
4 _____
5 _____
6 _____
7 _____
8 _____
9 _____

(b)
- inferior nasal concha 14
- inferior nasal meatus 12
- middle nasal concha 13
- middle nasal meatus 10
- nasal septum 11

10 _____
11 _____
12 _____
13 _____
14 _____

FIGURE 32.2 Nasal structures within nasal cavity.

2. Pharynx

The **pharynx** or throat is divided into three regions: the nasopharynx (*naso-* = nose), oropharynx (*oro-* = mouth), and laryngopharynx (*laryngo-* = larynx), corresponding to the anatomical structures nearby. The **nasopharynx** begins at the internal nares and ends at the soft palate. The **oropharynx** begins at the soft palate and extends to the level of the hyoid bone, and the **laryngopharynx** extends from the hyoid bone to the beginning of the esophagus. The **soft palate** is found in the pharynx as a posterior extension of the hard palate. The oval-shaped **uvula** dangles inferiorly as an extension of the soft palate. During swallowing, the soft palate pushes superiorly to close off the nasopharynx and to direct food toward the laryngopharynx.

Two **auditory tubes** (Eustachian tubes) have openings from the middle ear into the nasopharynx, and the single **pharyngeal tonsil** (adenoid) is located in its posterior wall. The oropharynx has two pairs of tonsils, the **palatine**

(palate) and **lingual** (tongue) tonsils. Note that the mucous membranes of the nose, paranasal sinuses, pharynx, and middle ear are connected by ducts, which explains why infections can spread to any adjacent areas.

Before Going to Lab

1 Label the three divisions of the pharynx in Figure 32.3(b).

2 Label the pharyngeal structures in Figure 32.3(c).

LAB ACTIVITY 3 The Pharynx

1 Identify the pharyngeal structures on a sagittal model or chart, or use the search text box in **Real Anatomy** (Respiratory) to find these structures. ■

(a)

FIGURE 32.3 The pharynx and associated structures.

(b) Divisions of the pharynx

- laryngopharynx 3
 (la-rin-go-FAIR-inks)
- nasopharynx 1
- oropharynx 2
 (oro-FAIR-inks)

1 _____

2 _____

3 _____

(c) Pharyngeal structures

(c)

- internal naris 10
- laryngopharynx 9
- lingual (LING-gwal) tonsil 13
- nasopharynx 5
- opening of auditory tube
 (Eustachian) 6
- oropharynx 8
- palatine tonsil 12
- pharyngeal (fair-IN-gee-al)
 tonsil 4
- soft palate 11
- uvula (YOU-view-la) 7

4 _____ 9 _____

5 _____ 10 _____

6 _____ 11 _____

7 _____ 12 _____

8 _____ 13 _____

FIGURE 32.3 **The pharynx and associated structures,** *continued.*

3. Larynx

The **larynx** or voice box connects the laryngopharynx with the trachea. This organ consists of nine hyaline cartilages and houses the vocal cords. The larynx contains three paired cartilages—the arytenoid, cuneiform, and corniculate cartilages—which are small cartilages located in the posterior wall of the larynx, and three single cartilages—the thyroid cartilage, cricoid cartilage, and epiglottis—which constitute the main body of the larynx. The largest cartilage, the **thyroid cartilage** (*thyroid* = shield-shaped), is seen anteriorly and has a prominence called the Adam's apple, which is made of hyaline cartilage. The **cricoid cartilage** (*cricoid* = ring-like), which is inferior to the thyroid cartilage, is larger on the posterior side than on the anterior side and consists of hyaline cartilage. As the term *cricoid* suggests, this is the only cartilage that is a complete ring in the larynx or trachea. The **glottis** is the opening that allows air into the larynx. The oval-shaped **epiglottis** (*epi-* = over; *-glottis* = tongue) is composed of elastic cartilage and closes over the glottis during swallowing. If you have food or drink in your mouth and start laughing, the glottis opens and you could aspirate these substances into the larynx with the air. Coughing forces the food or liquid

back into the laryngopharynx. Lateral to the glottis are two pairs of folds. The superior folds are the **ventricular** (little belly) **folds** (vestibular folds), or false vocal cords, while the **vocal folds** or true vocal cords are inferior and medial. The vocal folds attach via small muscles to the **arytenoid** (*arytenoid* = ladle-like) **cartilage.** The vocal folds, which contain elastic tissue, vibrate with phonation or speaking. If these folds become inflamed or have tumors, they cannot vibrate and hoarseness (laryngitis) results.

> #### Before Going to Lab
>
> **1** Label the laryngeal structures listed in Figure 32.4(c), (d), (e), and (f).

LAB ACTIVITY 4 The Larynx

1 Identify the laryngeal structures on a model or chart, or use the search text box in **Real Anatomy** (Respiratory) to find these structures.

2 Palpate your thyroid cartilage. Swallow and feel the thyroid cartilage move upward. ■

Epiglottis

Thyroid cartilage

Cricoid cartilage

Tracheal cartilage

(a) Anterior view

Epiglottis

Thyroid cartilage

Cricoid cartilage

(b) Posterior view

FIGURE 32.4 **The larynx and associated structures.**

(c) Anterior view

(d) Posterior view

(c)
- cricoid (CRY-koid) cartilage
- epiglottis
- thyroid cartilage

1 _____

2 _____

3 _____

(d)
- cricoid cartilage
- epiglottis
- thyroid cartilage

4 _____

5 _____

6 _____

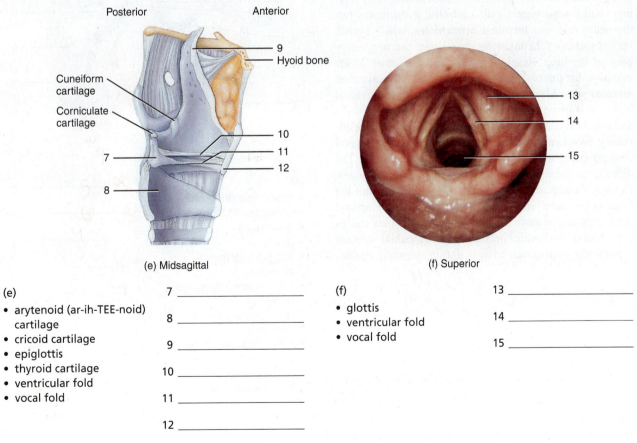

(e) Midsagittal

(f) Superior

(e)
- arytenoid (ar-ih-TEE-noid) cartilage
- cricoid cartilage
- epiglottis
- thyroid cartilage
- ventricular fold
- vocal fold

7 _____

8 _____

9 _____

10 _____

11 _____

12 _____

(f)
- glottis
- ventricular fold
- vocal fold

13 _____

14 _____

15 _____

FIGURE 32.4 **The larynx and associated structures,** *continued.*

4. Bronchial Tree

Just as a tree has a trunk, branches, twigs, and fruit, the lower airways have a trachea, bronchi, bronchioles, and alveoli. If the trachea and bronchi were inverted, they would resemble the branches of a tree, hence the name **bronchial tree.** The **trachea** (*trachea* = sturdy) or windpipe is a tube-like conduit that conducts air from the larynx to the bronchi. It is located anterior to the esophagus and can be palpated on the anterior surface of the neck. When palpating the trachea, horizontal tracheal cartilages that keep the airway open can be felt as bumps one on top of the other. **Tracheal cartilages** are C-shaped, and a smooth muscle, the **trachealis muscle,** connects the open sides of the C-shaped cartilage. The trachealis muscle is on the posterior surface of the trachea and allows expansion of the esophagus into the trachea during swallowing. The trachea ends at the **carina** (*carina* = keel of a boat) inferiorly and branches into a shorter right and longer left primary bronchus, each serving the corresponding lung. Each **primary (main) bronchus** branches into **secondary (lobar) bronchi** to supply a **lobe** of each lung. The right primary bronchus branches into three secondary bronchi, and the left primary bronchus branches into two secondary bronchi. The secondary bronchi further subdivide into **tertiary (segmental) bronchi** that enter **bronchopulmonary segments** within a lobe. Tertiary bronchi divide into **bronchioles,** each serving small compartments called **lobules.** Bronchioles further subdivide into **terminal bronchioles,** which branch into **respiratory bronchioles** that begin the respiratory zone of the lung. Respiratory bronchioles further divide into **alveolar ducts** that lead into clusters of alveoli called **alveolar sacs. Alveoli** bud off alveolar sacs like individual grapes in a grape cluster. Simple squamous alveolar cells and simple squamous blood capillary cells form the **respiratory membrane,** where gas exchange occurs. Several changes occur as the bronchi turn into bronchioles. The diameter of the airway diminishes, the epithelium changes, cartilage disappears early in the tertiary bronchi, and the bronchioles have only smooth muscle spiraled around them. Because smooth muscle in the bronchioles causes constriction, asthmatic inhalers with bronchial dilators cause the smooth muscle to relax during asthmatic attacks.

Before Going to Lab

1 Label the bronchial tree structures in Figure 32.5(a) and (b) and Figure 32.6.

LAB ACTIVITY 5 **Bronchial Tree**

1 Locate the bronchial tree structures on a lung model, or use the search text box in **Real Anatomy** (Respiratory) to find these structures.

2 Palpate the sternal angle to locate where the trachea divides into the primary (main) bronchi.

3 Observe the sheep pluck demonstration with the bronchial tree dissected. ■

(a) Anterior view

(a)
- bronchiole (BRON-key-ol) 10
- carina (ka-RYE-na) 3
- diaphragm 5
- larynx 6
- left primary (main) bronchus 8
- left tertiary (segmental) bronchus 9
- right primary (main) bronchus 2
- right secondary (lobar) bronchus 4
- trachea 1
- tracheal cartilage 7

1 _____
2 _____
3 _____
4 _____
5 _____
6 _____
7 _____
8 _____
9 _____
10 _____

FIGURE 32.5 **Bronchial tree.**

CLINICAL NOTE: *The airways can be clinically divided into three areas according to size:*

> **Large airways**—*trachea and bronchi*
>
> **Medium airways**—*bronchioles and terminal bronchioles*
>
> **Small airways**—*respiratory bronchioles, alveolar ducts, alveolar sacs, and alveoli*

Respiratory medicine uses these terms to identify the location of diseases. Bronchitis is a large airway disease, asthma is a middle airway disease, and emphysema is a small airway disease.

(b)

- alveolar (al-VEE-oh-lur) sac
- primary (main) bronchus
- respiratory bronchiole
- secondary (lobar) bronchus
- terminal bronchiole
- tertiary (segmental) bronchus

11 _____

12 _____

13 _____

14 _____

15 _____

16 _____

(b) Bronchial tree branching

FIGURE 32.5 Bronchial tree, *continued.*

- alveolar ducts
- alveolar sac
- alveoli
- respiratory bronchiole

1 _____

2 _____

3 _____

4 _____

FIGURE 32.6 Diagram of portion of lung lobule.

5. Lungs

The two **lungs** are divided into *three lobes* on the right side and *two lobes* on the left. The three right lobes are the **superior, middle,** and **inferior lobes,** and the left lobes are called the **superior** and **inferior lobes.** The rounded superior part of the lung is the **apex,** and the broader inferior part is the **base,** which rests on the diaphragm. The left lobe has a concave surface called the **cardiac notch,** which has the apex of the heart projecting into it. Each lung has a **hilum,** an area surrounded with pleura, where the bronchi, blood and lymphatic vessels, and nerves enter or exit the medial side of the lung. The lungs are in the thoracic cavity and are separated from each other by the heart and the mediastinum. **Parietal pleura** lines the thoracic cavity wall, and **visceral pleura** covers the surface of each lung.

The **pleural cavity** is the space between the two pleural layers that contains pleural fluid.

Before Going to Lab

1 Label the structures of the lung in Figure 32.7(a), (b), (c), and (d).

2 Label the parts of the pleura in Figure 32.8(a) and (b).

LAB ACTIVITY 6 Lungs and Pleura

1 Identify the lung structures on a model or chart, or use the search text box in **Real Anatomy** (Respiratory) to find these structures. ■

(a) Anterior right lung

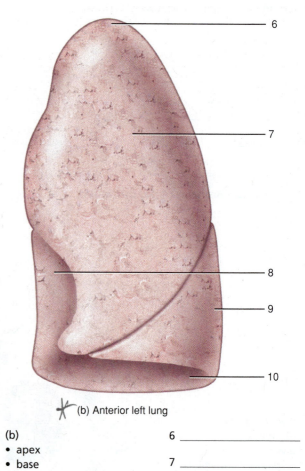

(b) Anterior left lung

(a)
- apex
- base
- inferior lobe
- middle lobe
- superior lobe

1 _____

2 _____

3 _____

4 _____

5 _____

(b)
- apex
- base
- cardiac notch
- inferior lobe
- superior lobe

6 _____

7 _____

8 _____

9 _____

10 _____

FIGURE 32.7 Lung structures.

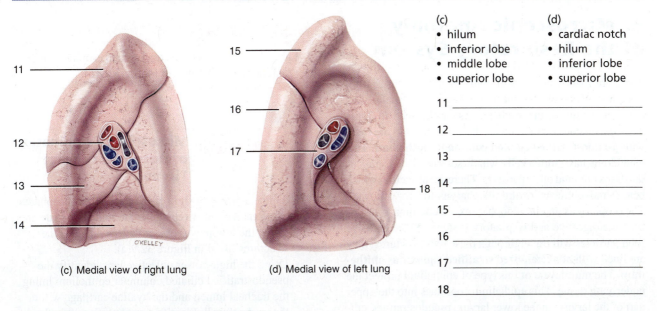

(c) Medial view of right lung

(d) Medial view of left lung

(c)
• hilum
• inferior lobe
• middle lobe
• superior lobe

(d)
• cardiac notch
• hilum
• inferior lobe
• superior lobe

11 _____

12 _____

13 _____

14 _____

15 _____

16 _____

17 _____

18 _____

FIGURE 32.7 **Lung structures,** *continued.*

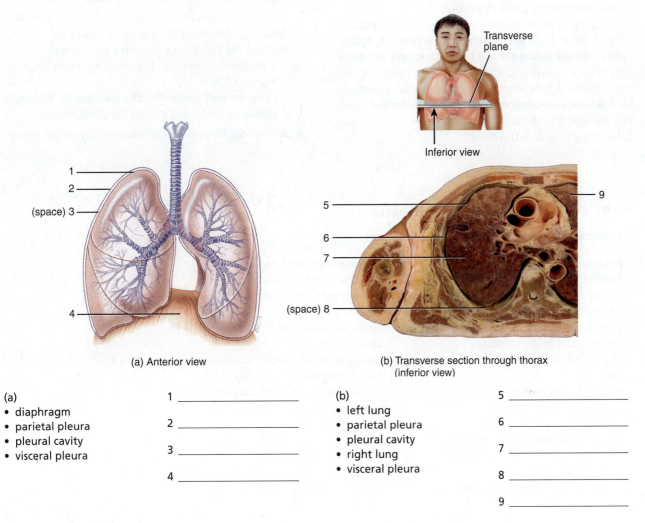

(a) Anterior view

(b) Transverse section through thorax
(inferior view)

(a)
• diaphragm
• parietal pleura
• pleural cavity
• visceral pleura

1 _____

2 _____

3 _____

4 _____

(b)
• left lung
• parietal pleura
• pleural cavity
• right lung
• visceral pleura

5 _____

6 _____

7 _____

8 _____

9 _____

FIGURE 32.8 **Pleura.**

B. Microscopic Anatomy of the Respiratory System

There are four basic types of epithelium in the respiratory system, with areolar connective tissue as the underlying tissue just beneath the epithelia. As can be seen in Table 32.1, the mucous membranes of the upper airways are lined with **pseudostratified ciliated columnar epithelium.** This epithelium has **goblet cells** which secrete viscous mucus that traps dust and other particles. There are also **cilia,** which beat to move mucus toward the pharynx to be swallowed. The oropharynx and laryngopharynx are common areas for both the digestive and respiratory systems. Since abrasive food particles will travel through these two structures, they are lined with nonkeratinized **stratified squamous epithelium.** The multilayers of this type of epithelium protect the underlying tissue. This epithelium continues into the upper part of the larynx. In the lower larynx, pseudostratified ciliated columnar epithelium resumes and extends through the trachea and primary (main) bronchi.

Deeper into the bronchial tree, the epithelium gradually changes from high **pseudostratified columnar epithelium** to **simple columnar** and finally to **simple cuboidal,** with an accompanying gradual loss of cilia. The respiratory zone structures (respiratory bronchioles to the alveoli) all have **simple squamous epithelium** that allows the highly vascularized alveoli to facilitate diffusion of respiratory gases, oxygen, and carbon dioxide.

Before Going to Lab

1 Label the structures of the trachea and esophagus in Figure 32.9(a) and the lung in Figure 32.9(b).

LAB ACTIVITY 7 Microscopic Examination of Respiratory Tract Tissues

1 Examine a prepared slide of the trachea and esophagus, or use **Real Anatomy** (Histology) to find these structures.
- Using the low-power objective lens, identify the structures listed in Figure 32.9(a).
- Using the high-power objective lens, identify the pseudostratified ciliated columnar epithelium lining the tracheal lumen and the hyaline cartilage within the tracheal wall. Identify a goblet cell within the epithelium.

2 Examine a prepared slide of lung tissue, or use **Real Anatomy** (Histology) to find these structures.
- Using the low-power objective lens, identify alveoli and, if present, the other structures listed in Figure 32.9(b).
- Using the high-power objective lens, identify the simple squamous epithelium forming the alveolar walls.

3 Answer the Discussion Questions with your lab group.

TABLE 32.1 Respiratory Epithelia

RESPIRATORY STRUCTURE	EPITHELIA
Nasal cavity	Pseudostratified ciliated columnar
Paranasal sinuses	Pseudostratified ciliated columnar
Nasopharynx	Pseudostratified ciliated columnar
Oropharynx	Stratified squamous
Laryngopharynx	Stratified squamous
Larynx (vocal folds and above)	Stratified squamous
Larynx (below vocal folds)	Pseudostratified ciliated columnar
Trachea	Pseudostratified ciliated columnar
Primary (main) bronchi	Pseudostratified ciliated columnar
Secondary (lobar) bronchi	Simple columnar with fewer cilia
Tertiary (segmental) bronchi	Simple columnar; no cilia
End of tertiary (segmental) bronchi	Simple cuboidal; no cilia
Bronchioles, including terminal bronchioles	Simple cuboidal; no cilia
Respiratory bronchioles	Begins as simple cuboidal; ends as simple squamous
Alveolar ducts, alveolar sacs, alveoli	Simple squamous

FIGURE 32.9 **Sectional views of the trachea and lung.**

(a) Transverse section of trachea and esophagus

(b) Sectional view of lung

(a)
- epithelial lining of trachea
- lumen of esophagus
- lumen of trachea
- tracheal cartilage
- trachealis muscle

1 _____
2 _____
3 _____
4 _____
5 _____

(b)
- alveolar ducts
- alveolar sacs
- alveoli
- respiratory bronchiole
- simple squamous epithelium

6 _____
7 _____
8 _____
9 _____
10 _____

DISCUSSION QUESTIONS
Microscopic Anatomy of Respiratory System

1 Which connective tissue type allows the lungs to expand and regain their original size and shape (recoil) during inhalation and exhalation?

2 Which connective tissue type in the bronchial tree provides the support to maintain an open airway?

3 Discuss the significance of the changing epithelia along the respiratory tree.

C. Dissection of Respiratory Structures

If you are dissecting a cat or fetal pig to observe respiratory system structures, refer to the appropriate dissection manual.

The respiratory system organs of a cat or fetal pig are similar to those of the human. This dissection will illustrate the structure of the larynx and trachea, the relationship of respiratory organs to other organs in the mediastinum, and the connective tissues surrounding these organs.

Real Anatomy, a virtual cadaver dissection, can be used to complement or substitute for animal dissection of respiratory structures.

Reviewing Your Knowledge

A. Air Flow

Trace the air flow through the respiratory system starting with the *external nares*. Number the structures 1 through 17.

_____ alveolar duct

_____ alveolar sac

_____ alveolus

_____ bronchiole

_____ external nares

_____ internal nares

_____ laryngopharynx

_____ larynx

_____ nasal cavity

_____ nasopharynx

_____ oropharynx

_____ primary (main) bronchus

_____ respiratory bronchiole

_____ secondary (lobar) bronchus

_____ terminal bronchiole

_____ tertiary (segmental) bronchus

_____ trachea

B. Structural Changes in the Respiratory Tree

Name three structural changes that occur in the bronchi as they branch into bronchioles.

1. _____

2. _____

3. _____

C. Structure and Function of the Respiratory System

Write the name of the structure described.

_____ **1.** Tubular airways that begin the respiratory zone

_____ **2.** Connects the laryngopharynx with the trachea

_____ **3.** Tube-like structure that conducts air from the larynx to the bronchi

_____ **4.** Closes over the glottis during swallowing

_____ **5.** Keep the trachea from collapsing

_____ **6.** Division of the bronchi that enter bronchopulmonary segments

_____ **7.** Last division of the conducting zone

_____ **8.** Conducts air from the nasopharynx to the laryngopharynx

_____ **9.** Small, round sacs where gas exchange occurs

_____ **10.** Small conduction airway that serves a lobule

D. Location and Function of Epithelial Tissue

Match the type of epithelium with its location in the respiratory system and its function.

Epithelial Tissue

 a. pseudostratified ciliated columnar d. simple squamous
 b. simple columnar e. stratified squamous
 c. simple cuboidal

Location

_____ **1.** Nasal cavity through nasopharynx

_____ **2.** Oropharynx through larynx above vocal cords

_____ **3.** Larynx below vocal cords through primary (main) bronchi

_____ **4.** Secondary (lobar) bronchi through tertiary (segmental) bronchi

_____ **5.** Bronchioles through beginning of respiratory bronchiole

_____ **6.** End of respiratory bronchiole through alveoli

Function

_____ **7.** Secretes mucus to trap and remove dust and debris

_____ **8.** Diffusion of respiratory gases

_____ **9.** Protects underlying tissues

Using Your Knowledge

A. Anatomy of the Respiratory System

Answer the following questions.

1. Explain how it is possible for a person to drink liquid and then have the liquid come out through the nose when the person laughs.

2. Explain how an infection in the nasopharynx can also result in an infection in the paranasal sinuses and/or the middle ear.

3. Asthma attacks are caused by smooth muscle spasms in the bronchial tree. These spasms can close the airway. What airway structures are closed? Why are these areas closed but not the rest of the bronchial tree?

4. Emphysema is characterized by destructive changes of alveolar walls resulting in large air-filled spaces due to overinflation. There is a loss of lung elasticity, decreased gas exchange, and large air-filled spaces.

 Pneumonia is an acute inflammation of the lungs with the alveoli and bronchioles becoming plugged with an exudate (fluid), and the alveolar cell membranes thickening.

 In Figure 32.10, identify which alveolar section came from a person with no lung disease (normal alveoli), with emphysema, or with pneumonia.

(a) _____ (b) _____ (c) _____

FIGURE 32.10 Normal and diseased alveolar sections.

Explain how the changes in alveolar structure cause a decrease in blood oxygen levels in:

5. Emphysema

6. Pneumonia

Identify the structures in Figures 32.11 and 32.12.

7 _____ (include right or left)

8 _____ (membrane)

FIGURE 32.11 Thoracic cavity, cross-section.

9 _____

10 _____

FIGURE 32.12 Bronchoscopic view of the larynx.

Pulmonary Ventilation

OBJECTIVES

1 Describe how respiratory muscle contraction causes thoracic volume changes

2 Explain how changes in thoracic and lung volumes and lung pressures result in pulmonary ventilation

3 Define and measure or calculate lung volumes and capacities

4 Discuss the relationship between blood carbon dioxide levels and pulmonary ventilation

MATERIALS

- models, charts, or use **Real Anatomy** (Muscular)
- bell jar model of lungs
- **Effect of Exercise on Lung Volumes and Capacities:** handheld dry spirometer, disposable mouthpieces, nose clip, 70% alcohol and cotton or alcohol wipes
- **Formation of Bicarbonate and Hydrogen Ions:** paper bag (lunch size), straws, 50-mL beakers, 100-mL beakers, 25-mL graduated cylinders, pH meter, squeeze bottle of deionized water, stopwatch or watch with second hand
- **PowerPhys Experiments:**
 - Respiratory Volumes
 - Regulation of Pulmonary Ventilation
- **Biopac Laboratory Guide Experiments:**
 - Effect of Exercise on Lung Volumes and Capacities
 - Comparison of Predicted and Measured FVC and FEV_1

The **respiratory system supplies oxygen needed** by body cells to produce adenosine triphosphate (ATP) for metabolism and removes carbon dioxide produced by metabolic reactions. Respiration involves three steps and requires the cardiovascular system to transport oxygen and carbon dioxide throughout the body.

1. **Pulmonary ventilation,** or breathing, is the movement of air between the atmosphere and the lungs that occurs when we inhale (inhalation) and exhale (exhalation).

2. **External respiration** is the movement of oxygen from the alveoli into pulmonary capillaries and carbon dioxide from pulmonary capillaries to the alveoli.

3. **Internal respiration** is the movement of oxygen from capillaries into body cells and carbon dioxide from body cells into capillaries.

A. Pulmonary Ventilation

During pulmonary ventilation, air moves from an area of higher pressure to an area of lower air pressure. Changes in air pressure in the lung (alveolar pressure) occur when lung volume changes. The relationship between pressure and volume is described by Boyle's Law, which states that the pressure of a gas in a closed container is inversely proportional to the volume of the container. Therefore, when lung volume increases, the pressure of the air inside decreases and air flows into the lungs. When lung volume decreases, the pressure of the air inside increases and air flows out of the lungs.

1. Changing Thoracic and Lung Volumes

When the diaphragm and other respiratory muscles contract or relax, they change the size of the thorax, which in turn changes lung volume. Normal inhalation is caused mainly by contraction of the diaphragm. The **diaphragm** is dome-shaped when relaxed and flattens when contracted. When the diaphragm flattens, the length of the thoracic cavity and its volume increase. During quiet inhalation, contraction of the **external intercostal muscles** further increases the width of the thoracic cage by raising the ribs. Contraction of the **sternocleidomastoid, scalenes,** and **pectoralis minor muscles** cause a greater increase in thoracic volume during forced inhalation by elevating the rib cage and sternum, resulting in a greater volume of air inhaled.

Normal exhalation is a passive process involving relaxation of the diaphragm and elastic recoil of the chest wall and lungs. The diaphragm becomes dome-shaped and decreases the length of the thoracic cavity and thoracic volume. In forced exhalation, contraction of the **internal intercostal muscles** depresses the rib cage. Contraction of the **abdominal muscles (external oblique, internal oblique, transverse abdominis,** and **rectus abdominis)**

pushes the diaphragm superiorly, further decreasing thoracic volume and resulting in a greater volume of exhaled air.

Before Going to Lab

1 Label the respiratory muscles in Figure 33.1. Refer to Exercise 14: Skeletal Muscles and Their Actions or your textbook.

LAB ACTIVITY 1 Respiratory Muscles and Volume Changes

1 Identify the respiratory muscles on a model or chart, or use the search text box in **Real Anatomy** (Respiratory) to find these structures.

2 Pronounce the muscle names as you point to them.

3 Complete Table 33.1 by circling the correct choice in the column "Effect on Thoracic Dimensions."

4 Examine the radiographs in Figure 33.2. Decide whether a radiograph was taken after inhalation or exhalation, and put your answer in the appropriate blank. ■

TABLE 33.1 Respiratory Muscle Functions

MUSCLE	FUNCTION	EFFECT ON THORACIC DIMENSIONS (circle correct choice)
External intercostals	Elevate ribs	1. Increase or decrease diameter
Internal intercostals	Depress ribs	2. Increase or decrease diameter
Sternocleidomastoids	Elevate the sternum	3. Increase or decrease diameter
Scalenes	Elevate 1st and 2nd ribs	4. Increase or decrease diameter
Pectoralis minors	Elevate 3rd, 4th, and 5th ribs	5. Increase or decrease diameter
Diaphragm	Flattens when contracted	6. Increase or decrease length
Abdominal muscles	Compress abdominal contents, increase abdominal pressure, and force diaphragm superiorly	7. Increase or decrease length

(a) Anterior superficial view (b) Anterior deep view

- diaphragm
- external intercostals
- external oblique
- internal intercostals
- internal oblique
- pectoralis (pek-tor-A-lis) minor

- rectus abdominis
- scalenes (SKAY-leens)
- sternocleidomastoid (ster-no-kli-doh-MAS-toid)
- transverse abdominis

1 _____ 6 _____

2 _____ 7 _____

3 _____ 8 _____

4 _____ 9 _____

5 _____ 10 _____

Linea alba

FIGURE 33.1 Respiratory muscles.

(a)

(b)

Indicate whether the view shows inhalation or exhalation.

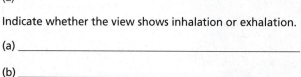

(a) _____

(b) _____

FIGURE 33.2 **Radiographs of the thorax taken either after inhalation or exhalation.**

2. Pressure Changes During Pulmonary Ventilation

As the thorax expands, the parietal pleura attached to the internal thoracic wall is pulled outward. The **pleural cavity** slightly increases in volume, causing a decrease in **intrapleural pressure,** the pressure between the pleural layers. The decrease in intrapleural pressure and the surface tension of pleural fluid cause the visceral pleura, and therefore the lungs, to be pulled outward. As the lungs increase in volume, the **alveolar (intrapulmonic) pressure** decreases to below atmospheric pressure, and air enters the lungs. When the thorax decreases in size the lungs recoil, causing intrapleural pressure and alveolar pressure to increase, and air leaves the lungs.

Before Going to Lab

1 Label the bell jar model lung in Figure 33.3.

LAB ACTIVITY 2 **Pressure Changes During Pulmonary Ventilation**

1 Using the model lung (bell jar model), observe what happens when the rubber diaphragm at the base of the model is domed or flattened.

2 Complete Table 33.2 by circling the correct choice in the column "Movement of Diaphragm."

3 Answer Discussion Questions with your lab group.

DISCUSSION QUESTIONS
Bell Jar (Model Lung) Demonstration

1 (a) In the model lung demonstration, the bell jar is filled with air. What is in the pleural cavity in the thorax? (b) Compare the relative size of the pleural space in the model lung to that in the human lung.

2 (a) In the model lung, is the air in the bell jar at atmospheric pressure or subatmospheric pressure? (b) Is the pressure in the pleural cavity atmospheric or subatmospheric?

3 (a) If the chest wall (including parietal pleura) is punctured, would the pressure in the pleural cavity increase or decrease? (b) Would lung size increase or decrease? (c) What is this condition called?

4 (a) In humans, is there one pleural cavity surrounding both lungs or two separate pleural cavities, one surrounding each lung? (b) Does this differ from the model lung?

5 In the model lung demonstration, air flows into the balloons through a rigid tube that divides once. Compare the rigidity and the number of branches of the conducting structures in the model to the rigidity and number of branches of the conducting structures in the human respiratory system.

6 In the model lung demonstration, the bell jar does not change in size. Describe the changes in the corresponding human respiratory system structures during inhalation and exhalation.

- bronchus
- diaphragm
- lungs covered with visceral pleura
- parietal pleura and thoracic wall
- pleural cavity
- trachea

1 _____
2 _____
3 _____
4 _____
5 _____
6 _____

FIGURE 33.3 Model lung (bell jar model).

TABLE 33.2 Bell Jar (Model Lung) Demonstration

OBSERVATIONS	MOVEMENT OF DIAPHRAGM	
	As Diaphragm Domes (circle correct choice)	*As Diaphragm Flattens* (circle correct choice)
Volume of air in bell jar	1. Increased or decreased	5. Increased or decreased
Pressure of air in bell jar	2. Increased or decreased	6. Increased or decreased
Balloon size	3. Inflated or deflated	7. Inflated or deflated
Direction of air flow	4. Into or out of balloons	8. Into or out of balloons

B. Lung Volumes and Capacities

Lung volumes and capacities are defined in Table 33.3. Notice that lung capacities are equal to 2 or more volumes. Both lung volume and lung capacity values vary according to gender, age, and height. Lung volume and some, but not all, capacity values vary with exercise.

1. Measuring and Calculating Lung Volumes and Capacities

A **spirometer** is an instrument used to measure lung volumes. There are different types of spirometers, depending on how they measure air volumes and capacities. Some spirometers measure only expiratory volumes and capacities, while others measure both inspiratory and expiratory volumes and capacities. In the following activity, directions are given for using a dry, handheld spirometer. *Exhale only into this type of spirometer*. It does not measure inspiratory volumes. Spirometers that measure inspiratory air volumes must be cleaned or supplied with new filters when used by a different person to prevent spread of infection. Your instructor will supply you with instructions if you are using a different type of spirometer instead of the dry, handheld spirometer.

Dry, handheld spirometers cannot be used to measure inspiratory volumes and capacities. However, IRV and IC can be calculated using values for TV, ERV, and VC.

Clinically, FRC is measured indirectly with special equipment, and RV is calculated using values for FRC and ERV. In this activity, RV will be calculated with equations that were developed using RV values from many individuals.

LAB ACTIVITY 3 **Experiment: Effect of Exercise on Lung Volumes and Capacities**

1 **Prediction:** With your lab group, predict how exercise will affect TV, IRV, ERV, VC, and TLC. Circle your answer (in italics) below.
 - During exercise TV will *increase, decrease,* or *stay the same.*
 - During exercise IRV will *increase, decrease,* or *stay the same.*
 - During exercise ERV will *increase, decrease,* or *stay the same.*
 - During exercise VC *will increase, decrease,* or *stay the same.*
 - During exercise TLC *will increase, decrease,* or *stay the same.*

2 **Materials:** Obtain the materials for Effect of Exercise on Lung Volumes and Capacities given in the Materials list.

3 **Data Collection:** Measure TV, ERV, and VC before and after exercise and record the values in Table 33.4.
 - Decide who will be the subject, who will prepare the spirometer, who will prepare and coach the subject, and who will be the timer and recorder. The subject should not have heart problems and should be in good health.

TABLE 33.3 Lung Volumes and Capacities

VOLUME OR CAPACITY	ABBREVIATION	DEFINITION	AVERAGE RESTING VALUES
Tidal volume	TV	The amount of air inhaled and exhaled during one normal breath	500 mL
Inspiratory reserve volume	IRV	Maximum amount of air that can be inhaled after a normal inhalation	3,100 mL
Expiratory reserve volume	ERV	Maximum amount of air that can be exhaled after a normal exhalation	1,200 mL
Residual volume	RV	Amount of air that remains in the lungs after a maximal exhalation	1,200 mL
Inspiratory capacity	IC	IC = TV + IRV	3,600 mL
Functional residual capacity	FRC	FRC = RV + ERV	2,400 mL
Vital capacity	VC	Maximum volume of air expelled after a maximal inhalation VC = IRV + TV + ERV	4,800 mL
Total lung capacity	TLC	TLC = IRV + TV + ERV + RV	6,000 ML

TABLE 33.4 **Measured Lung Volumes and Capacities**

AT REST	VALUE 1	VALUE 2	VALUE 3	AVERAGE VALUE
TV	N/A	N/A	N/A	
ERV				
VC				
AFTER EXERCISE	**VALUE 1**	**VALUE 2**	**VALUE 3**	**AVERAGE VALUE**
TV				
ERV				
VC				

Preparing the spirometer
- Wipe the nozzle of the spirometer with 70% alcohol.
- Place a clean, disposable mouthpiece on the nozzle.
- With use, water will condense inside the dial. Follow instructions included with your spirometer for removing condensation.

Preparing and coaching the subject
- Have the subject use a noseclip or fingers to close the nostrils to prevent air leaking out of the nose.
- When blowing into the spirometer, it should be held in a horizontal position with the face of the dial upward.
- Have subject practice breathing to measure TV, ERV, and VC. When **measuring TV,** the subject should breathe normally. When **measuring ERV,** the subject should take two or three breaths, then inhale normally and exhale as much air as possible. Verbally encourage the subject to exhale as much air as possible. When **measuring VC,** the subject should take two or three normal breaths, then completely fill the lungs with air, and exhale as much air as possible.
- Remind subject to *exhale only* into a handheld spirometer.

Measuring TV, ERV, and VC at rest
- Measure tidal volume (TV)—Reset the dial on the spirometer to 0. The subject should take two or three normal breaths and then inhale normally. The spirometer is then placed in the subject's mouth and the subject is instructed to exhale normally into the spirometer. Repeat this process *two more times,* and do **not reset** the dial before each measurement. Divide the volume by 3 to obtain the average TV. Record the average TV in Table 33.4.

- Measure expiratory reserve volume (ERV)—Reset the dial to 0. The subject should take two or three normal breaths, and then inhale normally and exhale normally. Without taking another breath, place the spirometer to the subject's mouth and have the subject exhale as much air as he or she can. Record the value for ERV in Table 33.4. Repeat this process two more times, *resetting the dial to 0 before each measurement.*
- Measure vital capacity (VC)—Reset the dial to 0. Have the subject take two or three normal breaths, then inhale as much air as possible. Place the spirometer to the subject's mouth and instruct the subject to exhale as much air as he or she can. Record the value for VC in Table 33.4. Repeat this process two more times, *resetting the dial to 0 before each measurement.*

Measuring TV, ERV, and VC after exercise
- The subject should not have heart problems and should be in good health.
- Have the subject run in place for 5 minutes.
- Measure tidal volume (TV)—Reset the dial on the spirometer to 0. Have the subject take two or three breaths, and then inhale normally. Place the spirometer to the subject's mouth and instruct the subject to exhale normally into the spirometer. Record the value for TV in Table 33.4. Repeat this process two more times, but this time *reset the dial before each measurement.*
- Measure expiratory reserve volume (ERV) as before.
- Measure vital capacity (VC) as before.

4 Clean up as directed by your instructor.

5 Data Analysis: Calculate average ERV and VC and record in Table 33.4. Calculate IRC, IC, FRC, and TLC and record values in Table 33.5.

Calculate average ERV and VC

- Add the individual ERVs and divide by 3. Record the average ERV in Table 33.4.
- Add the individual VCs and divide by 3. Record the average VC in Table 33.4.

Calculate IRV and IC using average TV, ERV, and VC values

- Use the equations for VC and IC in Table 33.3 to calculate IRV and IC. Rearrange the VC equation to formulate an equation to calculate IRV.
- Calculate the values and record them in Table 33.5.

Calculate RV, FRC, and TLC

- Use the appropriate equation in Table 33.6 to calculate your RV based on gender and age, and record the RV in Table 33.5.
- Calculate your FRC and TLC using the equations in Table 33.3 and record the value in Table 33.5.

6 Graphs: If graph paper is available, construct a bar graph that compares resting and exercising lung volumes and capacities.

7 Complete the Experimental Report with your lab group.

8 Complete **PowerPhys Experiment:** Respiratory Volumes.

9 Complete **Biopac Laboratory Guide Experiment:** Effect of Exercise on Lung Volumes and Capacities.

EXPERIMENTAL REPORT
Effect of Exercise on Lung Volumes and Capacities

Results: State how the various lung volumes and capacities changed with exercise.

Discussion:

1 Refer to Table 33.7 to answer this question. If TLC and RV remain constant, indicate whether an increase in TV would cause the following volumes and capacities to increase, decrease, or remain the same.
- IRV—increase, decrease, remain the same
- ERV—increase, decrease, remain the same
- IC—increase, decrease, remain the same
- VC—increase, decrease, remain the same

2 Discuss **why** TV changed during exercise.

3 Discuss **why** the change of TV during exercise either did or did not cause a change in the following lung volumes and capacities.
- IRV
- ERV
- IC
- FRC
- VC
- TLC

Conclusion: In one or two sentences, state how exercise affects lung volumes and capacities.

TABLE 33.5 Calculated Lung Volumes and Capacities

VOLUME OR CAPACITY	EQUATION	VALUE
IRV		
IC		
RV		
FRC		
TLC		

TABLE 33.6 Equations for Calculating Residual Volume

GENDER	AGE	EQUATION TO PREDICT RV (liters)
Male or female[1]	<19	RV = (0.029 × height in inches) − 0.9192
Female[2]	19–99	RV = (0.0813 × height in inches) + (0.009 × age in years) − 3.9
Male[2]	19–99	RV = (0.0686 × height in inches) + (0.017 × age in years) − 3.45

1. Polgar G, Promadhat V: *Pulmonary Function Testing in Children. Techniques and Standards*. Philadelphia: WB Saunders Co., 1971.

2. Goldman HI, Becklake MR: Respiratory function tests—normal values at median altitudes and the prediction of normal results. *Am Rev Tuberc* 79:457–467, 1959.

TABLE 33.7 Lung Volumes and Capacities

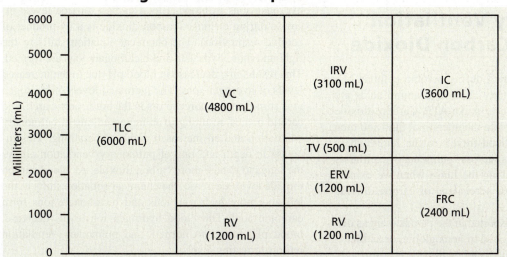

2. Timed Forced Expiratory Volumes

A timed forced expiratory volume is used to indicate whether someone has a condition in which air flow is obstructed or restricted. The subject is asked to inhale as much air as possible and then forcibly blow as much air as fast as possible into the spirometer. The total volume of forcibly and rapidly expelled air is the **forced vital capacity (FVC),** and the volume of air forcibly expelled in the first second is the **forced expiratory volume in 1 second (FEV$_1$).** These volumes and the ratio of FEV$_1$/FVC are used to determine if a person has an obstructive lung disease such as emphysema, chronic obstructive pulmonary disease (COPD), or asthma or a restrictive lung disease such as pneumonia or lung cancer.

LAB ACTIVITY 4 Measuring FEV$_1$ and FVC

1 The graph in Figure 33.4 illustrates the change in volume over time. The *x*-axis indicates time, and the *y*-axis indicates volume.

2 Use the values on the *y*-axis to determine the maximum volume indicated by the curved line. Record this FVC value in the space under the graph in Figure 33.4.

3 On the *x*-axis, find 1 second after the beginning of the forced exhale. Using a ruler, draw a vertical line that intersects the curve at this point. Find the corresponding volume on the *y*-axis, the FEV$_1$, and record this value in Figure 33.4.

4 Calculate the FEV$_1$/FVC ratio by dividing the FEV$_1$ by the FVC, and record this value in Figure 33.4.

5 Complete the **Biopac Laboratory Guide Experiment:** Comparison of Predicted and Measured FVC and FEV$_1$. ■

FVC = _____ L FEV$_1$ = _____ L FEV$_1$/FVC = _____

FIGURE 33.4 Timed forced expiratory volume.

C. Control of Pulmonary Ventilation: The Role of Carbon Dioxide

Carbon dioxide is formed during aerobic cellular respiration when glucose, fatty acids, and amino acids are chemically broken down to make ATP. Carbon dioxide then diffuses out of cells into the interstitial fluid and then into the bloodstream. Blood travels to the lungs where carbon dioxide diffuses across pulmonary capillaries and into the alveoli, leaving the lungs when we exhale. Exhaled air has a greater percentage of carbon dioxide than atmospheric air.

Carbon dioxide is transported in the bloodstream as dissolved carbon dioxide, bound to hemoglobin, or as **bicarbonate ions.** The majority of carbon dioxide is transported as bicarbonate ions. Bicarbonate ions (HCO_3^-) are formed from carbon dioxide as follows:

1. Carbon dioxide combines with water to form **carbonic acid.**

$$CO_2 + H_2O \rightleftharpoons H_2CO_3$$

2. Carbonic acid dissociates in water to form hydrogen ions and bicarbonate ions.

$$H_2CO_3 \rightleftharpoons H^+ + HCO_3^-$$

The chemical equation for the formation of bicarbonate and hydrogen ions from carbon dioxide and water is:

$$CO_2 + H_2O \rightleftharpoons H_2CO_3 \rightleftharpoons H^+ + HCO_3^-$$

The ratio of carbon dioxide to carbonic acid to bicarbonate ions is constant. If blood carbon dioxide levels increase (as occurs after exercise), then more carbonic acid and bicarbonate and hydrogen ions are formed. If carbon dioxide levels are reduced (as occurs in lungs after expiration), then more bicarbonate ions and hydrogen ions will form carbonic acid, which will dissociate back into carbon dioxide and water.

The pH of blood is determined by the concentration of hydrogen ions in blood. Blood carbon dioxide levels increase during exercise (carbon dioxide is a by-product of aerobic respiration). The chemical equation shifts to the right and more hydrogen and bicarbonate ions are formed. This results in a decrease in blood pH due to the increased levels of hydrogen ions. The increased levels of hydrogen ions stimulate chemoreceptors in the brain, aortic arch, and carotid bodies, resulting in an increased rate of pulmonary ventilation and an increased depth of breathing. This increase in depth and rate of pulmonary ventilation allows the lungs to exhale more carbon dioxide. As blood carbon dioxide levels decrease, the chemical equation shifts to the left and more hydrogen ions and bicarbonate ions form carbonic acid. The blood hydrogen ion levels decrease, blood pH returns to normal, and pulmonary ventilation returns to resting levels.

LAB ACTIVITY 5 **Formation of Bicarbonate and Hydrogen Ions**

1 **Prediction:** With your lab group, predict how exhaling into water after exercise will change the pH of the water. Circle your answer (in italics) below.
 - Exhaling into water after exercise will *increase, decrease,* or *not change* the pH.

2 **Materials:** Obtain the materials for Formation of Bicarbonate and Hydrogen Ions given in the Materials list.

3 **Data Collection:** Measure change in pH after exhaling into water while at rest and after exercise. Record your results in Table 33.8.
 - Decide who will be the subject, who will set up the experiment, who will be the timer, and who will measure and record. The subject should be in good health.

TABLE 33.8 Exhaled CO_2 Levels and Breathing Rate Before and After Exercise

PROCEDURE: BLOWING AIR INTO BEAKER	BREATHING RATE (breaths/min)	CONTROL pH	EXPERIMENTAL pH	% pH CHANGE	CLASS AVERAGE OF % pH CHANGE
At rest					
After exercise					

- Measure 20 mL of deionized water using a graduated cylinder and pour into a 50-mL beaker.
- Measure pH of the deionized water and record the value in Table 33.8 as control pH.
- Count the number of breaths the subject takes over a 30-second period, multiply by 2, and record the number of breaths/min in Table 33.8.
- Have the subject take a deep breath and, using a straw, blow air into the water for four 10-second exhalations. Inhale between exhalations.
- Measure the pH of the water and record the value in Table 33.8 as experimental pH.
- Discard water and rinse the beaker with deionized water.

Change in pH after exercise

- Measure 20 mL of deionized water using the graduated cylinder and pour into the 50-mL beaker.
- Measure pH of the deionized water and record the value in Table 33.8 as control pH.
- Have the subject run in place or climb steps for at least 3 minutes. During the last 30 seconds, count the number of breaths over this 30-second period, multiply by 2, and record the number of breaths/min in Table 33.8.
- As soon as the subject stops exercising, have the subject take a deep breath and, using a straw, blow air into the water for four 10-second exhalations. Inhale between exhalations.
- Immediately measure the pH of the water and record the value in Table 33.8 as experimental pH.
- Discard the water.

4 Clean up as directed by your instructor.

5 **Data Analysis:** Calculate the percentage change in pH for both procedures and record your values in Table 33.8.

- % change in pH = (control pH − experimental pH)/ control pH × 100
- Pool the class data and calculate average values for percent change in pH and breathing rate for both procedures, and record in Table 33.8.

6 **Graphs:** If graph paper is available, construct a bar graph that compares resting and exercising pH values.

7 Complete the Experimental Report with your lab group.

8 Complete **PowerPhys Experiment:** Regulation of Pulmonary Ventilation.

EXPERIMENTAL REPORT
Bicarbonate and Hydrogen Ion Formation

Results: State the pH values at rest and after exercise, and state the change in pH values after exercise.

Discussion:

- Explain how blowing air into the water caused the change in pH.

- Describe how hydrogen ion concentration and bicarbonate ion concentration change as pH changes.

- Describe why exercise caused the observed change in pH.

- Describe why exercise caused the observed change in breathing rate.

Conclusion: In one or two sentences, state how exercise causes pH changes, which, in turn, lead to respiratory changes.

Reviewing Your Knowledge

A. Respiratory Muscles

Match the respiratory muscles to the appropriate muscle function. Muscles may be used more than once.

_____ **1.** elevates 3rd, 4th, and 5th ribs

_____ **2.** elevates the sternum

_____ **3.** depresses ribs

_____ **4.** compresses abdominal contents and increases abdominal pressure

_____ **5.** elevates 1st and 2nd ribs

_____ **6.** elevates ribs

_____ **7.** main inspiratory muscle

_____ **8.**
_____ **9.** } muscles used in forced exhalation

a. abdominal muscles
b. diaphragm
c. external intercostals
d. internal intercostals
e. pectoralis minors
f. scalenes
g. sternocleidomastoids

B. Volume and Pressure Changes During Pulmonary Ventilation

Indicate whether the volume or pressure *increases* or *decreases* during inhalation and exhalation in Table 33.9.

TABLE 33.9 **Volume and Pressure Changes During Pulmonary Ventilation**

VOLUME OR PRESSURE	INHALATION	EXHALATION
Thoracic volume	1.	2.
Intrapleural cavity volume	3.	4.
Intrapleural pressure	5.	6.
Lung volume	7.	8.
Alveolar (intrapulmonic) pressure	9.	10.

C. Lung Volumes and Capacities

Match lung volumes and capacities with the appropriate definition or equation.

_____ **1.** equal to TV + IRV + ERV + RV

_____ **2.** equal to IC − TV

_____ **3.** equal to FRC − ERV

_____ **4.** equal to TV + IRV

_____ **5.** volume of air remaining in lungs after normal exhalation

_____ **6.** maximum amount of air that can be exhaled after a normal exhalation

_____ **7.** maximum amount of air that can be exhaled after a maximal inhalation

_____ **8.** equal to IC − IRV

a. expiratory reserve volume
b. functional residual capacity
c. inspiratory capacity
d. inspiratory reserve volume
e. residual volume
f. tidal volume
g. total lung capacity
h. vital capacity

D. Control of Pulmonary Ventilation: The Role of Carbon Dioxide

Circle True or False for the following questions. If false, underline and change the word(s) that are incorrect to make the statement true.

1. If blood carbon dioxide levels increase, then blood hydrogen ion levels will decrease. True or False

2. If blood carbon dioxide levels increase, then blood pH will decrease. True or False

3. Increasing blood carbon dioxide levels decreases breathing rate. True or False

Using Your Knowledge

A. Physiology of the Respiratory System

Write a short answer for the following questions.

1. A 32-year-old man presented to the emergency room with a pneumothorax. The right lung was collapsed, but the left lung was still inflated. Explain.

2. How does a pneumothorax affect lung volume and alveolar pressure when inspiratory muscles contract? How does this affect air flow?

3. The Valsalva maneuver is forced exhalation against a closed glottis. How would you demonstrate the Valsalva maneuver using the model lung?

A 42-year-old woman is breathing rapidly and deeply after exercise. Indicate whether each of the following volumes *increases*, *decreases*, or *stays the same* when compared to resting volumes.

Example: TLC: stays the same _____

4. RV: _____

5. TV: _____

6. IRV: _____

7. ERV: _____

8. A 5-year-old child announces to her parents (former A&P students) that she is going to hold her breath until she is allowed to watch more television. The parents are not worried. (a) Explain why the parents are not worried. (b) Explain how the child's blood carbon dioxide levels and blood hydrogen ion and bicarbonate ion levels change while she is holding her breath.

9. Women in the late nineteenth and early twentieth centuries wore whalebone corsets that severely restricted the respiratory and digestive systems. These corsets were worn to have a wasp-like waist. It was quite common for these women to faint—in fact, they had "fainting couches." Explain the physiological reason for their fainting problems.

10. When individuals are hyperventilating, they are told to breathe into a paper bag or into their cupped hands. (a) Explain why this increases blood carbon dioxide levels. (b) What will this do to the breathing rate?

Digestive System Structure and Function

OBJECTIVES

1 Identify the major gastrointestinal (GI) tract organs and accessory digestive organs

2 Describe the layers of the GI tract walls and the peritoneum

3 Describe the functions of each GI tract organ and accessory digestive organ

4 Identify the major microscopic structures of digestive organs

5 Dissect and identify the main digestive organs of a cat, fetal pig, or cadaver

MATERIALS

- human torso model, sagittal section of head model, or use **Real Anatomy** (Digestive)
- skull with teeth
- **Length of GI Tract and Transit Time:** 2"-wide roll of adding machine paper (19 ft per group), tape measure, yardstick, or ruler for each group
- compound microscope, lens paper, prepared slides: esophagus, stomach, small intestine, large intestine, pancreas, and liver, or use **Real Anatomy** (Histology)
- Dissections: preserved cat or fetal pig, dissection equipment, disposable gloves, safety glasses, and cat or fetal pig dissection manual
- **Real Anatomy:** Virtual Cadaver Dissection
- **PowerPhys Experiment:**
 - Effect of Dietary Fiber on Transit Time and Bile

The digestive system contains organs of the gastrointestinal (GI) tract and accessory digestive organs. The **GI tract,** or **alimentary** (*alimentary* = nourishment) **canal,** is a tube that extends from the mouth to the anus. The lumen of the GI tract opens to the external environment at either end, and anything inside the lumen is therefore considered to be external to the body. Functions of the digestive system include **ingestion** of food; **secretion** of digestive juices; **mixing** and **propulsion** of food; **digestion** of food into small, absorbable molecules; **absorption** of digestive products across the GI tract wall and into the body; and **defecation.** Digestion includes both **mechanical digestion,** mechanically breaking food into smaller pieces,

and **chemical digestion,** enzymatically breaking large molecules into smaller ones.

A. Organs of the Digestive System

The GI tract organs include the mouth, most of the pharynx, esophagus, stomach, small intestine (duodenum, jejunum, ileum), large intestine (colon or bowel), rectum, and anus. The **accessory digestive organs** (teeth, tongue, salivary glands, liver, gallbladder, and pancreas) assist in digestion.

Before Going to Lab

1 Label the digestive structures in Figure 34.1.

LAB ACTIVITY 1 | Organs of the Digestive System

1 Identify the digestive organs on a model or chart, or use the search text box in **Real Anatomy** (Digestive) to find these structures. ■

- anus
- esophagus
- gallbladder
- large intestine
- liver
- mouth (oral cavity)

- pancreas
- pharynx
- salivary glands
- small intestine
- stomach

1 _____

2 _____

3 _____

4 _____

5 _____

6 _____

7 _____

8 _____

9 _____

10 _____

11 _____

FIGURE 34.1 **Organs of the digestive system.**

B. Layers of the Gastrointestinal Tract and the Peritoneum

The esophagus, stomach, small intestine, and large intestine have a similar four-layered arrangement to their walls. From the interior (deep layer) to the exterior (superficial layer), the four layers are: the mucosa, submucosa, muscularis, and either serosa or adventita. The **mucosa** is a moist layer that includes the surface epithelium, connective tissue (lamina propria), and smooth muscle (muscularis mucosae). Inspect the inside of your cheek with your tongue. The wet, slick feeling is the mucosa that lines all the organs of the GI tract. The **submucosa** is areolar connective tissue, and the **muscularis** in all GI tract organs except the stomach is composed of two layers: an inner layer of circular smooth muscle and an outer layer of longitudinal smooth muscle. The stomach muscularis has three layers: an inner oblique, middle circular, and an outer longitudinal smooth muscle layer. **Adventitia** is the outermost layer of the esophagus, whereas the outermost layer of GI tract organs in the abdominopelvic cavity is **serosa**. Adventitia is comprised of areolar connective tissue and fibrous connective tissue, and anchors organs to other body structures. Serosa, the outermost layer of organs within body cavities, is the visceral layer of the serous membrane and reduces friction during organ movement.

The largest serous membrane in the body, the **peritoneum**, is a combination of simple squamous mesothelium (epithelium) with underlying connective tissue. As with all serous membranes, the **parietal peritoneum** lines the inner surface of the abdominopelvic wall and the **visceral peritoneum** (serosa) covers organs within the abdominopelvic cavity. **Peritoneal** (serous) **fluid** is found between the two peritoneal layers in the **peritoneal cavity**. Organs that lie outside the peritoneal cavity are called **retroperitoneal** (*retro-* = behind) **organs** (i.e., pancreas and kidneys). The peritoneum has large folds not found in other serous membranes that secure organs together and to the abdominal walls. Some of these peritoneal folds are the mesentery, mesocolon, falciform ligament, greater omentum, and lesser omentum. The **mesentery** holds the small intestine to the posterior abdominal wall, whereas the large intestine is similarly secured to the posterior abdominal wall by the **mesocolon**. The liver is bound to the anterior abdominal wall by the **falciform** (*falc-* = sickle-shaped) **ligament**. The **greater omentum** (*omentum* = fat skin) is a large, fatty-looking apron that folds back on itself and attaches to the transverse colon. It forms a pouch that hangs between the body wall and the anterior surface of the small intestine. Attached to the lateral curve of the stomach and duodenum, the greater omentum loosely covers the transverse colon and small intestine. The **lesser omentum** connects the medial curve of the stomach with the liver.

Before Going to Lab

1 Label the structures in Figure 34.2 and Figure 34.3(a)–(d).

LAB ACTIVITY 2 Layers of the GI Tract and the Peritoneum

1 Identify the digestive structures on a model or chart, or use the search text box in **Real Anatomy** (Digestive) to find these structures. ■

- mucosa (myu-KOH-sa)
- muscularis
- serosa (sir-OH-sa)
- submucosa

1 _____

2 _____

3 _____

4 _____

Lumen

1

2

3

4

FIGURE 34.2 **Layers of the gastrointestinal tract.**

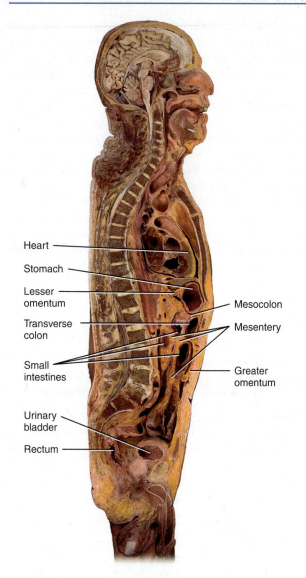

Heart

Stomach

Lesser
omentum

Transverse
colon

Small
intestines

Urinary
bladder

Rectum

Mesocolon

Mesentery

Greater
omentum

FIGURE 34.3 Peritoneal folds.

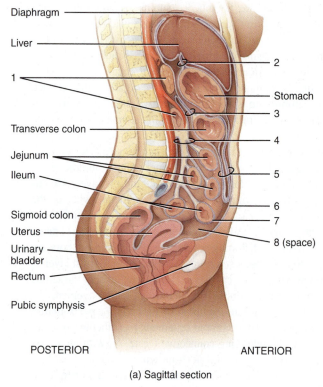

Diaphragm

Liver

1

Transverse colon

Jejunum

Ileum

Sigmoid colon

Uterus

Urinary
bladder

Rectum

Pubic symphysis

2

Stomach

3

4

5

6

7

8 (space)

POSTERIOR

ANTERIOR

(a) Sagittal section

- greater omentum (oh-MEN-tum)
- lesser omentum
- mesentery (MEH-sen-tary)
- mesocolon
- parietal peritoneum (per-ih-toe-NEE-um)
- peritoneal cavity
- retroperitoneal organs
- visceral peritoneum (serosa)

1 _____

2 _____

3 _____

4 _____

5 _____

6 _____

7 _____

8 _____

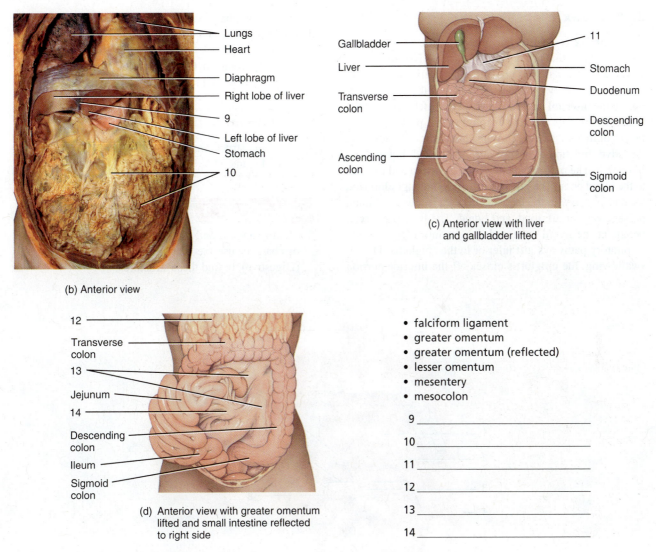

(b) Anterior view

(c) Anterior view with liver and gallbladder lifted

(d) Anterior view with greater omentum lifted and small intestine reflected to right side

- falciform ligament
- greater omentum
- greater omentum (reflected)
- lesser omentum
- mesentery
- mesocolon

9 _____

10 _____

11 _____

12 _____

13 _____

14 _____

FIGURE 34.3 Peritoneal folds, *continued*.

C. Gastrointestinal Tract Organs

1. Mouth

Food first enters the digestive system through the mouth. The **hard** and **soft palates** form the roof of the mouth, the tongue forms the floor, and the **cheeks** form the lateral walls. The anterior border is the **lips,** and the posterior border is the **fauces** (*fauces* = passages), which is the opening to the oropharynx. The mouth is also called the **oral** or **buccal** (*bucca-* = cheeks) **cavity.** The **vestibule** (*vestibule* = courtyard or entrance) is the space between the lips and teeth. The lips are attached to the **gingiva** (*gingiva* = gums) by the **labial frenulum** (*frenulum* = small bridle). The palatine bone and the palatine processes of the maxillae form the **hard palate,** and the soft palate (smooth muscle) is posterior to this. The **soft palate** forms the two arches that border the fauces, and also hangs down to form the oval process called the **uvula.** During swallowing, the soft palate keeps food from entering the nasal cavity. When you look into a mirror with your mouth wide open, you see two arches, the **palatoglossal arch** (anterior) and **palatopharyngeal arch** (posterior), with the uvula hanging down in between. The **palatine tonsils** are located between the two arches.

Digestive system functions that occur in the mouth include ingestion, secretion, mechanical and chemical digestion, mixing, and propulsion. The teeth break food into smaller pieces by a mechanical process called **mastication.** The **tongue** mixes food with saliva that is secreted into the mouth by the salivary glands, initiating chemical digestion. Movements of the tongue and soft palate propel food into the oropharynx during swallowing.

2. Pharynx

The **pharynx** is divided into three specific areas, named after structures that are in close proximity. The **nasopharynx,** the superior area, is posterior and inferior to the nose and has respiratory functions. This part of the pharynx begins at the **internal naris** and ends at the inferior end of the uvula. The middle area, the **oropharynx,** is posterior to the fauces of the oral cavity. This area has both respiratory and digestive functions because air, food, and liquid pass through it. The oropharynx begins at the uvula and extends to the hyoid bone. The inferior area is the **laryngopharynx,** located posterior to the larynx. This area is also a common passageway for air, food, and liquid. The laryngopharynx begins at the hyoid bone and divides into digestive and respiratory pathways just inferior to the epiglottis. During swallowing, the **epiglottis** closes off the inferior portion of the laryngopharynx, ensuring food and drink enter the digestive pathway and not the respiratory pathway.

The only digestive system function that occurs in the pharynx is propulsion. Contractions of the muscularis layer of the pharynx propel food to the esophagus.

Before Going to Lab

1 Label the structures of the mouth and pharynx in Figures 34.4 and 34.5.

LAB ACTIVITY 3 Mouth and Pharynx

1 Identify mouth and pharyngeal structures on a model or chart, or use the search text box in **Real Anatomy** (Digestive) to find these structures. ■

Anterior view

- cheek
- fauces
- gingivae (gums)
- hard palate
- inferior labial frenulum (FREN-u-lum)
- inferior lip
- palatine tonsil
- soft palate
- superior lip
- tongue
- uvula
- vestibule

1 _____ 7 _____

2 _____ 8 _____

3 _____ 9 _____

4 _____ 10 _____

5 _____ 11 _____

6 _____ 12 _____

FIGURE 34.4 The mouth.

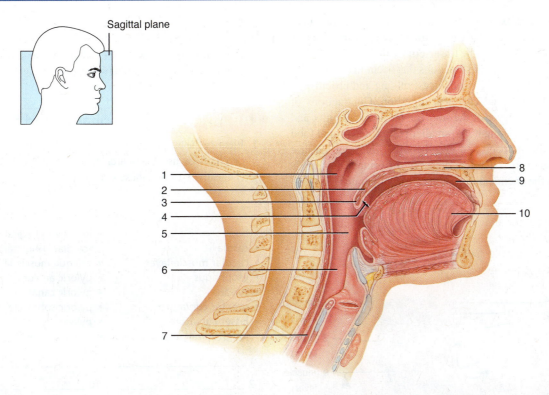

Sagittal plane

Sagittal section

FIGURE 34.5 **Mouth and pharynx.**

- esophagus
- fauces
- hard palate
- laryngopharynx (la-RIN-go-FAIR-inks)
- nasopharynx (NA-zo-fair-inks)
- oral or buccal cavity
- oropharynx (OR-o-fair-inks)
- soft palate
- tongue
- uvula (YOU-vu-la)

1 _____ 7 _____

2 _____ 8 _____

3 _____ 9 _____

4 _____ 10 _____

5 _____

6 _____

3. Esophagus

Food and drink move from the mouth into the oropharynx, laryngopharynx, and then the esophagus. The esophagus is a flattened muscular tube about 10 inches long that appears oval in cross-section but distends when food moves through it. The esophagus is posterior to the trachea and is bound with it by connective tissue. The **upper esophageal sphincter** controls the passage of food from the laryngopharynx into the esophagus. Food then continues through the **esophagus** and the **lower esophageal sphincter** (cardiac sphincter) into the stomach. The esophageal hiatus is the opening in the diaphragm through which the esophagus passes and is the site of a hiatal hernia.

The only digestive system function that occurs in the esophagus is propulsion, which moves food into the stomach. This is accomplished by **peristalsis,** a series of contractions and relaxations of the muscularis layer.

4. Stomach

The esophagus leads into the superior part of the stomach, which lies against the diaphragm superiorly. The J-shaped stomach has four areas: the cardia, fundus, body, and pylorus. The **cardia** is the first part of the stomach, inferior to the lower esophageal sphincter. The **fundus** is the elevated area to the left of the cardia that serves as a temporary holding area for food. The **body** is the largest, curved part of the stomach, whereas the **pylorus** continues to the right of the body after the curve. The pylorus is divided into two areas, the **pyloric antrum** (*antrum* = cave), which is next to the body, and the **pyloric canal,** which is nearer the pyloric sphincter. The **pyloric sphincter** is a smooth muscle that allows food to enter the first part of the small intestine, the duodenum. The stomach has large, conspicuous ridges in the mucosa called **rugae** that allow the stomach to be stretched when food and drink are present. Rugae disappear

as the stomach becomes fully distended. The lateral curve of the stomach is the **greater curvature,** and the smaller, medial curve is called the **lesser curvature.**

The digestive system functions that occur in the stomach include secretion of gastric juices, mixing and propulsion, and mechanical and chemical digestion. Contractions of the muscularis layer of the stomach mechanically mix and churn food, breaking it into smaller pieces, and propel food through the stomach and into the small intestine. Gastric juices begin the chemical digestion of protein and lipids and stop the chemical digestion of carbohydrates.

Before Going to Lab

1 Label the structures of the esophagus and stomach in Figure 34.6. Refer to cadaver figure.

LAB ACTIVITY 4 Esophagus and Stomach

1 Identify the esophagus and stomach structures on a model or chart, or use the search text box in **Real Anatomy** (Digestive) to find these structures. ■

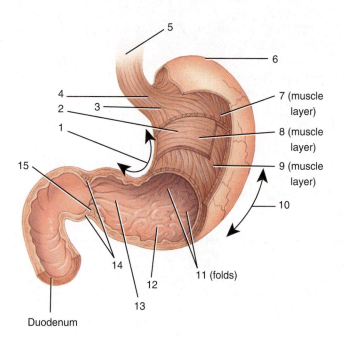

- body
- cardia
- circular muscle layer
- esophagus
- fundus
- greater curvature
- lesser curvature
- longitudinal muscle layer
- lower esophageal sphincter (cardiac sphincter)
- oblique muscle layer
- pyloric antrum
- pyloric canal
- pyloric sphincter
- pylorus
- rugae

1 _____	9 _____
2 _____	10 _____
3 _____	11 _____
4 _____	12 _____
5 _____	13 _____
6 _____	14 _____
7 _____	15 _____
8 _____	

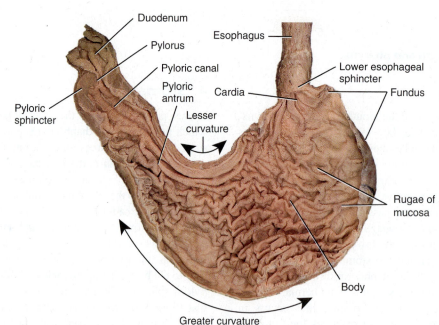

FIGURE 34.6 Esophagus and stomach.

5. Small Intestine

The small intestine is a long tube (about 10 feet long in a living person) divided into three sections: the duodenum, jejunum, and ileum (see Figure 34.7). The shortest section, the **duodenum,** resembles the shape of a "C" and receives food through the pyloric sphincter. The duodenum (*duodenum* = 12 fingers) is about 10 inches long, or approximately the width of 12 fingers. Secretions from the pancreas, liver, and gallbladder enter the duodenum at the hepatopancreatic ampulla, located about 4 inches inferior to the pyloric sphincter. The next section, the **jejunum** (*jejunum* = empty), is about 3 feet long and has a thicker layer of smooth muscle. It is named jejunum because on death it is found to be empty. The **ileum** (*ileum* = intestine), the last section of the small intestine, is about 6 feet long and ends at the **ileocecal sphincter** (valve), where it joins the large intestine. From the exterior, it is difficult to distinguish these three sections of the small intestine.

Digestive system functions of the small intestine include secretion, mixing and propulsion, mechanical and chemical digestion, and absorption. Most absorption occurs in the small intestines. Secretions of the pancreas, liver, and intestinal glands are deposited into the lumen of the small intestine. Local contractions of the muscularis layer of the small intestine mix and churn food by **segmentation** and propel it through the small intestine into the large intestine by peristalsis. Pancreatic and intestinal enzymes digest larger food molecules of carbohydrates, proteins, and lipids into smaller, absorbable molecules.

The small intestine has three special anatomical modifications for increasing the surface area available for absorption. The **circular folds** or **plicae circularis** are folds or ridges of the mucosa that can be observed with the unaided eye. The other two structural specializations, the **villi** and **microvilli,** can be observed only with the microscope and will be studied in the histology of the digestive system.

FIGURE 34.7 **Entire small intestine, longitudinal section.**

6. Large Intestine

The large intestine, also known as the colon (*colon* = food passage) or large bowel, is approximately 5 feet long and is composed of the cecum, colon, rectum, and anal canal. A blind pouch called the **cecum,** which is about 2 to 3 inches long, extends inferiorly from the large intestine at the ileocecal sphincter in the lower right quadrant. The slender **appendix,** which is lymphatic tissue, branches off the cecum and is also about 2 to 3 inches long. From the ileocecal junction, the **ascending colon** extends superiorly to the liver on the right side of the abdomen and makes a 90° turn to the left at the **right colic** (hepatic) **flexure.** The colon continues as the **transverse colon** until it reaches the spleen, where it makes another 90° turn at the **left colic** (splenic) **flexure.** The **descending colon** then continues inferiorly on the left side to an S-shaped curve called the **sigmoid colon** at the level of the iliac crest. **Teniae coli** are three longitudinal bands of smooth muscle that run along the length of the large intestine. Contractions of the teniae coli form pouches called **haustra. Epiploic appendages** are fat-filled pouches of visceral peritoneum that hang from the teniae coli.

Following the sigmoid colon, at about the third sacral vertebra level, is the **rectum.** The last segment of the colon is the **anal canal,** which has longitudinal **anal columns** or folds rich with blood vessels. The anal canal opens to the exterior at an opening called the **anus.** There are two sets of muscle sphincters in the anus: the **internal sphincter** of smooth muscle, which is involuntary, and the **external sphincter,** which is voluntary skeletal muscle.

The digestive system functions of the large intestine include secretion, mixing and propulsion, chemical digestion (by bacterial enzymes), absorption, and defecation. The large intestine absorbs a large amount of water, concentrating the undigested feces into a solid form. It is the site where bacteria complete digestion and produce vitamins. The large intestine also secretes mucus and eliminates fecal material from the body through the anus.

> **CLINICAL NOTE:** *A sigmoidoscopy views the lining of the sigmoid colon, whereas a colonoscopy views the lining of the entire colon.*

Before Going to Lab

1 Label the structures of the small and large intestines in Figure 34.8(a)–(c).

1 Identify the structures of the small and large intestines on a model or chart, or use the search text box in **Real Anatomy** (Digestive) to find these structures. ■

 (a) Anterior view

(a)
- ascending colon 2
- descending colon 7
- duodenum 1
- ileum 4
- jejunum 3
- rectum 8
- stomach 5
- transverse colon 6

1 _____

2 _____

3 _____

4 _____

5 _____

6 _____

7 _____

8 _____

FIGURE 34.8 Small intestine and large intestine.

(b) Large intestine, anterior view

(c) Frontal section of anal canal

(b)
- anal canal
- anus
- appendix
- ascending colon
- cecum (SEE-cum)
- descending colon
- epiploic (ep-ih-PLOH-ic) appendages
- haustra (HAW-struh)
- ileocecal (ileo-SEE-cal) sphincter
- ileum
- left colic (splenic) flexure
- rectum
- right colic (hepatic) flexure
- sigmoid colon
- teniae coli (TEE-nee-ee CO-lye)
- transverse colon

9 _____

10 _____

11 _____

12 _____

13 _____

14 _____

15 _____

16 _____

17 _____

18 _____

19 _____

20 _____

21 _____

22 _____

23 _____

24 _____

(c)
- anal canal
- anal column
- anus
- external anal sphincter
- internal anal sphincter
- rectum

25 _____

26 _____

27 _____

28 _____

29 _____

30 _____

FIGURE 34.8 **Small intestine and large intestine,** *continued*.

D. Length of GI Tract and Transit Time

The gastrointestinal (GI) tract is a tube that spans the distance between the mouth and the anus, with its sections varying in length and diameter. Most people don't realize how long this tube actually is until they see it stretched out its entire length. The lengths of the segments given here are those found in an average living adult and will obviously vary in length depending on the person's size. The GI tract length is noticeably longer in a cadaver, 30 ft (9 m), due to smooth muscle relaxation and the lack of muscle tone compared to the total length in a live adult of approximately 19 ft (5.8 meters).

The time it takes for food to pass through GI tract is called **transit time.** Transit time is typically 24–72 hours and is affected by a variety of factors including composition and volume of food, physical activity, emotional state, illness, and medication. Peristalsis moves food through the GI tract, and the velocity of peristalsis differs for each segment of the GI tract.

LAB ACTIVITY 6 **GI Tract Length and Transit Time**

1 Measurement of GI tract length
 - Obtain the materials for GI Tract Length and Transit Time from the Materials list.
 - Find a place to stretch out the paper.
 - Starting with the mouth, measure out and mark on the paper the lengths of each organ with the lengths given in Table 34.1.

 - Write the name of each organ in the appropriate segment of the paper.
 - In addition, mark the location of the following structures:
 - pyloric sphincter
 - hepatopancreatic ampulla
 - ileocecal sphincter
 - cecum (2.4 inches or 6 cm)
 - appendix (3 inches or 8 cm)

2 Calculate the *velocity of peristalsis* through each portion of the GI tract for organ transit times stated.
 - The time it takes for food to travel through each portion of the GI tract is indicated in Table 34.1.
 - Normally, organ transit times are stated as a range of values. In order to simplify this activity, a specific time within the range was chosen.
 - Velocity = length in inches/sec (or cm/sec) in GI tract organ.
 - Using the length of the organ and the organ transit time columns, calculate the *velocity of peristalsis* through each segment and write it in Table 34.1. Note that velocity will vary depending on composition and volume of food, physical activity, emotional state, illness, and medications.

3 Calculate your *total transit time*.
 - Because there isn't a way for you to calculate your own individual *organ transit times*, you will be watching the total time it takes for food to travel through your entire GI tract. This *total transit time* can be calculated at home using kernel corn or beets.
 - Decide whether you are going to eat kernel corn or beets, and avoid eating this food for a week.

TABLE 34.1 GI Tract Length and Transit Time

ORGAN	LENGTH	ORGAN TRANSIT TIME	PERISTALTIC VELOCITY
Mouth	2.5" (6.5 cm)	NA	NA
Pharynx	5" (13 cm)	1 sec	
Esophagus	10" (25 cm)	5 sec	
Stomach	10" (25 cm)	4 hours	
Small Intestine • duodenum–10" (25 cm) • jejunum–3' (1 m) • ileum–6' (2 m)	Total length is approx 10' (3 m)	6 hours	
Large Intestine • ascending, transverse descending and sigmoid colon	Total length is approx 5' (1.5 m)	16 hours	
Rectum	8" (20 cm)		
Total GI Tract Length	**Approximately 19 ft (5.8 m)**		

- Record the date and time that you eat the chosen food and monitor your stools for the appearance of the food. Beets will cause your stool to turn red.

 Date started: _____ Time started: _____

- Record the date and time that the food (corn kernel or red stool) first *starts* appearing in your stool, and the date and time that the food *stops* appearing your stool.

 Date food first appeared: _____

 Date food stopped appearing: _____

 Total transit time: _____ hours

4 Discuss with the class what surprised you or helped you to clarify concepts about the GI tract after doing this exercise.

5 Complete **PowerPhys Experiment:** Effect of Dietary Fiber on Transit Time and Bile. ◼

E. Accessory Digestive Organs

1. Salivary Glands

The mucous membranes of the mouth and tongue have many small salivary glands that secrete saliva. However, most of the saliva secreted into the oral cavity is secreted by the parotid, submandibular, and sublingual salivary glands. Saliva moistens food and contains enzymes that initiate chemical digestion of carbohydrates. It also contains enzymes for lipid digestion that become activated when they reach the acidic stomach environment.

The large **parotid** (*par-* = near; *ot-* = ear) **glands** are located anterior and inferior to the ears between the skin and the masseter muscle. Their salivary secretions reach the oral cavity via the **parotid ducts** that open near the second molar in the upper jaw. The **submandibular glands** are located in the posterior part of the mouth floor just medial to the mandible, deep to the mylohyoid muscle. The **submandibular ducts** open into the oral cavity lateral to the lingual frenulum. The **sublingual glands** are located under the tongue as the name suggests and are more medial and superior than the submandibular glands. There are several **lesser sublingual ducts** that open into the floor of the mouth; they are very small compared with the other main salivary gland ducts.

2. Tongue

Extrinsic muscles attach the tongue to bone and hold it in position, but they also allow movement to maneuver food in the mouth. The tongue is composed of skeletal muscle called **intrinsic muscles** that are used for speech and swallowing. The **lingual frenulum** attaches the tongue to the floor of the mouth. The dorsal and lateral surfaces of the tongue have numerous **papillae** (*papilla* = nipple-shape), many of which have taste buds.

3. Teeth

Teeth mechanically break up the food into small pieces, aided by the tongue, which manipulates the food in the oral cavity. Teeth are composed of three basic regions: the crown, neck, and root. The **crown** is the part that can be seen above the gums, whereas the **neck** is the constricted region at the gum line where the crown and root meet. The **root** is embedded in an alveolar bony socket of the maxilla or mandible and is held in place by the **periodontal** (*peri-* = around; *odont-* = tooth) **ligament** lining the socket. Most of the tooth is formed of **dentin,** a hardened, calcified connective tissue. **Enamel** covers and protects the dentin of the crown. Dentin of the root is covered with **cementum,** which attaches the tooth to the periodontal ligament. Surrounding each tooth is **gingiva** or gum tissue.

In the crown of the tooth deep to the dentin is a **pulp cavity** filled with connective tissue called **pulp** that contains nerves, blood vessels, and lymphatic vessels. **Root canals** are slender continuations of the pulp cavity that run the entire length of the root. The small opening at the base of each root, the **apical foramen,** is the location for nerves, blood vessels, and lymphatic vessels to enter and exit the tooth.

Adults have 32 permanent teeth, with each jaw containing four incisors, two cuspids (canines), four premolars (bicuspids), and six molars. **Incisors** cut food, **cuspids** tear food, and **premolars** and **molars** crush and grind food. Premolars have two cusps (pointed surfaces), and molars have four cusps. Incisors, cuspids, and the second premolars have one root, the first bicuspids and mandibular (lower) molars have two roots, and the maxillary (upper) molars have three roots.

Children's teeth, **deciduous** (*decidu-* = falling out) **teeth** or primary teeth, are temporary and are lost by the child between the ages of 6 and 12. They are replaced by **permanent** or **secondary teeth.** The deciduous set has 20 teeth, and the permanent set has 32 teeth.

From the midline moving laterally, the dental formula for *deciduous teeth* is:

$$\frac{\textbf{Upper teeth: } \text{2 incisors, 1 cuspid, 0 premolars, 2 molars}}{\textbf{Lower teeth: } \text{2 incisors, 1 cuspid, 0 premolars, 2 molars}} \times 2 = 20 \text{ teeth}$$

From the midline moving laterally, the dental formula for *permanent teeth* is:

$$\frac{\textbf{Upper teeth: } \text{2 incisors, 1 cuspid, 2 premolars, 3 molars}}{\textbf{Lower teeth: } \text{2 incisors, 1 cuspid, 2 premolars, 3 molars}} \times 2 = 32 \text{ teeth}$$

Before Going to Lab

1 Label the accessory digestive structures in Figure 34.9(a)–(c).

LAB ACTIVITY 7 The Teeth, Tongue, and Salivary Glands

1 Identify the accessory digestive structures on a model or chart, or use the search text box in **Real Anatomy** (Digestive) to find these structures.

2 Identify incisors, cuspids, premolars, and molars on a skull with teeth. ■

Second maxillary molar tooth

Tongue

Lingual frenulum

(a) Location of salivary glands

(a)
- openings of sublingual ducts
- opening of parotid duct (near secondary molar)
- parotid duct
- parotid gland
- sublingual gland
- submandibular duct
- submandibular gland

1 _____

2 _____

3 _____

4 _____

5 _____

6 _____

7 _____

FIGURE 34.9 **The salivary glands, tongue, and teeth.**

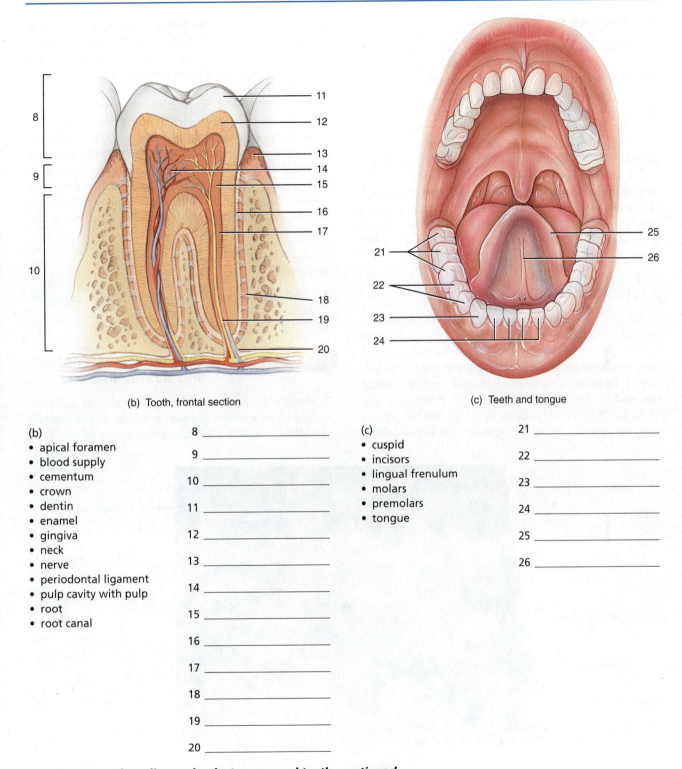

(b) Tooth, frontal section

(c) Teeth and tongue

(b)
- apical foramen
- blood supply
- cementum
- crown
- dentin
- enamel
- gingiva
- neck
- nerve
- periodontal ligament
- pulp cavity with pulp
- root
- root canal

8 _____

9 _____

10 _____

11 _____

12 _____

13 _____

14 _____

15 _____

16 _____

17 _____

18 _____

19 _____

20 _____

(c)
- cuspid
- incisors
- lingual frenulum
- molars
- premolars
- tongue

21 _____

22 _____

23 _____

24 _____

25 _____

26 _____

FIGURE 34.9 **The salivary glands, tongue, and teeth,** *continued.*

4. Pancreas

The **pancreas,** composed of a head, body, and tail, is located posterior to the stomach. The **head** is the expanded part of the pancreas that lies in the C-shape of the duodenum. The pancreatic **body** is in the middle, and the **tail** tapers to the left toward the spleen. The main **pancreatic duct** runs from the tail to the head of the pancreas and joins the common bile duct to form the **hepatopancreatic ampulla.** This ampulla opens into the duodenum at the **major duodenal papilla,** about 4 inches inferior to the pyloric sphincter. In the head of the pancreas, a smaller **accessory pancreatic duct** branches off the pancreatic duct and opens into the duodenum just superior to the hepatopancreatic ampulla. The pancreas secretes digestive enzymes and bicarbonate into the duodenum.

5. Liver and Gallbladder

The liver is a large organ that lies inferior to the diaphragm and touches the diaphragm all along the liver's superior border. The liver has two main lobes, the **right lobe** and **left lobe,** which are separated by the **falciform** (*falc-* = sickle-shaped) **ligament.** Two smaller lobes, the superior **caudate lobe** and the inferior **quadrate lobe,** lie on the same side of the falciform ligament as the right lobe and can be seen from the posterior view. The **right** and **left hepatic ducts** join to form the **common hepatic duct,** which takes bile to the cystic duct of the gallbladder. The main digestive function of the liver is to make bile. Bile emulsifies lipids (breaks large spheres of lipids into small ones), which prepares them for digestion in the small intestine.

The **gallbladder** is a sac made of smooth muscle that is located on the inferior surface of the liver between the right lobe and the quadrate lobe. The main function of the gallbladder is to concentrate and store bile. Bile leaves the gallbladder via the **cystic duct,** which joins with the **common hepatic duct** to form the **common bile duct.** The common bile duct joins the pancreatic duct in the pancreas to form the **hepatopancreatic ampulla,** which opens into the duodenum.

> ### Before Going to Lab
>
> **1** Label the accessory digestive structures in Figure 34.10(a)–(c).

LAB ACTIVITY 8 The Pancreas, Liver, and Gallbladder

1 Identify the structures of the pancreas, liver, and gallbladder on a model or chart, or use the search text box in **Real Anatomy** (Digestive) to find these structures. ■

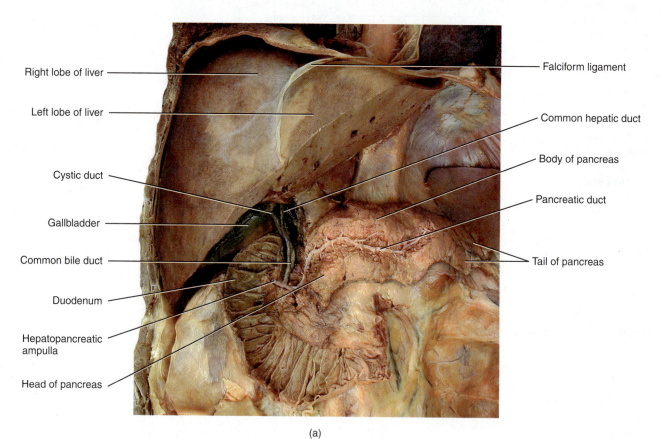

Right lobe of liver

Left lobe of liver

Cystic duct

Gallbladder

Common bile duct

Duodenum

Hepatopancreatic ampulla

Head of pancreas

Falciform ligament

Common hepatic duct

Body of pancreas

Pancreatic duct

Tail of pancreas

(a)

FIGURE 34.10 **Pancreas, liver, and gallbladder.**

(b) Anterior view

FIGURE 34.10 Pancreas, liver, and gallbladder, *continued*.

- accessory pancreatic duct 11
- body of pancreas 13
- common bile duct 10
- common hepatic duct 9
- cystic duct 2 .
- duodenum 4
- falciform ligament 6
- gallbladder 3
- head of pancreas 15
- hepatopancreatic (heh-PA-toe-pan- 5
 cree-A-tic) ampulla
- jejunum 16
- left hepatic duct 8
- left lobe of liver 7
- pancreatic duct 14
- right hepatic duct 1
- tail of pancreas 12

1 _____

2 _____

3 _____

4 _____

5 _____

6 _____

7 _____

8 _____

9 _____

10 _____

11 _____

12 _____

13 _____

14 _____

15 _____

16 _____

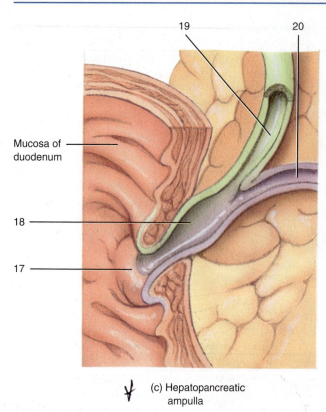

Mucosa of duodenum

(c) Hepatopancreatic ampulla

(b)
• common bile duct
• duodenal papilla
• hepatopancreatic ampulla
• pancreatic duct

17 _____

18 _____

19 _____

20 _____

Left hepatic vein

Inferior vena cava

Hepatic portal vein

Hepatic artery

(d) Liver, posterior view

(c)
• caudate lobe
• gallbladder
• left lobe
• quadrate lobe
• right lobe

21 _____

22 _____

23 _____

24 _____

25 _____

FIGURE 34.10 **Pancreas, liver, and gallbladder,** *continued.*

F. Histology of the Digestive System

1. Gastrointestinal Tract Organs

The walls of the GI tract organs are composed of four layers: the mucosa (epithelium, lamina propria, and muscularis mucosae), submucosa, muscularis, and either serosa (GI tract organs within abdominopelvic cavity) or adventitia (GI tract organs outside of abdominopelvic cavity). The epithelium of the mucosa changes along the digestive tract. Stratified squamous epithelium lines the mouth, oropharynx, laryngopharynx, and esophagus to protect underlying tissues. At the junction of the esophagus and stomach, the epithelium changes to simple columnar epithelium that lines the stomach, small intestine, and large intestine and reflects the absorptive and secretive functions of these organs. The epithelium changes again to stratified squamous epithelium in the anal canal, providing protection from the external environment.

The simple columnar epithelium of the **stomach** contains **goblet cells** that secrete mucus for stomach-lining protection. The gastric epithelium dips down into the underlying **lamina propria** to form narrow channels called **gastric pits.** The gastric pits lead into areas called **gastric glands** that are lined by three types of specialized secretory cells: mucous neck cells, chief cells, and parietal cells. **Mucous neck cells** secrete mucus that protects the stomach lining from being digested. **Chief cells** produce and secrete gastric lipase, which digests triglycerides, and pepsinogen, which is an inactive form of pepsin. **Parietal cells** produce hydrochloric acid and intrinsic factor. Hydrochloric acid converts pepsinogen into pepsin, which begins the digestion of protein. Intrinsic factor facilitates vitamin B_{12} absorption in the small intestine.

The **small intestinal** wall has an unusual feature in that the mucosa forms finger-like projections called **villi** (*villi* = tufts of hair) that greatly increase the surface area. In the center of each villus is the lamina propria with an arteriole, venule, blood capillary, and a lymphatic capillary called a **lacteal.** Absorptive cells of the **simple columnar epithelium** have **microvilli** on the surface facing the lumen, adding more surface area. When viewed with the compound microscope, the microvilli give the cell surface a brush-like appearance called the **brush border. Goblet cells** that secrete mucus are also found in the epithelium. Deep fissures in the intestinal mucosa are lined with **intestinal glands,** also called the **crypts of Lieberkühn.** In the submucosa of the duodenum, there are **duodenal (Brunner's) glands,** which secrete mucus to help neutralize the acidic stomach chyme.

2. Pancreas and the Liver

The **pancreas** has two distinct areas: the *exocrine portion* of glandular epithelial cells in clusters called **acini,** and the *endocrine portion* formed of **pancreatic islets (islets of Langerhans).** The acinar exocrine cells secrete the digestive enzymes pancreatic amylase, trypsin, and pancreatic lipase. Pancreatic amylase digests carbohydrates, trypsin digests proteins, and pancreatic lipase digests triglycerides (lipids).

The **liver** is composed of structural and functional units called **lobules,** which have **hepatocytes** (*hepato-* = liver) that produce bile. Hepatocytes are arranged in branching, flat plates around a **central vein.** The liver lobules have modified capillaries with large spaces called **sinusoids** where gases, nutrients, and wastes are exchanged with the hepatocytes. Special **fixed macrophages** in the sinusoids phagocytize dead or foreign matter from the blood. These fixed macrophages are called **reticuloendothelial (Kupffer's) cells.** The lobules are typically six-sided with portal triads present at each of the corners. A **portal triad** is composed of three structures: a branch of the **hepatic portal vein,** a branch of the **hepatic artery,** and a **bile canaliculus** (*canaliculus* = little canal).

Before Going to Lab

1 Label the microscopic structures of the esophagus, stomach, small intestine, pancreas, and liver in Figure 34.11(a)–(e).

LAB ACTIVITY 9 Histology of the Digestive System

1 Examine prepared slides of stomach, duodenum, pancreas, and liver using the scanning or low-power objective lens or use **Real Anatomy** (Histology). Identify the structures in Figure 34.11(a)–(e). ■

(a) Wall of the esophagus

(b) Fundic mucosa of the stomach

(a)
- adventitia
- circular layers of smooth muscle fibers
- lamina propria
- longitudinal layer of smooth muscle fibers
- mucosa
- muscularis mucosae
- stratified squamous epithelium
- submucosa

1 _____

2 _____

3 _____

4 _____

5 _____

6 _____

7 _____

8 _____

(b)
- gastric glands
- gastric pit
- lamina propria
- simple columnar epithelium

9 _____

10 _____

11 _____

12 _____

FIGURE 34.11 Histology of digestive organs.

(c) Duodenum of small intestine

LM 450×

(d) Histology of the pancreas

(c)
- duodenal (Brunner's) gland in submucosa
- goblet cell
- intestinal gland (crypts of Lieberkühn)
- lamina propria
- simple columnar epithelium
- villus

13 _____

14 _____

15 _____

16 _____

17 _____

18 _____

(d)
- acini (exocrine)
- pancreatic islet (islet of Langerhans)

19 _____

20 _____

LM 180×

(e) Histology of the liver lobule

(e)
- central vein
- hepatocytes
- sinusoids

21 _____

22 _____

23 _____

FIGURE 34.11 Histology of digestive organs, *continued*.

Blood flows through each liver lobule from branches of the hepatic artery and the portal vein to the **sinusoids.** The **hepatic artery** carries oxygen-rich blood, and the **hepatic portal vein** carries oxygen-poor blood plus nutrients from the digestive system. The sinusoids empty blood into the **central vein,** and blood then flows from the **central vein** to the **hepatic vein** and finally into the **inferior vena cava.** **Hepatocytes** produce and secrete **bile** first into narrow **bile canaliculi** and then into the **bile ducts,** part of the lobule triad. During this time, bile travels in an opposite direction from the blood in the triad and sinusoids. Bile eventually empties into the **hepatic ducts** and on to the **cystic duct** to be stored and concentrated in the **gallbladder.**

> **Before Going to Lab**
>
> **1** Label the liver lobule structures in Figure 34.12.

SAFETY NOTE: *Wear safety glasses and gloves when using preserved or fresh tissue. Wash hands thoroughly with soap and water when you are finished.*

G. Dissection of Digestive System

If you are dissecting a cat or fetal pig to observe digestive system structures, refer to the dissection manual.

In your dissection of the digestive system, you will find many organs and structures similar to the human. Pay attention to all the connective tissue, the greater omentum, and the mesentery.

Real Anatomy, a virtual cadaver dissection, can be used to complement or substitute for animal dissection of the digestive system.

- bile canaliculus
- bile duct
- branch of hepatic portal vein
- branches of hepatic artery
- central vein
- hepatic vein
- hepatocytes
- portal triad
- sinusoids

1 _____

2 _____

3 _____

4 _____

5 _____

6 _____

7 _____

8 _____

9 _____

FIGURE 34.12 **Flow of blood and bile through a liver lobule.**

Reviewing Your Knowledge

A. Layers of the Gastrointestinal Tract

Name the layers of the GI tract that are described.

_____ 1. The layer that contracts to churn food or move food along

_____ 2. Layer consisting of a membrane that lines a body cavity that opens to the exterior; overlays smooth muscle

_____ 3. Areolar connective tissue layer located deep to the mucosa

_____ 4. A serous membrane that is the external layer of a GI tract organ

B. Gastrointestinal Tract Organs

Write the name of the term that is described.

_____ 1. Its only function is propulsion

_____ 2. A continuous digestive tube from the mouth to the anus

_____ 3. Conducts both air and food

_____ 4. Primary site of nutrient absorption; is composed of 3 sections

_____ 5. Section of small intestine that receives bile, pancreatic secretions, and food from the stomach

_____ 6. Churns food and begins protein digestion

_____ 7. Has two sphincters that control elimination of feces from the body

_____ 8. Has regions called the cardia, fundus, body, and pylorus

_____ 9. Has regions called the cecum, colon, rectum, and anal canal

_____ 10. Receives secretions from salivary glands; mastication occurs here

C. Digestive Organ Structures

Write the name of the organ that is described.

_____ **1.** Finger-like extensions increasing surface area in the small intestine

_____ **2.** Folds in the gastric mucosa

_____ **3.** Permanent deep ridges in the small intestine mucosa

_____ **4.** Sphincter valve between the stomach and duodenum

_____ **5.** Serous membrane that lines the abdominal wall

_____ **6.** Serous membrane that covers the abdominal organs

_____ **7.** Sphincter that connects the small and large intestine

_____ **8.** Fluid that begins digestion of carbohydrates

_____ **9.** Bony plate between the mouth and nose

_____ **10.** Forms a brush border; extension of epithelial cells

_____ **11.** Keeps food and fluids from going up into the nasopharynx

_____ **12.** Area between the lips and teeth

_____ **13.** Region of the stomach where the lower esophageal sphincter meets the stomach

_____ **14.** Has regions called ascending, transverse, descending, and sigmoid

_____ **15.** Narrowed region of the stomach before the small intestine

_____ **16.** A series of gathered pouches in the large intestine

D. Accessory Digestive Organs and the Peritoneum

Write the name of the organ that is described.

_____ **1.** Mechanically breaks up food during mastication

_____ **2.** Secretes enzymes that digest carbohydrates, proteins, lipids, and nucleic acids

_____ **3.** Salivary glands located under the tongue with ducts that open in the floor of mouth

_____ **4.** Produces and secretes bile into ducts

_____ **5.** Peritoneal membrane that holds the small intestine to the posterior abdominal wall

_____ **6.** The secretions of this gland join with bile to enter the duodenum

_____ **7.** Peritoneal fold that holds the liver to the anterior abdominal wall

_____ **8.** Manipulates food in mastication

_____ **9.** Fatty, large fold of peritoneum covering the transverse colon and small intestine

_____ **10.** Largest salivary glands whose ducts open by the upper second molars

_____ **11.** Stores and secretes bile into the duodenum

_____ **12.** Peritoneal membrane that attaches the stomach and duodenum to the liver

_____ **13.** Salivary glands whose ducts open lateral to the lingual frenulum

14. Trace bile from its secretion to the gallbladder for storage and concentration, and then to the duodenum, listing the structures *in order,* using Figure 34.10.

(a) hepatocytes

(b) _____

(c) _____

(d) _____

(e) _____

(f) _____

(g) gallbladder

(h) _____

(i) _____

(j) _____

(k) duodenum

15. Trace blood from the hepatic portal vein through the liver to the inferior vena cava, listing the structures *in order,* using Figure 34.12.

(a) hepatic portal vein

(b) _____

(c) _____

(d) _____

(e) _____

(f) inferior vena cava

Using Your Knowledge

A. Digestive System Structure

1. A hiatal hernia or failure of the lower esophageal (cardiac) sphincter to close causes the stomach contents to back up into the esophagus. This causes the esophageal wall to have a burning sensation (heartburn) and, in serious cases, bleeding. What causes the burning sensation and bleeding to occur?

2. How does removal of the gallbladder affect digestion?

B. Identification of Digestive System Structures

Identify the structures numbered 3–6 on the X-ray of the human stomach shown in Figure 34.13.

3. _____

4. _____

5. _____

6. _____

FIGURE 34.13 **X-ray of the human stomach.**

Identify the structures numbered 7–10 on the X-ray of the lower GI tract, shown in Figure 34.14.

7. _____

8. _____ (sacs)

9. _____

10. _____

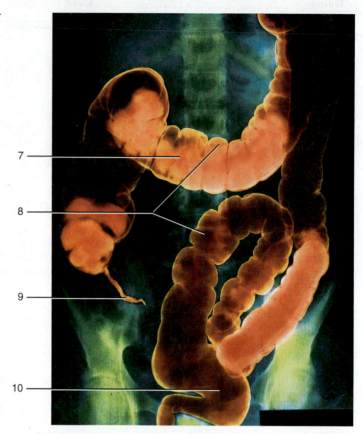

FIGURE 34.14 **X-ray of lower GI tract.**

Mechanical and Chemical Digestion

OBJECTIVES

1 Name the substrate for each of the following enzymes and the products produced from the enzyme-catalyzed reaction: amylase, protease, peptidase, and lipase

2 Describe the importance of mechanical digestion and chemical digestion

3 Explain the importance of pH on digestive enzyme function

4 Describe the action of bile salts on lipids

MATERIALS

- **Starch Digestion** (per group): 6 test tubes, test tube rack, test tube holder, wax marker, Parafilm, 37°C water bath, 1 saltine cracker (starch), and 1 dropper bottle each of salivary amylase, 0.5% HCl (hydrochloric acid), pH 7.4 buffer, distilled water

- **Starch Test** (per group): Lugol's iodine (IKI) in dropper bottle, 1 clean spot plate with 6 depressions, wax marker

- **Sugar Test** (per group): Benedict's reagent in dropper bottle, 250-mL beaker, hot plate

- **Starch Digestion in Mouth** (per person): $2\frac{1}{2}$-inch-square portions of *unsalted* soda crackers (saltine-type)

- **Bile Emulsification** (per group): bile salts, vegetable oil, medicine dropper, distilled water in dropper bottle, 2 test tubes, test tube rack, test tube holder, wax marker

- **PowerPhys Experiment:** Enzyme Activity

D igestion includes both mechanical digestion, mechanically breaking food into smaller pieces, and chemical digestion, the enzymatic breakdown of large molecules into smaller ones that can be absorbed into the bloodstream.

and forth movements of portions of the small intestine, that mix the chyme with digestive enzymes prior to peristaltic contractions moving the contents along. Mechanical digestion enhances the speed of chemical digestion by creating a greater surface area for digestive enzymes to do their work.

A. Mechanical Digestion

Mechanical digestion begins in the mouth and continues in the stomach and small intestine. Chewing or **mastication** breaks down large pieces of food into smaller ones that are swallowed. In the stomach, peristaltic waves mechanically churn and mix the food with gastric juices to enhance digestion. Churning continues in the small intestine by smooth muscle contractions called **segmentations,** or back

B. Chemical Digestion

Chemical digestion involves digestive enzymes breaking the bonds of the food macromolecules. The food we eat contains macromolecules of carbohydrates, lipids, and proteins. Because these large molecules cannot be absorbed by the gastrointestinal tract, enzymes secreted by various parts of the digestive system **catabolize** or break down large molecules into small ones that can be absorbed. **Starch,** a carbohydrate,

is digested into **disaccharides** (*di-* = two; *saccharide* = sugar) and **monosaccharides** (*mono-* = one) by the action of specific enzymes called **amylases.** Disaccharidases break down disaccharides into monosaccharides. **Lipids** are macromolecules that are catabolyzed or digested to **glycerol** and **fatty acids** by the enzyme **lipase. Proteases** digest **proteins** into **peptides** and **amino acids. Peptidases** digest **peptides** into amino acids. Monosaccharides, glycerol and fatty acids, and amino acids are small enough to be absorbed across the wall of the gastrointestinal tract and are used by cells for building new macromolecules or to provide energy (ATP). Table 35.1 contains digestive enzymes, substrates, and products.

Enzymes are typically large protein molecules produced by cells of the body that are special biological catalysts acting on specific **substrates** to produce **products.** Enzymes speed up the rate of chemical reactions between substrates without becoming chemically changed in the reaction. Enzymes work best on food substrates that are chewed and broken down mechanically into smaller particles with a greater surface area.

Each enzyme has environmental conditions, typically temperature and pH, that allow it to be the most active. Enzymes in our body function well at body temperature (37°C) but will denature and become inactive if temperature becomes too high. The pH of the digestive system changes from the mouth to the small intestine; therefore, the enzymes in these different areas have different optimal pHs.

In this exercise, we are using one specific enzyme, salivary amylase, to demonstrate digestion of carbohydrates into glucose. Whole pieces of cracker (starch) are compared with crumbled crackers to observe the importance of mechanical digestion. The activity of amylase will be observed at a pH similar to that of the oral cavity and a pH similar to the stomach to demonstrate the importance of pH on enzyme activity. Carbohydrate digestion begins in the mouth with a pH of about 7.4, but salivary amylase is denatured by the low pH in the stomach due to the presence of hydrochloric acid (HCl). Pancreatic buffers bring the pH back to 7.4, a favorable pH for pancreatic amylase, allowing carbohydrate digestion to continue in the small intestine.

SAFETY NOTE:
- *Wear safety glasses when heating test tubes and working with acids.*
- *Use a test tube clamp for handling heated test tubes.*

TABLE 35.1 Chemical Digestion: Enzymes, Substrates, and Products

ORGAN	ENZYMES PRESENT	SUBSTRATES	PRODUCTS
Mouth	Salivary amylase	Starches (polysaccharides)	Shorter polysaccharides and maltose (disaccharide)
	Lingual lipase (not active in mouth)	N/A	N/A
Esophagus	None	N/A	N/A
Stomach	Gastric lipase (limited role in adults)	Triglycerides (fats and oils)	Fatty acids and monoglycerides
	Pepsin (secreted as pepsinogen and converted to pepsin by HCL)	Proteins	Peptides
	Lingual lipase (activated by HCl; limited action)	Triglycerides (fats and oils)	Fatty acids and diglycerides
Small Intestine	Major Pancreatic Juice Enzymes		
	• Amylase	Starches (polysaccharides)	Disaccharides and shorter polysaccharides
	• Trypsin and other proteases	Proteins	Peptides
	• Lipase	Triglycerides (fats and oils)	Fatty acids and monoglycerides
Small Intestine	Brush Border Enzymes		
	Disaccharidases	Disaccharides	Monosaccharides
	Peptidases	Peptides and dipeptides	Amino acids

LAB ACTIVITY 1 Experiment: Digestion of Starch

1 Prediction: Circle the correct tube choice for each statement.
- The greatest amount of starch digestion will occur using: *whole cracker, crumbled cracker*
- The optimum pH for amylase to digest starch is: *7.0 (water), 7.4 buffer, acidic pH (HCl)*

2 Materials
- Materials: Obtain materials for Starch Digestion, Starch Test, and Sugar Test (see Materials list).

3 Data Collection:
- Read the procedures and decide who will do each step.

Digestion of starch by amylase

- Mark the tubes 1–6 with the wax pencil and put your initials on the upper part of the test tube for identification in the water bath. Place tubes in the test tube rack. Test tubes 1 and 2 are your controls to be used as comparisons with your results in tubes 3–6.
- In tubes 1, 3, and 5, place 1/8 of a piece of saltine cracker, uncrumbled. Add 12 drops of buffer to each tube and swirl it around.
- In tubes 2, 4, and 6, crumble 1/8 of a saltine cracker into very small pieces; crush with a mortar and pestle, if available. Add 12 drops of buffer to each tube and swirl it around.
- Swirl to mix the amylase solution before using. To tubes 3–6, add 3 drops of *amylase*.
- To tubes 5 and 6, add 3 drops of *0.5% hydrochloric acid (HCl)*.
- Add 6 drops of water to tubes 1 and 2, and 3 drops of water to tubes 3 and 4 to equal the liquid in tubes 5 and 6.
- Cover the test tubes with Parafilm and shake the tubes well to mix the contents.
- Remove the Parafilm and place the test tube holder with all 6 filled tubes in a 37°C water bath. Incubate for at least 30 minutes.

- Appoint one member of your group to be the timer for timing the tubes, and have this person gently shake the test tube rack periodically for best results.
- Proceed to the bile salts preparation setup in Lab Activity 3 while these tubes incubate.
- Start water boiling in the 250-mL beaker for the sugar test.

Determine the amount of starch in each tube

- Mark the depressions on the spot plates 1–6, matching the test tube identification.
- Obtain test tubes from the water bath and swirl to mix.
- Place 2 drops of each test tube solution in the appropriately marked depression on the spot plate.
- Place 1 drop of **Lugol's iodine (IKI)** solution (brown) on top of the sample drops.
- Observe the color and record this in Table 35.2. IKI turns *blue* or *black* depending on the amount of starch present and indicates a **positive starch test**. If there is **no color change,** this indicates a **negative starch test.**

brown (−)	dark blue (+++)
light blue (+)	black (++++)
medium blue (+ +)	

Determine the amount of sugar formed when starch is digested by amylase

- Place 3 drops of **Benedict's solution** (light blue) in the remaining solutions in all 6 tubes and swirl.
- Put all 6 tubes into the beaker when the water is boiling and time it for 3 minutes.
- Observe the color of the solution and record this in Table 35.2. A **positive sugar test** is green, yellow, orange, or red, depending on the amount of sugar present. If the color remains blue, you have a **negative sugar test.**

blue (−)	orange (+++)
green (+)	red (++++)
yellow (++)	

TABLE 35.2 Digestion of Starch by Salivary Amylase

	TUBE NO.					
	1	2	3	4	5	6
Contents added to test tubes	Whole cracker Buffer Water	Crumbled cracker Buffer Water	Whole cracker Buffer Amylase Water	Crumbled cracker Buffer Amylase Water	Whole cracker Buffer Amylase 0.5% HCl	Crumbled cracker Buffer Amylase 0.5% HCl
IKI test results for starch						
Benedict's test results for sugar						

4 Clean up as directed by your instructor.

5 Complete the Experimental Report with your lab group.

6 Complete **PowerPhys Experiment:** Enzyme Activity.

EXPERIMENTAL REPORT
Digestion of Starch

Results:

- List the test tubes that tested positive for sugar from highest amount of sugar to lowest.

- Identify any test tubes that tested negative for sugar.

- List the test tubes that tested positive for starch from lowest amount of starch to highest.

- Identify any test tubes that tested negative for starch.

- State the pH that was most effective for starch digestion.

Discussion:

- Compare the amounts of sugar and starch in all of the test tubes. Explain the differences.

- Discuss how the results would be different if the incubation time were longer.

- Identify which test tube mimicked the conditions in the mouth, esophagus, and the stomach.

Conclusion:

- State how the presence of an enzyme influences the digestion of starch to sugar.

- State how substrate size and pH influence enzyme activity. ■

LAB ACTIVITY 2 Starch Digestion in the Mouth

1 Obtain a cracker and break off two squares, each approximately a 1/2 inch square.

2 Place one square on your tongue (do not chew it and do not swallow it).

3 Time how long it takes for you to taste sweetness. _____ sec.

4 Swallow the cracker and rinse your mouth with water.

5 Place the other square in your mouth and chew it (do not swallow).

6 Time how long it takes for you to taste sweetness. _____ sec. .

7 Calculate time to taste sweetness for the chewed cracker as a percentage of the time to taste sweetness for the whole cracker. _____ %

8 Pool your results with the rest of your class and calculate the following averages:
- Average time to taste sweetness for whole cracker. _____ sec.
- Average time to taste sweetness for chewed cracker. _____ sec.
- Average % of time to taste sweetness of chewed cracker compared with whole cracker. _____ sec.

9 Explain the difference in average time to taste sweetness between the whole cracker and chewed cracker.

10 Compare your results with the class averages and explain any differences. Think how conditions in your mouth may differ from other students in the class. List all possible differences. ■

C. Bile Emulsification

Bile is not a digestive enzyme. **Bile** is composed of a mixture of substances, including **bile salts,** that aid in digestion by physically breaking up large aggregates of lipid molecules into smaller aggregates of lipid molecules with a greater surface area, a process called **emulsification.**

LAB ACTIVITY 3 Bile Emulsification

1 Read the procedure and decide who will do each step.

2 Obtain materials for Bile Emulsification (see Materials list).

3 Observe the ability of bile salts to emulsify lipids.
- Mark the tubes 1 and 2 with wax marker.
- To each test tube, add 12 drops of water and 2 drops of vegetable oil (a lipid).
- To test tube 2, add a pinch of bile salts.
- Cover the tubes with Parafilm and shake them to mix well.
- Let the tube contents stand for 15 minutes.
- Describe the contents of the 2 tubes:

 Tube 1 _____

 Tube 2 _____

4 Clean up as directed by your instructor.

5 Answer Discussion Questions with your lab group.

DISCUSSION QUESTIONS
Bile Emulsification

1 In which tube had the oil emulsified? _____

2 Is bile a digestive enzyme? _____ Explain.

3 Explain how emulsification increases chemical digestion of lipids by lipases.

■

Reviewing Your Knowledge

A. General Terminology

Write the term that matches the description.

absorption	macromolecule
amylase	product
enzyme	protease
lipase	substrate

_____ **1.** Large molecules resulting from anabolism

_____ **2.** A biological catalyst

_____ **3.** The molecule acted on by an enzyme

_____ **4.** Process that happens to nutrients after digestion

_____ **5.** Enzyme that hydrolyzes lipid

_____ **6.** Enzyme that hydrolyzes protein

_____ **7.** Enzyme that hydrolyzes starch

_____ **8.** The molecule produced as a result of an enzymatic process

B. Substrates and Products

Fill in the blank with the appropriate term.

1. Proteins are hydrolyzed into and absorbed as _____.

2. Lipids are hydrolyzed into and absorbed as _____ and _____.

3. Carbohydrates are hydrolyzed into and absorbed as _____.

4. Carbohydrate digestion begins in the _____.

5. Bile salts prepare lipids for digestion by a process called _____.

6. Lipid digestion begins in the _____.

7. Protein digestion begins in the _____.

C. Carbohydrate Digestion Exercise

Fill in the blank with the appropriate term.

1. The reagent used to test for the presence of starch was _____.

2. The reagent used to test for the presence of sugar was _____.

3. The three major food macromolecules digested in our GI tract are _____, _____, and _____.

4. Why were there 2 tubes in Lab Activity 1 that had no enzyme added to them?

5. The enzymes that digest carbohydrates work best at a(n) _____ pH.

6. The reason why we chew food (and crumbled the cracker in Lab Activity 1) is to

7. A positive test for the presence of sugar is indicated by which colors?

8. A positive test for the presence of starch is indicated by which colors?

Using Your Knowledge

A. Digestion

1–3. Enzymes that digest carbohydrates are produced by what organs?

1. _____

2. _____

3. _____

4–6. Enzymes that digest lipids are produced by what organs?

4. _____

5. _____

6. _____

7–9. Enzymes that digest proteins are produced by what organs?

7. _____

8. _____

9. _____

10. People who practice purging themselves after eating (bulimics) ruin the dentin of their teeth. What causes the dentin loss?

Urinary System Structure and Function

OBJECTIVES

1 Identify the organs of the urinary system
2 Describe the structure of urinary system organs
3 Describe the structure of the nephron and trace the path of filtrate through the nephron
4 Trace the path of urine from the collecting ducts to the exterior of the body
5 Describe the blood supply to the kidneys and trace the pathway of blood from the renal artery to the renal vein
6 Dissect a cat, fetal pig, or cadaver and identify the main urinary tract organs

MATERIALS

- male and female urogenital models, charts, or use **Real Anatomy** (Urinary)
- human kidney and nephron models; articulated skeleton
- Compound microscope, lens paper, prepared slides of the kidney and a transverse section through a ureter or urinary bladder
- Dissection: Fresh or preserved kidney, dissection equipment, disposable gloves, safety glasses
- Dissection: Preserved cat or fetal pig, dissection equipment, disposable gloves, safety glasses, dissection manual
- **Real Anatomy:** Virtual Cadaver Dissection

Urinary system organs include the kidneys, ureters, urinary bladder, and urethra. Urine is formed in the kidneys and flows through the ureters to the urinary bladder, which stores urine until it is eliminated from the body through the urethra.

A. Location and Gross Anatomy of Urinary System Organs

1. Kidneys

Kidneys are bean-shaped structures that are **retroperitoneal** (*retro* = behind), located between the abdominal wall and the peritoneum. They are found at waist level between the 12th thoracic vertebra and the 3rd lumbar vertebra. The concave surface of each kidney faces the vertebral column and contains a vertical fissure, the **renal hilum.** The ureter, renal arteries and veins, nerves, and lymphatics pass through the renal hilum.

The kidney is attached to the abdominal wall by the **renal fascia,** an outer layer of dense irregular connective tissue. The **renal capsule,** a thin fibrous membrane, covers the outer surface of the kidney. Between the renal fascia and the renal capsule is adipose tissue which forms the **adipose capsule** for protection and padding.

Within the kidney are three main regions: the cortex, medulla, and sinus. The **renal cortex,** a smooth area, is the most superficial region, whereas the **renal medulla** is deeper. The medulla contains cone-shaped **renal pyramids**

and extensions of the cortex, **renal columns,** that are between each pyramid. The base of each pyramid faces the cortex, and the **renal papilla** is the apex that is pointed toward the renal sinus. The cortex and medulla contain **nephrons,** the structural and functional units of the kidney that form urine. Urine drains into **papillary ducts** that exit through the openings in the renal papilla. The **renal sinus** is a space or cavity that is adjacent to the medulla and extends to the renal hilum. Within the renal sinus are minor and major calyces and the renal pelvis, which collect urine from the papillary ducts and deliver it to the ureters. The **minor calyces** are cup-like structures adjacent to the renal papillae, which receive urine from the papillary ducts. Several minor calyces drain into a major calyx. Each kidney has 8 to 18 minor calyces and 2 to 3 major calyces. The **major calyces** drain into the **renal pelvis,** which is continuous with the ureter. Blood vessels, lymphatics, and nerves also travel through the renal sinus. Adipose tissue fills the remaining space within the renal sinus.

Before Going to Lab

1 Label the urinary system structures in Figures 36.1, 36.2, and 36.3.

LAB ACTIVITY 1 Location and Structure of Kidneys

1 Identify the urinary system structures on a human torso model or chart, or use the search text box to find the structures in **Real Anatomy** (Urinary).

2 Use an articulated skeleton to identify which ribs help protect the kidneys.

3 Identify the structures from Figure 36.3 on a dissectible kidney model.

4 Answer the Discussion Questions with your lab group.

DISCUSSION QUESTIONS

1 Which kidney is lower? Identify the organ that forces this kidney into a lower position.

2 Which ribs protect the kidneys?

Diaphragm
Esophagus
Left adrenal (suprarenal) gland
Abdominal aorta
Interior vena cava
Rectum
Uterus

Right renal vein
Right renal artery

• right kidney
• right ureter
• urethra
• urinary bladder

1 _____

2 _____

3 _____

4 _____

 FIGURE 36.1 **Organs of the female urinary system.**

- adipose capsule
- left kidney
- peritoneum
- renal capsule
- renal fascia
- renal hilum
- right kidney

1 _____

2 _____

3 _____

4 _____

5 _____

6 _____

7 _____

Transverse section through
the abdomen (lying down and
viewed from the feet)

FIGURE 36.2 Location and coverings of the kidneys.

- major calyx (KAY-liks) 8
- minor calyx 7
- renal artery 9
- renal capsule 4
- renal column 3
- renal cortex 2
- renal hilum 11
- renal medulla 1
- renal papilla 6
- renal pelvis in renal sinus 12
- renal pyramid 5
- renal vein 10

1 _____

2 _____

3 _____

4 _____

5 _____

6 _____

7 _____

8 _____

9 _____

10 _____

11 _____

12 _____

Adipose
tissue

FIGURE 36.3 Internal structure of the kidney.

2. Ureters, Urinary Bladder, and Urethra

The **ureters** are narrow, 25- to 30-cm long muscular tubes located behind the peritoneum (retroperitoneal). The diameter of the ureters varies from 1 to 10 mm. The ureters descend toward the urinary bladder, curving medially as they approach the inferior portion of the bladder, and enter the posterior wall of the bladder at an oblique angle. Urine is propelled through the ureters by peristalsis, hydrostatic pressure, and gravity.

The **urinary bladder** is a hollow, muscular organ that distends to store urine. Its superior surface is covered by visceral peritoneum and is secured to the pelvic wall by parietal peritoneal folds. The smooth muscle within the wall of the urinary bladder is called the **detrusor muscle,** and the epithelial lining of the bladder forms folds or **rugae.** The inferior surface of the urinary bladder contains three openings that form a triangle called the **trigone.** The two posterior openings are the **ureteral openings,** whereas the anterior opening is the opening into the urethra, the **internal urethral orifice.** In males, the urinary bladder is anterior to the rectum and posterior to the pubic symphysis. In females, the urinary bladder is anterior to the vagina, inferior to the uterus, and posterior to the pubic symphysis.

The tube-like **urethra** carries urine from the internal urethral orifice to the **external urethral orifice,** the opening through which urine exits the body. The **internal urethral sphincter** is a layer of circular, involuntary smooth muscle that controls passage of urine into the urethra from the urinary bladder. Voluntary skeletal muscle within the urogenital diaphragm (deep muscles of perineum) forms the **external urethral sphincter** that permits the passage of urine to the external urethral orifice. During **micturition,** the detrusor muscle contracts and forces urine out of the urinary bladder, and the internal and external urethral sphincters relax to allow the passage of urine. In females, the urethra is short (4 cm), whereas the urethra in males is longer (15–20 cm). The male urethra has three regions: the **prostatic urethra** that passes through the prostate gland, the **membranous urethra** that passes through the urogenital diaphragm, and the **spongy (penile) urethra** that passes through the penis.

> **CLINICAL NOTE:** *BPH, or benign prostatic hypertrophy, is a noncancerous enlargement of the prostate gland that restricts the prostatic urethra and inhibits urine flow, causing urinary retention.*

Before Going To Lab

1 Label the urinary system structures in Figures 36.4 and 36.5(a) and (b).

LAB ACTIVITY 2 **Location and Gross Anatomy of the Ureters, Urinary Bladder, and Urethra**

1 Identify the urinary system structures from Figures 36.4 and 36.5 on urogenital models or charts, or use the search text box to find the structures in **Real Anatomy** (Urinary). ■

- detrusor muscle of urinary bladder 3
- external urethral orifice 9
- external urethral sphincter 8
- internal urethral orifice 6
- internal urethral sphincter 1
- trigone (TRY-goan) 5
- ureteral openings 4
- ureters 2
- urethra 7

1 _____ 6 _____

2 _____ 7 _____

3 _____ 8 _____

4 _____ 9 _____

5 _____

FIGURE 36.4 **Ureters, urinary bladder, and urethra in a female.**

(a)

- anus 2
- external urethral orifice 10
- internal urethral orifice 6
- membranous urethra 8
- prostatic urethra 7
- rectum 1
- spongy urethra 9
- ureter 3
- ureteral opening 5
- urinary bladder 4
- urogenital diaphragm 11

1 _____

2 _____

3 _____

4 _____

5 _____

6 _____

7 _____

8 _____

9 _____

10 _____

11 _____

(b)

- anus
- external urethral orifice
- internal urethral orifice
- rectum
- urethra
- urinary bladder
- urogenital diaphragm
- uterus
- vagina

12 _____

13 _____

14 _____

15 _____

16 _____

17 _____

18 _____

19 _____

20 _____

(a) Sagittal section through male pelvis

(b) Sagittal section through female pelvis

✶ FIGURE 36.5 **Comparison of the urinary bladder and urethra in males and females.**

B. Structure of the Nephron

The 1 million nephrons in each kidney filter the blood and form urine. Each **nephron** is composed of a renal corpuscle that filters blood and a renal tubule that modifies the filtrate to form urine. The **renal corpuscle,** which is located in the renal cortex, consists of a **glomerulus** (capillary network) and a **glomerular** (Bowman's) **capsule,** a cup-shaped epithelial membrane surrounding the glomerulus. Blood is filtered across a filtration membrane formed by the walls of the glomerulus and the glomerular capsule into the **capsular space** (glomerular cavity). The liquid, now called filtrate, drains into the renal tubule.

Each renal tubule is subdivided into three structural and functional sections. Starting at the glomerular capsule, these sections are proximal convoluted tubule, loop of Henle, and distal convoluted tubule. The **proximal convoluted tubule** is located in the renal cortex and connects to the **loop of Henle (nephron loop),** which dips down into the renal medulla. The third portion is the **distal convoluted tubule,** which is located in the cortex.

Distal convoluted tubules from several nephrons drain the liquid (now called urine) into **collecting ducts** that descend through the renal pyramids and merge to form larger **papillary ducts.** These ducts open into the renal papillae, and immediately drain urine into minor calyces, then into major calyces, the renal pelvis, ureters, and finally into the urinary bladder for storage.

The vasculature of the nephron is important to the formation of urine. The glomerulus is located between two arterioles—a larger-diameter **afferent arteriole** that delivers systemic blood to the glomerulus and a smaller-diameter **efferent arteriole** that receives blood from the glomerulus. The efferent arteriole then delivers blood to a second capillary bed, the **peritubular capillary network** that entwines the renal tubule.

There are two types of nephrons: cortical nephrons and juxtamedullary nephrons. **Cortical nephrons** have short loops of Henle that descend slightly into the medulla, whereas **juxtamedullary nephrons** have long loops of Henle that extend deep into the medulla. In addition to peritubular capillaries, the loops of Henle of juxtamedullary nephrons are supplied by **vasa recta,** long capillary loops that extend from the efferent arteriole.

Before Going to Lab

1 Label the renal tubule in Figure 36.6(a), and the blood vessels and renal tubule in Figure 36.6(b).

LAB ACTIVITY 3 **Structure of the Nephron**

1 Identify the structures of the nephron in Figure 36.6(a) and (b) on a nephron model or chart. ■

(a) Renal tubule of cortical nephron (b) Renal corpuscle and vasculature of the renal tubule

(a)

- ascending limb of loop of Henle 3
- collecting duct 6
- descending limb of loop of Henle 4
- distal convoluted tubule 5
- glomerular (glow-MER-u-lar) capsule 1
- papillary duct 7
- proximal convoluted tubule 2

1 _____

2 _____

3 _____

4 _____

5 _____

6 _____

7 _____

(b)

- 13 afferent arteriole
- 12 efferent arteriole
- 9 glomerular capsule
- 8 glomerulus
- 11 peritubular capillary
- 10 renal corpuscle

8 _____

9 _____

10 _____

11 _____

12 _____

13 _____

FIGURE 36.6 **Structure and vascular supply of a cortical nephron.**

C. Blood Flow Through the Kidneys

During rest, the renal arteries carry 20 to 25% of the cardiac output to the kidneys, and the renal veins return blood to the inferior vena cava. The pathway of blood through the kidneys is outlined below.

Renal artery

↓

Segmental arteries

↓

Interlobar arteries

↓

Arcuate arteries

↓

Interlobular arteries Renal vein

↓ ↑

Afferent arteriole Interlobar veins

↓ ↑

Glomerular capillaries Arcuate veins

↓ ↑

Efferent arterioles → Peritubular capillaries → Interlobular veins
 and/or vasa recta

- afferent arteriole
- arcuate artery
- arcuate vein
- efferent arteriole
- interlobar artery
- interlobar vein
- interlobular artery

- interlobular vein
- peritubular capillary
- renal artery
- renal vein
- segmental artery
- vasa recta

1 _____ 8 _____

2 _____ 9 _____

3 _____ 10 _____

4 _____ 11 _____

5 _____ 12 _____

6 _____ 13 _____

7 _____

FIGURE 36.7 **Blood supply of the kidney.**

Before Going to Lab

1 Label the blood vessels in Figure 36.7. Find the renal artery and label it first to follow the blood flow.

LAB ACTIVITY 4 **Blood Flow Through the Kidneys**

1 Identify the blood vessels on a model or chart.

2 Answer the Discussion Questions with your lab group.

Blood supply of the nephron

DISCUSSION QUESTIONS
Blood Flow Through the Kidney

Use Figure 36.7 to answer the following questions.

1 Identify the arteries and veins that travel through the renal sinus.

2 Identify the artery and vein that travel through the renal columns.

3 Identify the artery and vein that arch over the base of the renal pyramid between the cortex and medulla.

4 Identify the blood vessels located in the renal cortex.

D. Dissection of a Pig (or Sheep) Kidney

Pig and sheep kidneys are very similar to human kidneys. This dissection will illustrate the location of the calyces and renal pelvis within the renal sinus.

> **SAFETY NOTE:** *Wear gloves and safety glasses when using preserved or fresh tissue. Wash hands thoroughly with soap and water when you are done.*

LAB ACTIVITY 5 Kidney Dissection

1 Identify the boldface structures listed below on a pig or sheep kidney, using Figure 36.8 as a reference.

2 If you have a fresh kidney, carefully remove any fat on its exterior. If you are lucky, the adrenal glands may still be located on the superior surface. Adrenal glands are usually not present on preserved kidneys.

3 If you have a preserved kidney, rinse it to remove excess preservative.

4 Locate the **renal hilum** on the concave surface. Identify the **ureter, renal artery,** and **renal vein** that travel through the hilum. The renal artery will have a thicker wall than the renal vein.

5 Using a scalpel or knife, carefully cut along the frontal plane to separate the kidney into anterior and posterior sections.

6 Observe the cut surface. Using a needle probe, pull the thin **renal capsule** away from the surface of the kidney. The outer region of the kidney, the **cortex,** will be a slightly lighter color than the deeper **medulla.**

Cortex
Renal column
Renal pelvis

Pyramid
Renal papillae
Minor calyx
Major calyx
Renal capsule
Hilum

FIGURE 36.8 Frontal section of preserved triple-injected pig kidney.

7 **Glomeruli** may be observed in preserved kidneys that are injected with red and blue latex. The glomeruli will appear as red and blue dots within the cortex.

8 Within the medulla, extensions of light-colored cortical tissue, **renal columns,** are found between the darker, cone-shaped **renal pyramids.** Identify the **renal papillae** located at the apices of the renal pyramids.

9 Between the medulla and the renal hilus is a space called the **renal sinus.** Within the renal sinus are the **minor** and **major calyces,** and the **renal pelvis.** Adipose tissue, blood vessels, lymphatics, and nerves fill the remaining space within the renal sinus.

10 Clean up as directed by your instructor.

E. Dissection of the Urinary System

The urinary system organs of the cat and fetal pig are similar to those of the human. This dissection will illustrate the retroperitoneal location of the urinary system organs, the relationship of urinary system organs to other organs, and how well these organs are secured by connective tissues. Refer to the appropriate accompanying cat or fetal pig dissection manual for dissection instructions and figures.

Real Anatomy, a virtual cadaver dissection, can be used to complement or substitute for animal dissection of the urinary system.

F. Histology of the Urinary System

1. Kidneys

The **glomerular** (Bowman's) **capsule** is composed of a **visceral** and **parietal epithelial layer.** The **visceral layer,** which forms part of the filtration membrane, is adjacent to the glomerulus, and the **parietal layer** forms a funnel-like structure that collects the filtrate within the capsular space (glomerular cavity). Both layers are simple squamous epithelium; however, the **simple squamous epithelial** cells of the visceral layer have foot-like projections that wrap around glomerular capillary walls and these cells are named **podocytes** (*podo* = foot).

The epithelium lining the **proximal convoluted tubule** is **simple cuboidal epithelium.** The **microvilli** forming the **brush border** on the apical surface of these epithelial cells give a brush-like appearance.

The **descending limb of the loop of Henle (nephron loop)** is lined by **simple squamous epithelium.** The initial part of the ascending limb or **thin section of the ascending limb** is also lined by simple squamous epithelium, whereas the **thick section of the ascending limb of the loop of Henle (nephron loop)** is lined by **simple cuboidal** to low **columnar cells.**

The **distal convoluted tubule** is lined with **simple cuboidal epithelium.** These cells do not have a brush border and appear shorter than the simple cuboidal cells lining the proximal convoluted tubule. The final part of the distal convoluted tubule and the **collecting ducts** are lined with **simple cuboidal epithelium.**

> ### Before Going to Lab
>
> **1** Label the microscopic kidney structures in Figure 36.9(a) and (b). Notice that, with this stain, the distal convoluted tubules stain a lighter pink than the proximal convoluted tubules.

LAB ACTIVITY 6 Histology of the Kidney

1 Observe a prepared slide of the kidney.
- Using a low-power objective lens, identify **renal corpuscles** and **proximal and distal renal tubules** in the cortex and the **loops of Henle** and **collecting ducts** that extend into the renal medulla.
- Place a renal corpuscle in the center of the field of view, and using a high-power objective lens, identify the following structures: **simple squamous epithelium of the parietal layer of the glomerular capsule, capsular space,** and **glomerulus.**
- Return to a low-power objective lens, and place a section of the cortex with proximal and distal convoluted tubules in the center of the field of view. Switch to a high-power objective lens and identify the **simple cuboidal epithelium lining a proximal convoluted tubule** and **a distal convoluted tubule.**
- Return to a low-power objective lens and place a section of the renal medulla with collecting tubules in the center of the field of view. Switch to a high-power objective lens and identify the **epithelial lining of the collecting tubules.**

2 Answer Discussion Questions with your lab group.

DISCUSSION QUESTIONS
Histology of the Kidney

1 State the function of the simple squamous epithelium of the visceral layer of the glomerular capsule.

2 The epithelium lining of the proximal convoluted tubule has a brush border. Name the cell structure that forms the brush border and state its function.

(a) Survey photomicrograph of kidney cortex

LM 16×

LM 1225×

Lumen of proximal convoluted tubule

Lumen of distal convoluted tubule

4 5 6

(b) Renal corpuscle and proximal and distal convoluted tubules

(a)
- loop of Henle and collecting ducts
- proximal and distal convoluted tubules (outer cortex)
- glomerulus

1 _____

2 _____

3 _____

(b)
- parietal layer of glomerular capsule (simple squamous epithelium)
- capsular space
- glomerulus

4 _____

5 _____

6 _____

FIGURE 36.9 Sectional view of the kidney.

2. Ureters, Urinary Bladder, and Urethra

The ureters are composed of three layers: a mucosa that lines the lumen, a muscular layer called the muscularis, and an outer adventitial layer. The **mucosa** contains transitional epithelium, which can distend to accommodate urine, and an underlying lamina propria of areolar connective tissue. The **muscularis** layer contains smooth muscle. **Peristalsis,** the function of the muscularis layer, moves urine from the renal pelvis to the urinary bladder. The **adventitia** secures the ureters in place.

The urinary bladder contains the same layers as the ureter. The epithelium of the **mucosa** is transitional epithelium, and the **muscularis** contains smooth muscle. Contraction of the muscularis layer forces urine out of the urinary bladder, and the **adventitia** secures the bladder in place.

The urethra consists of a **mucosa** surrounded by a **muscularis** layer of circular smooth muscle. In both males and females, the epithelium of the mucosa changes along the course of the urethra. The epithelium is transitional near the urinary bladder, becomes stratified columnar or pseudostratified columnar distally, and changes to stratified squamous epithelium near the external urethral orifice.

Before Going to Lab

1 Label the urinary tract tissues in Figure 36.10.

LAB ACTIVITY 7 Histology of the Ureters, Urinary Bladder, and Urethra

1 Obtain a prepared microscopic slide of a transverse section through the ureter or urinary bladder or use **Real Anatomy** (Histology).
- Using a low-power objective lens, identify the **mucosa,** the **muscularis,** and the **adventitia.**
- Using the high-power objective lens, identify the **transitional epithelium** lining the lumen.

2 Answer the Discussion Question with your lab group.

DISCUSSION QUESTION
Histology of Ureters, Urinary Bladder, and Urethra

1 Explain why the epithelial layer changes to stratified squamous epithelium near the external urethral orifice.

• adventitia
• lamina propria
• lumen
• mucosa
• muscularis
• transitional epithelium

1 _____

2 _____

3 _____

4 _____

5 _____

6 _____

LM 40×

FIGURE 36.10 **Sectional view of the ureter.**

Reviewing Your Knowledge

A. Renal Structures

Write the renal structure that the phrase describes.

_____ **1.** Urine-forming structure of the kidney

_____ **2.** Region of the kidney deep to cortex; contains collecting ducts

_____ **3.** Extensions of renal cortex found in between renal pyramids

_____ **4.** Urine flowing through this structure drains into a minor calyx

_____ **5.** Located between renal fascia and renal capsule

_____ **6.** Apex of renal pyramid

_____ **7.** Urine flowing through this structure drains into the renal pelvis

_____ **8.** Vertical fissure in concave surface of kidney through which blood vessels and ureters pass

_____ **9.** Covers outer surface of kidney

_____ **10.** Dense irregular connective tissue that covers the adipose capsule and attaches the kidney to the abdominal wall

_____ **11.** Receives urine from the major calyces

_____ **12.** Space within kidney that is adjacent to renal medulla, contains calyces and renal pelvis

_____ **13.** Cup-like structure that is located in renal sinus that receives urine from openings of papillary ducts

_____ **14.** Cone-shaped structures located within the renal medulla

_____ **15.** Outermost region of the kidney, contains renal corpuscles

B. Ureters, Urinary Bladder, and Urethra

Write the name of the structure that the phrase describes.

_____ **1.** Detrusor muscle is the main muscle for this structure

_____ **2.** Region of male urethra that passes through the prostate gland

_____ **3.** Two openings in posterior urinary bladder wall

_____ **4.** Region of male urethra that passes through penis

_____ **5.** Area bounded by ureteral openings and internal urethral orifice

_____ **6.** Voluntary skeletal muscle in urogenital diaphragm that allows passage of urine to exterior of body

_____ **7.** Carries urine from renal pelvis to urinary bladder

_____ **8.** Urine is excreted through this opening

_____ **9.** Circular smooth muscle that involuntarily controls passage of urine from the urinary bladder to the urethra

_____ **10.** Anterior opening in urinary bladder that leads into urethra

_____ **11.** Region of male urethra that passes through urogenital diaphragm

C. The Nephron

Write the name of the structure the phrase describes.

_____ **1.** Blood vessel that delivers blood to glomerulus

_____ **2.** Blood from the efferent arteriole flows into this capillary bed

_____ **3.** Structure that surrounds glomerulus and collects filtrate

_____ **4.** Section of renal tubule that descends into medulla

_____ **5.** Capillary network within the renal corpuscle

_____ **6.** Structure composed of glomerulus and glomerular capsule

_____ **7.** Blood vessel that drains blood from glomerulus

_____ **8.** Capillary loops that extend from the efferent arteriole and run along loop of Henle (nephron loop) of juxtamedullary nephrons

D. Urine Formation and Flow

Trace the flow of filtrate and urine through the urinary system. Write the structures *in order*, starting with the glomerulus.

1. glomerulus

2. _____

3. _____

4. _____

5. _____

6. _____

7. _____

8. _____

9. _____

10. _____

11. _____

12. _____

13. _____

14. external urethral orifice

E. Blood Flow Through the Kidneys

Trace blood flow through the kidneys. Start at the *renal artery* and number the blood vessels in sequence.

_____ afferent arteriole

_____ arcuate artery

_____ arcuate vein

_____ efferent arteriole

_____ glomerulus

_____ interlobar artery

_____ interlobar vein

_____ interlobular artery

_____ interlobular vein

_____ peritubular capillary

___1___ renal artery

_____ renal vein

_____ segmental artery

Using Your Knowledge

Observe the X-ray in Figure 36.11 and identify the structures.

1. _____

2. _____

3. _____

4. _____

5. _____

FIGURE 36.11 Urinary system structures in the cat.

6. Is the urinary bladder full or partially voided? [Refer to Figure 2.5(d) for another view of the urinary bladder.]

7. Explain why pregnancy increases the frequency of urination.

8. Epithelial cells that reabsorb solutes are cuboidal, but epithelial cells involved in filtration are thin, squamous cells. The cuboidal cells contain more cytoplasm and organelles, especially mitochondria and rough endoplasmic reticulum (RER), than the squamous cells. Why do the cuboidal cells of the kidney tubules need more mitochondria and RER?

9. Define *ptosis* of the kidney.

10. Define *incontinence* and describe how this situation would be different in a child under 2 years old versus in an adult.

Urine Formation and Urinalysis

OBJECTIVES

1 Describe the function of the nephron
2 Describe the main components of the nephron
3 Explain the process of urine formation
4 Describe the major features of a urinalysis
5 Describe the composition and properties of normal urine
6 Compare and contrast the characteristics of normal and abnormal urine

MATERIALS

- **Filtration:** filter paper, funnel, ring stand with ring, 250-mL beaker, 5% powdered charcoal solution, and 1% copper sulfate solution
- **Urinalysis:** 100-mL beakers (4 per group), 50-mL beakers (4 per group), wax pencils, urine test strips, 1 normal and 3 abnormal artificial urines (obtained from a supply house or made with instructions on the Instructor Companion Site); slides and Sedi stain (if desired)
- **PowerPhys Experiment:** Influence of Fluid Intake on Urine Formation

The formation of urine by the 1 million nephrons in each kidney enables the body to remove metabolic wastes from blood and to maintain homeostasis by regulating water volume, concentration of ions in blood, and blood pH. A urinalysis is a simple test that can determine if urine formation is normal or abnormal. This test analyzes the volume of urine, and the physical, chemical, and microscopic composition of urine. In addition, the level of some metabolic wastes in the blood is assessed.

The two main components of the nephron are the renal corpuscle and the renal tubule. The subdivisions of the renal tubule are the proximal convoluted tubules, the loop of Henle, and the distal convoluted tubule. Many distal convoluted tubules empty filtrate into a collecting duct, which is not part of the nephron.

A. Function of the Nephrons

Urine is produced by the nephron and collecting ducts through three different processes—glomerular filtration, which occurs in the renal corpuscle, and tubular reabsorption and secretion, which occur in the renal tubules.

- **Filtration**—The first step in urine production. Water and most solutes in blood pass through the filtration membranes (glomerular capillary walls and visceral wall of glomerular capsule) into the capsular (Bowman's) space (glomerular cavity).
- **Tubular reabsorption**—Water and solutes cross the wall of the renal tubule, diffuse through the interstitial fluid, and return to blood by entering the peritubular capillaries or vasa recta.
- **Tubular secretion**—Solutes leave the peritubular capillaries or vasa recta, diffuse through the interstitial fluid, cross the wall of the renal tubule, and are eliminated in urine.

1. Function of the Renal Corpuscle

Each nephron has a renal corpuscle that is composed of a glomerulus and a glomerular capsule. In the nephron, blood plasma is filtered across the **filtration membrane** (glomerular capillary walls and the visceral wall of the glomerular capsule) into the capsular space (glomerular cavity). **Filtrate,** the liquid and solutes filtered into the capsular space, contains all the components of plasma except those that are too large to pass through the pores in the filtration membrane—the blood cells, platelets, and most plasma proteins. The rate of filtration across the glomerulus is determined by blood pressure (hydrostatic pressure). The greater the blood pressure, the greater the filtration rate. Also, filtrate moves by hydrostatic pressure into the first section of the renal tubule, the proximal convoluted tubule.

LAB ACTIVITY 1 Experiment: Filtration

In this activity, you will observe the **selectivity** of the filtration membrane and the effect of hydrostatic pressure on **filtration rate.**

1 **Prediction:** With your lab partners, predict which substances will pass through the filter: water, copper sulfate (blue), or charcoal.

_____ will pass through the filter.

_____ will not pass through the filter.

2 Read the procedure and decide who will do each step.

3 **Materials:** Obtain materials for **Filtration** (see Materials list).

4 **Data Collection:** Observe and record the substances that pass through a filter paper in the kidney filtration simulation. Record your data in Table 37.1.

TABLE 37.1 Filtration Results

TIME (sec)	#DROPS
15	
30	
45	
60	
90	

- Fold the circular filter paper in half twice, and open it to form a cone. Place the cone in the funnel and place the funnel in a ring on the ring stand. Put the beaker under the funnel (Figure 37.1).
- Stir or mix the solution composed of water, copper sulfate (blue), and powdered charcoal, and pour it into the funnel to just below the top of the filter paper. Watch so the edges of the filter paper do not capsize or that the solution doesn't go over the top.
- As soon as the liquid slows down enough to count individual drops, count the number of drops in 15, 30, 45, 60, and 90 seconds. Record your information in Table 37.1.

5 Clean up as directed by your instructor.

6 Complete the Experimental Report with your lab group.

Funnel

Filter paper

Filtrate

FIGURE 37.1 Filtration setup.

EXPERIMENTAL REPORT
Filtration

Results:

- State how filtration rate changed over time.

- Name the substance(s) that passed through the filter paper.

- Name the substance(s) that did not pass through the filter paper.

Discussion:

- Discuss why you saw a difference in the filtration rate during the activity.

- Identify the force that drives the filtration rate.

- Identify the cell structure that the filter paper represents.

- Compare the substances that were not filtered in this experiment to the substances that are not filtered in the kidney.

- Compare your results of this experiment with your predictions.

Conclusion:

Write a statement that postulates what substances can pass through the glomerulus of the kidney and how blood pressure affects the filtration rate.

2. Function of the Proximal Convoluted Tubule

The **proximal convoluted tubule** is the major site of **tubular reabsorption of water and solutes** from the filtrate. Proximal convoluted tubules reabsorb 100% of most organic solutes and nutrients (glucose, amino acids, lactic acid, water soluble vitamins); 65% of water, sodium ions, and potassium ions; 50% of chloride ions; and 80 to 90% of the bicarbonate ions. Water follows the solutes via osmosis.

The proximal convoluted tubule also **secretes hydrogen ions and urea** (nitrogenous waste) into the tubular fluid. The hydrogen ions secreted into the tubular fluid assist in the reabsorption of bicarbonate ions.

Because the amount of water reabsorbed is approximately equal to the amount of solutes reabsorbed (minus the solutes secreted into the tubule), the osmolarity (solute concentration) of the filtrate at the end of the convoluted tubule is the same as at the beginning of the convoluted tubule.

3. Function of the Loop of Henle (Nephron Loop)

The remaining filtrate travels on through the renal tubule to the **descending limb of the loop of Henle (nephron loop),** which is permeable to water but not to solutes. As the filtrate travels through the descending limb of the loop of Henle (nephron loop) toward the medulla of the kidney, the high solute content (higher osmolarity) of the interstitial fluid in the renal medulla causes 15% of the original filtered **water to be reabsorbed by osmosis.** This causes the osmolarity of the filtrate to increase.

The **ascending limb of the loop of Henle (nephron loop)** is impermeable to water but permeable to solutes. Therefore, the opposite occurs. As the filtrate travels toward the cortex, the decrease in solute content (lower osmolarity) of the interstitial fluid causes **reabsorption of Na^+ and Cl^-.** The osmolarity of the filtrate decreases and becomes more dilute than the proximal convoluted tubule filtrate as it heads toward the distal convoluted tubule.

4. Function of the Distal Convoluted Tubule

As the filtrate enters the **distal convoluted tubule,** which is located in the cortex of the kidney, a total of 80% of water has been reabsorbed into the blood. In this segment of the tubule, further **Na⁺ and Cl⁻ are reabsorbed** into the blood and a small amount of **water follows the solute by osmosis.** The filtrate in the DCT is more dilute than in the PCT.

5. Function of the Late Distal Tubules and Collecting Ducts

When the filtrate reaches the collecting ducts, 90 to 95% of the water and solutes present in the original filtrate have been removed. The filtrate is now more dilute than in the proximal convoluted tubule. In the late distal tubules and collecting ducts, there is **additional reabsorption of Na⁺** and **Cl⁻** and the **secretion of K⁺.**

Hormones affect reabsorption and secretion in the late distal tubules and collecting ducts. **ADH** determines whether a concentrated (higher osmolarity) or dilute (lower osmolarity) urine is produced. High levels of ADH increase water reabsorption, producing concentrated urine. When ADH levels are low, dilute urine is produced. **Aldosterone** acts on the collecting ducts, increasing the reabsorption of Na⁺ and Cl⁻ and the secretion of K⁺.

LAB ACTIVITY 2 **Function of the Nephron**

1 With your lab group, write the **number** of the structure shown in Figure 37.2 next to the phrase that describes its function. A number may be used more than once.

2 With your lab group, complete Table 37.2 by comparing the osmolarity of the filtrate in each portion of the renal tubule or collecting duct with the osmolarity of the filtrate in the glomerular capsule. Write in one of the following choices: *no change, increased, decreased, change dependent on presence of ADH and/ or aldosterone.*

3 Complete **PowerPhys Experiment:** Influence of Fluid Intake on Urine Formation.

_____ additional reabsorption of solutes and water

_____ ADH increases reabsorption of water, additional reabsorption and secretion of solutes

_____ aldosterone increases reabsorption of Na⁺ and Cl⁻ and secretion of K⁺

_____ filtration

_____ major site of tubular reabsorption of water and solutes; secretion of solutes

_____ Na⁺ and Cl⁻ reabsorbed but not water

_____ water reabsorbed but not solutes

FIGURE 37.2 **Structure and function of a cortical nephron.**

TABLE 37.2 Comparison of Filtrate Osmolarity in Renal Tubule and Glomerular Capsule

PART OF RENAL TUBULE	OSMOLARITY OF FILTRATE IN RENAL TUBULE
End of proximal convoluted tubule	
End of descending limb of the loop of Henle	
End of ascending limb of the loop of Henle	
Distal convoluted tubule	
Collecting duct	

B. Urinalysis

A **urinalysis** is an analysis of the physical, chemical, and microscopic characteristics of urine and a measure of urine volume. The normal characteristics of urine are listed in Table 37.3. **Urine volume** varies depending on the water content of the body and decreases when body fluid volume is low. ADH secreted by the posterior pituitary gland acts on the collecting ducts to stimulate reabsorption of water from the filtrate. If body fluid volume is high, ADH secretion is inhibited, the collecting ducts do not reabsorb water from the filtrate, and dilute urine is excreted. The **specific gravity of urine** is the weight of a volume of urine divided by the weight of the same volume of distilled water. Urine weight per volume is higher than distilled water because of the presence of solutes in urine. The more solutes present in urine, the higher the specific gravity. Normal urine contains 95% water and 5% solutes. The solutes found in normal urine include **electrolytes** (sodium, potassium, chloride, and other ions), **urea** (formed from breakdown of amino acids), **creatinine** (formed from breakdown of creatine phosphate), **uric acid** (formed from breakdown of nucleic acids), and **metabolic end products** of hormones and other substances. Although drugs are not a normal solute in urine, they will be excreted in urine if present in the bloodstream.

Selected abnormal components of urine are listed in Table 37.4. Your textbook may include a more extensive list. The presence of these substances in urine indicates an abnormality in metabolism or kidney function.

TABLE 37.3 Normal Characteristics of Urine

CHARACTERISTIC	DESCRIPTION
Volume	One to two liters per 24 hours; considerable variation in normal volume
Color	Yellow or amber color; urine color is darker in concentrated urine
Turbidity	Transparent in freshly voided urine; will turn cloudy after standing; microbes, pus, epithelial cells, or crystals may cause cloudiness in fresh urine
Odor	Aromatic when fresh; ammonia-like after standing because of breakdown of urea to ammonia by bacteria
pH	Normal range is 4.6–8.0, with an average of 6.0; high-protein diets produce an acidic urine; vegetarian diets produce an alkaline urine
Specific gravity	Normal range is 1.001–1.035; low specific gravity represents dilute urine; higher values represent a concentrated urine

TABLE 37.4 Selected Abnormal Constituents of Urine

ABNORMALITY	POSSIBLE CAUSE
Glucosuria (glucose in urine)	Caused by diabetes mellitus but sometimes may be caused by stress (epinephrine stimulates glycogen breakdown); diabetes mellitus and stress result in high blood glucose levels and therefore high levels of glucose in the filtrate; glucose transporters cannot work fast enough to reabsorb all glucose from filtrate.
Hematuria (erythrocytes in urine)	Caused by inflammation of urinary system organs, irritation by kidney stones, kidney disease, trauma to urinary system organs, or polyps or tumors within the urinary system.
Pyuria (white blood cells in urine)	Caused by an infection in the kidney, ureter, urinary bladder, or urethra.
Albuminuria (excess albumin in urine)	Results from an increase in filtration membrane permeability caused by high blood pressure, kidney trauma, disease, or inflammation; trace amounts of albumin in urine is normal.
Ketonuria (ketone bodies in urine)	Caused by a metabolic condition called ketosis; ketosis occurs when cells do not have enough glucose to completely break down fatty acids; ketosis can occur from starvation, low-carbohydrate diets, or untreated diabetes mellitus.
Casts (hardened clumps of material formed by protein precipitation and/or cell agglutination in renal tubules)	The following conditions may result in cast formation: albuminuria, abnormally acidic urine, and highly concentrated urine.
Calculi or kidney stones (insoluble salt crystals)	Calculi can form anywhere within the kidney tubules, ureters, urinary bladder, or urethra and can cause considerable pain as they pass through the lumens of urinary system organs. Some causes of calculi are dehydration and ingesting too much of the salts that form calculi.
Urinary tract infection (microbes)	Presence of bacteria or other microorganisms indicate the presence of a urinary tract infection (UTI); normal urine is sterile (no microorganisms present); however, microbes present on the skin surrounding the external urethral orifice may contaminate the urine sample if the urine sample is not carefully obtained.

LAB ACTIVITY 3 Experiment: Urinalysis

1 **Prediction:** Predict the normal urine values for each of the following characteristics by circling your choice (in italics).
- Color: *yellow* or *red*
- Turbidity: *clear* or *cloudy*
- pH range: *3.0–4.0* or *4.6–8.0* or *8.5–9.5*
- Presence of glucose: *negative* or *positive*
- Presence of protein: *negative, trace,* or *positive*
- Presence of ketone bodies: *negative, trace,* or *positive*
- Presence of hemoglobin: *negative, trace,* or *positive*
- Presence of leukocytes: *negative, trace,* or *positive*

2 **Materials:** Obtain materials for Urinalysis (from Materials list).

3 **Data Collection:** Determine color, turbidity, specific gravity, pH, and presence or absence of glucose, protein, ketone bodies, hemoglobin, or leukocytes and record the results in Table 37.5.
- Label each of four 100-mL beakers as Urine Sample #1, 2, 3, or 4.
- Put 50 mL from each of the 4 stock artificial urine samples into each appropriately labeled 100-mL

beaker. You will have 1 normal (control) and 3 abnormal urine samples.
- Look at the urine samples to determine the color and turbidity. Record the color and turbidity in Table 37.5.
- Using a urinometer, measure the specific gravity for each urine sample using instructions provided by your instructor. Do not discard urine; use it for the next test.
- Urine test strips contain small squares of different reagent paper that change color when they contact specific reagents (chemicals). Use a urine test strip to measure the pH and presence or absence of glucose, albumin (protein), hemoglobin (blood), leukocytes, and ketone bodies. Follow the instructions on the test strip container and record the results in Table 37.5.
- If desired by your instructor, view one or more microscope slides of real urine that has been centrifuged. The sediment is mounted on a slide and stained with Sedi-stain for better viewing of the sediment constituents (Figure 37.3).

4 Clean up as directed by your instructor.

5 Complete the Experimental Report with your lab group.

TABLE 37.5 Urinalysis Results

CHARACTERISTIC	SAMPLE 1	SAMPLE 2	SAMPLE 3	SAMPLE 4
Color				
Turbidity				
Specific gravity				
pH				
Glucose				
Albumin (protein)				
Blood (hemoglobin)				
Ketone bodies				
Casts				
Microbes				

EXPERIMENTAL REPORT
Urinalysis

Results:

For each urine sample, state whether the urine sample is normal or abnormal. For each abnormal sample, specify which characteristic of the urinalysis was abnormal.

Urine Sample #1:

Urine Sample #2:

Urine Sample #3:

Urine Sample #4:

Discussion:

For each abnormal urine sample, name the abnormality and give the possible cause. Refer to Table 37.4

Urine Sample #_____
 Abnormality:
 Cause:

Urine Sample #_____
 Abnormality:
 Cause:

Urine Sample #_____
 Abnormality:
 Cause:

Conclusion:

Give an overall statement regarding the characteristics of normal urine.

FIGURE 37.3 **Microscopic components of urinary sediments.**
(a) waxy cast and granular cast, (b) granular cast, (c) hyaline cast, (d) hemoglobin cast, (e) uric acid crystals,
(f) calcium oxide crystals, (g) triple phosphate crystals, (h) tyrosine crystals

Reviewing Your Knowledge

A. Functions of the Nephron

Write name of the part of the nephron that has the following function.

_____ **1.** Reabsorption of an additional 15% of water by osmosis. Impermeable to solutes.

_____ **2.** Hormones determine if urine is concentrated.

_____ **3.** Filters blood and forms filtrate.

_____ **4.** Reabsorption of majority of water and solutes from filtrate by active and passive processes. Also secretion occurs here.

_____ **5.** Reabsorption of more sodium, chloride, and about 10 to 15% more water.

_____ **6.** Reabsorption of sodium and chloride decreases the osmolarity of filtrate. Nearly impermeable to water.

B. Hormones and Urine Formation

Write the term the phrase describes.

_____ **1.** Hormone that increases sodium and chloride ion reabsorption and potassium secretion

_____ **2.** Process of moving substances from renal tubule into the peritubular capillary

_____ **3.** ADH acts mainly at this portion of the nephron

_____ **4.** Process of moving substances from the peritubular capillary into renal tubule

_____ **5.** Process occurring across wall of glomerular capillary and visceral layer of glomerular capsule

_____ **6.** The act of voiding (emptying) the urinary bladder

_____ **7.** Hormone that increases water reabsorption in the late distal convoluted tubule and collecting duct

C. Urinalysis

Answer the following questions with a short answer.

1. What is the normal volume of urine output per day in an adult?

2. What is the normal color of urine?

3. What could make urine cloudy?

4. What is the normal specific gravity of urine?

5. What makes urine have a higher specific gravity than distilled water?

6. Name five normal solutes found in urine.

7. Normal urine contains _____ % water and _____ % solutes.

8. Is normal daytime urine output dilute or concentrated?

9. What is the normal pH of urine?

10. What is pyuria?

11. What is ketonuria?

12. What is hematuria?

13. What is glucosuria?

14. True or false. Normal urine has some bacteria in it even if the urine sample was taken properly.

Using Your Knowledge

A. Urine Formation

1. Explain how a kidney stone blocking urine flow through the ureter eventually may decrease filtrate formation in the glomerular capsule. (*Hint:* hydrostatic pressure.)

2. Explain how inflammation of the glomerular capsule causes hematuria.

3. Compare the contents of the filtrate in the glomerular capsule and the contents of urine in the renal pelvis. (Yes, you can answer this question without research.)

B. Urinalysis

For each of the urine samples in Table 37.6, indicate whether the urine is normal or abnormal. If abnormal, name the abnormality and indicate the probable cause.

4. Urine A

5. Urine B

6. Urine C

7. Urine D

8. Which urine has the

 (a) lowest concentration of solutes?

 (b) highest concentration of solutes?

9. Which urines are:

 (a) acidic?

 (b) basic?

10. Which specific gravity value is the most concentrated? Urine: _____

TABLE 37.6 Urinalysis Results

MEASUREMENT	URINE A	URINE B	URINE C	URINE D
Color	Yellow	Slightly yellow	Amber	Red
Turbidity	Clear	Clear	Clear	Turbid
pH	6.0	6.5	4.0	7.5
Specific gravity	1.01	1.001	1.04	1.02
Glucose	Negative	Positive	Negative	Negative
Red blood cells	Negative	Negative	Negative	Positive
White blood cells	Negative	Negative	Negative	Positive
Albumin	Negative	Negative	Negative	Negative
Ketone bodies	Negative	Positive	Negative	Negative
Casts	Negative	Negative	Positive	Negative
Calculi	Negative	Negative	Positive	Negative
Microbes	Negative	Negative	Negative	Positive

Male Reproductive System Structure and Function

The male reproductive system has testes that produce **gametes** called **sperm (spermatozoa)**, ducts that store and transport sperm, and accessory sex glands that secrete fluid that, along with sperm, form **semen.** Supporting reproductive structures are the **scrotum** that houses the testes and the **penis** that transfers sperm into the female vagina.

A. Male Reproductive Organs

1. Scrotum and Testes

The **scrotum** (scrotal sac) is a pouch of loose skin that is suspended from the root of the penis outside the abdominopelvic cavity. The scrotum is internally divided into two portions by a central **scrotal septum** (partition) with one testis in each section. The **dartos muscle** (*dartos* = skinned) is within the scrotal septum and scrotal wall. This smooth muscle wrinkles and raises the scrotum when contracted. The **cremaster muscles** (*cremaster* = hanging)

are continuations of the internal oblique muscles and are located in the spermatic cords superior to the testes, one on each side of the groin. The contraction or relaxation of the cremaster muscles occurs when it is cold or hot, respectively, to adjust the distance of the testes from the body wall. Being on the exterior of the body, the scrotum is provided with a cooler temperature, 2 to 3°C lower than body temperature for normal sperm development.

The **testes (testicles** or **male gonads)** are oval-shaped glands that are covered with a white fibrous capsule called the **tunica albuginea** (*tunic* = coat or covering; *albus* = white). The tunica albuginea is dense fibrous connective tissue that extends into the testis to form **septa,** partitions that divide the testes into small compartments or **lobules.** External to this coat is an extension of the peritoneum, called the **tunica vaginalis,** and **spermatic fascia** composed of loose connective tissue.

The testes produce and release male hormones called androgens, primarily testosterone; they also produce and release sperm, which will travel through a series of tubular structures before being ejaculated from the body. **Sperm** are produced by a process called **spermatogenesis** in tiny **seminiferous tubules** that are located within the **lobules** of the testes. Seminiferous tubules form **straight tubules** that carry the sperm into a network of tubules called the **rete testis** (*rete* = network). **Efferent ducts** connect the rete testis to the **epididymis** (*epi-* = on or upon; *didymus* = pair), the first part of the male duct system.

Before Going to Lab

1 Label the structures of the testes and scrotum in Figures 38.1 and 38.2.

LAB ACTIVITY 1 Testes and Scrotum

1 Locate the structures of the testes and scrotum on a model or chart, or use the search text box to find the structures in **Real Anatomy** (Reproductive). ■

- cremaster (kree-MAS-ter) muscle
- dartos (DAR-tohs) muscle
- scrotum (SCROH-tum)
- tunica albuginea (TUNE-ih-kah ahl-byoo-GIN-ee-uh) of testis

1 _____

2 _____

3 _____

4 _____

Internal oblique muscle

Spermatic cord

Transverse section of penis

1

2

3

4

Tunica vaginalis

Spermatic fascia

FIGURE 38.1 Anterior view of the testes and scrotum.

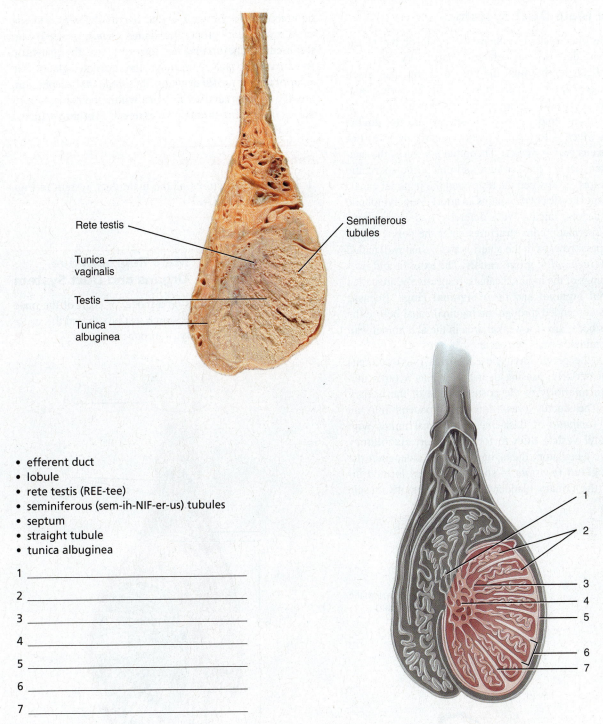

- efferent duct
- lobule
- rete testis (REE-tee)
- seminiferous (sem-ih-NIF-er-us) tubules
- septum
- straight tubule
- tunica albuginea

1 _____

2 _____

3 _____

4 _____

5 _____

6 _____

7 _____

FIGURE 38.2 **Sagittal section of the testis showing seminiferous tubules.**

2. The Male Duct System

The male duct system includes the ducts of the testes, epididymis, ductus (vas) deferens, ejaculatory duct, and urethra. Each **epididymis,** the site of sperm maturation and storage, curves over the posterior of the corresponding testis. Peristaltic contractions of smooth muscle in the wall of the epididymis move sperm into the **ductus deferens** (*duct* = to lead; *deferens* = carrying away) or **vas deferens** (*vas* = vessel). The initial portion of the ductus deferens leaves the scrotum and enters the spermatic cord that travels between the testes and the inguinal canal.

Each **spermatic cord** contains an artery, vein, lymphatic vessels, nerves, ductus (vas) deferens, and a cremaster muscle. Spermatic cord structures enter the pelvic cavity through passageways in the anterior abdominal wall called **inguinal** (*inguinal* = groin) **canals.** The exterior and interior openings of the inguinal canals, respectively, are called **superficial inguinal** and **deep inguinal rings.** Because the testes descended through the inguinal canal before the male was born, this is a weaker area in the abdominal wall where inguinal hernias can occur.

Within the pelvic cavity, each ductus (vas) deferens travels posteriorly, passing over a ureter as it turns medially and inferiorly to the posterior side of the urinary bladder. The ductus (vas) deferens broadens into an **ampulla** (*ampulla* = flask-like bottle) that unites with the **seminal vesicle duct** to form the short **ejaculatory duct.** Two ejaculatory ducts, one on either side, enter the **prostate gland** (*prostate* = standing before) located inferior to the urinary bladder. Peristaltic smooth muscle

contractions in the wall of the ductus deferens and the ejaculatory ducts propel the **semen** (*semen* = seed) into the urethra. The urethra has three regions: the **prostatic urethra** that passes through the prostate gland, the **membranous urethra** within the urogenital diaphragm, and the **spongy (penile) urethra** within the penis. Semen leaves the urethra through the **external urethral orifice.**

Before Going to Lab

1 Label the structures of the male duct system in Figures 38.3 and 38.4.

LAB ACTIVITY 2 — Male Reproductive Organs and Duct System

1 Identify the structures on a model or chart of the male duct system, or use the search text box to find the structures in **Real Anatomy** (Reproductive). ■

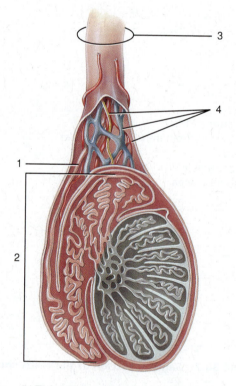

(b) The epididymis and ductus deferens

- blood vessels and nerves 1 _____
- ductus (vas) deferens
 (DUK-tus DEAF-er-ens) 2 _____
- epididymis
 (ep-ih-DID-ih-miss) 3 _____
- spermatic (sper-MAT-ic)
 cord 4 _____

(a) Lateral view of testis and associated structures

Spermatic cord

Body of epididymis

Head of epididymis

Tail of epididymis

Testis

Tunica vaginalis

FIGURE 38.3 The testis and epididymis.

POSTERIOR

ANTERIOR

Seminal vesicle

Ductus (vas) deferens

Prostatic urethra

Ejaculatory duct

Membranous urethra

Rectum

Bulb of penis

Bulbospongiosus muscle

Urinary bladder (opened)

Prostate

Pubic symphysis

Corpora cavernosum penis

Corpus spongiosum penis

Spongy (penile) urethra

Testis

Glans penis

- ampulla (AM-pyoo-la) of ductus deferens
- ductus (vas) deferens in inguinal canal
- ejaculatory duct
- epididymis
- external urethral orifice
- membranous urethra
- penis
- prostate (PRAH-state) gland
- prostatic (prah-STA-tic) urethra
- scrotum
- spongy (penile) urethra
- testis

1 _____

2 _____

3 _____

4 _____

5 _____

6 _____

7 _____

8 _____

9 _____

10 _____

11 _____

12 _____

FIGURE 38.4 **Male reproductive organs and duct system, sagittal section.**

B. Accessory Structures

1. Accessory Sex Glands and Semen

Three types of accessory glands—the seminal vesicles, prostate gland, and bulbourethral (Cowper's) glands—make and secrete most of the fluid portion of the semen. To assist sperm movement, the testes secrete a small amount of fluid before the addition of fluid from these three glands.

The **seminal vesicles** (*seminal* = pertaining to semen) are sac-like glands posterior to the bladder, flanking either side. When sperm cells reach the end of the **ampulla** (expanded end) of the ductus deferens, additional fluid is added from the seminal vesicle through the **seminal vesicle duct.** This alkaline fluid constitutes about 60% of the semen volume and neutralizes the acidic male urethra and female vagina because sperm motility is adversely affected by an acidic environment. It also contains fructose, an energy source for sperm, and prostaglandins that stimulate sperm motility and muscular contractions in the female reproductive tract.

The **prostate gland** surrounds the prostatic urethra inferior to the urinary bladder. Its milky secretion has several proteolytic (*proteo-* = protein; *lytic* = to break apart) enzymes, including prostate-specific antigen (PSA). This slightly acidic secretion, constituting approximately 25% of the semen volume, increases sperm viability and motility.

The prostate gland increases in size during male development and is known to enlarge after age 45, a condition known as BPH or benign prostatic hypertrophy. BPH can constrict the prostatic urethra causing difficulty in urination and ejaculation.

The smallest accessory glands are the paired **bulbourethral (Cowper's) glands** (*bulbus* = swollen root) that are located on either side of the membranous urethra within the urogenital diaphragm, a skeletal muscle. These small glands secrete an alkaline secretion that neutralizes the acidic male urethra and secrete mucus that lubricates the urethra and glans penis.

> ### Before Going to Lab
>
> **1** Label the accessory sex gland structures in Figures 38.5 and 38.6.

LAB ACTIVITY 3 Male Reproductive Organs and Accessory Sex Glands

1 Identify the structures of the accessory sex glands on a model or chart, or use the search text box to find the structures in **Real Anatomy** (Reproductive). ■

- ampulla of ductus deferens 6
- bulbourethral (buhl-boh-your-REETH-ruhl) gland 8
- ductus (vas) deferens 5
- ejaculatory duct 7
- membranous urethra 4
- prostate gland 2
- prostatic urethra 3
- seminal vesicle (SEM-ih-nul VES-sih-kul) 1
- spongy (penile) urethra 9

1 _____

2 _____

3 _____

4 _____

5 _____

6 _____

7 _____

8 _____

9 _____

FIGURE 38.5 **Male reproductive organs and accessory sex glands, posterior view.**

- bulbourethral gland
- prostate gland
- seminal vesicle

1 _____

2 _____

3 _____

FIGURE 38.6 **Accessory male sex glands, sagittal section.**

2. Penis

The **penis** is a cylindrical shaft that contains the distal portion of the urethra, a passageway for urine and for the ejaculation of semen. The penis is located medial and anterior to the scrotum and testes and contains the spongy (penile) urethra. The penis is composed of three main parts: a body, a root, and the glans penis. The **body of the penis** contains three separate cylinders of erectile tissue, two dorsolateral **corpora cavernosa penis** (*corpor-* = body; *cavernosa* = hollow) and one midventral **corpus spongiosum penis,** all containing blood sinuses. Fibrous connective tissue, fascia, and skin surround all three cylinders. The spongy (penile) urethra is located posteriorly in the corpus spongiosum penis.

The **glans penis** (*glans* = acorn) is the distal, expanded portion of the corpus spongiosum that is mostly covered with **prepuce** (foreskin) in an uncircumcised male. **Circumcision** removes the prepuce, exposing the glans penis. The **external urethral orifice** is the external opening of the urethra in the glans penis.

Before Going to Lab

1 Label the penile structures in Figure 38.7(a) and (b).

LAB ACTIVITY 4 The Penis

1 Identify the penile structures on a model or chart, or use the search text box to locate the structures in **Real Anatomy** (Reproductive).

(a) Frontal section

Urinary bladder

Urogenital diaphragm

(a)

- bulbourethral gland
- corpora cavernosa (core-PORE-ah cav-er-NOH-suh) penis
- corpus spongiosum (CORE-pus spun-gee-OH-sum) penis
- external urethral orifice
- glans penis
- membranous urethra
- prepuce (PREE-pyoos)
- prostate gland
- prostatic urethra
- spongy (penile) urethra

1 _____

2 _____

3 _____

4 _____

5 _____

6 _____

7 _____

8 _____

9 _____

10 _____

Ventral

(b) Transverse section

(b)

- corpora cavernosa penis
- corpus spongiosum penis
- spongy (penile) urethra

11 _____

12 _____

13 _____

Deep dorsal vein
Dorsal artery
Corpora cavernosa penis
Deep artery of penis
Spongy (penile) urethra
Corpus spongiosum penis

(c) Cadaver transverse section

FIGURE 38.7 **Penis.**

LAB ACTIVITY 5 Function of Male Reproductive Organs

1 With your lab group, write the number of the structure shown in Figure 38.8 next to the phrase that describes its function. ■

_____ carries sperm from epididymis to ejaculatory duct

_____ copulatory organ; passageway for urine and sperm

_____ organ that produces sperm

_____ secretes 60% of semen volume; alkaline pH; prostaglandins

_____ secretes 25% of semen volume; increases viability and motility of sperm

_____ secretes mucus for lubrication of glans; alkaline pH

_____ site of storage and maturation of sperm

FIGURE 38.8 Function of the male reproductive organs.

C. Microscopic Anatomy of the Testes

1. Histology of the Testis and Spermatogenesis

Spermatogenesis (*spermato-* = sperm; *genesis* = origin) or the formation of haploid sperm with half the number of chromosomes, occurs in the walls of numerous seminiferous tubules of each testis. This process takes about 2 to 2½ months for each sperm. Between the seminiferous tubules of the testes are triangular-shaped spaces called **interstitial** (*interstitial* = stand between) **spaces** that contain **Leydig cells** or **interstitial endocrinocytes** (endocrine cells). Leydig cells, when stimulated by luteinizing hormone from the anterior pituitary, produce and secrete hormones called androgens, which are primarily **testosterone.**

Spermatogenesis begins with **spermatogonia,** diploid (46 or 2*n* chromosomes) stem cells located in the outer portion of the seminiferous tubule wall just beneath the basement membrane, which surrounds the seminiferous tubules. Spermatogonia divide to form primary spermatocytes (diploid) and more spermatogonia stem cells. Each division moves the cells closer to the lumen of the tubule. **Primary spermatocytes** with 46 chromosomes enlarge and their nucleus divides in meiosis I to form two haploid (*n*) secondary spermatocytes, each with 23 *double* chromosomes. The two **secondary spermatocytes** divide by a second nuclear division, meiosis II, to form four haploid (*n*) **spermatids,** each with 23 *single* chromosomes. Spermatids go through a drastic change in shape and composition in a process called **spermiogenesis,** the making of sperm. **Sertoli cells** surround the cells undergoing spermatogenesis, providing the chemical environment for this process and assisting the cells to move from the basement membrane to the lumen.

2. Sperm

Sperm are also called **sperm cells** or **spermatozoa.** Mature sperm have lost most of their cytoplasm and cellular organelles, and do not survive in the female reproductive tract more than a few days after ejaculation. Sperm are composed of a head, midpiece, and a tail. The **head** consists mostly of the nucleus with an acrosome (*acro-* = atop) covering it. The **acrosome** is a vesicle filled with hydrolytic enzymes that allow sperm to penetrate the female oocyte for fertilization. The **midpiece** contains many mitochondria that are needed to produce energy (ATP) for the tail once the sperm cells are ejaculated. The **tail** is a flagellum that performs a whip-like motion, propelling the sperm.

Before Going to Lab

1 Label the microscopic structures in Figures 38.9(a) and (b), 38.10, and 38.11.

LAB ACTIVITY 6 **Histology of the Testis, Spermatogenesis, and Sperm**

1 Observe a prepared slide of the testis, or use **Real Anatomy** (Histology).
- Using low power, identify the walls of the **seminiferous tubules, tails of sperm** in the lumen, and **interstitial spaces.**
- Using high power, identify the **basement membrane, spermatogonia, spermatocytes, spermatids,** and **sperm.**

2 Observe a prepared slide of sperm.
- Using a lower power, locate the sperm.
- Using high power, identify the **head, midpiece,** and **flagellum.**

3 Answer the Discussion Question with your lab group.

DISCUSSION QUESTION
Spermatogenesis

1 How many sperm are produced in meiosis from one primary spermatocyte? _____

D. Dissection of Male Reproductive System

If you are dissecting a cat or fetal pig, consult the appropriate dissection manual. You will probably be dissecting the urinary system along with the reproductive system. Although these two systems are not identical to humans, there are several similarities, and the differences make an interesting study.

(a)

- primary spermatocyte (spur-MAH-toe-site)
- secondary spermatocyte
- spermatid (SPERM-uh-tid)
- spermatogonium (spur-ma-toe-GOHN-ee-um)
- spermatozoa (spur-ma-toe-ZOH-uh)

1 _____

2 _____

3 _____

4 _____

5 _____

(a) Events of spermatogenesis

(b)

- basement membrane 13
- early spermatid 8
- late spermatid 7
- Leydig (interstitial) cells 14
- primary spermatocyte 11
- secondary spermatocyte 9
- Sertoli cells 10
- spermatozoa (sperm cell) 6
- spermatogonium 12

6 _____

7 _____

8 _____

9 _____

10 _____

11 _____

12 _____

13 _____

14 _____

(b) Transverse section through a seminiferous tubule

FIGURE 38.9 Spermatogenesis.

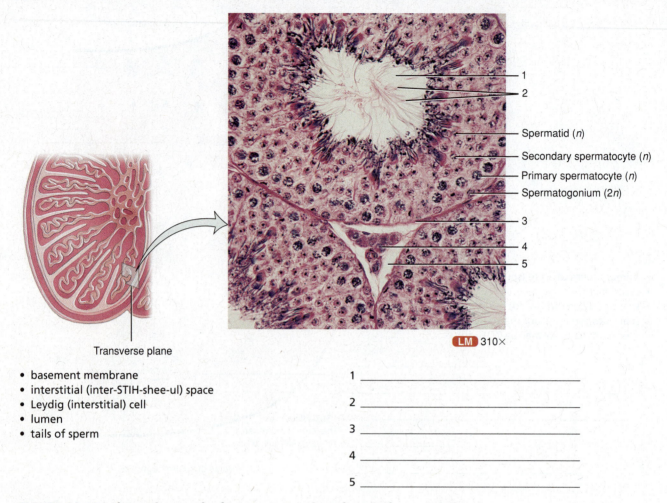

Transverse plane

1 _____
2 _____
Spermatid (*n*)
Secondary spermatocyte (*n*)
Primary spermatocyte (*n*)
Spermatogonium (2*n*)
3 _____
4 _____
5 _____

LM 310×

- basement membrane
- interstitial (inter-STIH-shee-ul) space
- Leydig (interstitial) cell
- lumen
- tails of sperm

1 _____

2 _____

3 _____

4 _____

5 _____

FIGURE 38.10 **Photomicrograph of transverse section of seminiferous tubules.**

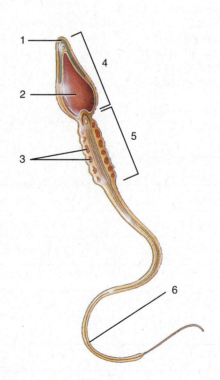

- acrosome (AK-row-sohm)
- head
- midpiece
- mitochondria
- nucleus
- tail (flagellum)

1 _____

2 _____

3 _____

4 _____

5 _____

6 _____

FIGURE 38.11 **Sperm cell.**

Reviewing Your Knowledge

A. Male Reproductive Organ Structure and Function

Write the name of the structure that the phrase describes.

_____ **1.** Secretes 60% of semen volume; alkaline pH; prostaglandins

_____ **2.** Organ that produces sperm

_____ **3.** Part of urethra within the urogenital diaphragm

_____ **4.** Site of storage and maturation of sperm

_____ **5.** Secretes 25% of semen volume; increases viability and motility of sperm

_____ **6.** Located in prostate; formed by union of ducts of the ductus deferens and seminal vesicle

_____ **7.** Covers and protects testes

_____ **8.** Contracts to pull testes close to body; located in spermatic cord

_____ **9.** Last section of urethra located in penis

_____ **10.** Tube located in both the spermatic cord and pelvic cavity for transporting sperm

_____ **11.** Contracts to wrinkle scrotum; located in scrotal wall

_____ **12.** Ejaculatory duct opens into this part of urethra

_____ **13.** Copulatory organ; passageway for urine and semen

_____ **14.** Secretes mucus for lubrication of glans; alkaline pH

B. Male Duct System

List the following ducts *in order* (1–11), tracing sperm from their production to ejaculation.

_____ ampulla of ductus deferens

_____ ductus (vas) deferens

_____ epididymis

_____ efferent ducts

_____ ejaculatory duct

_____ membranous urethra

_____ prostatic urethra

_____ rete testis

_____ seminiferous tubules

_____ spongy (penile) urethra

_____ straight tubules

C. Spermatogenesis and Sperm Cells

Write the name of the structure that the phrase describes.

_____ **1.** Part of sperm that contains chromosomes

_____ **2.** Haploid cells formed when secondary spermatocytes undergo meiosis II

_____ **3.** Cells that secrete testosterone

_____ **4.** Diploid stem cells

_____ **5.** Process that produces haploid sperm from diploid stem cells

_____ **6.** Cells that surround spermatocytes and spermatids and provide chemical environment for spermatogenesis

_____ **7.** Space within testes where Leydig cells are located

_____ **8.** Diploid cells that undergo meiosis I to form two haploid secondary spermatocytes

_____ **9.** Process that transforms spermatids into sperm

_____ **10.** Part of sperm that produces ATP

_____ **11.** Propels the sperm

Using Your Knowledge

A. Male Reproductive System

Write a short answer for the following questions.

1. Can sperm move in reverse as well as forward? Explain.

2. (a) Which men would have a higher testicular temperature, those who wear briefs or those who wear boxer shorts?

 (b) Which would be more likely to cause sterility? Explain.

3. The male reproductive system has both exocrine and endocrine functions. Name one endocrine secretion.

Name four exocrine secretions.

4. _____

5. _____

6. _____

7. _____

8. To perform a vasectomy, does the doctor have to surgically enter the abdominal cavity? Explain.

9. A vasectomy involves cutting the ductus deferens. Does a vasectomy impede testosterone levels, having a penile erection, or having an ejaculation? Explain each situation.

10. Many body builders who take high doses of anabolic steroids have undesirable side effects. Name four side effects.

Female Reproductive System Structure and Function

OBJECTIVES

1 Describe the structure and function of the female reproductive organs

2 Identify the location of the female reproductive organs

3 Describe oogenesis

4 Trace an oocyte from its production until implantation

5 Identify the selected structures on the microscopic slides of the ovary and uterus

6 Name the organ and cells that produce the female hormones and describe the functions of the hormones

7 Dissect a cat, fetal pig, or cadaver and identify the main female reproductive structures

MATERIALS

- models or charts of the female reproductive system, or use **Real Anatomy** (Reproductive)

- compound microscopes, lens paper, prepared slides of the ovary and uterus, or **Real Anatomy** (Histology)

- Dissection: preserved cats or fetal pigs, dissection equipment, disposable gloves, safety glasses, cat or fetal pig dissection manual

The female reproductive system has ovaries that produce **gametes (oocytes),** uterine tubes to transport the oocytes, a **uterus** to protect and nourish the developing embryo and fetus, a **vagina, external genitalia,** and **mammary glands.**

A. Female Reproductive Organs

1. Ovaries and Uterine Tubes

The paired, oval-shaped **gonads** or **ovaries** (*ovary* = egg receptacle) are the size of unshelled almonds (about 1 inch long) and are located on either side of the uterus.

Oocytes (eggs or gametes) are produced in the ovary and are released during a process called **ovulation.** The

hormones **estrogen** and **progesterone** are produced by the ovaries. The ovaries are anchored in place by three paired ligaments: the broad ligaments, the ovarian ligaments, and the suspensory ligaments. The sheet-like **broad ligament** of the uterus is a double layer of peritoneum that attaches to each ovary and also secures the uterine tube in place. The **ovarian ligament** on the superior medial side of each ovary attaches the ovary to the uterus. The **suspensory ligament** on the superior lateral side of each ovary attaches the ovary to the pelvic wall in the lumbar region and surrounds the ovarian artery and ovarian vein.

Paired **uterine (fallopian) tubes** or **oviducts** curve around the ovaries to receive and transport the oocyte to the superior part of the uterus. There are three main regions of the uterine tube: the infundibulum (with fimbriae), ampulla, and isthmus. The funnel-shaped **infundibulum** composes the distal portion of the uterine tube and has finger-like structures projecting from its expanded end called **fimbriae.** During ovulation, the oocyte is projected toward the infundibulum, and the beating of the fimbriae forms a current that sweeps the oocyte into the infundibulum. The oocyte then enters the longest region of the uterine tube, the **ampulla,** which constitutes about two-thirds of the length of the uterine tube. Fertilization of the oocyte by a sperm cell usually occurs in the ampulla. The third region, the **isthmus,** is proximal to and opens into the superior

lateral part of the uterus. The shorter isthmus region has a narrower lumen and thicker walls than the rest of the uterine tube.

The lumen of the uterine tubes is lined with **ciliated columnar epithelia.** The fertilized ovum (**zygote**) is moved through the uterine tubes by the beating of the cilia and by peristaltic, smooth muscle contractions of the uterine tube walls. The zygote reaches the uterus about 1 week after ovulation.

Before Going to Lab

1 Label the structures of the ovaries and uterine tubes in Figure 39.1(b).

LAB ACTIVITY 1 **Ovaries and Uterine Tubes**

1 Identify the structures of the ovaries and uterine tubes on a urogenital model or chart, or use the search text box to locate the structures in **Real Anatomy** (Reproductive). ■

(a) Superior view

FIGURE 39.1 **Ovaries, uterine tubes, and uterus.**

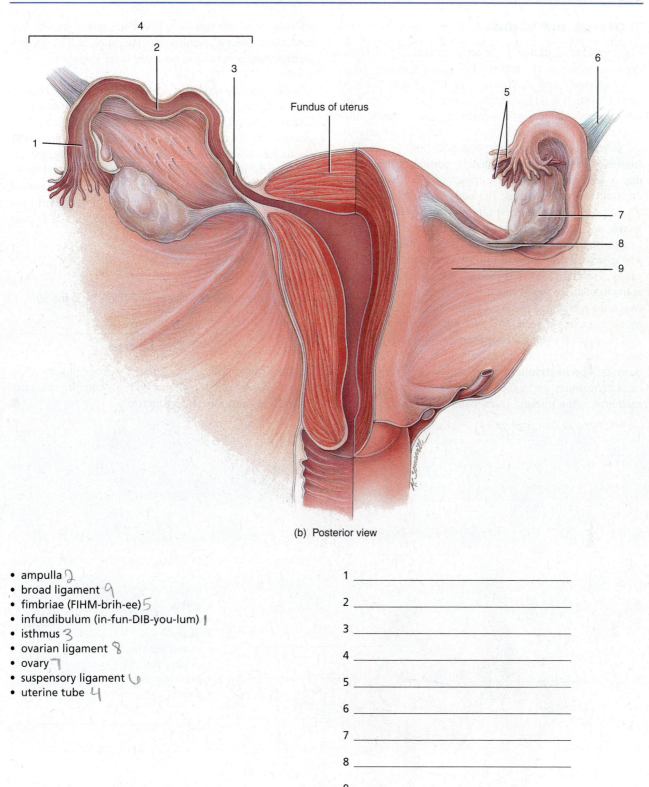

Fundus of uterus

(b) Posterior view

- ampulla 2
- broad ligament 9
- fimbriae (FIHM-brih-ee) 5
- infundibulum (in-fun-DIB-you-lum) 1
- isthmus 3
- ovarian ligament 8
- ovary 7
- suspensory ligament 6
- uterine tube 4

1 _____

2 _____

3 _____

4 _____

5 _____

6 _____

7 _____

8 _____

9 _____

FIGURE 39.1 **Ovaries, uterine tubes, and uterus,** *continued.*

2. Uterus and Vagina

The pear-shaped **uterus** is posterior to the urinary bladder and anterior to the rectum. The uterus normally projects superiorly and anteriorly over the urinary bladder (anteflexion). It is a little smaller than the size of your fist, but obviously larger in a recently pregnant female and smaller after menopause. Externally, three main regions of the uterus can be seen: the fundus, body, and cervix. The **fundus** is the dome-shaped superior portion of the uterus that is wider and meets the entrance of the uterine tubes on both sides of the uterus. The **body** starts inferior to the openings of the uterine tubes, is the largest region of the uterus, and decreases in size to meet the narrower cervix inferiorly. The space within the uterus is the **uterine cavity.** The narrowed **cervix** opens into the vagina inferiorly. The cavity within the cervix is the **cervical canal** that begins at the **internal os** (*os* = mouth), a constricted opening between the uterus and the cervix, and ends at the **external os,** the opening between the cervix and the vagina.

The uterus is a thick-walled organ with three layers: the perimetrium, the myometrium, and the endometrium. The outer, thin **perimetrium** (*metr-* = uterus) is part of the visceral peritoneum, also called the serosa. The middle **myometrium** is the thickest layer, composed of three layers of smooth muscle that have fibers going in three different directions: oblique, circular, and longitudinal. The thin **endometrium** is the inner, vascular layer lining the uterine cavity.

Although the uterus is secured in its position by four paired ligaments, only two of these ligaments are easily found on models: the broad ligaments and the round ligaments. The large, flat **broad ligaments** attach the uterus to the lateral pelvic walls. The **round ligaments,** located between the two layers of the broad ligaments, begin below the uterine tubes and attach the uterus to the labia majora of the external genitalia. The ligaments of the uterus keep it in an **anteflexion** (*ante-* = forward) position as it leans forward over the urinary bladder.

Before Going to Lab

1 Label the uterine structures in Figures 39.2 and 39.3.

LAB ACTIVITY 2 The Uterus

1 Identify the uterine structures on a urogenital model or chart, or use the search text box to locate the structures in **Real Anatomy** (Reproductive). ■

- cervix
- myometrium of uterus
- ovary
- round ligament
- urinary bladder
- uterine cavity
- vagina

1 _____

2 _____

3 _____

4 _____

5 _____

6 _____

7 _____

FIGURE 39.2 Female reproductive organs, sagittal section.

Layers of uterus
1
2
3
4
5
6
7
8
9
10
11
12
13
14

- body of uterus 12
- broad ligament 11
- cervical canal 7
- cervix (SIR-vex) 8
- endometrium (endo-MEE-tree-um) 2
- external os 13
- fundus 10
- internal os 6
- myometrium (my-oh-MEE-tree-um) 3
- perimetrium (peri-MEE-tree-um) 4
- round ligament 5
- rugae of vagina 9
- uterine cavity 1
- vagina 14

1 _____
2 _____
3 _____
4 _____
5 _____
6 _____
7 _____
8 _____
9 _____
10 _____
11 _____
12 _____
13 _____
14 _____

FIGURE 39.3 **Posterior view of the female reproductive organs.**

3. External Genitalia

The external female genitalia are referred to as the **vulva** (*vulva* = to wrap around) or **pudendum** (pyoo-DEN-duhm) (*pudens* = modest). The vulva consists of the mons pubis, labia majora, labia minora, clitoris, and vestibule. The **mons** (*mons* = mountain) **pubis** is an anterior pad of adipose tissue beneath the skin that covers the pubic symphysis. Posterior to this are two longitudinal folds of adipose tissue covered with skin and pubic hair called the **labia majora** (labium majus, sing.). Medial to these larger folds are smaller paired longitudinal folds of skin without pubic hair called the **labia minora** (labium minus, sing.). The **clitoris,** a cylindrical erectile tissue, is posterior to the mons pubis and anterior to the external urethral orifice. The **prepuce,** an anterior extension of the labia minora, covers the body of the clitoris like a hood but leaves part of the clitoris exposed. The **vestibule** is the area medial to the paired labia minora and contains the external urethral orifice, vaginal orifice, hymen (if present), and openings of glands that secrete mucus for lubrication. The **hymen** (*hymen* = membrane) is a thin membrane around the perimeter of the vaginal orifice that partially blocks the orifice.

Before Going to Lab

1 Label the external genitalia structures in Figures 39.4(b) and 39.5.

LAB ACTIVITY 3 **External Genitalia**

1 Identify the external genitalia structures on a urogenital model or use the search text box to locate the structures in **Real Anatomy** (Reproductive). ■

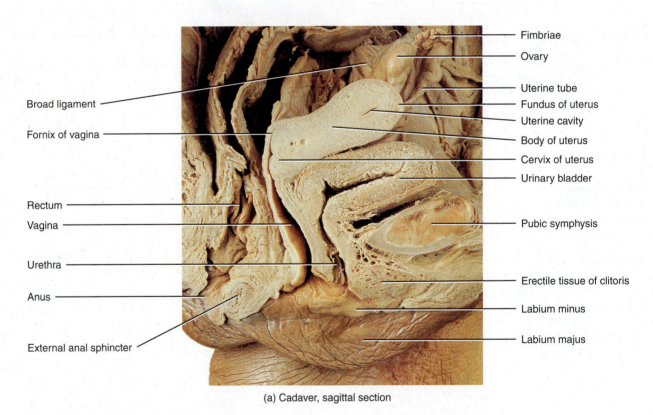

(a) Cadaver, sagittal section

FIGURE 39.4 **Female external genitalia.**

(b) Sagittal section

- anus
- clitoris (KLIH-tor-us)
- external urethral orifice
- labium majus (LAY-bee-um MAY-jus)
- labium minus
- mons pubis (mohnz PYOO-bis)
- pubic symphysis (PYOO-bik SYM-fih-sis)
- vaginal orifice

1 _____

2 _____

3 _____

4 _____

5 _____

6 _____

7 _____

8 _____

✱ **FIGURE 39.4 Female external genitalia, *continued*.**

- anus
- clitoris
- external urethral orifice
- hymen (HIGH-men)
- labia majora
- labia minora
- mons pubis
- prepuce (PREE-pyoos)
- vaginal orifice

1 _____

2 _____

3 _____

4 _____

5 _____

6 _____

7 _____

8 _____

9 _____

FIGURE 39.5 Female external genitalia, inferior view.

4. Mammary Glands

Although **mammary** (*mamma-* = breast) **glands** are derivatives of the skin and are categorized as modified sweat glands, they belong to the reproductive system because in females they secrete milk for the offspring. The mammary glands in both males and females lie anterior to the pectoralis major and serratus anterior muscles. Female breasts are enlarged compared with those of the male due to a greater amount of adipose tissue that surrounds the mammary glands. The main external structures of the breasts are the darkened **areola** and the raised **nipple** with **lactiferous duct openings** (*lact-* = milk). The internal structures are the alveoli, lobules, lobes, mammary ducts, lactiferous sinuses, lactiferous ducts, and adipose tissue. **Alveoli** (mammary glands) are milk-producing glands that are arranged in clusters called **lobules.** Several lobules make up a **lobe.** Milk is secreted from alveoli into **mammary ducts** that drain one lobe and then into dilated **lactiferous sinuses** that merge to form lactiferous ducts. Milk is ejected through **lactiferous ducts** in the nipple.

> ### Before Going to Lab
>
> **1** Label the mammary gland structures in Figure 39.6(a) and (b).

LAB ACTIVITY 4 Mammary Glands

1 Identify the mammary gland structures on a model or chart. ◼

Pectoralis major

Intercostal muscles

Rib

(a) Sagittal section

(b) Anterior view, partially sectioned

- adipose tissue
- areola (ah-REE-oh-la)
- lactiferous (lak-TIF-er-us) duct
- lactiferous sinus
- lobule with alveoli
- mammary duct
- nipple

1 _____

2 _____

3 _____

4 _____

5 _____

6 _____

7 _____

FIGURE 39.6 Mammary glands.

LAB ACTIVITY 5 **Function of Female Reproductive Structures**

1 With your lab group, write the number of the organ shown in Figure 39.7 next to the phrase that describes its function. Numbers may be used more than once.

2 Discuss the pathway of a secondary oocyte from its production until implantation in the uterus with your lab partners.

_____ Layer of uterus that sheds during menstruation

_____ Produces secondary oocytes

_____ Transports secondary oocytes toward uterus

_____ Usual site of fertilization

_____ Beat together to bring ovulated oocyte into uterine tube

_____ Implantation of zygote and development of fetus occurs here

_____ Produces estrogen and progesterone

_____ Narrow region of uterine tube that opens into uterus

FIGURE 39.7 Function of the female reproductive structures.

B. Oogenesis and Microscopic Anatomy of the Ovary and Uterus

1. Oogenesis

Oogenesis (*oo-* = egg) is the formation of haploid ova with half the number of chromosomes that occurs in the ovary. Early in fetal development, diploid ($2n$) stem cells called **oogonia** are produced, which divide to create millions of germ cells, most of which degenerate.

Meiosis I

Some of the **oogonia** develop into **primary oocytes** ($2n$) that begin meiosis I before birth but stop in prophase I. At puberty, several primary oocytes complete meiosis I each month, forming two haploid cells of unequal sizes, each with 23 pairs of chromosomes. The much larger cell with most of the cytoplasm and organelles is called the **secondary oocyte.** The smaller one is called the **first polar body,** which may or may not survive to undergo the meiosis II division. Polar bodies are mostly DNA or genetic material and will eventually degenerate.

Meiosis II

The secondary oocyte begins meiosis II but stops in metaphase II prior to ovulation. If the ovulated secondary oocyte is fertilized, it will complete meiosis II and form a larger haploid ovum and a **second polar body.** When the nucleus from the sperm (n) unites with the nucleus of the ovum (n), a diploid single-celled **zygote** ($2n$) is formed.

LAB ACTIVITY 6 **Oogenesis**

1 With your lab group, label the structures and phases in Figure 39.8.

2 Answer the Discussion Questions with your lab group.

DISCUSSION QUESTIONS
Oogenesis

1 Compare the size of the secondary oocyte with the first polar body.

2 Compare the numbers of gametes produced from one stem cell by oogenesis and spermatogenesis.

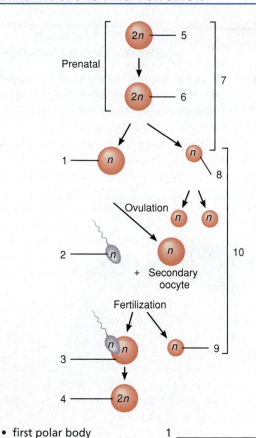

- first polar body
- meiosis I
- meiosis II
- oogonium (oh-oh-GOHN-ee-um)
- ovum
- primary oocyte (OH-oh-site)
- second polar body
- secondary oocyte
- sperm cell
- zygote

1 _____
2 _____
3 _____
4 _____
5 _____
6 _____
7 _____
8 _____
9 _____
10 _____

FIGURE 39.8 **Oogenesis.**

2. Histology of the Ovary

The regions of the ovary from superficial to deep are the **germinal epithelium** (simple epithelium), cortex, and medulla. The stringy looking **medulla** contains connective tissue, blood vessels, and nerves. The **cortex** contains follicles (oocytes surrounded by cells) in various stages of development. **Primordial follicles,** the smallest follicles, each contain a **primary oocyte** surrounded by a single layer of cells called **follicular cells.** These follicles mature into **primary follicles** that have **granulosa cells** surrounding the follicular cells; follicular cells secrete **estrogen.** As the primary follicle matures, the granulosa cells secrete follicular fluid and become secondary follicles. **Secondary follicles** contain an antrum (space within granulosa cell layers) that is filled with follicular fluid. Each month one secondary follicle develops into a **mature** or **graafian follicle.** This follicle, which contains a secondary oocyte, is the largest follicle and has the greatest amount of fluid in its antrum. When a graafian follicle expels the secondary oocyte during ovulation, the remaining follicular tissue becomes a **corpus luteum** (*corpus* = body; *luteum* = yellow), which secretes **progesterone.** When this structure is no longer functional, it disintegrates and a scar-like **corpus albicans** (*alba* = white) is formed. The corpus albicans doesn't remain in the ovary very long and also disintegrates. Because of this, neither structure may be seen in your slides of the ovary.

Before Going to Lab

1 Label the microscopic ovarian structures in Figure 39.9(a)–(d).

LAB ACTIVITY 7 Histology of the Ovary

1 Examine a prepared slide of the ovary or use **Real Anatomy** (Histology).

- Hold the ovary slide up to the light and determine whether or not you have a section of a complete ovary. If you do not have a complete ovary, some of the structures may not be present.
- Using low power, identify the **medulla, follicles** of various sizes, and the **hilum.**
- Search for **primordial follicles,** a **primary follicle, secondary follicle,** and a mature (**graafian**) **follicle.**
- Center the mature follicle, and observe the **secondary oocyte** with its **granulosa cells** surrounding it, forming the **corona radiata.**

2 Answer the Discussion Questions with your lab group.

DISCUSSION QUESTIONS
Histology of the Ovary

1 Which follicle types contain fluid?

2 Describe the difference in appearance between a corpus luteum and a corpus albicans.

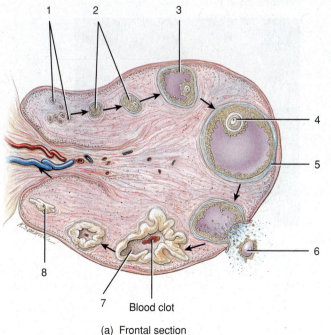

(a) Frontal section

(a)
- corpus albicans (KOR-pus AL-bih-kanz) 8
- corpus luteum (LOO-tee-um) 7
- graafian (GRAF-ee-uhn) (mature) follicle 5
- ovulated secondary oocyte 6
- primary follicles 2
- primordial (pry-MORE-dee-uhl) follicles 1
- secondary follicle 3
- secondary oocyte 4

1 _____
2 _____
3 _____
4 _____
5 _____
6 _____
7 _____
8 _____

Blood clot

FIGURE 39.9 Histology of the ovary.

Germinal epithelium

9

10

11

12

Corpus luteum

LM 30×

(b) Ovarian cortex

(b)
• primary follicle
• primordial follicles
• primary oocyte in primary follicle
• primary oocyte in secondary follicle

9 _____

10 _____

11 _____

12 _____

13 Germinal epithelium Theca folliculi 14

15 16 LM about 200×

(c) Primordial and primary follicles

(c)
• granulosa cells
• ovarian cortex
• primary oocyte in primary follicle
• primordial follicles

13 _____

14 _____

15 _____

16 _____

17

18

19

LM 70×

(d) Secondary follicle

(d)
• follicular fluid in antrum
• granulosa cells (corona radiata)
• primary oocyte

17 _____

18 _____

19 _____

Ovary

Secondary oocyte

LM 30×

(e) Ovulation of a secondary oocyte from a graafian follicle

FIGURE 39.9 **Histology of the ovary,** *continued.*

3. Histology of the Uterus and Uterine Tube

The endometrium of the uterus has epithelium facing the uterine cavity with **endometrial (uterine) glands** lying deep to the epithelium. The endometrium is further divided into the stratum functionalis and the stratum basalis. The epithelial layer and the endometrial glands together form the **stratum functionalis** (functional layer). Each month estrogen and progesterone stimulate the stratum functionalis to greatly increase in size in preparation to nourish a zygote. The stratum functionalis is the layer that is sloughed off in menstruation if implantation does not occur. The **stratum basalis,** the layer that remains after menstruation is completed for that month, gives rise to the new stratum functionalis the following month.

Before Going to Lab

1 Label the microscopic uterine structures in Figure 39.10.

LAB ACTIVITY 8 Histology of the Uterus

1 Observe a prepared slide of the uterus.
- Using scanning power or low power, observe a prepared slide of the uterus to identify the **endometrium** and the **myometrium.** You will have to move the slide to see both layers. If your slide is taken from a uterus that is in the secretory phase, it will have a thick endometrium.
- Using high power, identify the **simple columnar epithelium** lining the lumen, **endometrial glands, stratum functionalis,** and **stratum basalis.** ■

C. Dissection of Female Reproductive System

If you are dissecting a cat or fetal pig, consult the appropriate dissection manual. You will probably be dissecting the urinary system along with the reproductive system. Although these two systems are not identical to humans, there are several similarities, and the differences make an interesting study.

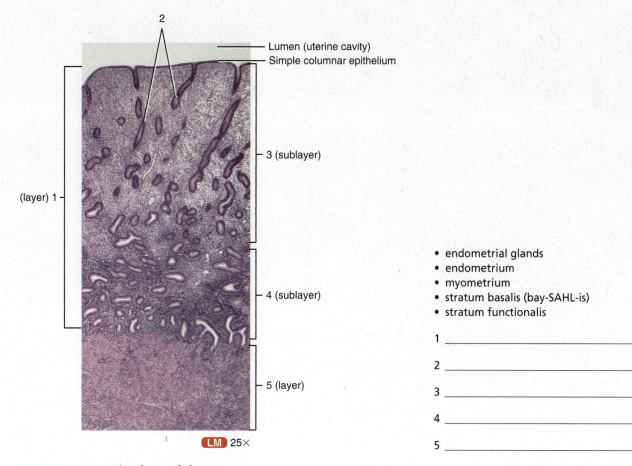

- endometrial glands
- endometrium
- myometrium
- stratum basalis (bay-SAHL-is)
- stratum functionalis

1 _____

2 _____

3 _____

4 _____

5 _____

FIGURE 39.10 Histology of the uterus.

Reviewing Your Knowledge

A. The Ovaries, Uterine Tubes, and Uterus

Write the name of the structure that the phrase describes.

_____ **1.** Domed portion of uterus

_____ **2.** Transports secondary oocyte toward uterus

_____ **3.** Usual site of fertilization

_____ **4.** Found on end of infundibulum of uterine tubes

_____ **5.** Opening between uterus and cervix

_____ **6.** Largest portion of uterus

_____ **7.** Implantation of zygote and development of fetus occurs in this organ

_____ **8.** Narrower portion of uterus; connects with vagina

_____ **9.** Produces secondary oocytes and hormones

_____ **10.** Middle layer of uterus

_____ **11.** Sheet-like ligament attaches uterus to lateral body wall

_____ **12.** Layer of uterus that sheds during menstruation

_____ **13.** Ligament attaches uterus to labia majora

B. Vagina, External Genitalia, and Mammary Gland

Write the name of the structure that the phrase describes.

_____ **1.** Area between the labia minora with openings of urethra and vagina

_____ **2.** Folds in vagina

_____ **3.** Erectile tissue

_____ **4.** Tissue that pads mons pubis

_____ **5.** Milk-producing glands

_____ **6.** Openings in the nipple

_____ **7.** Organ for intercourse; canal for childbirth and menstruation

_____ **8.** Another name for external genitalia

C. Histology of Ovary

Number in correct order (1–6) from the earliest to the latest stage of development.

_____ corpus albicans

_____ corpus luteum

_____ mature (graafian) follicle

_____ primary follicle

_____ primordial follicle

_____ secondary follicle

D. Pathway of a Secondary Oocyte

Number the structures (1–7) for the pathway of a secondary oocyte from the ovary to the uterus.

_____ ampulla of uterine tube

_____ body of uterus

_____ fimbriae of uterine tube

_____ fundus of uterus

_____ infundibulum of uterine tube

_____ isthmus of uterine tube

_____ ovary

Using Your Knowledge

A. Female Reproductive System

Write a short answer for the following questions.

1. An ovariectomy is a surgical procedure that removes one or both ovaries, and a hysterectomy is a procedure that removes the uterus. State whether ovulation and menstruation will occur after:

 (a) A bilateral (both ovaries) ovariectomy

 (b) A hysterectomy

2. A hysterectomy can cause a protrusion of the urinary bladder through the wall of the vagina. Based on your knowledge of anatomy, why would a hysterectomy cause this problem?

3. Vasectomies and female tubal ligations both involve severing and tying a tube to prevent a pregnancy. Which procedure is less invasive surgically? Explain.

4. An ectopic pregnancy is a pregnancy somewhere other than in the uterus. Explain how a pregnancy could occur in the abdominal cavity.

5. Some women notice a small blood spot shortly after ovulation. Explain why ovulation can cause this.

6. Endometriosis is a condition in which the endometrial tissue grows on the outside of the uterus and in other locations in the pelvic cavity. Explain how the endometrial tissue travels to these locations.

7. A female who has contracted gonorrhea (bacterial infection) through intercourse can develop scarring of the uterine tubes. Explain how an ectopic pregnancy may then result.

8. **(a)** Compare the amount of cytoplasm and number of organelles in ova to that in sperm.

(b) Explain why this difference is important to their function.

9. A homologous (*homo-* = same; *logos* = relation) structure is an organ that has the same embryonic origin. Using your textbook, look up which male structure is homologous to the clitoris.

10. Explain why it would be advisable for a patient to have an empty bladder when having a pelvic exam.

Human Development

OBJECTIVES

1 Describe the major events that occur during embryonic development, and give the time period during which they occur

2 Describe the process of fertilization and early cleavage to the blastocyst stage

3 Describe the process of implantation and placenta formation

4 Name the three primary germ layers and the body systems that are formed by each

5 Name the embryonic membranes and give the importance of each

6 Describe the structure and function of the placenta and umbilical cord

7 Describe the major developmental events that occur during fetal development, and give the time period during which they occur

MATERIALS

- **Fertilization, Cleavage, and Blastocyst Formation:** compound microscope, lens paper, depression microscope slides, sea urchin egg suspension, sea urchin sperm suspension, medicine droppers, Vaseline, toothpicks

- compound microscopes, lens paper, prepared slides of sea urchin embryonic development (early and late cleavage)

- models or charts of human embryo development

- models or charts of human fetal development

- model or chart of pregnant human torso

- fresh or preserved placenta or placenta model

- Demonstration: preserved pregnant cat or other animal

Fertilization is the process by which the nuclei of an egg cell and a sperm cell, each with 23 chromosomes, fuse to form a single cell with 46 chromosomes. The fertilized ovum, which is a single cell called a **zygote** (*zygote* = a pair or yoked), is the beginning stage of human life. After fertilization, the zygote undergoes a phenomenal series of rapid mitotic divisions that result in an organized ball of cells called an **embryo** (*en-* = in; *-bryein* = to grow). The developing human embryo continues to grow, and its cells differentiate (specialize), forming organs. By 9 weeks, the developing human is called a **fetus** (*fetus* = fruitful, unborn offspring). During this stage, all organs continue to develop and grow, taking on the appearance of adult organs.

A. Embryonic Period

The embryonic period is the time from fertilization through the 8th week of development. The major events during the embryonic period are fertilization, cleavage, blastocyst formation, implantation, formation of the three primary germ layers, and formation of embryonic membranes.

1. Fertilization

After ovulation, the secondary oocyte is swept into the uterine tube by the motion of fimbriae in the distal part of the uterine tube, the infundibulum. The secondary oocyte is surrounded by two layers, the corona radiata and the zona pellucida. The **corona radiata** (*corona* = crown) is the outer layer of cells, and the **zona pellucida** (*pellucida* = allowing light to pass through) is the translucent layer of glycoprotein just beneath the corona radiata and exterior to the plasma membrane. Sperm must break through both of these layers to reach the oocyte. After one sperm cell penetrates the plasma membrane, the zona pellucida is enzymatically hardened so other sperm may not enter. The secondary oocyte now completes the second meiotic division and is called an **ovum.** The 26-chromosome sperm cell then fertilizes the 26-chromosome ovum, and the two fuse to form a one-celled **zygote** with 46 chromosomes. Fertilization usually occurs within the ampulla of the uterine tube.

2. Cleavage

The rapid series of early mitotic divisions of the zygote are referred to as **cleavage** (*cleavage* = to split into two or more). The one-celled zygote divides into two cells that are called **blastomeres** (*blasto-* = to sprout, germ; *mere-* = part) 24 to 30 hours after fertilization. The two blastomeres further divide into 4, 8, 16, and 32 cells. Although there is an increase in the number of cells because of rapid mitosis, each cell becomes increasingly smaller in size so that the entire ball of cells remains approximately the same size as the original zygote. By the 4th day, the 32-cell stage is a solid ball of cells called a **morula** (*morula* = raspberry) that passes through the isthmus of the uterine tube and into the uterine cavity.

3. Blastocyst Formation

By the 5th day, the solid morula develops a hollow cavity called the **blastocyst cavity (blastocele),** and the developing zygote is now called a **blastocyst** (*blasto-* = to sprout, germ; *cyst* = bag). Blastocyst cells begin the process of differentiation by forming two different cell regions, the inner cell mass and the trophoblast. The **inner cell mass** is a group of cells in the interior of the blastocyst that will develop into the embryo and form the developing human body. The **trophoblast** (*troph-* = to nourish) is the superficial layer of cells that form the wall of the blastocyst, which will burrow into the endometrium and subsequently form the fetal side of the placenta.

> ### Before Going to Lab
>
> **1** Label the embryonic structures in Figure 40.1.

> **NOTE TO THE INSTRUCTOR:** *You may want to fertilize several eggs an hour or more before class and have the students examine those zygotes for cleavage while waiting for their slides.*

LAB ACTIVITY 1 Fertilization, Cleavage, and Blastocyst Formation

1 Examine the models or charts of fertilization, cleavage, and early embryologic development.

2 Fertilization, Cleavage, and Blastocyst Formation:

Observe fertilization
- Obtain materials for Fertilization and Blastocyst Formation (see Materials list).
- Using a medicine dropper, transfer one drop of sea urchin egg-cell suspension to the well of a depression slide.
- Examine the drop under a low-power objective lens and focus on one egg cell. Now go to high-power and observe the egg cell.
- Add a drop of sea urchin sperm-cell suspension to the egg-cell suspension on your slide.
- Observe the sperm cells with high-power lens to see them cluster around the egg cells. The eggs secrete chemicals that attract the sperm.
- Watch the egg cell for the rapid appearance of a **fertilization membrane.** This membrane appears as a clear halo and is formed when a sperm cell has penetrated the egg cell. Because of its small size, you will not actually see the nucleus of the sperm.

Observe cleavage and blastocyst formation

- Using a toothpick, place a thin line of Vaseline around the edge of the depression and gently press a clean coverslip over the Vaseline and depression. This will prevent water evaporation.
- The first cleavage should occur in about 45 to 60 minutes. **It is important to keep the slide at 72°F or 22°C.** You may need to use an incubator set at that temperature.

- Use low power to view the slide approximately every 30 minutes. After the first mitotic division, examine the slide about every 15 to 20 minutes.
- Find 2-, 4-, 8-, and 16-cell stages depending on lab time available.

3 Examine the prepared slides of sea urchin embryonic development, both early and late cleavage. Find 2-, 4-, 8-, and 16-cell stages.

4 Describe the processes of fertilization and cleavage to the blastocyst stage to your lab partners. ◼

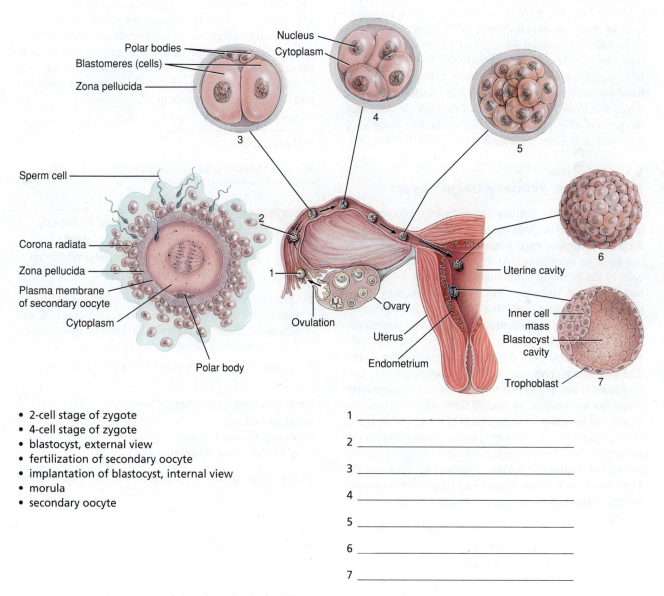

- 2-cell stage of zygote
- 4-cell stage of zygote
- blastocyst, external view
- fertilization of secondary oocyte
- implantation of blastocyst, internal view
- morula
- secondary oocyte

1 _____

2 _____

3 _____

4 _____

5 _____

6 _____

7 _____

FIGURE 40.1 First week of development—fertilization to blastocyst.

4. Implantation

At approximately the 7th day, the blastocyst sheds the zona pellucida and usually begins **implantation,** or insertion of the blastocyst into the wall of the uterus. This process starts when the blastocyst adheres to the wall of the uterus at approximately the 6th day after fertilization. Implantation typically occurs in the posterior wall of the fundus or body of the uterus with the inner cell mass facing the endometrium. When this occurs, the endometrium of the uterus is in its secretory phase with a rich blood supply. The trophoblast secretes enzymes that allow the blastocyst to penetrate and bury itself in the endometrium. A hormone, **human chorionic gonadotropin (hCG),** is secreted by the trophoblast to maintain secretion of estrogen and progesterone by the corpus luteum of the ovary. The endometrium is now called the **decidua** (*decida* = falling off), and the basal portion (next to the embryo) is the **deciduas basalis,** which will form the maternal side of the placenta. Although the process of implantation started in the first week after fertilization, it is not completed until around the 14th day.

5. Formation of the Three Primary Germ Layers

About 8 days after fertilization, cells of the inner mass begin to differentiate into two different layers of cells that form the **bilaminar embryonic disc** (*bi-* = two; *lamina* = flat layer or membrane). At approximately the 15th day, cells of the bilaminar embryonic disc further differentiate and form three layers of cells. This structure, now called the **trilaminar embryonic disc,** consists of three **primary germ layers,** the outer ectoderm, the middle mesoderm, and the inner endoderm. All tissues, organs, and body systems will arise from these three primary germ layers. The **ectoderm** gives rise to the epidermis, glands, nervous tissue, and parts of the eye and ear; the **mesoderm** gives rise to cartilage, bones, all three muscle types, and blood; and the **endoderm** gives rise to the epithelial lining of the digestive and respiratory systems, and other internal organs. The tissues and organs are all growing rapidly during the embryonic period. Around the 22nd day the heart starts beating and by the end of the 8th week, the embryo takes on a human appearance and brain waves can be detected.

6. Formation of Embryonic (Extraembryonic) Membranes

The thin, protective membrane that surrounds and protects the embryo is called the **amnion.** This transparent sac is formed around the 8th day after fertilization. The cavity located between the amnion and the embryo is called the **amniotic cavity. Amniotic fluid** fills the amniotic cavity, cushioning and protecting the embryo from temperature extremes and trauma. The amnion is also known as the "bag of waters."

The **yolk sac,** formerly called the blastocyst cavity, is another embryonic membrane that lies outside the embryo and also begins to develop around the 8th day. The bilaminar disc is now located between the yolk sac and the amniotic cavity. During the 2nd and 3rd weeks, while the placenta is being established, the yolk sac, although small, gives nourishment to the embryo.

The **chorion** (*chorion* = a skin or membrane), a membrane that surrounds all the other embryonic membranes, develops into the embryonic portion of the placenta. Cells of the chorion form the **connecting (body) stalk,** which connects the trophoblast and the bilaminar embryonic disc. The connection stalk will become the umbilical cord.

The fourth embryonic membrane, the **allantois** (*allas* = sausage; *-eidos* = resemblance), is an extension of the yolk sac and extends into the connecting stalk of the embryo. It contains blood vessels that will become the umbilical arteries and the umbilical vein.

Before Going to Lab

1 Label the embryonic structures in Figure 40.2(b).

LAB ACTIVITY 2 — Implanation, Primary Germ Layers, and Embryonic Membrane Development

1 Observe models or charts of the 4 embryonic membranes and locate the **amnion, yolk sac, chorion,** and **allantois.** Refer to Figure 40.4 later in this manual. State the function of each embryonic membrane.

2 Answer Discussion Questions with your lab group.

DISCUSSION QUESTIONS
Implantation, Primary Germ Layers, and Embryonic Membranes

1 How does the blastocyst implant into the endometrium?

2 What body structures are produced from the three primary germ layers?

3 Discuss the difference among the four embryonic membranes and the importance of each.

Endometrium of uterus:

Endometrial stroma

Simple columnar epithelium

Endometrial gland

Chorion

Amnion
Amniotic cavity

Bilaminar embryonic disc

Blastocyst cavity

Uterine cavity
Maternal blood vessels

(a) Implantation of blastocyst, about 8 days after fertilization

Endometrium of uterus:
Simple columnar epithelium
Endometrial stroma

Sinusoid

Lacunae

1

Chorion

2
3

Lacunar network

4

Uterine cavity

Endometrial gland (right) and sinusoid (left) emptying into lacunar network

- amnion
- amniotic cavity
- bilaminar embryonic disc
- yolk sac

1_____

2_____

3_____

4_____

(b) Development of embryonic membranes, about 12 days after fertilization

FIGURE 40.2 Second week of development—embryonic membrane development.

7. Placenta and Umbilical Cord

At birth the placenta (*placenta* = flat cake) is about 8 inches in diameter and approximately 1 inch to 1½ inches thick. The **placenta** is disc-shaped and dark burgundy in color. It almost has the appearance of raw meat. This is the structure in the uterus through which the developing embryo and fetus obtain nutrients and get rid of wastes. The placenta has an embryonic side (chorionic villi) and a maternal side (decidua basalis). During the 3rd week of growth, the chorion forms **chorionic villi** (*villus* = hair-like projection) with blood vessels that lengthen and protrude into the **decidua basalis** of the endometrium. The chorionic villi lie in **intervillus spaces** surrounded by **lacunae** (*lacuna* = a pit, hollow space) that are filled with maternal blood. The **umbilical cord** is formed from the connecting stalk and is the connection between the placenta and the embryo (and later the fetus) and contains 2 arteries and 1 vein. The umbilical arteries carry oxygen-poor blood and wastes from the fetus to the placenta. The umbilical vein carries oxygen-rich blood and nutrients obtained from the intervillous spaces (containing maternal blood) in the placenta to the fetus. When the baby is born, the placenta separates from the uterus and is called the **afterbirth.** The area where the umbilical cord was attached to the baby is called the **umbilicus** (navel).

> ### Before Going to Lab
>
> **1** Label the structures of the placenta and umbilical cord in Figures 40.3 and 40.4.

LAB ACTIVITY 3 Placenta and Umbilical Cord

1 Identify the structures of the placenta and umbilical cord on a model or chart of the placenta.

2 Observe the demonstration of a fresh or preserved placenta, if available.

- Observe the maternal surface, which is spongy with a cobblestone appearance. This side of the placenta is textured due to the fetal chorionic villi, which extensively branched into the endometrium.

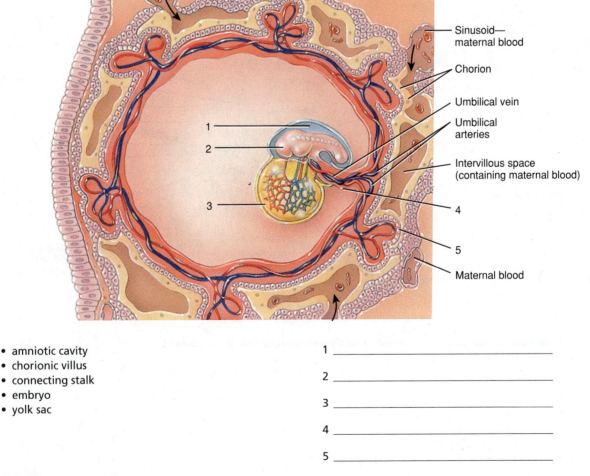

- amniotic cavity
- chorionic villus
- connecting stalk
- embryo
- yolk sac

1 _____

2 _____

3 _____

4 _____

5 _____

FIGURE 40.3 **Development of chorionic villi in a 21-day embryo.**

- Now observe the smooth, shiny **fetal surface** of the placenta, which is somewhat translucent. The umbilical cord is attached to this side of the placenta, generally in the center.
- The umbilical cord contains 2 arteries and 1 vein, which are surrounded by mucous connective tissue called Wharton's jelly. The shiny layer on the exterior of the umbilical cord is a layer of the amnion.

3 Observe the demonstration of a preserved pregnant cat or other animal, if available. Identify the **chorion, amnion, embryo, placenta,** and **umbilical cord.**

4 Answer Discussion Questions with your lab group.

DISCUSSION QUESTIONS
Placenta and Umbilical Cord

1 Create a sentence including the following three terms: *fetus, placenta, umbilical cord.*

2 How is the placenta formed? What are the embryonic and maternal parts of the placenta?

3 Why do the umbilical arteries carry oxygen-poor blood and the umbilical vein carries oxygen-rich blood?

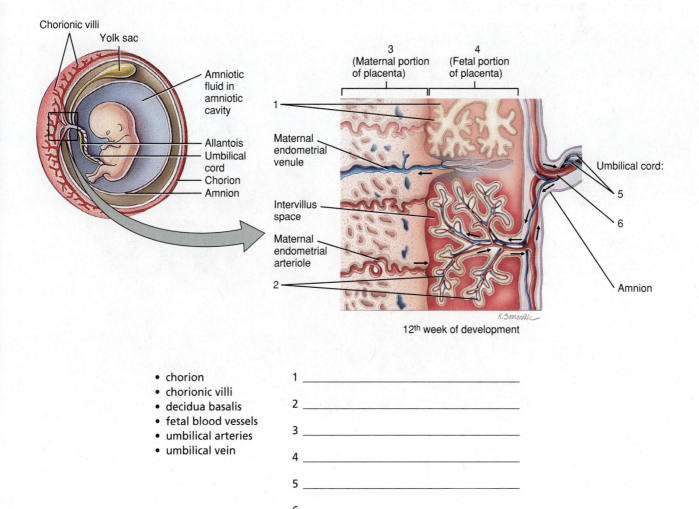

12th week of development

- chorion
- chorionic villi
- decidua basalis
- fetal blood vessels
- umbilical arteries
- umbilical vein

1 _____

2 _____

3 _____

4 _____

5 _____

6 _____

FIGURE 40.4 Placenta and umbilical cord.

B. Fetal Period

The fetal period is from week 9 to week 38. By the 9th week all organ systems are present in the fetus, and the fetal head is about half of the size of the body. The fetus doubles in size between the 12th and 20th fetal weeks, doubles again between the 16th and the 24th weeks, and more than doubles between the 20th and 38th weeks. The uterus expands as the fetus grows, and the mother's abdominal organs are shifted from their normal locations. Although at 8 weeks, toward the end of the embryonic period, the developing embryo looks like a human, the 16-week fetus takes on even more of a human appearance.

Before Going to Lab

1 Examine the drawings in Figure 40.5 and the photographs in Figure 40.6.

LAB ACTIVITY 4 **Fetal Development**

1 Examine the fetal development models and describe the position of the fetus in the uterus.

2 Answer Discussion Questions with your lab group.

DISCUSSION QUESTIONS
Fetal Development

1 At what stage can you first identify the:
- upper and lower limbs
- future auditory meatus
- eye
- fingers and toes

2 Does the position of the fetus in the uterus change very much toward the end of the 38th week?

| 4 | 8 | 12 | 16 | 20 | 24 | 28 | 32 | 36 | (weeks) |

FIGURE 40.5 **Fetal development.**

(a) 20-day embryo

Neural plate
Neural groove
Cut edge of amnion
Somite
Yolk sac
Primitive streak

(b) 24-day embryo

Developing brain
Heart prominence
Developing spinal cord
Somite

(c) 32-day embryo

Pharyngeal arches
Lens placode
Heart prominence
Upper limb bud
Tail
Lower limb bud

(d) 44-day embryo

Otic placode
Developing nose
Upper limb
Lower limb
Umbilical cord

FIGURE 40.6 Fetal development.

Ear

Eye

Nose

Upper limb

Umbilical cord

Lower limb

(e) 52-day embryo

Ear

Nose

Upper limb

Rib

Lower limb

Eye

Yolk sac

Umbilical cord

Placenta

(f) Ten-week fetus inside amniotic sac in uterus

Ear

Eye

Nose

Mouth

Upper limb

Umbilical cord

Lower limb

(g) Thirteen-week fetus

Ear

Eye

Nose

Mouth

Upper limb

Lower limb

(h) Twenty-six-week fetus

FIGURE 40.6 **Fetal development,** *continued.*

Reviewing Your Knowledge

A. Embryonic Development

Give the name of the structure that the phrase describes.

_____ 1. The name of the ovulated structure prior to fertilization

_____ 2. The blastocyst structure that becomes the embryo

_____ 3. The anatomical structure where fertilization normally takes place

_____ 4. The stage of the embryo that implants in the uterine wall

_____ 5. After fertilization, implantation normally occurs on this day

_____ 6. The membrane that forms the embryonic side of the placenta

_____ 7. The structure that is formed after the morula stage

_____ 8. The structure that the blastocyst sheds before implantation

_____ 9. The structure that forms the maternal side of the placenta

_____ 10. The name of the thin, protective membrane that encompasses the embryo

_____ 11. The side of the placenta that is smooth and shiny

_____ 12. The number of umbilical arteries in the umbilical cord

_____ 13. The number of umbilical veins in the umbilical cord

_____ 14. This placental structure has villi

_____ 15. The number of primary germ layers that form all body organs

_____ 16. Sustains the embryo with nutrients and removes wastes

_____ 17. The name of the female gamete

_____ **18.** The name of the male gamete

_____ **19.** The inner layer of the uterus where implantation occurs

_____ **20.** The name of the fertilized egg

_____ **21.** Primary germ layer that becomes the outer skin and nervous system

_____ **22.** Primary germ layer that gives rise to epithelial lining of digestive and respiratory organs

_____ **23.** Primary germ layer that produces bone, muscle, cartilage, and blood

B. Fetal Development

Give the name of the structure that the phrase describes.

_____ **1.** The "bag of waters" that breaks at birth.

_____ **2.** The fetal period lasts from week _____ to week _____.

_____ **3.** By week _____ all organ systems are present.

_____ **4.** Toward the end of the embryonic period, week _____, the developing embryo looks like a human.

_____ **5.** The fetus first doubles in size between weeks _____ and _____.

_____ **6.** The fetus doubles in size again between weeks _____ and _____.

_____ **7.** The fetus more than doubles between weeks _____ and _____.

Using Your Knowledge

A. Human Development

Write a short answer for each of the following questions.

1. The acrosome of a sperm contains hydrolytic enzymes. Using your knowledge, what is the purpose of these enzymes?

2. Ectopic pregnancies occur in the uterine tubes or pelvic cavity. Explain how this can occur.

3. Amniocentesis is a test conducted on fetal cells found in amniotic fluid to determine whether the fetal cells have genetic abnormalities. Amniocentesis is usually conducted at 12 weeks. Amniotic fluid is collected by inserting a needle into the mother's abdomen. List the structures the needle must go through to collect the amniotic fluid.

4. Are the cells obtained for culture from amniocenteses most likely derived from ectoderm, mesoderm, or endoderm? Explain your answer.

5. Chorionic villi sampling (CVS) is a test to determine whether fetal cells have genetic abnormalities and can be performed as early as 8 weeks. Chorionic villi cells are removed using a small tube that is passed through the cervix. These cells are cultured and undergo chromosomal analysis. Explain why chorionic villi cells will determine whether the fetus has genetic abnormalities.

6. Drinking alcohol during pregnancy can lead to fetal alcohol syndrome, a condition that can cause heart defects, damage to the central nervous system, deformed limbs, slow growth, and abnormal facial features. Explain how the ingested alcohol is transferred from mother to fetus.

7. Early pregnancy tests used antibodies to detect the presence of human chorionic gonadotropin (hCG) in maternal urine. Explain how hCG, produced by the chorion, gets in maternal urine.

8. Human embryonic stem cell research uses stem cells cultured from the blastula. These cells are pluripotent, meaning that these stem cells can develop many different types of human cells. Identify the part of the blastula that they are using and explain your choice.

9. During an ultrasonography, the doctor observes that all four chambers of the fetal heart are developed normally and that blood can be observed passing from the right atrium to the left atrium. Explain why it is normal for blood to pass from the right to left atrium in the fetal heart.

10. Placenta previa (*previa* = before or in front of) is a condition in which either part or the entire placenta implants in the inferior part of the uterus near the cervix. Explain why this can cause spontaneous abortion or premature birth.

Heredity

OBJECTIVES

1 Understand the terms and language of genetics
2 Be able to proficiently use the Punnett square
3 Complete monohybrid crosses with dominant-recessive inheritance, incomplete dominance, co-dominance, and sex-linked inheritance
4 Describe the roles genetics and the environment play in determining phenotypes

Heredity is the transfer of genetic traits, such as hair or eye color, from parents to offspring, generation after generation. **Genetics** is the science that studies the principles and patterns governing heredity. Geneticists can study the complete set of an organism's genes (located on chromosomes) called a **genome.** The complete set of genes and alleles for all humans that is studied in genetics is called the **human gene pool.** Geneticists have discovered that environmental influences like nutrition play a very critical part in our appearance (height, weight, etc.) and whether or not we will inherit a genetic disease.

A. Basic Genetic Language

Each chromosome contains one **DNA** (deoxyribonucleic acid) molecule. **Genes** are short segments of DNA that are the functional or heritable units of DNA. Each gene is

situated at a specific location on a chromosome and provides a template for the cell to make one or more proteins.

Each human cell contains the same 46 chromosomes, 22 pairs of **homologous** (*homo-* = same; *logos* = proportion) chromosomes called **autosomes** and 2 **sex chromosomes.** The homologous chromosomes look identical in size and shape under the microscope and have the same genes at identical locations. There are two types of human sex chromosomes, an X chromosome and a Y chromosome. The X and Y chromosomes not only look different in size and shape, but they also carry genes for different characteristics.

Each gene has two or more alternate forms, called **alleles,** which occupy the same **locus** (location) on homologous chromosomes. The different alleles code for different proteins and therefore will result in the expression of different traits. For example, there are two alleles pertaining to dimples. One allele of this gene codes for dimples (*D*) and the other allele is for no dimples (*d*). If the pair of alleles of a particular gene are identical (i.e., *DD* or *dd*),

the individual is **homozygous** (*homo-* = same; *zygous* = yoked) for that trait. If the alleles are different (*Dd*), the person is **heterozygous** (*hetero-* = different). If alleles are heterozygous, one allele is usually completely expressed and is called **dominant,** while the other allele is not expressed and is called **recessive.** Note that a capital letter is used to designate a dominant allele, which is written first, and a lowercase letter is used for a recessive allele. The particular genetic composition of a person is the **genotype** (*DD*, *Dd*, or *dd*), and the expressed physical traits that are observable are called the **phenotype** (dimples or no dimples).

LAB ACTIVITY 1 Genotypes and Phenotypes

1 For each genotype, indicate whether it is homozygous (hom) or heterozygous (het).

_____ *Zz* _____ *bb* _____ *MM*

_____ *Tt* _____ *cc* _____ *AA*

2 For each genotype, indicate the phenotype:

The freckles allele (*F*) is dominant to the no freckles allele (*f*). The attached earlobes allele (*E*) is dominant to the unattached earlobes allele (*e*).

ff _____ *Ee* _____

FF _____ *EE* _____

Ff _____ *ee* _____

3 For each phenotype, list the genotype(s):

Widow's peak (*W*) is dominant to a straight hairline (*w*).

_____ widow's peak (homozygous)

_____ widow's peak (heterozygous)

_____ straight hairline

B. Monohybrid Crosses and the Punnett Square

Gregor Mendel was the first person to systematically study, record, analyze, and publish information regarding the patterns of inheritance. Mendel's first rule of inheritance, developed while studying inheritance in garden peas, was the **principle of segregation.** He showed that when gametes are produced, the 2 units (genes) for a characteristic separate from each other and the sperm and ovum get only one unit each. For instance, if the genotype for one parent was *Dd*, as the sperm or ovum is formed, each sperm or ovum gets only a '*D*' or a '*d*.'

An uncomplicated way to view patterns of inheritance for a single pair of alleles is by using a diagram called the **Punnett square.** This diagram is used to predict possible genotypes and therefore phenotypes of the offspring from two parents. In a **monohybrid cross,** the inheritance of only one trait is studied using a single pair of alleles. In a **dihybrid** or **trihybrid cross,** there are two and three traits, respectively, that are studied. All of the genetic activities in this exercise are monohybrid crosses. The presence or absence of dimples is used as the trait in the next activity.

LAB ACTIVITY 2 Dimples or No Dimples

1 Let *D* = dimples and *d* = no dimples. The genotype of the female is homozygous dominant (*DD*) and the male is homozygous recessive (*dd*).

2 In Figure 41.1, look at the genotype of the parents that displays the **diploid** state (2*n*). There are 2 sets of chromosomes, one inherited from the father and one from the mother. According to the first principle of inheritance, the two alleles for any trait will separate during meiosis. Therefore, each '*D*' or each '*d*' is now **haploid** (*n*), with half the number of chromosomes (alleles) present in the ova or sperm.

3 Put the alleles of the sperm on the side of the square, and the alleles of the ova on top. In order to have a pattern, always put the same parent on the same axis.

4 Combine the letters of the sperm and ova in the squares that now represent the diploid genotype(s) of their offspring.

5 In Figure 41.1, indicate the number of offspring with each **genotype** and the percent chance offspring has the genotype.

6 In Figure 41.1, indicate the number of offspring with each **phenotype** and the percent chance offspring has the phenotype.

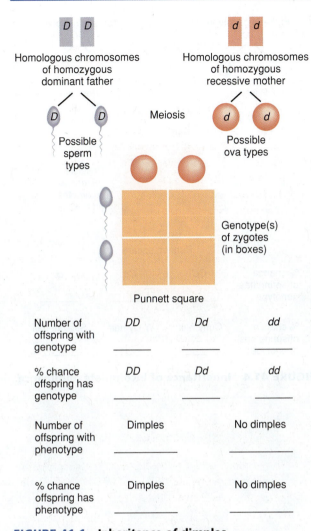

Homologous chromosomes of homozygous dominant father

D *D*

Possible sperm types

Meiosis

Homologous chromosomes of homozygous recessive mother

d *d*

Possible ova types

Genotype(s) of zygotes (in boxes)

Punnett square

Number of offspring with genotype	*DD* ___	*Dd* ___	*dd* ___
% chance offspring has genotype	*DD* ___	*Dd* ___	*dd* ___
Number of offspring with phenotype	Dimples ___		No dimples ___
% chance offspring has phenotype	Dimples ___		No dimples ___

FIGURE 41.1 Inheritance of dimples.

FIGURE 41.2(a) Tongue roller.

FIGURE 41.2(b) Non-tongue roller.

LAB ACTIVITY 3 Tongue Rolling or No Tongue Rolling

1 Some people can roll the tongue into a U-shape [Figure 41.2(a)], while others do not have the ability to roll the tongue [Figure 41.2(b)]. Let R = tongue roller, r = non-tongue roller. Cross a heterozygous tongue-rolling mother and father.
 • In Figure 41.3, choose the dominant and recessive alleles for the sperm and the ova and write the alleles on the top and left side of the Punnett square.
 • Combine the letters of the sperm and ova in the squares. These letters represent all of the possible diploid genotypes of the offspring.

2 In Figure 41.3, indicate the number of offspring with each **genotype** and the percent chance offspring has the genotype.

3 In Figure 41.3, indicate the number of offspring with each **phenotype** and the percent chance offspring has the phenotype. ■

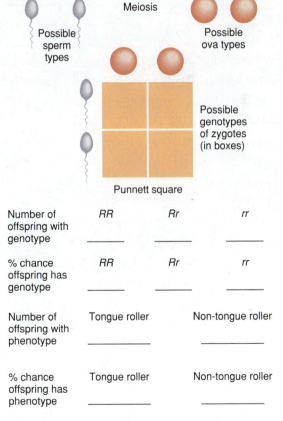

Father's genotype *Rr*

Possible sperm types

Meiosis

Mother's genotype *Rr*

Possible ova types

Possible genotypes of zygotes (in boxes)

Punnett square

Number of offspring with genotype	*RR* ___	*Rr* ___	*rr* ___
% chance offspring has genotype	*RR* ___	*Rr* ___	*rr* ___
Number of offspring with phenotype	Tongue roller ___		Non-tongue roller ___
% chance offspring has phenotype	Tongue roller ___		Non-tongue roller ___

FIGURE 41.3 Inheritance of tongue rolling.

C. Incomplete Dominance

A few human traits are inherited because neither allele for a gene is dominant over the other and the offspring show an intermediate phenotype or blending of the two characteristics. This type of inheritance is called **incomplete dominance.** An example of this is curly, wavy, or straight hair. Curly hair (*CC*) is incompletely dominant over straight hair (*cc*), therefore the heterozygous genotype (*Cc*) produces wavy hair, a phenotype intermediate between curly and straight hair.

LAB ACTIVITY 4 **Curly, Wavy, and Straight Hair**

1 Cross a mother with curly hair with a father with straight hair in Figure 41.4.

2 Calculate the percentages of each genotype and phenotype of the offspring from the Punnett square. ■

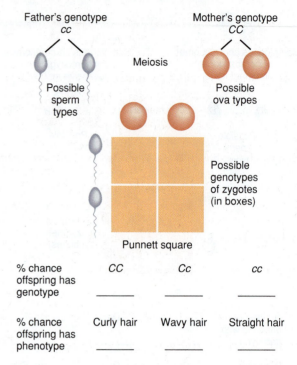

	CC	*Cc*	*cc*
% chance offspring has genotype	_____	_____	_____
% chance offspring has phenotype	Curly hair	Wavy hair	Straight hair
	_____	_____	_____

FIGURE 41.4 Inheritance of incomplete dominance.

D. Codominance in Monohybrid Crosses

Codominance is a pattern of inheritance in which two different alleles of a gene are equally expressed in a heterozygote and no blending of the characteristic occurs. There are three alleles for the ABO human blood type system: *A*, *B*, and *O*. The *A* and *B* alleles are codominant and both are dominant over the *O* allele. The *A* allele causes an A antigen to appear on the surface of red blood cells, the *B* allele causes a B antigen to appear on the surface of red blood cells, and the *O* allele results in no antigen on the surface of the red blood cell. Because each of us inherits only two of these alleles, there are six possible genotypes: *AA*, *AB*, *AO*, *BB*, *BO*, and *OO*. This means that there are only four phenotypes for blood type: A, B, AB, and O.

LAB ACTIVITY 5 **ABO Blood Type**

1 Cross a father with blood type AO and a mother with BO blood type in Figure 41.5.

2 Calculate the percentages of each genotype and phenotype of the offspring from the Punnett square.

3 Cross a father with AB blood type with a mother with O blood type in Figure 41.6.

4 Calculate the percentages of each genotype and phenotype of the offspring from the Punnett square. ■

% chance offspring has genotype	AO	BO	AB	OO
	___	___	___	___

% chance offspring has phenotype	A	B	AB	O
	___	___	___	___

FIGURE 41.5 **Codominance with ABO blood types.**

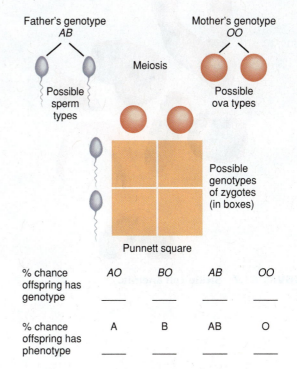

% chance offspring has genotype	AO	BO	AB	OO
	___	___	___	___

% chance offspring has phenotype	A	B	AB	O
	___	___	___	___

FIGURE 41.6 **Codominance with ABO blood types.**

Sickle-cell anemia is another example of codominance. Two alleles are present that are involved in the production of hemoglobin (Hb), a protein that carries oxygen in red blood cells. The sickle-cell allele is designated *S* and produces an abnormal Hb. Individuals who are *SS* produce only the abnormal Hb. When oxygen levels in the red blood cells are low (i.e., at high altitudes or during exercise), the sickle-cell Hb crystallizes within red blood cells, causing the red blood cells to become sickle- or crescent-shaped (Figure 41.7). The crescent-shaped sickle cell clumps and forms clots in small blood vessels. Untreated individuals with sickle-cell anemia often die at an early age. The normal Hb allele is *A*, and individuals who are *AA* have normal Hb and normal-shaped red blood cells. Heterozygous individuals have the **sickle-cell trait** (but not the disease); half of their Hb is normal and half is sickle-cell Hb. These individuals are usually healthy, and only a small number experience some symptoms of sickle-cell disease. Heterozygous individuals are also resistant to malaria. Therefore, in regions of the world where malaria is a problem, heterozygous individuals have an advantage and more individuals with the *SA* genotype survive compared to individuals with the *AA* genotype.

LAB ACTIVITY 6 Sickle-Cell Anemia

1 The abbreviation for hemoglobin is Hb. Let $Hb^S Hb^S$ = sickle-cell anemia, $Hb^A Hb^A$ = normal Hb (unaffected), and $Hb^S Hb^A$ = sickle-cell trait.

2 Using Figure 41.8, cross a mother with sickle-cell anemia and father with the sickle-cell trait.

3 Calculate the percentages of each genotype and phenotype of the offspring from the Punnett square. ■

FIGURE 41.7 Sickle-cell anemia.

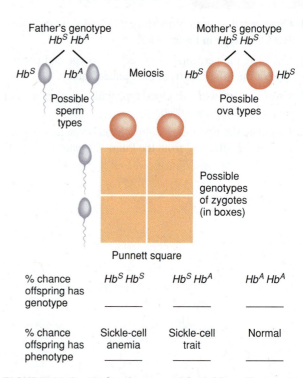

FIGURE 41.8 Codominance with sickle-cell anemia.

E. Sex-Linked Inheritance

The majority of genes are carried on **autosomes,** chromosomes that code for traits not directly involved in determining the sex of the individual. There are 22 pairs of autosomes and 2 **sex chromosomes** (X and Y) for each person (Figure 41.9). Females have two X chromosomes (XX) while males have one X and one Y (XY). The X chromosome is as large as many of the autosomes, but the Y chromosome is about one-third of the size of the X chromosome and is missing genes present on the X chromosome. The genes carried *only* on the X chromosome are called **sex-liked genes** and do not produce female characteristics. Sex-linked genes have recessive alleles that cause baldness, hemophilia, and color blindness. For a woman to express these phenotypes, she must have two recessive alleles, one on each X chromosome. However, because the male has only one X chromosome and these sex-linked genes are not present on the Y chromosome, the presence of only one recessive allele will give a male the recessive phenotype. In sex-linked inheritance, the terms *homozygous* and *heterozygous* in the male do not apply. It is interesting to note that with sex-linked inheritance, the father cannot pass the condition to his son. The son can receive only a Y chromosome from his father and an X chromosome from his mother.

FIGURE 41.9 **Autosomes and sex chromosomes.**

LAB ACTIVITY 7 Sex-Linked Inheritance

1 Let X^C = dominant allele for **normal** vision and X^c = recessive allele for **color blindness.** Indicate the phenotype of the following alleles for recessive sex-linked color blindness.

$X^C X^C$_____ $X^C X^c$_____ $X^c X^c$_____

$X^C Y$_____ $X^c Y$_____

2 Using Figure 41.10, cross a heterozygous female carrying the allele for color blindness (a carrier) with a male with normal vision.

3 Calculate the percentages of each genotype and phenotype of the offspring from the Punnett square.

4 Answer Discussion Questions with your lab group.

DISCUSSION QUESTIONS
Sex-Linked Inheritance

1 What percentage of girls, if any, would be a carrier for the color-blind allele?

2 What percentage of boys, if any, would be a carrier for the color-blind allele?

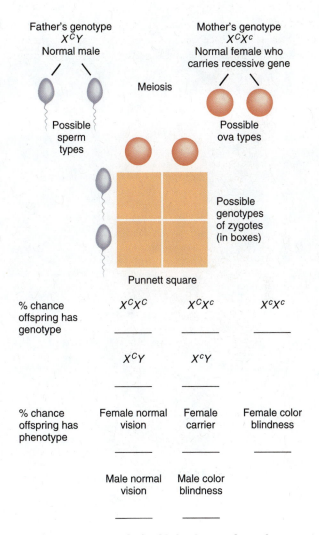

FIGURE 41.10 **Sex-linked inheritance for color blindness.**

Reviewing Your Knowledge

A. Basic Genetic Language

Write the name of the term that describes the phrase.

_____ **1.** Two different alleles of a gene are both expressed with no blending in a heterozygous individual

_____ **2.** Non-identical chromosomes that determine male or female

_____ **3.** The functional units of DNA; short segments of chromosomes

_____ **4.** The allele that is expressed in a heterozygous individual

_____ **5.** Two or more alternate forms or variations of a gene

_____ **6.** The genetic composition for a trait

_____ **7.** Individual having identical alleles for a gene

_____ **8.** 22 pairs of chromosomes that determine most of the body traits

_____ **9.** The allele that is not expressed in heterozygous alleles

_____ **10.** The physical or functional expression of a genetic trait

_____ **11.** Possessing all of the normal number of chromosomes

_____ **12.** Possessing half of the normal number of chromosomes

13. Explain what a "carrier" for an allele of a gene means.

14. Distinguish between a phenotype and a genotype.

15. How do we obtain two copies of each gene?

B. Monohybrid Dominant-Recessive Inheritance

1. Unattached earlobes (*E*) are dominant over attached earlobes (*e*) [Figure 41.11(a) and (b)]. If a woman with unattached earlobes and a man with attached earlobes have children, what percentage of their children has the possibility of having unattached earlobes if the mother is homozygous for the trait?

FIGURE 41.11(a) Unattached earlobes.

FIGURE 41.11(b) Attached earlobes.

2. What percentage of the offspring in question 1 above has the possibility of having unattached earlobes if the mother is heterozygous for the trait?

3. For questions 1 and 2, could the phenotype of the children born allow you to determine whether the mother is homozygous or heterozygous for the trait? Explain.

C. Incomplete Dominance Inheritance

1. If a woman with wavy hair and a man with curly hair have children, calculate the percentages of each genotype and phenotype of the offspring from the Punnett square.

2. If a woman with straight hair and a man with wavy hair have children, calculate the percentages of each genotype and phenotype of the offspring from the Punnett square.

D. Monohybrid Codominant Inheritance

1. If a father with type O blood and a mother with AB blood have children, calculate the percentages of each genotype and phenotype of the offspring from the Punnett square.

2. If a mother with sickle-cell anemia and a father with the sickle-cell trait have children, calculate the percentages of each genotype and phenotype of the offspring from the Punnett square.

E. Sex-Linked Inheritance

1. Cystic fibrosis is due to an inherited recessive gene. The recessive allele produces a defective version of a protein. Perform a cross with a heterozygous (carrier) mother and a father who has a normal gene for the trait. Let C = the normal allele and c = cystic fibrosis allele. Calculate the percentages of each genotype and phenotype of the offspring from the Punnett square.

Using Your Knowledge

Answer the following questions with a short answer.

1. Why are marriages between blood relatives prohibited in the United States and other countries?

2. If a phenotypic ratio for a single trait is 3:1 (75%:25%) in the offspring, what genotypes would you propose the parents might be? Draw a Punnett square to explain.

3. The allele for baldness is on the X chromosome only. Can any sons obtain the allele for baldness from the father with the allele? Draw a Punnett square to explain.

4. Why are men more likely than women to be color blind? Explain.

5. Can a son be a carrier for an X-linked trait? Explain.

6. Could a father who has blood type A and a mother who has blood type B possibly have a child who has blood type O? Draw a Punnett square to explain.

7. Huntington's disease is due to inheriting a lethal dominant allele that affects the nervous system and causes early dementia around age 40. Cross a heterozygous mother who has the trait with a normal father. Draw a Punnett square to explain.
 (a) What are the percentages and types of genotype?

 (b) What are the percentages and types of phenotype?

8. Why does the lethal dominant allele of Huntington's disease stay in the gene pool?

9. If there is a possibility of four different genotypes in the offspring that result from a dominant-recessive inheritance, and if the couple has four children, would they definitely have children with one expressing each genotype? Explain.

10. List the 6 possible genotypes of the four blood types. Identify the parental genotypes that could produce children with blood type AO.

Answer Key

(Answers to Discussion Questions can be found on the Instructor Companion Site)

Exercise 1

Figure 1.1

1. cranial
2. facial
3. cervical
4. acromial
5. axillary
6. brachial
7. antecubital
8. antebrachial
9. carpal
10. digital
11. palmar
12. femoral
13. patellar
14. crural
15. tarsal
16. digital
17. frontal
18. orbital
19. otic
20. buccal
21. nasal
22. oral
23. mental
24. sternal
25. mammary, pectoral, or thoracic
26. umbilical
27. coxal
28. inguinal
29. manual
30. pubic
31. pedal
32. occipital
33. scapular
34. vertebral
35. olecranal
36. gluteal
37. popliteal
38. sural
39. fibular (peroneal)
40. plantar
41. lumbar
42. calcaneal

Lab Activity 2 Directional Terms

1. a. anterior
 b. medial and inferior
 c. medial
 d. distal
 e. ipsilateral
 f. superficial

Figure 1.2

1. superior
2. posterior or dorsal
3. anterior or ventral
4. proximal
5. distal
6. inferior

Figure 1.3

1. frontal plane
2. oblique plane
3. transverse plane
4. parasagittal plane
5. midsagittal plane
6. longitudinal section
7. cross-section

Figure 1.4

a. midsaggital
b. transverse
c. frontal

Exercise 2

Figure 2.1

1. Cardiovascular System
 aorta
 heart
 thoracic aorta
 abdominal aorta
 inferior vena cava
2. Digestive System
 esophagus
 stomach
 large intestine
 small intestine
 liver
 pancreas
3. Endocrine System
 pancreas
4. Integumentary System
 none
5. Lymphatic System
 spleen
6. Muscular System
 diaphragm
7. Nervous System
 none
8. Reproductive System
 none
9. Respiratory System
 trachea
 bronchus
 diaphragm
 lung
10. Skeletal System
 none
11. Urinary System
 kidney
 ureter
 urinary bladder

Lab Activity 1 Identification of Organs on Torso

1. a, 2. b, 3. d, 4. c,
5. d, 6. b, 7. b, 8. a,
9. b, 10. b

Figure 2.2

1. cranial cavity
2. vertebral canal
3. thoracic cavity
4. diaphragm
5. abdominal cavity
6. pelvic cavity

Figure 2.3

1. parietal
2. visceral

Lab Activity 6 Abdominopelvic Quadrants and Regions

Quadrants

3. a. RLQ
 b. all four quadrants
 c. mainly RUQ
 d. LLQ and RLQ
 e. LUQ
 f. all four quadrants
 g. LUQ
 h. mainly LUQ

Regions

5. a right inguinal (iliac)
 b. epigastric
 c. left inguinal (iliac)
 d. hypogastric (pubic)
 e. left hypochondriac
 f. epigastric

Exercise 3

Lab Activity 3 Using the Microscope

12. When you move the slide to the left, the object as seen through the ocular lens moves to the right.
 When you move the slide to the right, the object as seen through the ocular lens moves to the left.
 When you move the slide forward (away from you), the object as seen through the ocular lens moves backward (toward you).
 When you move the slide backward (toward you), the object as seen through the ocular lens moves forward (away from you).

Exercise 4

Lab Activity 2 Cell Specialization

2. a. long, cylindrical; many nuclei
 b. columnar; cilia, nucleus
 c. irregular shape with processes that extend from cell body; nucleus
 d. oval with a flagellum and is tapered for movement; nucleus and flagellum
 e. biconcave; no nucleus
 f. round; nucleus
 g. columnar with microvilli; nucleus

Figure 4.4

a. interphase
b. prophase
c. metaphase
d. anaphase
e. telephase

Exercise 5

Figure 5.2

1. hypotonic solution
2. hypertonic solution

Lab Activity 4 Experiment: Osmosis in Living RBCs

1. 0.9% saline solution—isotonic; 5% saline solution—hypertonic; distilled water—hypotonic

Figure 5.3

(a) shape: normal: solution: isotonic
(b) shape: swollen: solution: hypotonic
(c) shape: crenated: solution: hypertonic

Exercise 6

Figure 6.3

1. cytoplasm
2. nucleus
3. plasma membrane

Figure 6.4

1. nucleus of simple squamous epithelial cell in visceral layer of pleura
2. simple squamous epithelium
3. connective tissue
4. nucleus of simple squamous epithelial cell in alveolar wall

Figure 6.5

1. simple cuboidal epithelium
2. apical surface of simple cuboidal epithelial cell
3. lumen of kidney tubule
4. nucleus of simple cuboidal epithelial cell

Figure 6.6

1. microvilli on apical surface of simple columnar epithelial cell
2. nucleus of simple columnar epithelial cell
3. simple columnar epithelium
4. connective tissue

Figure 6.7

1. nucleus of squamous epithelial cell
2. stratified squamous epithelium
3. nucleus of epithelial cell in basal layer of epithelium
4. connective tissue

Figure 6.8

1. nucleus of transitional epithelial cell in apical layer

2. nucleus of transitional cell in basal layer
3. transitional epithelium
4. connective tissue

Figure 6.9

1. nucleus of ciliated columnar epithelial cell
2. cilia on apical surface of columnar epithelial cell
3. pseudostratified ciliated columnar epithelium
4. connective tissue

Figure 6.11

1. collagen fiber
2. elastic fiber
3. connective tissue cells

Figure 6.12

1. reticular fiber
2. reticulocyte

Figure 6.13

1. nucleus of adipocyte
2. lipid storage area

Figure 6.14

1. fibroblast
2. collagen fiber bundle

Figure 6.15

1. fibroblast
2. collagen fiber bundles running in different directions
3. parallel collagen fiber bundles

Figure 6.16

1. fibroblast
2. individual elastic fiber
3. bundle of elastic fibers

Figure 6.17

1. extracellular matrix
2. lacuna
3. nucleus of chondrocyte

Figure 6.18

1. lacuna
2. nucleus of chondrocyte
3. elastic fibers

Figure 6.19

1. collagen fibers
2. lacuna
3. chondrocytes

Figure 6.20

1. lamella
2. canaliculus
3. lacuna
4. central canal

Figure 6.21

1. trabecula
2. osteocyte

Figure 6.22

1. red blood cells
2. nucleus of white blood cell
3. platelet

Figure 6.24

1. width of individual muscle fiber
2. nucleus
3. striation

Figure 6.25

1. width of cardiac muscle fiber
2. nucleus
3. branches of cardiac muscle fiber
4. intercalated discs

Figure 6.26

1. nucleus of smooth muscle fiber in cross-section
2. nucleus of smooth muscle fiber in longitudinal section

Figure 6.28

1. processes
2. cell body of multipolar neuron
3. nucleus

Exercise 7

Figure 7.1

1. epidermis
2. papillary layer of dermis
3. reticular layer of dermis
4. hypodermis
5. dermal papillae

Figure 7.2

1. dermal papillae
2. epidermis
3. papillary layer of dermis
4. reticular layer of dermis
5. hypodermis

Figure 7.3

1. stratum corneum
2. stratum lucidum
3. stratum granulosum

4. stratum spinosum
5. stratum basale
6. dermis

Figure 7.4

1. hair shaft
2. hair root
3. sebaceous gland
4. arrector pili muscle
5. hair follicle
6. hair bulb
7. eccrine sweat gland
8. papilla of hair
9. apocrine sweat gland

Figure 7.5

1. sebaceous gland
2. hair follicle
3. hair root
4. hair bulbs
5. papilla of hair

Figure 7.6(a & b)

1. free edge
2. nail body
3. lunula
4. eponychium
5. nail matrix
6. nail root
7. eponychium
8. lunula
9. nail body
10. free edge
11. hyponychium

Exercise 8

Figure 8.1

1. proximal epiphysis
2. diaphysis
3. distal epiphysis
4. articular cartilage
5. epiphyseal line
6. compact bone
7. medullary cavity
8. yellow marrow
9. periosteum
10. nutrient artery
11. endosteum
12. spongy bone

Figure 8.2(a & b)

1. spongy bone
2. trabeculae of spongy bone covered with endosteum
3. concentric lamellae
4. blood vessels

5. canaliculus
6. lacuna
7. osteocyte
8. periosteum
9. central canal
10. perforating canal
11. compact bone
12. trabeculae covered with endosteum
13. osteocyte in lacuna
14. interstitial lamellae

Figure 8.3(a, b, & c)

1. proximal epiphysis
2. spongy bone
3. metaphysis
4. compact bone
5. medullary cavity
6. concentric lamella
7. lacuna
8. central canal
9. canaliculi
10. osteocyte
11. trabecula

Exercise 9

Figure 9.1

1. parietal bone
2. coronal suture
3. squamous suture
4. temporal bone
5. lambdoid suture
6. occipital bone
7. frontal bone
8. sphenoid bone
9. ethmoid bone
10. lacrimal bone
11. nasal bone
12. zygomatic bone
13. maxilla
14. mandible

Figure 9.2

1. frontal bone
2. coronal suture
3. parietal bone
4. sagittal suture
5. lambdoid suture
6. occipital bone

Figure 9.3

1. maxilla
2. vomer
3. sphenoid bone
4. occipital bone
5. zygomatic bone
6. palatine bone
7. temporal bone

Figure 9.4

1. parietal bone
2. occipital bone
3. frontal bone
4. ethmoid bone
5. sphenoid bone
6. temporal bone
7. lambdoid suture

Figure 9.5

1. parietal bone
2. sphenoid bone
3. ethmoid bone
4. lacrimal bone
5. inferior nasal concha
6. vomer
7. frontal bone
8. temporal bone
9. nasal bone
10. zygomatic bone
11. maxilla
12. mandible

Figure 9.6

1. frontal bone
2. sphenoid bone
3. zygomatic bone
4. ethmoid bone
5. lacrimal bone
6. maxilla

Figure 9.7

1. supraorbital foramen
2. orbit of eye
3. inferior orbital fissure
4. perpendicular plate of ethmoid
5. mental foramen of mandible
6. supraorbital margin
7. superior orbital fissure
8. middle nasal concha

Figure 9.8

1. external auditory meatus
2. mastoid process
3. lacrimal fossa in lacrimal bone
4. zygomatic process of temporal bone
5. condylar process
6. coronoid process
7. ramus of mandible
8. body of mandible

Figure 9.9

1. palatine process of maxilla
2. palatine bone
3. hard palate
4. pterygoid processes
5. foramen ovale
6. jugular foramen
7. occipital condyle

8. hypoglossal foramen
9. mandibular fossa
10. foramen lacerum
11. carotid foramen (canal)
12. stylomastoid foramen
13. mastoid process
14. foramen magnum
15. external occipital protuberance

Figure 9.10

1. crista galli
2. olfactory foramina
3. cribriform plate
4. sella turcica
5. foramen ovale
6. internal auditory meatus
7. foramen magnum
8. lesser wing of sphenoid
9. optic foramen
10. greater wing of sphenoid
11. foramen rotundum
12. foramen lacerum
13. jugular foramen

Figure 9.11

1. frontal sinus
2. ethmoidal sinus
3. sphenoidal sinus
4. maxillary sinus

Figure 9.12

1. perpendicular plate of ethmoid
2. septal cartilage
3. vomer

Figure 9.13(b & c)

1. anterolateral (sphenoidal) fontanel
2. posterolateral (mastoid) fontanel
3. anterior (frontal) fontanel
4. posterior (occipital) fontanel

Figure 9.15

1. cervical vertebrae
2. thoracic vertebrae
3. lumbar vertebrae
4. sacrum
5. coccyx
6. cervical curve
7. thoracic curve
8. lumbar curve
9. sacral curve

Figure 9.17(b, c, d)

1. superior articular facet
2. transverse foramen
3. transverse process
4. lamina
5. dens (odontoid process)
6. spinous process

7. body
8. transverse process
9. bifurcated spinous process
10. pedicle

Figure 9.18(a & b)

1. transverse process
2. facet for articular part of tubercle
3. superior articular facet
4. superior demifacet
5. superior articular facet
6. facet for articular part of tubercle
7. slanted spinous process
8. superior demifacet
9. inferior demifacet

Figure 9.19(a & b)

1. pedicle
2. superior articular process
3. transverse process
4. vertebral foramen
5. body
6. hatchet-shaped spinous process
7. inferior articular facet

Figure 9.20(a & b)

1. intervertebral foramen
2. nucleus pulposus
3. annulus fibrosus
4. intervertebral disc
5. herniation
6. nucleus pulposus
7. annulus fibrosus

Figure 9.21(a & b)

1. sacral ala
2. base of sacrum
3. sacral promontory
4. sacral foramen
5. coccyx
6. sacral canal
7. superior articular facet
8. auricular surface (for sacroiliac joint)
9. sacral hiatus

Figure 9.22

a. scoliosis
b. kyphosis
c. lordosis

Figure 9.23

1. suprasternal notch (jugular notch)
2. manubrium
3. body of sternum
4. xiphoid process
5. sternum
6. costal cartilage
7. sternal angle

8. true ribs
9. floating ribs
10. false ribs

Figure 9.24(a & b)

1. facet for articular part of tubercle
2. tubercle
3. articular part of tubercle
4. head of rib
5. tubercle
6. intervertebral foramen
7. inferior demifacet of vertebra
8. head of rib
9. superior demifacet of vertebra
10. body of rib

Exercise 10

Figure 10.2(a, b, & c)

1. acromial end
2. sternal end
3. acromion (acromial process)
4. coracoid process
5. glenoid cavity or fossa
6. lateral (axillary) border
7. subscapular fossa
8. (medial) vertebral border
9. supraspinous fossa
10. infraspinous fossa
11. acromion (acromial process)
12. spine of scapula
13. glenoid cavity or fossa

Figure 10.3(a & b)

1. greater tubercle
2. intertubercular sulcus (groove)
3. lesser tubercle
4. lateral epicondyle
5. capitulum
6. head
7. anatomical neck
8. deltoid tuberosity
9. coronoid fossa
10. medial epicondyle
11. trochlea
12. olecranon fossa
13. medial epicondyle
14. lateral epicondyle

Figure 10.4(a & b)

1. head of radius
2. radial tuberosity
3. styloid process of radius
4. olecranon (process)
5. trochlear notch (semilunar)
6. coronoid process
7. radial notch
8. styloid process of ulna

9. ulna
10. radius

Figure 10.5

1. carpals
2. metacarpals
3. proximal phalanx V
4. middle phalanx V
5. distal phalanx V

Figure 10.6

1. ilium
2. posterior superior iliac spine
3. posterior inferior iliac spine
4. greater sciatic notch
5. ischium
6. ischial spine
7. lesser sciatic notch
8. obturator foramen
9. ischial tuberosity
10. iliac crest
11. anterior superior iliac spine
12. anterior inferior iliac spine
13. acetabulum
14. pubis

Figure 10.7(a & b)

1. iliac crest
2. ilium
3. ischial spine
4. pelvic brim
5. pubic symphysis
6. false pelvis
7. true pelvis
8. pubis
9. ischial spine
10. false pelvis
11. sacroiliac joint
12. sacrum
13. coccyx
14. true pelvis
15. pubis

Figure 10.8(a & b)

1. head of femur
2. greater trochanter
3. neck
4. lesser trochanter
5. medial epicondyle
6. medial condyle
7. linea aspera
8. lateral epicondyle
9. lateral condyle

Figure 10.9

1. lateral condyle
2. head of fibula
3. fibula
4. lateral malleolus

5. medial condyle
6. tibial tuberosity
7. tibia
8. medial malleolus

Figure 10.10(a & b)

1. calcaneus
2. talus
3. proximal phalanx II
4. middle phalanx II
5. distal phalanx II
6. tarsals
7. metatarsals
8. phalanges
9. calcaneus
10. tibia
11. lateral malleolus of fibula
12. talus
13. tarsals
14. metatarsals
15. phalanges

Exercise 11

Figure 11.1(a & b)

1. synovial and diarthrosis
2. synovial and diarthrosis
3. cartilaginous and amphiarthrosis
4. synovial and diarthrosis
5. fibrous and amphiarthrosis
6. fibrous and synarthrosis
7. cartilaginous and amphiarthrosis
8. cartilaginous and amphiarthrosis
9. synovial and diarthrosis
10. synovial and diarthrosis
11. fibrous and synarthrosis

Figure 11.2(b)

1. articular bone
2. fibrous membrane
3. synovial membrane
4. articular capsule
5. synovial cavity (contains synovial fluid)
6. articular cartilage

Figure 11.3(a)

1. articular cartilage of femur
2. lateral (fibular) collateral ligament
3. lateral meniscus
4. posterior cruciate ligament
5. anterior cruciate ligament
6. medial meniscus
7. medial (tibial) collateral ligament

Figure 11.4(a, b, c, d, e, & f)

1. flexion
2. extension

3. hyperextension
4. hyperextension
5. extension
6. flexion
7. flexion
8. extension
9. flexion
10. extension
11. hyperextension
12. hyperextension
13. extension
14. flexion
15. flexion
16. extension

Figure 11.5(a, b, c, d, e, & f)

1. abduction
2. adduction
3. circumduction
4. abduction
5. adduction
6. circumduction
7. adduction
8. abduction
9. abduction
10. adduction

Figure 11.6(a & b)

1. rotation
2. lateral rotation
3. medial rotation

Figure 11.7(a, b, c, d, & e)

1. elevation
2. depression
3. protraction
4. retraction
5. pronation
6. supination

Figure 11.8(a & b)

1. inversion
2. eversion
3. dorsiflexion
4. plantar flexion

Exercise 12

Figure 12.1(a & b)

1. perimysium
2. epimysium
3. fascicle
4. endomysium
5. muscle fiber
6. myofibril
7. perimysium
8. endomysium
9. muscle fiber
10. sarcolemma
11. myofibril
12. filament

Figure 12.2(a & b)

1. endomysium
2. skeletal muscle fiber in cross-section
3. fascicle
4. perimysium
5. width of skeletal muscle fiber
6. nucleus
7. striation

Figure 12.3(a, b, & c)

1. triad
2. sarcolemma
3. myofibril
4. T-tubule
5. terminal ciserna of sarcoplasmic reticulum
6. sarcoplasmic reticulum
7. sarcomere
8. thin filament
9. thick filament
10. actin
11. tropomyosin
12. troponin
13. myosin heads
14. myosin tails

Figure 12.5

1. Z disc
2. M line
3. H zone
4. I band
5. A band

Figure 12.7

1. axon terminal
2. synaptic end bulb
3. synaptic vesicle with acetylcholine
4. synaptic cleft
5. motor end plate
6. receptors

Figure 12.8

1. skeletal muscle fibers
2. axon terminal with synaptic end bulbs
3. motor neuron axon

Exercise 13

Figure 13.1

1. slow oxidative fiber
2. fast glycolytic fiber
3. fast oxidative-glycolytic fiber

Lab Activity 4 Isotonic and Isometric Contractions

1. Isotonic contraction occurs when flexing the knee. Isometric contraction occurs when maintaining the knee in a flexed position.
2. Concentric. Extend the leg slowly to original position.

Lab Activity 5 Record Frog Gastrocnemius Muscle Contraction

1. Force = 1.5 g; latent period = 10 msec; contraction period = 25 msec; relaxation period = 45 msec

Exercise 14

Figure 14.1(a, b, & c)

1. orbicularis oculi
2. orbicularis oris
3. platysma
4. frontalis
5. zygomaticus minor
6. zygomaticus major
7. occipitalis
8. frontalis
9. orbicularis oculi
10. zygomaticus major
11. buccinator
12. orbicularis oris
13. masseter
14. temporalis
15. lateral pterygoid
16. medial pterygoid
17. orbicularis oris
18. buccinator

Figure 14.2(a, b, & c)

1. splenius capitis
2. sternocleidomastoid
3. trapezius
4. levator scapulae
5. scalene
6. digastric, anterior belly
7. mylohyoid
8. stylohyoid
9. digastric, posterior belly
10. omohyoid
11. sternocleidomastoid
12. sternohyoid
13. mylohyoid
14. scalenes
15. levator scapulae

Figure 14.3(a & b)

1. deltoid
2. pectoralis major

3. pectoralis minor
4. serratus anterior
5. external oblique
6. rectus abdominis

Figure 14.4(a & b)

1. trapezius
2. deltoid
3. latissimus dorsi
4. levator scapulae
5. rhomboid minor
6. supraspinatus
7. infraspinatus
8. teres minor
9. rhomboid major
10. teres major

Figure 14.5(a & b)

1. biceps brachii
2. subscapularis
3. teres major
4. brachialis
5. triceps brachii, long head
6. triceps brachii, lateral head
7. triceps brachii, medial head

Figure 14.6(a, b, c, & d)

1. pronator teres
2. flexor carpi radialis
3. brachioradialis
4. palmaris longus
5. flexor carpi ulnaris
6. supinator
7. flexor digitorum profundus
8. flexor pollicis longus
9. extensor carpi radialis longus
10. flexor carpi ulnaris
11. extensor carpi radialis brevis
12. extensor carpi ulnaris
13. extensor digiti minimi
14. extensor digitorum
15. supinator
16. extensor pollicis longus

Figure 14.7(a, b, & e)

1. iliacus
2. psoas major
3. tensor fasciae latae
4. rectus femoris (cut)
5. vastus intermedius
6. vastus lateralis
7. pectineus
8. adductor longus
9. gracilis
10. sartorius
11. vastus medialis
12. adductor brevis
13. gracilis
14. adductor magnus

15. gluteus medius
16. piriformis
17. biceps femoris
18. gluteus maximus
19. semitendinosus
20. semimembranosus

Figure 14.8(a, b, c, & d)

1. fibularis (peroneus) longus
2. tibialis anterior
3. extensor digitorum longus
4. fibularis (peroneus) brevis
5. gastrocnemius
6. soleus
7. fibularis (peroneus) longus
8. extensor digitorum longus
9. tibialis anterior
10. gastrocnemius
11. soleus
12. flexor hallucis longus
13. flexor digitorum longus

Exercise 15

Figure 15.1(a & b)

1. frontalis muscle
2. zygomaticus major muscle
3. mental protuberance
4. supraorbital margin
5. nasal bone
6. body of mandible
7. temporalis muscle
8. occipitalis muscle
9. external occipital protuberance
10. mastoid process
11. ramus of mandible
12. temporomandibular joint
13. masseter muscle covered by parotid gland

Figure 15.1(c)

1. common carotid artery
2. sternocleidomastoid muscle
3. clavicle
4. hyoid bone
5. thyroid cartilage
6. cricoid cartilage
7. suprasternal (jugular) notch

Figure 15.1(d)

1. sternocleidomastoid
2. levator scapulae
3. trapezius
4. scalenes

Figure 15.2(a)

1. second rib (deep to muscle)
2. body of sternum

3. xiphisternal joint
4. xiphoid process of sternum
5. serratus anterior muscle
6. manubrium of sternum
7. sternal angle
8. pectoralis major muscle
9. ribs (deep to muscle)
10. costal margin

Figure 15.2(b)

1. rectus abdominis muscle
2. linea semilunaris
3. external oblique muscle
4. iliac crest
5. McBurney's point
6. linea alba
7. tendinous intersection
8. umbilibus
9. anterior superior iliac spine

Figure 15.2(c)

1. tendinous insertion
2. McBurney's point
3. serratus anterior muscle
4. rectus abdominis muscle
5. external oblique muscle
6. iliac crest
7. anterior superior iliac spine

Figure 15.2(d)

1. trapezius muscle
2. acromion
3. supraspinatus
4. infraspinatus muscle
5. teres major muscle
6. erector spinae muscle
7. spinous processes of thoracic vertebrae
8. vertebral border of scapula

Figure 15.2(e)

1. deltoid muscle, lateral fibers
2. deltoid muscle, posterior fibers
3. triangle of ascultation
4. spinous process of vertebra
5. trapezius muscle
6. teres major muscle
7. vertebral border of scapula
8. latissimus dorsi
9. spinous process of lumbar vertebra

Figure 15.2(f)

1. iliac crest
2. posterior sperior iliac spine
3. coccyx
4. greater trochanter
5. sacrum
6. gluteus medius muscle
7. gluteus maximus muscle
8. ischial tuberosity

Figure 15.2(g)

1. sacrum
2. greater trochanter of the femur
3. coccyx
4. gluteal cleft
5. iliac crest
6. posterior superior iliac spine
7. gluteus maximus muscle

Figure 15.3(a)

1. acromion
2. spine of scapula
3. deltoid muscle (lateral fibers)
4. acromioclavicular joint
5. clavicle
6. greater tubercle of humerus

Figure 15.3(b)

1. acromion
2. deltoid muscle (lateral fibers)
3. biceps brachii muscle
4. triceps brachii muscle
5. lateral epicondyle of humerus

Figure 15.3(c)

1. biceps brachii muscle
2. tendon of biceps brachii muscle
3. olecranon of ulna
4. medial epicondyle of humerus
5. triceps brachii muscle
6. groove for brachial artery

Figure 15.3(d)

1. median cubital vein
2. biceps brachii muscle
3. cephalic vein
4. brachioradialis muscle
5. cubital fossa
6. basilic vein

Figure 15.4(a)

1. sartorius muscle
2. adductor longus muscle
3. gracilis muscle
4. rectus femoris muscle
5. vastus lateralis muscle
6. vastus medialis muscle
7. lateral condyle of femur
8. lateral condyle of tibia
9. patellar ligament
10. medial condyle of femur
11. patella
12. medial condyle of tibia
13. tibial tuberosity
14. tibialis anterior muscle
15. anterior border of tibia (shin)
16. fibularis (peroneus) longus muscle

17. medial malleolus of tibia
18. lateral malleolus of fibula
19. tendon of extensor hallucis longus muscle
20. tendons of extensor digitorum longus muscle

Figure 15.4(b)

1. biceps femoris tendon
2. gastrocnemius muscle (medial and lateral heads)
3. soleus muscle
4. calcaneus
5. tendon of semitendinosus
6. biceps femoris muscle
7. semitendinosus and semimembranosus muscles

Exercise 16

Table 16.1

1. Schwann cells - PNS
2. ependymal cells - CNS
3. microglia - CNS
4. astrocytes - CNS
5. satellite cells - PNS
6. oligodendrocytes - CNS

Figure 16.1(a & b)

1. oligodendrocyte
2. microglial cell
3. astrocytes
4. ependymal cell
5. satellite cell
6. Schwann cell

Figure 16.2(a & b)

1. processes
2. cell body
3. dendrites
4. axon
5. axon hillock
6. cell body
7. nucleus

Figure 16.3

1. axon terminals
2. Schwann cell
3. node of Ranvier
4. axon collateral
5. axon or soma
6. dendrites
7. cell body
8. axon hillock
9. trigger zone (initial segment)
10. myelin sheath

Figure 16.4(a & b)

1. process
2. satellite cells
3. nucleus
4. neuron cell body
5. axon
6. dendrites
7. neuron cell body

Figure 16.5(a & b)

1. unipolar
2. multipolar
3. bipolar
4. sensory neuron
5. interneuron
6. motor neuron

Figure 16.6(a & b)

(a) myelinated
 1. axon
 2. myelin sheath
 3. Schwann cell cytoplasm
 4. node of Ranvier
(b) unmyelinated
 5. axons
 6. Schwann cell cytoplasm

Figure 16.7

1. neurolemma
2. myelin sheath
3. node of Ranvier
4. axon

Figure 16.8

1. white matter in brain and spinal cord
2. gray matter in brain and spinal cord

Before Going to Lab Synapses

1. axoaxonic
2. axodendritic
3. axosomatic
4. dendrites on postsynaptic neurons receive neurotransmitter at synapse
5. graded potential sent toward trigger zone
6. action potential started on postsynaptic neuron
7. action potential travels down axon
8. action potential reaches axon terminal

Exercise 17

Figure 17.1

1. dura mater
2. pia mater

3. epidural space
4. subarachnoid space
5. web-like projection of arachnoid mater

Figure 17.2(a)

1. cervical enlargement
2. lumbar enlargement
3. conus medullaris
4. cauda equina
5. filum terminale

Figure 17.3

1. spinal nerve
2. posterior (dorsal) root ganglion
3. posterior (dorsal) root
4. posterior median sulcus
5. central canal
6. anterior (ventral) root
7. anterior median fissure

Figure 17.4

1. anterior white column
2. anterior gray horn
3. lateral white column
4. lateral gray horn
5. posterior gray horn
6. posterior white column
7. gray commissure

Figure 17.5

1. posterior (dorsal) root
2. lateral white column
3. central canal
4. posterior median sulcus
5. posterior white column
6. posterior gray horn
7. posterior (dorsal) root ganglion
8. anterior (ventral) root
9. anterior gray horn
10. anterior white column
11. gray commissure
12. anterior median fissure

Exercise 18

Figure 18.1

1. spinal nerve
2. epineurium
3. fascicle
4. perineurium
5. myelin sheath
6. axon
7. endoneurium

Figure 18.2

1. perineurium
2. axon

3. myelin sheath
4. endoneurium

Figure 18.3(a & b)

1. spinal nerve
2. anterior (ventral) ramus
3. posterior (dorsal) ramus
4. rami communicantes
5. sympathetic ganglion
6. intervertebral foramen
7. anterior (ventral) ramus
8. posterior (dorsal) ramus
9. rami communicantes

Figure 18.4

1. sacral plexus
2. lumbar plexus
3. brachial plexus
4. cervical plexus
5. cervical nerves
6. thoracic nerves
7. lumbar nerves
8. sacral nerves
9. coccygeal nerve

Figure 18.5

1. axillary nerve
2. musculocutaneous nerve
3. median nerve
4. ulnar nerve
5. radial nerve
6. phrenic nerve

Figure 18.6

1. femoral nerve
2. pudendal nerve
3. obturator nerve
4. sciatic nerve
5. tibial nerve
6. common fibular (peroneal) nerve

Exercise 19

Table 19.1

• somatic, spinal nerve reflex
• autonomic, cranial nerve reflex
• autonomic, cranial nerve reflex
• somatic, spinal nerve reflex

Figure 19.1

1. integrating center
2. sensory neuron axon
3. sensory neuron cell body
4. sensory receptor
5. effector
6. motor neuron axon
7. motor neuron cell body

Figure 19.2

1. receptor
2. sensory neuron
3. integrating center for reciprocal innervation
4. integrating center for patellar reflex arc
5. motor neuron for patellar reflex arc
6. motor neuron for reciprocal innervation
7. effector for patellar reflex arc
8. effector for reciprocal innervation

Exercise 20

Figure 20.1

1. cerebrum
2. cerebellum
3. diencephalon
4. brain stem

Figure 20.2(a & b)

1. corpora quadrigemina
2. cerebral peduncle
3. pons
4. medulla oblongata
5. midbrain
6. midbrain
7. pons
8. medulla oblongata
9. spinal cord

Figure 20.3(a & b)

1. cerebellar hemispheres
2. vermis
3. folia
4. superior colliculus
5. inferior colliculus
6. arbor vitae (white matter)
7. cerebellar cortex (gray matter)
8. pons
9. medulla oblongata

Figure 20.4(a, b, & c)

1. thalamus
2. hypothalamus
3. pineal gland (part of epithalamus)
4. diencephalon
5. mammillary body
6. pineal gland
7. thalamus
8. intermediate mass of thalamus
9. hypothalamus
10. infundibulum
11. pituitary gland
12. lateral ventricles
13. thalamus
14. third ventricles
15. hypothalamus

Figure 20.5(b)

1. internal capsule (projection fibers)
2. cerebral cortex (gray matter)
3. cerebral white matter
4. corpus callosum (commissural fibers)
5. fornix (association fibers)
6. basal nuclei (gray matter)

Figure 20.6

1. central sulcus
2. postcentral gyrus
3. parietal lobe
4. occipital lobe
5. transverse fissure
6. precentral gyrus
7. frontal lobe
8. insula
9. temporal lobe (cut)

Figure 20.7

1. Broca's speech area
2. primary gustatory area
3. primary motor area
4. central sulcus
5. primary somatosensory area
6. primary visual area
7. Wernicke's area
8. primary auditory area

Figure 20.9(a & b)

1. subarachnoid space
2. arachnoid villus
3. falx cerebri
4. white matter
5. superior sagittal sinus
6. parietal bone
7. dura mater
8. arachnoid mater
9. pia mater
10. cerebral cortex

Figure 20.10(a, b, & c)

1. lateral ventricle
2. choroid plexus
3. cerebral aqueduct
4. fourth ventricle
5. arachnoid villus
6. third ventricle
7. subarachnoid space
8. lateral ventricles
9. interventricular foramen
10. third ventricle
11. cerebral aqueduct
12. fourth ventricle
13. central canal of spinal cord
14. superior sagittal sinus
15. arachnoid villus

16. subarachnoid space
17. lateral ventricle
18. choroid plexus
19. third ventricle
20. cerebral aqueduct
21. fourth ventricle
22. central canal

Exercise 21

Figure 21.1

1. olfactory
2. optic
3. oculomotor
4. trochlear
5. trigeminal
6. abducens
7. facial
8. vestibulocochlear
9. glossopharyngeal
10. vagus
11. accessory
12. hypoglossal

Figure 21.2

1. trigeminal (mandibular branch)
2. trigeminal (maxillary branch)
3. hypoglossal
4. vestibulocochlear
5. glossopharyngeal, vagus, accessory
6. olfactory
7. optic
8. facial
9. oculomotor, trochlear, abducens, trigeminal (ophthalmic branch)

Exercise 22

Figure 22.1

1. right sympathetic trunk ganglia
2. splanchnic nerves
3. left sympathetic trunk ganglia
4. prevertebral ganglia

Figure 22.2

1. otic ganglion
2. ciliary ganglion
3. pterygopalatine ganglion
4. submandibular ganglion

Figure 22.3

1. interneuron
2. axon of sensory neuron (afferent)
3. sensory receptor
4. axon of preganglionic motor neuron
5. autonomic ganglion
6. axon of postganglionic motor neuron
7. visceral effector

Exercise 23

Figure 23.1(a & b)

1. type I cutaneous mechanoreceptor
2. corpuscle of touch (Meissner's)
3. type II cutaneous mechanoreceptor
4. hair root plexus
5. laminated corpuscle (Pacinian)
6. muscle spindle
7. tendon organ

Figure 23.2(a & b)

1. third-order neuron
2. second-order neuron
3. first-order neuron
4. third-order neuron
5. second-order neuron
6. first-order neuron

Exercise 24

Figure 24.1(a & b)

1. conjunctival fold
2. palpebral conjunctiva
3. bulbar conjunctiva
4. lacrimal gland
5. lacrimal canals
6. lacrimal sac
7. nasolacrimal duct

Figure 24.2(a & b)

1. superior oblique
2. superior rectus
3. lateral rectus
4. medial rectus
5. inferior oblique
6. inferior rectus

Table 24.1

Inferior oblique—moves eye superiorly and laterally
Inferior rectus—moves eye inferiorly
Lateral rectus—moves eye laterally
Medial rectus—moves eye medially
Superior oblique—moves eye inferiorly and laterally
Superior rectus—moves eye superiorly

Figure 24.3

1. scleral venous sinus
2. ciliary muscle
3. ciliary process
4. ciliary body
5. choroid
6. sclera
7. retina

8. cornea
9. pupil
10. iris
11. suspensory ligament (zonular fibers)
12. ora serrata

Figure 24.4

1. anterior chamber
2. posterior chamber
3. anterior cavity
4. scleral venous sinus
5. vitreous chamber
6. lens

Figure 24.6

1. macula lutea
2. central fovea
3. blood vessel
4. optic disc

Figure 24.7(a & b)

1. optic disc
2. central retinal vein
3. optic nerve
4. central retinal artery
5. central fovea
6. cornea
7. pupil
8. iris
9. ciliary body
10. lens
11. choroid
12. sclera
13. retina
14. optic disc
15. optic nerve

Figure 24.8(a & b)

1. optic nerve fibers (axons)
2. ganglion cell layer
3. bipolar cell layer
4. photoreceptor layer
5. neural portion of retina
6. pigmented epithelium of retina
7. ganglion cell layer
8. bipolar cell layer
9. photoreceptor layer
10. neural portion of retina
11. pigmented epithelium of retina

Figure 24.13

1. lobule
2. auricle
3. helix
4. external ear
5. malleus
6. incus

7. stapes attached to oval window
8. middle ear
9. internal ear
10. auditory tube
11. tympanic membrane
12. external auditory canal

Figure 24.14

1. anterior semicircular canal
2. posterior semicircular canal
3. lateral semicircular canal
4. ampulla of semicircular canal and duct
5. utricle
6. saccule
7. oval window
8. membranous semicircular duct
9. vestibule
10. round window
11. cochlea
12. cochlear duct

Figure 24.15(a & b)

1. scala vestibuli
2. scala tympani
3. vestibular membrane
4. basilar membrane
5. cochlear duct
6. spiral organ of Corti
7. vestibular membrane
8. tectorial membrane
9. basilar membrane
10. hair cells
11. cochlear duct

Figure 24.16(a & b)

1. otoliths
2. otolithic membrane
3. stereocilia in hair bundle
4. hair cell
5. vestibular branches of vestibulocochlear nerve
6. cupula
7. hair bundle
8. hair cell

Figure 24.17(a & b)

1. olfactory bulb
2. cribriform plate
3. olfactory tract
4. olfactory nerve
5. olfactory receptor cell
6. olfactory hair

Figure 24.18(a, b, & c)

1. vallate (circumvallate) papilla
2. fungiform papilla
3. foliate papilla
4. filiform papilla
5. vallate papillae

6. filiform papillae
7. fungiform papillae
8. taste bud
9. taste pore
10. gustatory hairs
11. gustatory receptor cell
12. sensory axons

Exercise 25

Figure 25.1

1. pineal gland
2. parathyroid glands
3. adrenal glands
4. pancreas
5. hypothalamus
6. pituitary gland or hypophysis
7. thyroid gland
8. thymus
9. ovaries
10. testes

Figure 25.2(a, b, & c)

1. posterior pituitary or neurohypophysis
2. hypothalamus
3. infundibulum
4. anterior pituitary or adenohypophysis
5. hypothalamus
6. infundibulum
7. anterior pituitary
8. posterior pituitary
9. posterior pituitary
10. anterior pituitary

Figure 25.3(a & b)

1. right lobe of thyroid gland
2. thyroid cartilage of larynx
3. isthmus of thyroid gland
4. trachea
5. left lobe of thyroid gland
6. left lobe of thyroid gland
7. left parathyroid glands
8. right parathyroid glands
9. trachea
10. right lobe of thyroid gland
11. isthmus of thyroid gland

Figure 25.4(a & b)

1. parathyroid gland
2. thyroid gland
3. follicular cell
4. parafollicular cell or C cell
5. colloid-filled follicle

Figure 25.6(a, b, & c)

1. right adrenal gland
2. left adrenal gland
3. kidney

4. capsule
5. adrenal cortex
6. adrenal medulla
7. capsule
8. zona glomerulosa
9. zona fasciculata
10. zona reticularis
11. adrenal medulla
12. adrenal cortex

Figure 25.8(a & c)

1. head of pancreas
2. body of pancreas
3. tail of pancreas
4. exocrine or acini cells
5. pancreatic islet

Table 25.1
Anterior pituitary

1. human growth hormone
2. thyroid-stimulating hormone
3. follicle-stimulating hormone
4. luteinizing hormone
5. prolactin
6. adrenocorticotropic hormone
7. melanocyte-stimulating hormone

Posterior pituitary

1. anti-diuretic hormone
2. oxytocin

Thyroid gland

1. thyroxine
2. triiodothyronine
3. calcitonin

Parathyroid glands

parathyroid hormone

Adrenal cortex

1. aldosterone
2. cortisol
3. androgens

Adrenal medulla

1. epinephrine
2. norepinephrine

Pancreas

1. insulin
2. glucagon

Ovaries

1. estrogen
2. progesterone

Testes

testosterone

Pineal Gland

melatonin

Thymus

thymosin

Exercise 26

TABLE 26.7

Summary of ABO Blood Group Interactions

Antigen on RBCs: A, B, A and B; neither A nor B

Antibody in plasma: anti-B, anti-A, none, anti-A and anti-B

Compatible donor: A, O; B, O; A, B, AB, O; O

Incompatible donor: B, AB; A, AB; none; A; B; AB

Exercise 27

Figure 27.2(a & b)

1. right auricle
2. coronary sulcus
3. right ventricle
4. left auricle
5. anterior interventricular sulcus
6. left ventricle
7. apex of heart
8. left auricle
9. coronary sulcus
10. left ventricle
11. adipose tissue
12. posterior interventricular sulcus
13. right auricle
14. right ventricle

Figure 27.3(b & c)

1. superior vena cava
2. right pulmonary artery
3. ascending aorta
4. pulmonary trunk
5. right pulmonary veins
6. inferior vena cava
7. aortic arch
8. ligamentum arteriosum
9. left pulmonary artery
10. descending aorta
11. left pulmonary veins
12. aortic arch
13. ligamentum arteriosum
14. left pulmonary artery
15. left pulmonary veins
16. coronary sinus
17. superior vena cava

18. ascending aorta
19. right pulmonary artery
20. right pulmonary veins
21. inferior vena cava

Figure 27.4(b)

1. superior vena cava opening
2. pulmonary (semilunar) valve
3. right atrium
4. coronary sinus opening
5. inferior vena cava opening
6. tricuspid valve
7. right ventricle
8. trabeculae carneae
9. left pulmonary vein openings
10. left atrium
11. aortic (semilunar) valve
12. bicuspid valve (mitral)
13. chordae tendineae
14. interventricular septum
15. papillary muscle
16. left ventricle

Before Going to Lab Systemic and Pulmonary Circulation: Blood vessels with oxygen-poor blood

a. pulmonary arteries
b. pulmonary trunk
c. venae cavae

Blood vessels with oxygen-rich blood

d. aorta
e. pulmonary veins

Trace blood flow

1. right atrium
2. tricuspid valve
3. right ventricle
4. pulmonary (semilunar) valve
5. pulmonary trunk
6. pulmonary arteries
7. pulmonary capillaries
8. pulmonary veins
9. left atrium
10. bicuspid (mitral) valve
11. left ventricle
12. aortic (semilunar) valve
13. aorta
14. systemic arteries
15. systemic capillaries
16. systemic veins
17. venae cavae and coronary sinus
18. right atrium

Figure 27.6(c & d)

1. marginal branch
2. small cardiac vein
3. right coronary artery

4. left coronary artery
5. circumflex branch
6. anterior interventricular branch (LAD)
7. great cardiac vein
8. middle cardiac vein
9. coronary sinus
10. posterior interventricular branch

Figure 27.7(a & b)

1. fibrous pericardium
2. diaphragm
3. fibrous pericardium
4. parietal layer of serous pericardium
5. pericardial cavity
6. visceral layer of serous pericardium (epicardium)
7. endocardium
8. myocardium

Figure 27.9

1. cardiac muscle fiber
2. intercalated discs
3. branching cardiac fibers
4. nucleus

Lab Activity 8: Dissection of a Sheep Heart

10. These three cusps are half-moon shaped, and chordae tendinae are not associated with the pulmonary semilunar valve.
12. The bicuspid valve has 2 cusps. The bicuspid valve looks similar to the tricuspid valve but they differ in the number of cusps. The bicuspid and tricuspid valves have chordae tendinae and papillary muscles. The left ventricle has a smaller number of papillary muscles, but these papillary muscles are larger. The wall of the left ventricle is thicker to pump blood through the systemic circulation.

Exercise 28

Figure 28.1

1. Purkinje (conduction) fibers in right ventricle
2. right bundle branch
3. AV bundle (of His)
4. atrioventricular (AV) node
5. sinoatrial (SA) node
6. left bundle branch
7. Purkinje (conduction) fibers in left ventricle

Figure 28.3

1. P wave
2. P-Q interval
3. QRS complex
4. S-T segment
5. T wave
6. Q-T interval

Figure 28.4

2. (a) 3 (number of P waves in 30 squares) \times 10 = 30 beats/min; bradycardia
(b) 6 (number of P waves in 30 squares) \times 10 = 60 beats per min; NSR
(c) 12 (number of P waves in 30 squares) \times 10 = 120 beats/min; tachycardia

Exercise 29

Before Going to Lab: Blood Vessel Structure

1. Figure 29.1(b) is vasoconstricted.

Figure 29.2

1. endothelium
2. basement membrane
3. internal elastic lamina
4. tunica interna
5. smooth muscle
6. external elastic lamina
7. tunica media
8. tunica externa

Figure 29.3

1. endothelium
2. basement membrane
3. tunica interna
4. valve
5. tunica media (smooth muscle)
6. tunica externa

Figure 29.4

1. tunica media of vein
2. tunica interna of vein
3. lumen of vein
4. tunica externa of vein
5. lumen of artery
6. tunica interna of artery
7. tunica media of artery
8. tunica externa of artery

Figure 29.5

1. capillary
2. arteriole

Figure 29.6(a, b, & c)

1. intercellular cleft
2. vesicles involved in transcytosis
3. tight junctions
4. fenestrations

Exercise 30

Figure 30.1(b)

1. right common carotid
2. brachiocephalic trunk
3. ascending aorta
4. left common carotid
5. left subclavian
6. aortic arch
7. thoracic aorta

Figure 30.2(a & b)

1. internal carotid
2. vertebral
3. external carotid
4. common carotid
5. brachiocephalic trunk
6. anterior cerebral
7. middle cerebral
8. internal carotid
9. posterior communicating
10. posterior cerebral
11. basilar
12. vertebral
13. anterior communicating

Figure 30.3(b)

1. subclavian
2. axillary
3. brachial
4. radial
5. ulnar

Figure 30.4

1. superior mesenteric
2. right common iliac
3. right internal iliac
4. right external iliac
5. common hepatic
6. splenic
7. celiac trunk
8. left renal
9. abdominal aorta
10. inferior mesenteric

Figure 30.5

1. left gastric
2. celiac trunk
3. common hepatic
4. splenic

Figure 30.6(a & b)

1. common iliac
2. external iliac
3. internal iliac
4. femoral
5. popliteal
6. anterior tibial
7. posterior tibial
8. fibular (peroneal)
9. dorsal artery of the foot (dorsalis pedis)

Figure 30.7

1. superior vena cava
2. coronary sinus
3. inferior vena cava

Figure 30.8

1. vertebral
2. internal jugular
3. external jugular
4. subclavian
5. brachiocephalic

Figure 30.9(a & b)

1. basilic
2. cephalic
3. median cubital
4. subclavian
5. axillary
6. brachial veins
7. radial veins
8. ulnar veins

Figure 30.10

1. right brachiocephalic
2. left brachiocephalic
3. superior vena cava
4. azygos
5. hepatic veins
6. right suprarenal
7. right renal
8. right gonadal
9. inferior vena cava
10. right common iliac
11. right internal iliac
12. right external iliac
13. accessory hemiazygos
14. hemiazygos
15. left suprarenal
16. left renal
17. left gonadal
18. left common iliac
19. left internal iliac
20. left external iliac

Figure 30.11

1. hepatic
2. hepatic portal

3. superior mesenteric
4. splenic
5. inferior mesenteric

Figure 30.12(a & b)

1. external iliac
2. femoral
3. anterior tibial
4. small saphenous
5. great saphenous
6. femoral
7. popliteal
8. great saphenous
9. posterior tibial
10. small saphenous
11. fibular (peroneal)

Figure 30.13

1. right pulmonary artery
2. right pulmonary vein
3. left pulmonary artery
4. pulmonary trunk
5. left pulmonary vein

Figure 30.14(a & b)

1. foramen ovale
2. ductus venosus
3. umbilical vein
4. umbilical arteries
5. placenta
6. ductus arteriosus
7. ligamentum arteriosum
8. fossa ovalis
9. ligamentum venosum
10. ligamentum teres
11. medial umbilical ligaments

Exercise 31

Figure 31.1(a & b)

1. tissue cell
2. interstitial fluid
3. blood capillary
4. lymphatic capillary
5. interstitial fluid
6. tissue cell
7. lymph in lymphatic capillary

Figure 31.2(a, b, & c)

1. right lymphatic duct
2. lymphatic vessel in arm
3. thoracic duct in thorax
4. cisterna chyli
5. lymphatic vessel in leg
6. thoracic duct draining into subclavian vein
7. area drained by right lymphatic duct
8. area drained by thoracic duct

9. right jugular trunk
10. right subclavian trunk
11. right lymphatic duct
12. cisterna chyli
13. thoracic duct
14. thoracic (left lymphatic) duct in thorax

Figure 31.3(a & b)

1. pharyngeal tonsil
2. palatine tonsil
3. lingual tonsil
4. submandibular lymph node
5. cervical lymph node
6. thymus
7. intestinal lymph node
8. appendix (vermiform)
9. red bone marrow
10. axillary lymph node
11. mammary lymph node
12. thoracic lymph node
13. spleen
14. Peyer's patches (aggregated lymphatic follicles)
15. iliac lymph node
16. inguinal lymph node

Lab Activity 2: Flow of Lymph

1. blood capillaries
2. interstitial fluid
3. lymphatic capillaries
4. lymphatic vessels
5. inguinal lymph nodes
6. lymphatic vessels
7. iliac lymph nodes
8. lumbar trunk
9. cisterna chyli
10. thoracic duct
11. junction of jugular and subclavian veins
12. brachiocephalic vein
13. superior vena cava
14. heart

Figure 31.5(a, b, & c)

1. trabecular sinus
2. afferent vessels
3. valve
4. subcapsular sinus
5. cortex
6. medulla
7. medullary sinus
8. efferent vessels
9. hilum
10. capsule
11. trabecula
12. cortex
13. medulla
14. capsule
15. lymphatic nodule
16. reticulocyte
17. reticular fiber

Figure 31.7(b)

1. capsule
2. cortex of lobule
3. medulla of lobule
4. trabecula of lobule

Figure 31.8(a & b)

1. splenic artery
2. splenic vein
3. hilum
4. capsule
5. white pulp
6. trabecula

Lab Activity 6: Cells Involved in Body Defenses

1. e
2. e
3. e
4. c
5. e
6. e
7. a, b, d
8. e
9. e
10. e
11. e
12. e

Exercise 32

Figure 32.1(b)

1. pharynx
2. right primary or main bronchus
3. lungs
4. nose
5. nasal cavity
6. larynx
7. trachea

Figure 32.2(a & b)

1. internal naris
2. superior nasal concha or turbinate
3. middle nasal concha or turbinate
4. inferior nasal concha or turbinate
5. superior nasal meatus
6. middle nasal meatus
7. inferior nasal meatus
8. external naris
9. hard palate
10. middle nasal meatus
11. nasal septum
12. inferior nasal meatus
13. middle nasal concha
14. inferior nasal concha

Figure 32.3(b & c)

1. nasopharynx
2. oropharynx

3. laryngopharynx
4. pharyngeal tonsil
5. nasopharynx
6. opening of auditory tube
7. uvula
8. oropharynx
9. laryngopharynx
10. internal naris
11. soft palate
12. palatine tonsil
13. lingual tonsil

Figure 32.4(c, d, e, & f)

1. epiglottis
2. thyroid cartilage
3. cricoid cartilage
4. epiglottis
5. thyroid cartilage
6. cricoid cartilage
7. arytenoid cartilage
8. cricoid cartilage
9. epiglottis
10. ventricular fold
11. vocal fold
12. thyroid cartilage
13. ventricular fold
14. vocal fold
15. glottis

Figure 32.5(a & b)

1. trachea
2. right primary (main) bronchus
3. carina
4. right secondary (lobar) bronchus
5. diaphragm
6. larynx
7. tracheal cartilage
8. left primary (lobar) bronchus
9. left tertiary (segmental) bronchus
10. bronchiole
11. primary (main) bronchus
12. secondary (lobar) bronchus
13. tertiary (segmental) bronchus
14. terminal bronchiole
15. respiratory bronchiole
16. alveolar sac

Figure 32.6

1. respiratory bronchiole
2. alveolar ducts
3. alveolar sac
4. alveoli

Figure 32.7(a, b, c, & d)

1. apex
2. superior lobe
3. middle lobe
4. inferior lobe
5. base

6. apex
7. superior lobe
8. cardiac notch
9. inferior lobe
10. base
11. superior lobe
12. hilum
13. middle lobe
14. inferior lobe
15. superior lobe
16. inferior lobe
17. hilum
18. cardiac notch

Figure 32.8(a & b)

1. visceral pleura
2. parietal pleura
3. pleural cavity
4. diaphragm
5. visceral pleura
6. parietal pleura
7. right lung
8. pleural cavity
9. left lung

Figure 32.9(a & b)

1. lumen of esophagus
2. trachealis muscle
3. lumen of trachea
4. epithelial lining of trachea
5. tracheal cartilage
6. respiratory bronchiole
7. alveolar ducts
8. simple squamous epithelium
9. alveoli
10. alveolar sacs

Exercise 33

Table 33.1

1. External intercostals—increase diameter
2. Internal intercostals—decrease diameter
3. Sternocleidomastoids—increase diameter
4. Scalenes—increase diameter
5. Pectoralis minors—increase diameter
6. Diaphragm—increases length
7. Abdominal muscles—decrease length

Figure 33.1(a & b)

1. sternocleidomastoid
2. scalenes
3. pectoralis minor
4. rectus abdominis
5. external oblique
6. internal oblique
7. transverse abdominis
8. internal intercostals
9. external intercostals
10. diaphragm

Figure 33.2(a & b)

a. exhalation
b. inhalation

Figure 33.3

1. parietal pleura and thoracic wall
2. trachea
3. bronchus
4. lungs covered with visceral pleura
5. pleural cavity
6. diaphragm

Table 33.2

1. decreased
2. increased
3. deflated
4. out of balloons
5. increased
6. decreased
7. inflated
8. into balloons

Figure 33.4

FVC = 3.1 L
FEV_1 = 1.9 L
FEV_1/FVC ratio = 0.61

Exercise 34

Figure 34.1

1. salivary glands
2. pharynx
3. esophagus
4. liver
5. gallbladder
6. large intestine
7. mouth (oral cavity)
8. stomach
9. pancreas
10. small intestine
11. anus

Figure 34.2

1. mucosa
2. submucosa
3. serosa
4. muscularis

Figure 34.3(a, b, c, & d)

1. retroperitoneal organs
2. lesser omentum
3. mesocolon
4. mesentery
5. greater omentum
6. parietal peritoneum
7. visceral peritoneum (serosa)
8. peritoneal cavity

9. falciform ligament
10. greater omentum
11. lesser omentum
12. greater omentum (reflected)
13. mesocolon
14. mesentery

Figure 34.4

1. hard palate
2. soft palate
3. uvula
4. cheek
5. vestibule
6. superior lip
7. fauces
8. palatine tonsil
9. tongue
10. gingivae (gums)
11. inferior labial frenulum
12. inferior lip

Figure 34.5

1. nasopharynx
2. soft palate
3. uvula
4. fauces
5. oropharynx
6. laryngopharynx
7. esophagus
8. hard palate
9. oral or buccal cavity
10. tongue

Figure 34.6

1. lesser curvature
2. body
3. cardia
4. lower esophageal sphincter (cardiac sphincter)
5. esophagus
6. fundus
7. longitudinal muscle layer
8. circular muscle layer
9. oblique muscle layer
10. greater curvature
11. rugae
12. pyloric antrum
13. pyloric canal
14. pyloric sphincter
15. pylorus

Figure 34.8(a, b, & c)

1. duodenum
2. ascending colon
3. jejunum
4. ileum
5. stomach
6. transverse colon
7. descending colon
8. rectum
9. right colic (hepatic) flexure

10. ascending colon
11. ileum
12. ileocecal sphincter
13. cecum
14. appendix
15. rectum
16. transverse colon
17. left colic (splenic) flexure
18. descending colon
19. teniae coli
20. epiploic appendages
21. haustra
22. sigmoid colon
23. anal canal
24. anus
25. rectum
26. anal canal
27. internal anal sphincter
28. external anal sphincter
29. anal column
30. anus

Figure 34.9(a, b, & c)

1. parotid gland
2. parotid duct
3. opening of parotid duct (near secondary molar)
4. openings of sublinguinal ducts
5. sublinguinal gland
6. submandibular duct
7. submandibular gland
8. crown
9. neck
10. root
11. enamel
12. dentin
13. gingiva
14. blood supply
15. pulp cavity with pulp
16. cementum
17. root canal
18. periodontal ligament
19. nerve
20. apical foramen
21. molars
22. premolars
23. cuspid
24. incisors
25. tongue
26. lingual frenulum

Figure 34.10(b, c, & d)

1. right hepatic duct
2. cystic duct
3. gallbladder
4. duodenum
5. hepatopancreatic ampulla
6. falciform ligament
7. left lobe of liver
8. left hepatic duct
9. common hepatic duct

10. common bile duct
11. accessory pancreatic duct
12. tail of pancreas
13. body of pancreas
14. pancreatic duct
15. head of pancreas
16. jejunum
17. duodenal papilla
18. hepatopancreatic ampulla
19. common bile duct
20. pancreatic duct
21. quadrate lobe
22. left lobe
23. caudate lobe
24. right lobe
25. gallbladder

Figure 34.11(a, b, c, d, & e)

1. stratified squamous epithelium
2. lamina propria
3. muscularis mucosae
4. mucosa
5. submucosa
6. circular layer of smooth muscle fibers
7. longitudinal layer of smooth muscle fibers
8. adventitia
9. gastric pit
10. simple columnar epithelium
11. lamina propria
12. gastric glands
13. simple columnar epithelium
14. goblet cell
15. villus
16. lamina propria
17. intestinal (crypts of Lieberkühn) gland
18. duodenal (Brunner's) gland in submucosa
19. acini (exocrine)
20. pancreatic islet (islet of Langerhans)
21. hepatocytes
22. sinusoids
23. central vein

Figure 34.12

1. bile canaliculus
2. hepatic vein
3. central vein
4. sinusoids
5. portal triad
6. bile duct
7. hepatocytes
8. branch of hepatic portal vein
9. branches of hepatic artery

Exercise 36

Figure 36.1

1. right kidney
2. right ureter

3. urinary bladder
4. urethra

Figure 36.2

1. peritoneum
2. renal fascia
3. right kidney
4. renal capsule
5. adipose capsule
6. renal hilum
7. left kidney

Figure 36.3

1. renal medulla
2. renal cortex
3. renal column
4. renal capsule
5. renal pyramid
6. renal papilla
7. minor calyx
8. major calyx
9. renal artery
10. renal vein
11. renal hilum
12. renal pelvis in renal sinus

Figure 36.4

1. internal urethral sphincter
2. ureters
3. detrusor muscle of urinary bladder
4. ureteral openings
5. trigone
6. internal urethral orifice
7. urethra
8. external urethral sphincter
9. external urethral orifice

Figure 36.5(a & b)

1. rectum
2. anus
3. ureter
4. urinary bladder
5. ureteral opening
6. internal urethral orifice
7. prostatic urethra
8. membranous urethra
9. spongy urethra
10. external urethral orifice
11. urogenital diaphragm
12. rectum
13. anus
14. uterus
15. urinary bladder
16. internal urethral orifice
17. urethra
18. urogenital diaphragm
19. external urethral orifice
20. vagina

Figure 36.6(a & b)

1. glomerular capsule
2. proximal convoluted tubule
3. ascending limb of loop of Henle
4. descending limb of loop of Henle
5. distal convoluted tubule
6. collecting duct
7. papillary duct
8. glomerulus
9. glomerular capsule
10. renal corpuscle
11. peritubular capillary
12. efferent arteriole
13. afferent arteriole

Figure 36.7

1. afferent arteriole
2. efferent arteriole
3. vasa recta
4. peritubular capillary
5. interlobular artery
6. arcuate artery
7. interlobar artery
8. segmental artery
9. renal artery
10. renal vein
11. interlobar vein
12. arcuate vein
13. interlobular vein

Figure 36.9(a & b)

1. proximal and distal convoluted tubules (outer cortex)
2. glomerulus
3. loop of Henle and collecting ducts
4. glomerulus
5. parietal layer of glomerular capsule (simple squamous epithelium)
6. capsular space

Figure 36.10

1. lamina propria
2. transitional epithelium
3. lumen
4. mucosa
5. muscularis
6. adventitia

Exercise 37

Figure 37.2

1. filtration
2. major site of tubular reabsorption of water and solutes; secretion of solutes
3. water reabsorbed but not solutes
4. Na^+ and Cl^- reabsorbed but not water
5. additional reabsorption of solutes and water

6. aldosterone increases reabsorption of Na$^+$ and Cl$^-$ and secretion of K$^+$; also ADH increases reabsorption of water, additional reabsorption and secretion of solutes

Table 37.2

1. no change
2. increased
3. decreased
4. decreased
5. Change dependent on presence of ADH and/or aldosterone

Exercise 38

Figure 38.1

1. cremaster muscle
2. dartos muscle
3. scrotum
4. tunica albuginea of testis

Figure 38.2

1. efferent duct
2. seminiferous tubules
3. straight tubule
4. rete testis
5. tunica albuginea
6. lobule
7. septum

Figure 38.3

1. ductus (vas) deferens
2. epididymis
3. spermatic cord
4. blood vessels and nerves

Figure 38.4

1. ampulla of ductus deferens
2. ejaculatory duct
3. prostate gland
4. membranous urethra
5. epididymis
6. testis
7. scrotum
8. ductus (vas) deferens (in inguinal canal)
9. prostatic urethra
10. spongy (penile) urethra
11. penis
12. external urethral orifice

Figure 38.5

1. seminal vesicle
2. prostate gland
3. prostatic urethra
4. membranous urethra
5. ductus (vas) deferens

6. ampulla of ductus deferens
7. ejaculatory duct
8. bulbourethral gland
9. spongy (penile) urethra

Figure 38.6

1. seminal vesicle
2. prostate gland
3. bulbourethral gland

Figure 38.7(a & b)

1. prostate gland
2. prostatic urethra
3. membranous urethra
4. corpus spongiosum penis
5. corpora cavernosa penis
6. spongy (penile) urethra
7. glans penis
8. bulbourethral gland
9. prepuce
10. external urethral orifice
11. corpora cavernosa penis
12. corpus spongiosum penis
13. spongy (penile) urethra

Lab Activity 5: Function of Male Reproductive Organs

1. secretes 60% of semen volume; alkaline pH; prostaglandins
2. secretes 25% of semen volume; increases viability and motility of sperm
3. secretes mucus for lubrication of glans; alkaline pH
4. site of storage and maturation of sperm
5. organ that produces sperm
6. copulatory organ; passageway for urine and sperm
7. carries sperm from epididymis to ejaculatory duct

Figure 38.9(a & b)

1. spermatogonium
2. spermatozoa
3. spermatid
4. secondary spermatocyte
5. primary spermatocyte
6. spermatozoa (sperm cell)
7. late spermatid
8. early spermatid
9. secondary spermatocyte
10. Sertoli cells
11. primary spermatocyte
12. spermatogonium
13. basement membrane
14. Leydig (interstitial) cells

Figure 38.10

1. lumen
2. tails of sperm

3. basement membrane
4. Leydig (interstitial) cell
5. interstitial space

Figure 38.11

1. acrosome
2. nucleus
3. mitochondria
4. head
5. midpiece
6. tail (flagellum)

Exercise 39

Figure 39.1(b)

1. infundibulum
2. ampulla
3. isthmus
4. uterine tube
5. fimbriae
6. suspensory ligament
7. ovary
8. ovarian ligament
9. broad ligament

Figure 39.2

1. uterine cavity
2. myometrium of uterus
3. vagina
4. ovary
5. round ligament
6. cervix
7. urinary bladder

Figure 39.3

1. uterine cavity
2. endometrium
3. myometrium
4. perimetrium
5. round ligament
6. internal os
7. cervical canal
8. cervix
9. rugae of vagina
10. fundus
11. broad ligament
12. body of uterus
13. external os
14. vagina

Figure 39.4(b)

1. anus
2. vaginal orifice
3. pubic symphysis
4. mons pubis
5. clitoris
6. external urethral orifice

7. labium minus
8. labium majus

Figure 39.5

1. labia majora
2. labia minora
3. hymen
4. anus
5. mons pubis
6. prepuce
7. clitoris
8. external urethral orifice
9. vaginal orifice

Figure 39.6(a & b)

1. lobule with alveoli
2. mammary duct
3. lactiferous sinus
4. lactiferous duct
5. adipose tissue
6. areola
7. nipple

Lab Activity 5: Function of Female Reproductive Organs

1. Beat together to bring ovulated oocyte into uterine tube
2. Usual site of fertilization
3. Narrow area of uterine tube that opens into uterus
4. Transports secondary oocytes toward uterus
5. Produces secondary oocytes; produces estrogen and progesterone
6. Layer of uterus that sheds during menstruation; implantation of zygote and development of fetus occurs here

Figure 39.8

1. secondary oocyte
2. sperm cell
3. ovum
4. zygote
5. oogonium
6. primary oocyte
7. meiosis I
8. 1st polar body
9. 2nd polar body
10. meiosis II

Figure 39.9(a, b, c, & e)

1. primordial follicles
2. primary follicles
3. secondary follicle
4. secondary oocyte
5. graafian (mature) follicle
6. ovulated secondary oocyte
7. corpus luteum
8. corpus albicans
9. primordial follicles

10. primary follicle
11. primary oocyte in primary follicle
12. primary oocyte in secondary follicle
13. primordial follicles
14. ovarian cortex
15. primary oocyte in primary follicle
16. granulosa cells
17. follicular fluid in antrum
18. primary oocyte
19. granulosa cells

Figure 39.10

1. endometrium
2. endometrial glands
3. stratum functionalis
4. stratum basalis
5. myometrium

Exercise 40

Figure 40.1

1. secondary oocyte
2. fertilization of secondary oocyte
3. 2-cell stage of zygote
4. 4-cell stage of zygote
5. morula
6. blastocyst, external view
7. implantation of blastocyst, internal view

Figure 40.2(b)

1. yolk sac
2. amnion
3. amnionic cavity
4. bilaminar embryonic disc

Figure 40.3

1. amniotic cavity
2. embryo
3. yolk sac
4. connecting stalk
5. chorionic villus

Figure 40.4

1. chorionic villi
2. fetal blood vessels
3. decidua basalis
4. chorion
5. umbilical arteries
6. umbilical vein

Exercise 41

Lab Activity 1 Genotypes and Phenotypes

1. het=Zz, Tt; hom=bb, MM, cc, AA

2. freckles=FF, Ff; no freckles=ff; attached earlobes=EE, Ee; unattached earlobes=ee
3. homozygous widow's peak =WW; heterozygous widow's peak =Ww; straight hairline=ww

Figure 41.1

DD=0; Dd=4; dd=0
DD=0%; Dd=100%; dd=0%
dimples=4; no dimples=0
dimples=100%; no dimples=0%

Figure 41.3

RR = 1; Rr=2; rr=1
RR=25%; Rr=50%; rr=25%
Tongue-roller=3; non-tongue roller=1
Tongue-roller=75%; non-tongue roller=25%

Figure 41.4

CC=0%; Cc=100%, cc=0%
Curly hair=0%; wavy hair=100%; straight hair=0%

Figure 41.5

AO=25%; BO=25%; AB=25%; OO=25%
A=25%; B=25%; AB=25%; O=25%

Figure 41.6

AO=50%; BO=50%; AB=0%; OO=0%
A=50%; B=50%; AB=0%; O=0%

Figure 41.8

Hb^SHb^S=50%; Hb^SHb^A=50%; Hb^AHb^A=0%
Sickle-cell anemia=50%; sickle-cell trait=50%; Normal=0%

Lab Activity 7 Sex-Linked Inheritance

1. X^CX^C=normal; X^CX^c=normal; X^cX^c=color blind; X^CY=normal; X^cY=color blind

Figure 41.10

X^CX^C=25%; X^CX^c=25%; X^cX^c=0%; X^CY=25%; X^cY=25%
Female normal=25%; Female carrier = 25%; female color blind=0%; male normal=25%; male color blind=25%

Appendix A: Word Roots

WORD ROOT	DEFINITION	NOUNS AND/OR ADJECTIVES FORMED FROM WORD ROOT AND A SUFFIX	WORDS FORMED BY COMBINING WORD ROOT WITH ANOTHER WORD ROOT, PREFIX AND/OR SUFFIX
abdomin-	abdomen	**Noun:** abdominis **Adjective:** abdominal	**abdomin**opelvic
acromi-	high point of shoulder	**Noun:** acromion (sing.) **Adjective:** acromial	**acromi**oscapular
axill-	armpit	**Noun:** axilla (sing.) and axillae (pl.) **Adjective:** axillary	**axill**ofemoral
brachi-	arm	**Noun:** brachium (sing.) and brachia (pl.) **Adjective:** brachial	**brachi**ocephalic, ante**brachi**um
bucc-	cheek	**Noun:** bucca (sing.), buccae (pl.) **Adjective:** buccal	**bucc**inator
calcane-	heel	**Noun:** calcaneus (sing.) and calcanei (pl.) **Adjective:** calcaneal	**calcane**ofibular
carp-	wrist bones	**Noun:** carpus (sing.) and carpi (pl.) **Adjective:** carpal	**carp**ectomy
cervic-	neck or neck-like structure	**Noun:** cervix (sing.) and cervices (pl.) **Adjective:** cervical	**cervic**othorax, **cervic**ectomy, **cervic**itis
cox-	hip	**Noun:** coxa (sing.) and coxae (pl.) **Adjective:** coxal	**cox**algia
crani-	cranium	**Noun:** cranium (sing.), crania (pl.) **Adjective:** cranial	**crani**osacral, **crani**ofacial
cubit-	elbow or ulna	**Noun:** cubitus (sing.), cubiti (pl.) **Adjective:** cubital	ante**cubit**al
digit-	finger or toe	**Noun:** digit and digitus (sing.) and digiti (pl.) **Adjective:** digital	**digit**ation
dors-	back	**Noun:** dorsum (sing.) and dorsa (pl.) **Adjective:** dorsal	**dors**iflexion
faci-	face	**Noun:** facies (sing.) **Adjective:** facial	**faci**oplasty
fibul-	lateral leg	**Noun:** fibula **Adjective:** fibular	**fibul**ocalcaneal
front-	forehead	**Noun:** frons (sing.) **Adjective:** frontal	**front**onasal
glute-	buttocks	**Noun:** gluteus **Adjective:** gluteal	**glute**ofemoral
lumb-	lower back or loin	**Noun:** lumbus (sing.) and lumbi (pl.) **Adjective:** lumbar	**lumb**ago, **lumb**osacral
ment-	mind or chin	**Noun:** mentum (sing.) **Adjective:** mental	**ment**alis muscle, **ment**al foramen

WORD ROOT	DEFINITION	NOUNS AND/OR ADJECTIVES FORMED FROM WORD ROOT AND A SUFFIX	WORDS FORMED BY COMBINING WORD ROOT WITH ANOTHER WORD ROOT, PREFIX AND/OR SUFFIX
nas-	nose	**Noun:** nasus (sing.) **Adjective:** nasal	**nas**opharynx, **nas**osinusitis
occipit-	back of head	**Noun:** occipitis, occiput (sing.) **Adjective:** occipital	**occipit**ofrontal, **occipit**omental
olecran-	proximal projection of ulna	**Noun:** olecranon (sing.) and olecrana (pl.) **Adjective:** olecranal	**olecran**arthropathy
or-	mouth	**Noun:** oris (sing.) **Adjective:** oral	**or**opharynx
orbit-	eye	**Noun:** orbit (sing), orbitae (pl.) **Adjective:** orbital	**orbit**onasal
ot-	ear	**Adjective:** otic	**ot**itis, **ot**olaryngology
palm -	flexor surface of hand	**Noun:** palm and palma (sing.) and palmae (pl.) **Adjective:** palmar	**palm**ature
patell-	kneecap	**Noun:** patella (sing.) and patellae (pl.) **Adjective:** patellar	**patell**ectomy
pector-	breast or chest	**Noun:** pectus (sing.) and pectora (pl.) **Adjective:** pectoral	**pect**oralgia
ped-	foot or child	**Noun:** pes (sing.) and pedes (pl.) **Adjective:** pedal	**ped**icle, **ped**ometer
pelvi-	pelvis, basin	**Noun:** pelvis (sing.) and pelves (pl.) **Adjective:** pelvic	**pelvi**operitonitis, abdomino**pelvic**
plant-	sole of foot	**Noun:** planta (sing.) and plantae (pl.) **Adjective:** plantar	**plant**arflexion, plantaris
pub-	pubis or genital area	**Noun:** pubis (sing.) **Adjective:** pubic	**pub**erty
scapul-	scapula (shoulder bone)	**Noun:** scapula (sing.) and scapulae (pl.) **Adjective:** scapular	**scapul**ectomy
stern-	chest	**Noun:** sternum (sing.) **Adjective:** sternal	**stern**ocleidomastoid
tars-	ankle bones or edge of eyelid	**Noun:** tarsus (sing.) and tarsi (pl.) **Adjective:** tarsal	**tars**ometatarsal, tibio**tarsal**
thorac-	chest	**Noun:** thorax (sing.) and thoraces (pl.) **Adjective:** thoracic	**thorac**olumbar
umbilic-	navel	**Noun:** umbilicus (sing.) and umbilici (pl.) **Adjective:** umbilical	**umbilic**ation
vertebr-	vertebra, spinal column	**Noun:** vertebra (sing.) and vertebrae (pl.) **Adjective:** vertebral	**vertebr**oarterial

Appendix B: Skeletal Muscle Origins and Insertions

TABLE B1 Muscles of the Head and Neck

MUSCLE	ORIGIN	INSERTION
MUSCLES OF FACIAL EXPRESSION		
Occipitofrontalis muscle		
Frontalis	Galea (epicranial) aponeurosis	Skin superior to supraorbital margin
Occipitalis	Occipital bone and mastoid process of temporal bone	Galea (epicranial) aponeurosis
Orbicularis oculi	Medial wall of orbit	Skin around margin of orbit
Zygomaticus minor	Zygomatic bone	Upper lip
Zygomaticus major	Zygomatic bone	Skin at angle of mouth and orbicularis oris
Orbicularis oris	Muscle fibers surrounding mouth	Skin at corner of mouth
Platysma	Fascia covering deltoid and pectoralis major	Mandible, muscles around angle of mouth, and skin of lower face
MUSCLES OF MASTICATION		
Superficial Muscles		
Temporalis	Temporal bone	Coronoid process and ramus of mandible
Masseter	Maxilla and zygomatic arch	Ramus and angle of mandible
Deep Muscles		
Buccinator	Alveolar processes of maxilla and mandible	Orbicularis oris
Lateral pterygoid	Greater wing of sphenoid and lateral part of pterygoid process of sphenoid bone	Mandibular condyle; temporomandibular joint (TMJ)
Medial pterygoid	Medial part of pterygoid process of sphenoid bone and maxilla bone	Ramus and angle of mandible
MUSCLES THAT MOVE THE HEAD AND NECK		
Anterior and Lateral Muscles		
Sternocleidomastoid	Sternum and clavicle	Mastoid process of temporal bone
Scalenes	Transverse processes of vertebrae C_3 through C_7	Ribs 1 and 2
Posterior Muscles		
Splenius capititis	Ligamentum nuchae and spinous processes of vertebrae C_7 through T_4	Occipital bone and mastoid process of temporal bone
Trapezius (superior portion)	Occipital bone, ligamentum nuchae, and spines of vertebrae C_7 through T_{12}	Clavicle, acromion (process) of scapula, and spine of scapula
MUSCLE THAT ONLY MOVES THE SCAPULA		
Levator scapulae	Vertebrae C_1–C_5	Superior vertebral border of scapula

TABLE B1 (Continued)

MUSCLE	ORIGIN	INSERTION
MUSCLES THAT MOVE HYOID BONE		
Suprahyoid muscles		
Digastric		
Anterior belly	Inner surface of inferior border of mandible	Body of hyoid bone
Posterior belly	Mastoid process of temporal bone	Body of hyoid bone
Stylohyoid	Styloid process of temporal bone	Body of hyoid bone
Mylohyoid	Inner surface of mandible	Body of hyoid bone
Infrahyoid Muscles		
Omohyoid	Superior border of scapula	Body of hyoid bone
Sternohyoid	Manubrium and medial end of clavicle	Body of hyoid bone

TABLE B2 Muscles of the Anterior Trunk

MUSCLE	ORIGIN	INSERTION
MUSCLES THAT MOVE THE ARM AT THE SHOULDER JOINT		
Superficial muscles		
Deltoid	**Anterior fibers:** acromial end of clavicle **Lateral fibers:** acromion of scapula **Posterior fibers:** spine of scapula	Deltoid tuberosity of humerus
Pectoralis major	Clavicle (clavicular head), sternum, and costal cartilages of ribs 2–6 (sternocostal head)	Greater tubercle and intertubecular sulcus of humerus
Deep muscle		
Subscapularis	Subscapular fossa of scapula	Lesser tubercle of humerus
MUSCLES THAT MOVE THE SCAPULA		
Serratus anterior	Ribs 1–9	Vertebral border of scapula and inferior angle of scapula
Pectoralis minor	Ribs 2–5, 3–5, or 2–4	Coracoid process of scapula
MUSCLES THAT MOVE THE ABDOMINAL WALL		
Superficial muscles		
Rectus abdominis	Pubic crest and pubic symphysis	Cartilage of ribs 5–7 and the xiphoid process
External oblique	Ribs 5–12	Iliac crest and linea alba
Deep muscles		
Internal oblique	Iliac crest, inguinal ligament, and thoracolumbar fascia	Cartilage of ribs 9–12 and linea alba
Transversus abdominis	Iliac crest, inguinal ligament, lumbar fascia, and cartilage of ribs 7–12	Xiphoid process, linea alba, and pubis
MUSCLES USED IN BREATHING		
Deep muscles		
External intercostals	Inferior border of rib superior to muscle	Superior border of rib inferior to muscle
Internal intercostals	Superior border of rib inferior to muscle	Inferior border of rib superior to muscle
Diaphragm	Xiphoid process of sternum, costal cartilages of inferior six ribs, and lumbar vertebrae	Central tendon

TABLE B3 **Muscles of the Posterior Trunk**

MUSCLE	ORIGIN	INSERTION
MUSCLES THAT MOVE THE ARM AT THE SHOULDER JOINT		
Superficial muscles		
Deltoid	**Posterior fibers:** spine of scapula	Deltoid tuberosity of humerus
Infraspinatus	Infraspinous fossa of scapula	Greater tubercle of humerus
Teres minor	Inferior lateral border of scapula	Greater tubercle of humerus
Teres major	Inferior angle of scapula	Intertubercular sulcus of humerus
Latissimus dorsi	Spines of T_7–T_{12} vertebrae, lumbar vertebrae, crests of sacrum and ilium, ribs 9–12	Intertubercular sulcus of humerus
Deep muscles		
Supraspinatus	Supraspinous fossa of scapula	Greater tubercle of humerus
MUSCLES THAT MOVE THE SCAPULA		
Superficial muscle		
Trapezius	Occipital bone, ligamentum nuchae, and spines of vertebrae C_7–T_{12}	Clavicle, acromion of scapula, and spine of scapula
Deep muscles		
Levator scapulae	Vertebrae C_1–C_5	Superior vertebral border of scapula
Rhomboideus minor	Spines of vertebrae C_7 and T_1	Vertebral border of scapula superior to spine
Rhomboideus major	Spines of vertebrae T_2–T_5	Vertebral border of scapula inferior to spine
MUSCLES THAT MOVE THE VERTEBRAL COLUMN		
Erector spinae consists of iliocostalis, longissiumus, and spinalis groups	Large group of muscles with origins on ribs, iliac crest, and cervical, thoracic, and lumbar vertebrae	Skull, ribs, and cervical, thoracic, and lumbar vertebrae

TABLE B4 **Muscles of the Arm**

MUSCLE	ORIGIN	INSERTION
ANTERIOR SURFACE		
Biceps brachii	**Long head:** tubercle superior to glenoid cavity of scapula	Radial tuberosity of radius
	Short head: coracoid process of scapula	
Brachialis	Anterior surface of distal humerus	Ulnar tuberosity and coronoid process of ulna
POSTERIOR SURFACE		
Triceps brachii	**Long head:** inferior to glenoid cavity of scapula	All three heads insert on the olecranon of ulna
	Medial head: posterior humerus inferior to groove for radial nerve	
	Lateral head: posterior surface of lateral humerus superior to radial groove	

TABLE B5 Muscles of the Forearm

MUSCLE	ORIGIN	INSERTION
ANTERIOR SURFACE		
Superficial muscles (lateral to medial)		
Brachioradialis	Distal lateral border of humerus	Superior to styloid process of radius
Pronator teres	Medial epicondyle of humerus and coronoid process of ulna	Midlateral surface of radius
Flexor carpi radialis	Medial epicondyle of humerus	2nd and 3rd metacarpals
Palmaris longus	Medial epicondyle of humerus	Flexor retinaculum and palmar aponeurosis
Flexor carpi ulnaris	Medial epicondyle of humerus and superior posterior border of humerus	Pisiform, hamate, and base of 5th metacarpal
Flexor digitorum superficialis	Medial epicondyle of humerus, coronoid process of ulna, and anterior surface of radius	Middle phalanx of each finger
Deep muscles		
Supinator	Lateral epicondyle of humerus and ridge near radial notch of ulna (supinator crest)	Lateral surface of proximal one-third of radius
Flexor pollicis longus	Anterior surface of radius and interosseous membrane	Base of distal phalanx of thumb
POSTERIOR SURFACE		
Superficial muscles (medial to lateral)		
Extensor carpi ulnaris	Lateral epicondyle of humerus and posterior ulna	5th metacarpal
Extensor digitorum	Lateral epicondyle of humerus	distal and middle phalanges of each finger
Extensor carpi radialis longus and brevis	**Brevis:** lateral epicondyle of humerus	3rd metacarpal
	Longus: lateral supracondylar ridge of humerus	2nd metacarpal
POSTERIOR SURFACE		
Deep muscles		
Supinator	Lateral epicondyle of humerus and ridge near radial notch of ulna	Lateral surface of proximal one-third of radius
Extensor pollicis longus	Posterior surface of middle of ulna, and interosseous membrane	Distal phalanx of thumb

TABLE B6 Muscles of the Thigh

MUSCLE	ORIGIN	INSERTION
ANTERIOR SURFACE *(lateral to medial)*		
Tensor fasciae latae	Iliac crest	Iliotibial tract (fascia)
Quadriceps femoris		Patella (quadriceps tendon) and tibial tuberosity (patellar ligament)
Rectus femoris	Anterior inferior iliac spine	
Vastus lateralis	Greater trochanter and linea aspera of femur	
Vastus medialis	Linea aspera of femur	
Vastus intermedius	Anterior and lateral surface of body of femur	
Sartorius	Anterior superior iliac spine	Medial surface of body of tibia
Iliopsoas		
Psoas major	Psoas major: Transverse processes and bodies of lumbar vertebrae	Psoas major and iliacus insert together into lesser trochanter of femur
Iliacus	Iliacus: Iliac fossa	
Pectineus	Superior ramus of pubis	Pectineal line of femur between lesser trochanter and linea aspera
Adductor longus	Pubic crest and pubic symphysis	Linea aspera of femur
Adductor magnus	Inferior ramus of pubis and ischium to ischial tuberosity	Linea aspera of femur
Adductor brevis	Inferior ramus of pubis	Proximal linea aspera of femur
Gracilis	Pubic symphysis and pubic arch	Medial surface of body of tibia
POSTERIOR SURFACE		
Gluteus maximus	Iliac crest, sacrum, and coccyx	Greater trocanter of femur and iliotibial tract
Gluteus medius	Ilium	Greater trochanter of femur
Gluteus minimus	Ilium	Greater trochanter of femur
Piriformis	Anterior sacrum	Superior border of greater trochanter of femur
Hamstrings		
Biceps femoris	**Long head:** ischial tuberosity **Short head:** linea aspera of femur	Head of fibula and lateral condyle of tibia
Semitendinosus	Ischial tuberosity	Medial surface of proximal part of tibia
Semimembranosus	Ischial tuberosity	Medial condyle of tibia

TABLE B7 Muscles of the Leg and Foot

MUSCLE	ORIGIN	INSERTION
ANTERIOR AND LATERAL SURFACE		
Superficial muscles		
Tibialis anterior	Lateral condyle and body of tibia and interosseous membrane	1st metatarsal and 1st (medial) cuneiform (tarsal bone)
Extensor digitorum longus	Lateral condyle of tibia, anterior fibula, and interosseous membrane	Middle and distal phalanges 2–5
Fibularis (peroneus) longus	Lateral condyle of tibia and head and body of fibula	1st metatarsal and 1st (medial) cuneiform
POSTERIOR SURFACE		
Superficial muscles		
Gastrocnemius	Lateral and medial condyles of femur	Calcaneus via calcaneal tendon
Soleus	Medial tibia and head of fibula	Calcaneus via calcaneal tendon
POSTERIOR SURFACE		
Deep muscles		
Flexor digitorum longus	Posterior tibia	Distal phalanges 2–5
Flexor hallucis longus	Inferior fibula	Distal phalanx of great toe

Appendix C: Measurements

U.S. Customary System

PARAMETER	UNIT	RELATION TO OTHER U.S. UNITS	SI (METRIC) EQUIVALENT
Length	inch	1/12 foot	2.54 centimeters
foot	12 inches	0.305 meter	
yard	36 inches	0.914 meter	
mile	5,280 feet	1.609 kilometers	
Mass	grain	1/1,000 pound	64.799 milligrams
dram	1/16 ounce	1.772 grams	
ounce	16 drams	28.350 grams	
pound	16 ounces	453.6 grams	
ton	2,000 pounds	907.18 kilograms	
Volume (Liquid)	ounce	1/16 pint	29.574 milliliters
pint	16 ounces	0.473 liter	
quart	2 pints	0.946 liter	
gallon	4 quarts	3.785 liters	
Volume (Dry)	pint	1/2 quart	0.551 liter
quart	2 pints	1.101 liters	
peck	8 quarts	8.810 liters	
bushel	4 pecks	35.239 liters	

International System (SI)

UNIT	BASE UNITS QUANTITY	SYMBOL	PREFIX	PREFIXES MULTIPLIER	SYMBOL
meter	length	m	tera-	$10^{12} = 1,000,000,000,000$	T
kilogram	mass	kg	giga-	$10^{9} = 1,000,000,000$	G
second	time	s	mega-	$10^{6} = 1,000,000$	M
liter	volume	L	kilo-	$10^{3} = 1,000$	k
mole	amount of matter	mol	hecto-	$10^{2} = 100$	h
			deca-	$10^{1} = 10$	da
			deci-	$10^{-1} = 0.1$	d
			centi-	$10^{-2} = 0.01$	c
			milli-	$10^{-3} = 0.001$	m
			micro-	$10^{-6} = 0.000,001$	μ
			nano-	$10^{-9} = 0.000,000,001$	n
			pico-	$10^{-12} = 0.000,000,000,001$	p

Temperature Conversion

FAHRENHEIT (F) TO CELSIUS (C)
$°C = (°F − 32) ÷ 1.8$

CELSIUS (C) TO FAHRENHEIT (F)
$°F = (°C × 1.8) + 32$

U.S. to SI (Metric) Conversion

WHEN YOU KNOW	MULTIPLY BY	TO FIND
inches	2.54	centimeters
feet	30.48	centimeters
yards	0.91	meters
miles	1.61	kilometers
ounces	28.35	grams
pounds	0.45	kilograms
tons	0.91	metric tons
fluid ounces	29.57	milliliters
pints	0.47	liters
quarts	0.95	liters
gallons	3.79	liters

SI (Metric) to U.S. Conversion

WHEN YOU KNOW	MULTIPLY BY	TO FIND
millimeters	0.04	inches
centimeters	0.39	inches
meters	3.28	feet
kilometers	0.62	miles
liters	1.06	quarts
cubic meters	35.32	cubic feet
grams	0.035	ounces
kilograms	2.21	pounds

Photo Credits

All photos in this *Laboratory Manual* 5e are by Mark Nielsen with the following exceptions:

Front Cover

Illustration by Bryan Christie Design

Exercise 1

Figure 1.7a: Simon Fraser/Photo Researchers*; Figure 1.7b: Alfred Pasieka/Photo Researchers

Exercise 2

Figure 2.5a: CNRI/Phototake; Figure 2.5b: © Benedet/Phototake; Figure 2.5c: Geoff Tompkinson/Photo Researchers; Figure 2.5d: CNRI/Photo Researchers

Exercise 3

Figure 3.1: Dan Ahn/iStockphoto; Figure 3.3: Bruce Iverson/Photo Researchers

Exercise 4

Figure 4.1b: Steve Gschmeissner/Photo Researchers; Figure 4.2a, 4.2b, 4.2f: Courtesy Michael Ross, University of Florida; Figure 4.2c: Carolina Biological Supply Company/Phototake; Figure 4.2d: M. Abbey/Photo Researchers; Figure 4.2e: Robert Becker, Ph.d./Custom Medical Stock Photo, Inc.; Figure 4.4a, 4.4b, 4.4c, 4.4d, 4.4e: Courtesy Michael Ross, University of Florida; Figure 4.5: Don Fawcett/Photo Researchers

Exercise 5

Figure 5.3a, 5.3b, 5.3c: David Phillips / Photo Researchers

Exercise 6

Figure 6.4a, 6.4b, 6.5a, 6.5b, 6.6a, 6.6b, 6.7a, 6.7b, 6.8a, 6.8b, 6.9a, 6.9b, 6.11, 6.12, 6.13, 6.14, 6.15, 6.16, 6.17, 6.18, 6.19, 6.20, 6.21, 6.22, 6.24, 6.25, 6.26a, 6.26b, 6.30-6, 6.30-17, 6.31-1, 6.31-3: Courtesy Michael Ross, University of Florida; Figure 6.28: Jan Leesma/Custom Medical Stock Photo, Inc.; Figure 6.29-13, 6.29-14, 6.29-15, 6.30-4, 6.30-5, 6.30-8, 6.30-10, 6.3-12, 6.31-4: Ed Reschke; Figure 6.29-12, 6.30-1, 6.30-2, 6.30-3, 6.30-9, 6.30-11, 6.30-13, 6.30-14, 6.30-15, 6.30-16, 6.30-18: Biophoto Associates/Photo Researchers; Figure 6-30-7: Andrew Kutzman; Figure 6.31-2 ISM/Phototake

Exercise 7

Figure 7.2, 7.3, 7.5: Courtesy Michael Ross, University of Florida; Figure 7.7a, 7.7c: Photo Researchers; Figure 7.7b: David Becker/Photo Researchers

Exercise 8

Figure 8.4: Lester V. Bergman/Corbis Images; Figure 8.5: PDSN/Phototake; Figure 8.6a: Mehau Kulyk/Photo Researchers; Figure 8.6b: Scott Camazine & Sue Trainor/Photo Researchers; Figure 8.7a: Photo Researchers; Figure 8.7b: P. Motta/Photo Researchers

Exercise 9

Figure 9.26a: Photo Researchers; Figure 9.26b: AFIP/Photo Researchers; Figure 9.27a, 9.27b: Courtesy Chaim J. Margolin

Exercise 10

Figure 10.13a: Larry Mulvehill/Photo Researchers; Figure 10.13b: © Robert Destefano/Alamy; Figure 10.13c: Photo Researchers; Figure 10.14a: Kevin Dodge/Masterfile; Figure 10.14b: Dick Luria/Photo Researchers

Exercise 11

Figure 11.3a, 11.3c, 11.3d, 11.9c: Dissection Shawn Miller, Photograph Mark Nielsen

Exercise 12

Figure 12.2a, 12.2b, 12.8: Courtesy Michael Ross, University of Florida; Figure 12.2c: MedImage/Photo Researchers; Figure 12.5: Dr. Clara Franzini-Armstrong; Figure 12.6a, 12.6b, 12.6c, 12.9-11, 12.9-12, 12.9-13 Courtesy Hiroyouki Sasaki, Yale E. Goldman and Clara Franzini-Armstrong

Exercise 13

Figure 13.1: Biophoto Associates/Photo Researchers

Exercise 14

Figure 14.1a, 14.1b, 14.1c, 14.2, 14.4, 14.5a, 14.7a, 14.7b, 14.8a, 14.8b: Dissection Shawn Miller; Photograph Mark Nielsen; Figure 14.3b, 14.3c, 14.3d, 14.3e, 14.5b, 14.7e: Dissection Nathan Mortensen and Shawn Miller, Photograph Mark Nielsen

Exercise 15

Figure 15.1a, 15.1b, 15.1c, 15.2d, 15.2f, 15.3a, 15.3b, 15.3e, 15.3f: © John Wiley & Sons, Inc.; Figure 15.3a: Andy Washnik / © John Wiley & Sons, Inc.

Exercise 16

Figure 16.2a, 16.4a, 16.7: Courtesy Michael Ross, University of Florida; Figure 16.2b: Jan Leesma/Cutom Medical Stock Photo, Inc.; Figure 16.4b: Cavallini James/BSIP/Phototake; Figure 16.8a: © Glauberman/Photo Researchers; Fig. 16.8b: ISM/Phototake

* Photo Researchers is now named Science Source.

Index